T0251272

# Fungi
## in Ecosystem Processes
Second Edition

# MYCOLOGY SERIES

Editor
## J. W. Bennett
Professor
Department of Plant Biology and Pathology
Rutgers University
New Brunswick, New Jersey

Founding Editor
## Paul A. Lemke

# Fungi
## in Ecosystem Processes
### Second Edition

## John Dighton
Rutgers University Pinelands Field Station, New Lisbon,
New Jersey, USA

**CRC Press**
Taylor & Francis Group
Boca Raton   London   New York

CRC Press is an imprint of the
Taylor & Francis Group, an **informa** business

Front Cover: Photos copyright John Dighton and compiled by Mike Anderson, Emotive New Media Ltd. Used with permission. All rights reserved.

CRC Press
Taylor & Francis Group
6000 Broken Sound Parkway NW, Suite 300
Boca Raton, FL 33487-2742

First issued in paperback 2021

© 2016 by Taylor & Francis Group, LLC
CRC Press is an imprint of Taylor & Francis Group, an Informa business

No claim to original U.S. Government works

Version Date: 20160127

ISBN 13: 978-1-03-209803-6 (pbk)
ISBN 13: 978-1-4822-4905-7 (hbk)

This book contains information obtained from authentic and highly regarded sources. Reasonable efforts have been made to publish reliable data and information, but the author and publisher cannot assume responsibility for the validity of all materials or the consequences of their use. The authors and publishers have attempted to trace the copyright holders of all material reproduced in this publication and apologize to copyright holders if permission to publish in this form has not been obtained. If any copyright material has not been acknowledged please write and let us know so we may rectify in any future reprint.

Except as permitted under U.S. Copyright Law, no part of this book may be reprinted, reproduced, transmitted, or utilized in any form by any electronic, mechanical, or other means, now known or hereafter invented, including photocopying, microfilming, and recording, or in any information storage or retrieval system, without written permission from the publishers.

For permission to photocopy or use material electronically from this work, please access www.copyright.com (http://www.copyright.com/) or contact the Copyright Clearance Center, Inc. (CCC), 222 Rosewood Drive, Danvers, MA 01923, 978-750-8400. CCC is a not-for-profit organization that provides licenses and registration for a variety of users. For organizations that have been granted a photocopy license by the CCC, a separate system of payment has been arranged.

**Trademark Notice:** Product or corporate names may be trademarks or registered trademarks, and are used only for identification and explanation without intent to infringe.

Publisher's Note
The publisher has gone to great lengths to ensure the quality of this reprint but points out that some imperfections in the original copies may be apparent.

**Visit the Taylor & Francis Web site at**
**http://www.taylorandfrancis.com**

**and the CRC Press Web site at**
**http://www.crcpress.com**

# Contents

# List of Figures

# Preface

Since the writing of the first edition of *Fungi in Ecosystem Processes*, much more literature has been published on many of the subjects I attempted to cover; thus, this edition is an update of the material content of the first. During the intervening years between editions, I was privileged to work with Peter Oudemans and Jim White in editing the third edition of *The Fungal Community* (2005) and have developed additional mycological courses with them. In this edition, therefore, I have expanded the section on plant pathogens, invasive species, and insect fungal interactions and have added a new section on fungi in the built environment. I am grateful to my undergraduate and graduate students for their interest in discussing fungal interactions with their environment, my research group for carrying many of my oddball ideas into research questions, and especially to a group of graduate students (Kristi Troster, Jonelle Scardino, and Alex Duffy) for their help in gathering new material for this edition as part of their independent studies.

The main thrust of this book follows the great discussions I had as a graduate student and beyond with Alan Rayner, who led me to believe that fungi rule the world. My view of this has changed little, and the more I delve into the world of mycology the greater the fascination I have with their diversity of function and interactions with other organisms. I hope to impart a small part of this enthusiasm to the readers of this book.

## REFERENCE

Dighton, J., J. F. White Jr., and P. Oudemans. (eds.) 2005. *The Fungal Community: Its Organization and Role in the Ecosystem.* CRC Press. 936 pp.

# Author

**John Dighton** earned his MSc in Ecology from Durham University, Durham, UK, and PhD from London University, London, UK. After a brief spell of teaching high school, he worked for 15 years for the Institute of Terrestrial Ecology, Merlewood, UK (Natural Environment Research Council), where he worked on ectomycorrhizal fungi, forest soil ecology, forest nutrition, and impacts of pollutants on fungi. He moved to the United States and started working with Rutgers University, New Brunswick, New Jersey, to run their Pinelands Field Station in the New Jersey pine barrens. He has continued his research in forest soil ecology and mycology and the impacts of pollutants. He teaches courses in mycology and soil ecology at two campuses of the university. He has published more than 100 scientific papers, serves on the editorial board of *Soil Biology and Biochemistry*, *Fungal Biology*, and *Fungal Ecology*, and has edited books on soil and mycology-related topics.

# CHAPTER 1

# Introduction

## 1.1 WHY FUNGI?

Fungi are a group of organisms that cannot fix energy and nutrients directly (but see Tugay et al., 2006; Dadachova et al., 2007; Dadachova and Casadevall, 2008), so they use the energy stored in plant and animal biomass to create their own growth. Fungi are a key group of organisms that interact with other organisms and the abiotic environment to regulate ecosystem processes. In his introductory chapter to *The Fungal Community: Its Organization and Role in the Ecosystem* (Carroll and Wicklow, 1992), Alan Rayner speaks of the importance of fungi in ecosystems in the following four major functions: (1) As decomposers, fungi drive the global carbon cycle. (2) As mycorrhizal symbionts, they form absorptive accessories to roots, linking the activities of separate plants and underpinning primary production in forests, heathlands, and grasslands. (3) In lichens, they clothe what might otherwise be bare parts of the planet. (4) As parasites, they regulate population dynamics of their hosts.

The pervasiveness of fungi in, for example, woodlands, suggests that "rather than being on the margins of forest life, fungi are...central to it, interconnecting and influencing the lives and deaths of plants and animals in countless, diverse and often surprising ways. To disregard them is to misunderstand the system" (Rayner, 1992). In his presidential address to the British Mycology Society, he showed how fungal linkages are inextricable and fundamental to the processes occurring in the ecosystem and to the sustainability of the system (Rayner, 1998). Here, fungi are able to share resources and span a range of spatial and temporal scales to mediate flows of energy and materials (Morris and Robertson, 2005). These interactions are described by Rayner (1998), wherein he regards the forest as a single organism held together by the functions and structure of fungi, rather than a collection of individuals. Much of these ideas are also expressed in the work of Trappe and Luoma (1992), who suggest that fungi are the "ties that bind" components of the ecosystem together in a literal and functional sense.

There are many textbooks describing fungi and their ecology (e.g., Cooke and Rayner, 1984; Dix and Webster, 1995; Webster and Webber, 2007). However, many of these books take a fungal view of ecology by describing the habitat in which certain species are found or the general physiological processes effected by selected species. It is the intent of this book to take a broader view of ecosystems, highlighting the major processes that occur within them and focusing on the importance of fungi in these processes. Indeed, their structure and life form make them ideal "ecosystem engineers" (*sensu* Lawton and Jones, 1995; Lavelle, 1997), a term usually reserved for animals, which have major physical effects on the environment. In this sense, we may consider fungi as forming the plumbing of the ecosystem as they are capable of regulating spatial and temporal flows of nutrient and energy through their extensive and, often, long-lived mycelial and rhizomorphic networks extending over tens or hundreds of meters (Smith et al., 1992). Although thought of as microorganisms, fungi may form extensive mycelia. Far from being ephemeral, many of these long-lived fungi are capable of continuing their function almost indefinitely and, by virtue of their continued growth, extend these functions into new areas. Their activity affects the local environment around their hyphae, at the scale of micrometers, yet their macroscale impact on the processes in the ecosystem are large (Anderson, 1995; Morris and Robertson, 2005). The complex and diverse modes of spore dispersal, together with the perennial nature of their mycelial networks, make fungi ubiquitous in all ecosystems.

So, why should we focus on fungi, rather than the more charismatic megafauna? Why should we be interested in some seemingly ephemeral mushroom that appears as a nuisance on a neatly cut lawn where a tree used to be, only to die and reappear a few months later? Why should we be interested in that mold on a piece of discarded fruit peel or in that spot on the leaf of our apple tree? It is exactly for the reasons that fungi are not very visible components of the ecosystem that they are overlooked as significant organisms in the environment. It is because the effects of fungi may be highly significant, that we need to explore the role of fungi in ecosystem processes.

Fungi are ubiquitous; they seem to be able to survive in almost all habitats. They occur in terrestrial, aquatic (Suberkropp and Chauvet, 1995), and marine (Hyde et al., 1998; Raghukumar, 2012) ecosystems, and in many extreme environments. Fungi can be active in Arctic and Antarctic conditions where the production of specific sugars allows their function to be maintained at below freezing temperatures (Robinson, 2001; Ludley and Robinson, 2008). Oligotrophic fungi (organisms that can survive in very nutrient poor environments) have been isolated and cultured from glass, where they survive by scavenging nutrients and carbon from the air (Wainwright et al., 1997; Bergero et al., 1999) and can cause significant damage to glass (Rodrigues et al., 2014). Chemotrophic fungi (organisms surviving on chemical nutrients alone) have been found in aircraft fuel lines (Hendey, 1964), deep oceans (Nagano and Nagahara, 2012; Xu et al., 2014) and hypersaline environments (Cantrell et al., 2006). Some fungal species, which are regarded as extremophiles (organisms found in extreme environmental conditions), have been isolated from the walls of the reactor room at Chernobyl, after the nuclear explosion of 1986 (Zhdanova et al., 2000).

## 1.2 WHAT ARE FUNGI?

We will not dwell on the taxonomy and structure of fungi as these topics are adequately discussed in other textbooks (Alexopoulus and Mims, 1979; Webster and

Weber, 2007). However, we will review some of the key features of the fungal body and its physiology that allow fungi to make an important contribution to ecosystem processes. The taxonomy of fungi is constantly under debate, and there are continual changes in the nomenclature of the hierarchical category under which species should be organized (Hibbett and Taylor, 2013; Money, 2013).

The mycelial portion of the fungus consists of hyphae. These hyphae, which are absent or rudimentary in the Chytridomycetes and yeasts, are a filamentous assemblage of tubular cells in which continuity is maintained between adjacent cells by the absence of cross-cell walls (septa) or a septum perforated by a pore. Thus, the hyphae develop a continual cytoplasmic connectivity between adjacent cells. Hyphae average 5–6 µm in diameter and grow by wall extension at the tip (Chiu and Moore, 1996; Moore, 1999; Moore and Novak Frazer, 2002). Because they have a narrow diameter and long length, fungal hyphae present a large surface area, relative to volume, to the environment around them. This property allows fungi to optimize the absorption of degradation products of simple carbohydrates and mineral nutrients that are derived from the action of extracellular enzymes produced by the fungi.

Fungal hyphae may grow independently or coalesce to form larger and structured assemblages called rhizomorphs or strands. These linear structures are larger and more robust than individual hyphae, and have been developed for long-distance transport of water and nutrients (Duddridge et al., 1980; Cairney, 1992; Lindahl and Olsson, 2004). Nutrient and carbon translocation has important implications for maintenance of functional continuity in a heterogeneous environment (Boddy, 1999), where resources can be reallocated within the fungal mycelium from areas of storage or excess to actively growing or functioning regions. As a result of the coenocytic arrangement of the hyphal network, there is the possibility of movement of resources within the hyphae from areas of high resource availability (sources) to areas of low availability or sites of resource demand (sinks). The movement of resources between sources and sinks is known as translocation and has been described by Jennings (1976, 1982) and reviewed by Cairney (1992), Boddy (1999), and Lamour et al. (2007). This attribute of the fungal mycelium allows for the movement of resources over short (millimeters) to long (meters) distances within the fungal mycelium, thereby reducing heterogeneity within ecosystems and connecting parts of the ecosystem in both space and time.

The temporal component, immobilization, of this activity is as important as the spatial component. As mycelia grow, there is incorporation of new carbon and mineral nutrients into the biomass of the advancing hyphal front and to more proximal biomass by translocation and immobilization. While the fungal mycelium is alive and active, much of this material remains bound to structural components or in the cytoplasm. However, upon the death of more proximal parts of the mycelium, materials incorporated into biomass may either be retranslocated from dying to living components or released into the environment via decomposition and mineralization processes. The duration of incorporation of material into biomass is regarded as an immobilization phase, where the material is unavailable for use by other organisms. The duration of this immobilization phase is dependent on the turnover time of the organism. In comparison with bacteria, which have turnover times of hours or days, some higher fungi may have hyphal turnover times of weeks or years. Hence, fungi may be important long-term accumulators of materials and, thereby, effect temporal changes in the availability of materials in the environment. Thus, the fungal mycelium is a sessile system of indeterminate growth. Andrews (1992) discusses the structure advantages and disadvantages of these "modular" organisms, and some of their attributes are listed in Table 1.1.

In their discussion of the development of fungal mycelia, Rayner et al. (1986) discuss the mycelial network as a branching linear organ designed as a means of entry into and exit from a resource by a fungus. The direction of growth and degree and direction of branching of the hyphae appear to be apically controlled. The location of nutrient supply is "sensed" by the hyphae using appropriate transporters whose regulation is by appropriate transcription factors activated by external nutrient concentration (Rutherford, 2011). The degree of branching is controlled by feedback mechanisms between the resource quality into which the hyphae are growing and the physiology (growth and metabolic activity) of the hyphae. Thus, the concept of "fast effuse" and "slow dense" was coined (Rayner et al., 1986; Cooke and Rayner, 1984) to describe the patterns of hyphal growth thought favorable (nutrient rich) media and unfavorable (nutrient-poor) media (Ritz, 1995; Rayner, 1996) (Figure 1.1). Fast growth occurs when few resources are available and slow dense growth occurs when there are abundant resources to be utilized and the fungus increases hyphal surface area to maximize enzyme release and end product absorption. The fact that the fungal hyphae appear to be ultimate fractal organisms (Ritz and Crawford, 1990) led Rayner (1996) to revise his discussion of hyphal growth in terms of nonlinear systems, fractals, and continuity of form over spatial scales. In terms of the role of fungi in ecosystem processes, fungal hyphae and rhizomorphs have resource exploitation patterns that relate to the quality of exploitable resources and the spatial

**Table 1.1    Some Attributes of Unitary and Modular Organisms**

| Attribute | Unitary (Discrete) Organisms | Modular (Nondiscrete) Organisms |
|---|---|---|
| Branching | Generally nonbranched | Generally branched |
| Mobility | Mobile; active | Nonmobile, sessile |
| Germ plasm | Segregated from soma | Not segregated |
| Growth pattern | Noniterative, determinate | Iterative, determinate |
| Internal age structure | Absent | Present |
| Reproductive value | Increases with age, then declines | Increases with delayed or no senescence |
| Environmental effects | Relatively minor role in development | Relatively major role in sessile forms |
| Effects on environment | Discrete, local effects increasing environmental heterogeneity | Discrete effects increasing heterogeneity and/or internal translocation allowing a soothing of environmental heterogeneity |
| Examples | Higher animals and plants | Fungi, bryozoans, corals, clonal plants |

Source: Modified from Andrews, J.H., *The Fungal Community: Its Organization and Role in the Ecosystem*, ed. Carrol, G.C., Wicklow, D.T., 119–145, Marcel Dekker, New York, 1992.

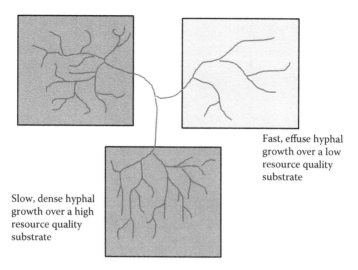

Fast, effuse hyphal
growth over a low
resource quality
substrate

Slow, dense hyphal
growth over a high
resource quality
substrate

**Figure 1.1** Diagrammatic representation of "fast effuse" and "slow dense" fungal hyphal growth over resources of contrasting quality. (After Ritz, K., *FEMS Microbiol. Ecol.*, 16, 269–280, 1995; Rayner, A.D.M., *Fungi and Environmental Change*, ed. Frankland, J.D., 317–341, Cambridge University Press, Cambridge, 1996.)

distribution of these resources (Boddy, 1999; Lindahl and Olsson, 2004; Boswell and Davidson, 2012).

To perpetuate themselves, fungi can either grow as clonal organisms or reproduce by sexual or asexual reproduction or both. In either case, new offspring can arise from spores, which are produced as dispersal agents. The production of spores often leads to the production of specialized fruiting structures on which the spores are borne. In the case of the higher fungi, Basidiomycotina and Ascomycotina, these fruiting bodies are large and visible. Because the production of spores demands energy and additional nutrients, these fruiting structures are a sink for internally translocated carbon and nutrients. Therefore, the fruiting bodies are an ideal food source for grazing animals (Claridge and Trappe, 2005), which, in turn, are important spore dispersers (Piattoni et al., 2014). In terms of fruiting structures specifically and mycelia in general, fungi are important constituents of food webs as they support secondary production (production of grazing animal biomass) in ecosystems. The functional role of fungi in a terrestrial ecosystem is schematically represented in Figure 1.2, showing the interactions between above- and belowground components of the ecosystem, plants, animals, and the abiotic environment.

## 1.3 WHAT ARE ECOSYSTEMS AND ECOSYSTEM FUNCTIONS?

The world is divided into biomes that are usually delineated by climatic constraints. Within these biomes, a number of ecosystems can exist and are usually characterized by the dominant vegetation (e.g., deciduous forest, coniferous forest,

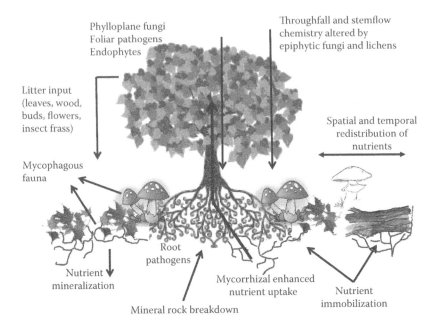

Phylloplane fungi
Foliar pathogens
Endophytes

Throughfall and stemflow
chemistry altered by
epiphytic fungi and lichens

Litter input
(leaves, wood,
buds, flowers,
insect frass)

Spatial and temporal
redistribution of
nutrients

Mycophagous
fauna

Nutrient
mineralization

Root
pathogens

Mycorrhizal enhanced
nutrient uptake

Nutrient
immobilization

Mineral rock breakdown

**Figure 1.2   (See color insert.)** Schematic representation of some of the roles of fungi in terrestrial ecosystems. (After Dighton, J., Boddy, L., *Nitrogen, Phosphorus and Sulphur Cycling in Temperate Forest Ecosystems*, ed. Boddy, L., 269–298, Cambridge University Press, Cambridge, 1989.)

grassland, tundra). Each ecosystem consists of communities of organisms that coexist and interact with each other and with the environment. The precise details of the composition of the organisms within an ecosystem depend on the geographical location, climate, and stage of evolution of that community. Thus, ecosystems can vary in the space–time continuum in that there is geographical separation of different ecosystems and that ecosystem community composition can change over time, resulting in changes in resources available to the fungal community.

Definitions and interpretations of an ecosystem can be found in many basic ecology textbooks (e.g., Ricklefs and Miller, 2000), but the definition of an ecosystem means different things to different people. Odum (1959) proposed a definition of an ecosystem as "any area of nature that includes living organisms and nonliving substances that interact to provide an exchange of materials between the living and nonliving parts is an ecological system or ecosystem." In their simplest form, ecosystems can be represented by a general energetic model in which there is a flux of kinetic energy into and between the potential energy stored within organisms in the ecosystem and a net loss of energy to the environment as a result of energy dissipative activities of the organisms (Odum, 1968). Heal and Dighton (1986) describe ecosystems in thermodynamic terms where energy flow is controlled by interactions and feedback mechanisms between the component organisms to maintain the homeostasis or stability of the system. However the ecosystem is described, there is general agreement that ecosystems consist of four main components into which fungi can become intimately involved:

1. Abiotic substances—the basic inorganic and organic compounds of the environment
2. Producers—autotrophic organisms that are capable of manufacturing food from the abiotic components of the environment by fixing energy from the sun (photo-autotrophs) or from the chemical environment (chemoautotrophs)
3. Consumers—heterotrophic organisms that are unable to fix their own energy, and derive energy and nutrient by consumption of other organisms
4. Decomposers or saprotrophs—heterotrophic organisms that break down the complex compounds of dead organisms

Within an ecosystem, these organisms occupy a niche, which is defined as "The position or status of an organism within its community and ecosystem resulting from the organism's structural adaptations, physiological responses and specific behavior (inherited and/or learned)" (Odum, 1959), or "Any dynamic segment of space, time, and energy whose occupation by biological entities is conditional on the possession of particular attributes, or combinations of attributes, by those entities" (Rayner, 1992). These definitions require that the organisms exist within the constraints of environmental factors, interactions with other organisms, and the physiological attributes of the organism to fit itself for its function within the ecosystem. Fungi are no exception, and their diversity provides adequate abilities for them to live in a variety of niches. As modular organisms, they are able to change their niche breadth by altering resource and environment within the niche (Swift et al., 1979).

The general concept of how fungi are engaged in ecosystem processes is outlined in Table 1.2, with a suggestion of the major functional groups involved in each process.

One of the difficulties in relating the role of fungi to ecosystem processes is the vast difference in spatial scale over which ecosystem processes (centimeters to kilometers) and fungal hyphae work (micrometers to meters). We think of ecosystems as being defined by climatic zones and vegetation communities that operate at the scale of tens or hundreds of square kilometers. The fungal hyphum, however, is of the order of 5–10 µm in diameter, and their immediate effect on the environment is restricted to a few micrometers on each side of these hyphae (Oberle-Kilic et al., 2013). The influence of fungi, as hyphal formers, allows them to exploit large volumes of their environment and, together with their intimate associations with both plants and animals, allows their effects to be seen at much larger scales. The interpolation of information between scales of resolution is, however, still one of the great challenges that we face, despite the emergence of hierarchy theory in ecology (O'Neill et al., 1991) and the development of models linking biogeochemical functioning of microbial communities with global processes (Schimel and Gulledge, 1998). A discussion of these problems in relation to fungi suggests that we need more information on the contribution of fungi to processes, the connection between hyphal networks, hyphal function, and the frequently assessed spore population, the relationship between results from molecular data and hyphal activity, and the interactions between fungi and other organisms in the ecosystem (Morris and Robertson, 2005).

**Table 1.2  Ecosystem Services Provided by Fungi**

| Ecosystem Service | | Fungal Functional Group |
|---|---|---|
| Soil formation | Rock dissolution | Lichens<br>Saprotrophs<br>Mycorrhizae |
| | Particle binding | Saprotrophs<br>Mycorrhizae |
| Providing fertility for primary production | Decomposition or organic residues and nutrient mineralization | Saprotrophs<br>(Ericoid and ectomycorrhizae) |
| | Soil stability (aggregates) | Saprotrophs<br>Arbuscular mycorrhizae |
| Primary production | Direct production | Lichens |
| | Nutrient accessibility | Mycorrhizae<br>Endophytes |
| | Plant yield | Mycorrhizae<br>Pathogens |
| | Defense against pathogens | Mycorrhizae<br>Endophytes<br>Saprotrophs |
| | Defense against herbivory | Endophytes |
| Plant community structure | Plant–plant interactions | Mycorrhizae<br>Pathogens |
| Secondary production | As a food source | Saprotrophs<br>Mycorrhizae |
| | Population/biomass regulation | Pathogens |
| Modification of pollutants | | Saprotrophs<br>Mycorrhizae |
| Carbon sequestration and storage | | Mycorrhizae<br>(Saprotrophs) |
| Decay of human structures and artifacts | | Saprotrophs |

*Note:* Fungal groups in parentheses are regarded as being of lesser importance in that function.

## 1.4  SPECIFIC ECOSYSTEM SERVICES CARRIED OUT BY FUNGI

In this book, I attempt to highlight some of the important ecosystem services that fungi are involved with. In particular, we will consider the role of fungi in the following major processes:

1. Primary production
   a. Making nutrients available for plant growth
   b. Enhancing nutrient uptake in the rhizosphere
2. Secondary production
   a. Providing food for both vertebrate and invertebrate animals
   b. Other fungal/faunal interactions
3. Population and community regulation by plant and animal pathogens
4. Interactions between fungi and human activities
   a. Pollution
   b. The built environment

Primary production is the activity of the autotrophs in the ecosystem that are intimately linked to fungi in the form of mycorrhizae or saprotrophs providing soil fertility. Autotrophs are mostly vascular plants, bryophytes, algae, and bacteria. The process of building more autotrophic biomass is dependent on the availability of light and carbon dioxide in the atmosphere for photosynthesis, or—in the context of chemoautotrophic bacteria—the availability of electron donors in specific chemical bonds. Plant growth requires a readily available source of mineral nutrients (particularly nitrogen, phosphorus, and potassium) from soil. Provision of these nutrients comes partially from minerals contained in the bedrock that has been eroded and incorporated into the soil profile and partly from recycling from dead plant and animal remains. As lichens, fungi are components of biological crust communities, and as ectomycorrhizae, fungi are involved in breaking rocks apart to develop the mineral component of soil (Burford et al., 2003; Belnap and Lange, 2005; Finlay et al., 2009). Other nutrients are recycled within the ecosystem and become available to plant roots by decomposition of dead plant and animal remains (Dighton, 2007). This decomposition is carried out by saprotrophic fungi in conjunction with bacteria and soil fauna. The nutrients contained in the dead material are attacked by exoenzymes from bacteria and fungi, and it is the "leakiness" of their activities—by inefficient recapture of the end products of enzyme action—that allows a proportion of the nutrients to become available to other organisms through the process of nutrient mineralization. These subjects are addressed in Chapter 2.

The uptake of nutrients by plant roots is mediated in about 90% of plant species by mycorrhizal associations of roots with fungi (Smith and Read, 1997; Bonfante and Genre, 2010). The ability of different groups of mycorrhizae to access inorganic and organic forms of nutrients and to protect the plant from microbial and faunal pathogens in the rhizosphere (Whipps, 2004; Sikes et al., 2009) is the subject of Chapter 3.

The organisms that comprise the biotic component of ecosystems coexist in communities. The interactions between members of the community can involve competition for available resources (e.g., food, light, and space) or competition among trophic levels in producer–consumer and predator–prey food web interactions (for a review of communities, see Morin, 2011). Within the context of population regulation of plants or animals, fungi play an important role as either food (promoting growth and health) or as pathogens (reducing growth and health).

As ubiquitous organisms that often occur at times when plant food is less available, fungi themselves play an important and direct role within food webs in ecosystems by being consumed by both vertebrate and invertebrate animals (Claridge and Trappe, 2005) and as such being a considerable industry providing human food. This food source is important for the growth of populations of many faunal taxa, and access to fungi as an alternate food supply may have implications on a species' competitive ability, thus influencing the development of faunal communities in an ecosystem Chapter 4).

In contrast, other fungal interaction with plants can be seen in the form of pathogens, which are detrimental to plant growth and fitness (Termorshuizen, 2014); mycorrhizae, which help plants obtain nutrients and provide defense against pathogens; and endophytes, which provide defense against herbivory and improve nutrient levels in the plant (Rudgers and Clay, 2005). Fungal pathogens of both invertebrate and vertebrate animals also cause decline in populations or health of a population of animals. Interactions with invertebrates, especially insects, can be complex as fungi may be pathogens, symbionts, or just hitching a ride from one place to another. Fungi have been implicated in the population decline of amphibians (Lipps, 1999) and, more recently, bats (Blehert, 2012; Eskew and Todd (2013). These interactions are discussed in more detail in Chapter 5.

The increasingly important role of humans on the landscape in most ecosystems around the world adds another dimension to the role of fungi in the environment. There are detrimental effects on the communities of fungi, as well as their physiology and biochemistry, through the toxicity of chemicals released by human activities. However, fungi are often tolerant to toxic elements or have abilities to sequester these elements in forms that do not interfere with their metabolism, allowing fungi to accumulate toxic metals in their biomass (e.g., Shen et al., 2013). In contrast, fungi have a great propensity to decompose both inorganic metal and organic pollutants, improving the environment in polluted areas (Dellamatrice et al., 2005; Fomina et al., 2008). Because of the relatively short generation times of microbes, particularly bacteria, fungi and actinomycetes are able to more rapidly evolve resistance to disturbance than other "higher" organisms. Thus, the effects of human interactions in the environment can induce adaptations to new conditions, allowing fungi to coexist with humans.

In the built environment, there are potential negative health effects of living with fungi and their abilities to produce copious airborne spores and toxic volatile metabolites (Crook and Burton, 2010). Fungi can play an important role in the destruction of our cultural heritage—that is, books, manuscripts, artwork (Sterflinger, 2010), as well as the structures in which we live (Schmidt, 2007). These interactions are discussed in Chapter 6.

## 1.5 CONCLUSIONS

In summary, the intent of the following information is to report on and speculate from the research that has been done since Jack Harley presented his presidential address to the British Ecological Society, entitled "Fungi in Ecosystems" (Harley, 1971). In this address, Harley states that "it is clear from recent work that the magnitude of their [fungal] intervention in nutrient and energy cycling may be very great," and he presents a diagrammatic scheme of fungal intervention in the carbon cycle (Figure 1.3). However, Harley cautions that there are two difficulties in determining the roles of fungi in ecosystems. First, there are difficulties in determining the identity (species) of fungi in their vegetative state and, second, of being able to predict the true physiology of fungi in the natural environment from the studies conducted with fungal cultures in the laboratory. The former problem is being overcome by the development of molecular tools to identify fungal species, and the second is being addressed by careful manipulation studies in the field and the use of isotope tracer

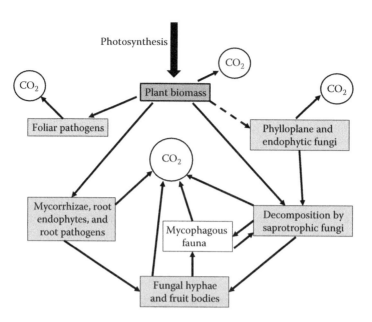

**Figure 1.3**   Broad outline of the role of fungi in carbon cycling in the ecosystem. Dashed line represents partial carbon transfer to the endophytic fungal component. (Redrawn from Harley, J.L., *J. Ecol.*, 59, 653–668, 1971.)

and natural abundance isotope studies. However, the links between these studies and the ecophysiology of fungi are still in their infancy. These studies are further hampered by our lack of understanding of the physiology of the majority of fungi in the world. Hawksworth (1991) reported that, at that time, we had identified some 69,000 fungal species worldwide. He then predicted the possible total number of fungi based on current knowledge of fungal species associated with plants and animals, and the rate of discovery of new plant and animal species. He arrived at a potential fungal diversity of approximately 1.5 million species (Hawksworth, 2001). Recent literature suggests that this is an underestimation, and based on molecular methods the assessment is up to about 5 million species (O'Brien et al., 2005; Blackwell, 2011). The molecular taxonomic age has increasingly revealed that old taxonomy has given different names to sexual and asexual stages of the same species and pleomorphic species problems are being untangled, resulting in a push to develop a new type of taxonomy not based on Linnean principles (Hibbett and Taylor, 2013) and using molecular bar-coding to identify individuals (Money, 2013).

For a functional perspective, we will still need to be able to characterize fungi on a functional taxonomic basis, irrespective of their true taxonomy and phylogeny as the biology, biochemistry, and physiology of most fungal species are still relatively unknown. The use of functional genomics is still in its infancy, and it is proposed that this is an area of fruitful development to link taxonomy to functionality in the fungal kingdom (Lee and Dighton, 2013).

The debate about the role of species diversity and ecosystem function has been started by investigation of plant community structure (Tilman et al., 1996; Naeem et al., 1994, 1996). In the fungal context, there is probably a greater need to separate out functional diversity from taxonomic diversity (Zak et al., 1994), as numerous species across a number of higher taxonomic categories are capable of carrying out similar functions. It is only recently that researchers have started to investigate the functional diversity of microbial communities (Ritz et al., 1994) and started linking this functional diversity to the ecosystem level functions of nutrient and energy flow in, primarily, terrestrial ecosystems (de Ruiter et al., 1998; Ekschmitt and Griffiths, 1998; Hodkinson and Wookey, 1999; Baxter and Dighton, 2001). It is with this level of ignorance, a high degree of extrapolation of function between species, and confronting the problems of transcending spatial scales from the micro- to landscape level (Friese et al., 1997) and temporal scales of seconds to decades that we will proceed to explore the role of fungi in ecosystem processes and amplify the sentiments of Rayner (1998)!

## REFERENCES

Alexopoulos, C. J. and C. W. Mims. 1979. *Introductory Mycology*. Third ed. Chichester, UK: John Wiley.

Anderson, J. M. 1995. Soil organisms as engineers: Microsite modulation of macroscale processes. In *Linking Species and Ecosystems*, ed. C. G. Jones and J. H. Lawton, 94–106, New York: Chapman Hall.

Andrews, J. H. 1992. Fungal life-history strategies. In *The Fungal Community: Its Organization and Role in the Ecosystem*, ed. G. C. Carrol and D. T. Wicklow. 119–145, New York: Marcel Dekker.

Baxter, J. W. and J. Dighton. 2001. Ectomycorrhizal diversity alters growth and nutrient acquisition of gray birch (*Betula populifolia* Marshall) seedlings in host–symbiont culture conditions. *New Phytol* 152:139–149.

Belnap, J. and O. L. Lange. 2005. Lichens and microfungi in biological crusts: Community structure, physiology, and ecological functions. In: *The Fungal Community: Its Organization and Role in the Ecosystem*, ed. J. Dighton, J. F. White, and P. Oudemans, 117–138. Boca Raton, FL: Taylor & Francis.

Bergero, R., M. Girlanda, G. C. Varese, D. Intilli, and A. M. Luppi. 1999. Psychooligotrophic fungi from Arctic soils of Franz Joseph Land. *Polar Biol.* 21:361–368.

Blackwell, M. 2011. The fungi: 1, 2, 3… 5.1 million species? *Am. J. Bot.* 98:426–438.

Blehert, D. S. 2012. Fungal disease and the developing story of bat white-nose syndrome. *PLoS Pathogens* 8:1–3 e1002779. doi:10.1371/journal. ppat.1002779.

Boddy, L.1999. Saprotrophic cord-forming fungi: Meeting the challenge of heterogenous environments. *Mycologia* 91:13–32.

Bonfante, P. and A. Genre. 2010. Mechanisms underlying beneficial plant-fungus interactions in mycorrhizal symbiosis. *Nature Comm.* 48. doi:10.1038/ncmms1046.

Boswell, G. P. and F. A. Davidson. 2012. Modelling hyphal networks *Fung. Biol. Rev.* 26:30–38.

Burford, E., M. Kierans, and G. M. Gadd. 2003. Geomycology: Fungi in mineral substrata. *Mycologist* 17:98–107.

Cairney, J. W. G. 1992. Translocation of solutes in ectomycorrhizal and saprotrophic rhizomorphs. *Mycol. Res.* 96: 135–141.

Cantrell, S. A., L. Castillas-Martínez, and M. Molina. 2006. Characterization of fungi from hypersaline environments and solar salterns using morphological and molecular techniques. *Mycol. Res.* 110:962–970.

Carroll, G. C. and D. T. Wicklow. (Eds.) 1992. *The Fungal Community: Its Organization and Role in the Ecosystem.* New York: Marcel Dekker, 976 pp.

Chiu, S. W. and D. Moor. (Eds) 1996. *Patterns in Fungal Development.* Cambridge, UK: CUP, 226 pp.

Claridge, A. W. and J. M. Trappe. 2005. Sporocarp mycophagy: Nutritional, behavioral, evolutionary and physiological aspects. In: *The Fungal Community: Its Organization and Role in the ecosystem* (3rd Edition), ed. J. Dighton, J. F. White, and P. Oudemans, 599–611. Boca Raton, FL: Taylor & Francis.

Cooke, R. C. and A. D. M. Rayner. 1984. *Ecology of Saprotrophic Fungi.* London: Longman.

Crook, B. and N. V. Burton. 2010. Indoor moulds, Sick Building Syndrome and building related illness. *Fung. Biol. Rev.* 24:106–113.

Dadachova, E., R. A. Bryan, X. Huang et al. 2007. Ionizing radiation changes the electronic properties of melanin and enhances the growth of melanized fungi. PLoS ONE 2(5): e457. doi:10.1371/jornal.pone.0000457.

Dadachova, E. and A. Casadevall. 2008. Ionizing radiation: How fungi cope, adapt, and exploit with the help of melanin. *Curr. Opin. Microbiol.* 11:1–7.

de Ruiter, P. C., A-M. Neutel, and J. C. Moore. 1998. Biodiversity in soil ecosystems: The role of energy flow and community stability. *Appl. Soil Ecol.* 10:217–228.

Dellamatrice, P. M., R. T. R. Montiero, H. M. Kamida, N. L. Nogueira, M. L. Rossi, and C. Blaise. 2005. Decolorization of municipal effluent and sludge by *Pleurotus sajor-caju* and *Pleurotus ostreatus. World J. Microbiol. Biotechnol.* 21:1363–1369.

Dighton, J. 2007. Nutrient cycling by saprotrophic fungi in terrestrial habitats. In: *The Mycota IV Environmental and Microbial Relationships*, 2nd Edn., ed. C. P. Kubicek and I. S. Druzhinina, 287–300. Berlin: Springer-Verlag.

Dighton, J. and L. Boddy. 1989. Role of fungi in nitrogen, phosphorus and sulphur cycling in temperate forest ecosystems. In: *Nitrogen, Phosphorus and Sulphur Cycling in Temperate Forest Ecosystems*, ed. L. Boddy, R. Marchant, and D. J. Read, 269–298. Cambridge: Cambridge University Press.

Dix, N. J. and J. Webster. 1995. *Fungal Ecology.* London: Chapman and Hall.

Duddridge, J. A., A. Malibari, and D. J. Read. 1980. Structure and function of mycorrhizal rhizomorphs with special reference to their role in water transport. *Nature* 287:834–836.

Ekschmitt, K. and B. S. Griffiths. 1998. Soil biodiversity and its implications for ecosystem functioning in a heterogeneous and variable environment. *Appl. Soil Ecol.* 10:201–215.

Eskew, E. A. and B. D. Todd. 2013. Parallels in amphibian and bat declines from pathogenic fungi. *Emerging Infectious Diseases* 19:379–385.

Finlay, R., H. Wallander, M. Smits et al. 2009. The role of fungi in biogenic weathering in boreal forest soils. *Fung. Biol. Rev.* 23:101–106.

Fomina, M., J. M., Charnock, S. Hillier, R. Alvarez, F. Livens, and G. M. Gadd. 2008. Role of fungi is the biogeochemical fate of depleted uranium. *Curr. Biol.* 18: R375–R377.

Friese, C. F., S. J. Morris, and M. F. Allen. 1997. Disturbance in natural ecosystems: Scaling from fungal diversity to ecosystem functioning. In: *The Mycota IV: Environmental and Microbial Relationships*, ed. D. T. Wicklow and B. Soderstrom, 47–63. Berlin: Springer Verlag.

Harley, J. L. 1971. Fungi in ecosystems. *J. Ecol.* 59:653–668.

Hawksworth, D. L. 1991. The fungal dimension of biodiversity: Magnitude, significance and conservation. *Mycol. Res.* 95:641–655.

Hawksworth, D. L. 2001. The magnitude of fungal diversity: The 1.5 million species estimate revisited. *Mycol. Res.* 105:1422–1432.

Heal, O. W. and J. Dighton. 1986. Nutrient cycling and decomposition of natural terrestrial ecosystems. *Microfloral and Faunal Interactions in Natural and Agro-Ecosystems*, ed. M. J. Mitchell and J. P. Nakas, 14–73. Martinus Nijhoff/Dr. W. Junk.

Hendey, N. I. 1964. Some observations on *Cladosporium resinae* as a fuel contaminant and its possible role in the corrosion of aluminium fuel tanks. *Trans. Br. Mycol. Soc.* 47:467–475.

Hibbett, D. S. and J. W. Taylor. 2013. Fungal systematics: Is a new age of enlightenment at hand? *Nat. Rev. Microbiol.*, pp. 1–5, doi:10.1038/nrmicro2942.

Hodkinson, I. D. and P. A. Wookey. 1999. Functional ecology of soil organisms in tundra ecosystems: Towards the future. *Appl. Soil Ecol.* 11:111–126.

Hyde, K. D., E. B. Gareth Jones, E. Leano, S. B. Pointing, A. D. Poonyth, and L. L. P. Vrijmoed. 1998. Role of fungi in marine ecosystems. *Biodiver. Conserv.* 7:1147–1161.

Jennings, D. H. 1976. Transport and translocation in filamentous fungi. In: *The Filamentous Fungi*, ed. J. E. Smith and D. R. Berry Vol. 2, 32–64. London: Edward Arnold.

Jennings, D. H. 1982. The movement of *Serpula lacrimans* from substrate to substrate over nutritionally inert surfaces. In: *Decomposer Basidiomycetes: Their Biology and Ecology*, ed. J. C. Frankland, J. N. Hedger, and M. J. Swift, 91–108. Cambridge: Cambridge University Press.

Lamour, A., A. J. Termorshuizen, D. Volker, and M. Jeger. 2007. Network formation by rhizomorphs of *Armillaria lutea* in natural soil: Their description and ecological significance. *FEMS Microbiol. Ecol.* 62:222–232.

Lavelle, P. 1997. Faunal activities and soil processes: Adaptive strategies that determine ecosystem function. *Adv. Ecol. Res.* 27:93–132.

Lawton, J. H. and C. G. Jones. 1995. Linking species and ecosystems: Organisms as ecosystem engineers. In: *Linking Species and Ecosystems*, ed. C. G. Jones and J. H. Lawton, 141–150. New York: Chapman Hall.

Lee, K. and J. Dighton. 2013. Advancement of functional genomics of a model species of *Neurospora* and its use for ecological genomics of soil fungi. In: *Genomics of Soil- and Plant-Associated Fungi*, ed. B. A. Horwitz et al., 29–44. Berlin: Springer Verlag.

Lindahl. B. D. and S. Olsson. 2004. Fungal translocation—Creating and responding to environmental heterogeneity. *Mycologist* 18:79–88.

Lipps, K. R. 1999. Mass mortality and population declines of anurans at an upland site in western Panama. *Conserv. Biol.* 15:117–125.

Ludley, K. E. and C. H. Robinson. 2008. Decomposer Basidiomycota in Arctic and Antarctic ecosystems. *Soil Biol. Biochem.* 40:11–29.

Money, N. P. 2013. Against naming fungi. *Fung. Biol.* 117:463–465.

Moore, D. (Ed.) 1999. *Fungal Morphogenesis*. New York: CUP, 486 pp.

Moore, D. and L. A. Novak Frazer. 2002. *Essential Fungal Genetics*. Berlin: Springer Verlag, 357 pp.

Morin, P. J. 2011. *Community Ecology*, Second Edition. Oxford, UK: Wiley Blackwell, 407 pp.

Morris, S. J. and G. P. O. Robertson. 2005. Linking function between scales of resolution. In: *The Fungal Community: Its Organization and Role in the Ecosystem*, ed. J. Dighton, J. F. White and P. Oudemans, 13–26. Boca Raton, FL: Taylor & Francis.

Naeem, S., K. Hakamsson, J. H. Lawton, M. J. Crawley, and L. J. Thompson. 1996. Biodiversity and plant productivity in a model assemblage of plant species. *Oikos* 76:259–264.

Naeem, S., L. J. Thompson, S. P. Lawler, J. H. Lawton, and R. M. Woodfin. 1994. Declining biodiversity can alter the performance of ecosystems. *Nature* 365:734–737.

Nagano, Y. and T. Nagahama. 2012. Fungal diversity in deep-sea extreme environments. *Fung. Ecol.* 5:463–471.

O'Brien, H. E., J. L. Parrent, J. A. Jackson, J.-M. Moncalvo, and R. Vilgalis. 2005. Fungal community analysis by large-scale sequencing of environmental samples. *Appl. Environ. Microbiol.* 71:5544–5550.

Oberle-Kilic, J., J. Dighton, and G. Arbuckle-Keil. 2013. Atomic force microscopy and micro-ATR–FT-IR imaging reveals fungal enzyme activity at the hyphal scale of resolution. *Mycology* 4:44–53.

Odum, E. P. 1959. *Fundamentals of Ecology*. Philadelphia: Saunders.

Odum, E. P. 1968. *Fundamentals of Ecology*. Philadelphia: Saunders.

O'Neill, E. G., R. V. O'Neill, and R. J. Norby. 1991. Hierarchy theory as a guide to mycorrhizal research on large-scale problems. *Environ. Pollut.* 73:271–284.

Piattoni, F., A. Amicucci, M. Iotti, F. Ori, V. Stocchi, and A. Zambonelli. 2014. Viability and morphology of *Tuber aestivum* spores after passage though the gut of *Sus scrofa*. *Fung. Ecol.* 9:52–60.

Raghukumar, C. (Ed.) 2012. *Biology of Marine Fungi*. Berlin: Springer Verlag, 334 pp.

Rayner, A. D. M. 1992. Introduction. In *The Fungal Community: Its Organization and Role in the Ecosystem*, ed. G. C. Carroll and D. T. Wicklow, xvii–xxiv. New York: Marcel Dekker.

Rayner, A. D. M. 1996. Has chaos theory a place in environmental mycology. In *Fungi and Environmental Change*, ed. J. C. Frankland, N. Magan, and E. M. Gadd, 317–341. Cambridge University Press.

Rayner, A. D. M. 1998. Fountains of the forest—The interconnectedness between trees and fungi. *Mycol. Res.* 102:1441–1449.

Rayner, A. D. M., K. A. Powell, W. Thompson, and D. H. Jennings. 1986. Morphogenesis of vegetative organs. In *Developmental Biology of Higher Fungi*, ed. D. Moore, L. A. Casselton, D. A. Wood, and J. C. Frankland, 249–279. Cambridge: Cambridge University Press.

Ricklefs, R. E. and G. L. Miller. 2000. *Ecology*. New York: W. H. Freeman and Co.

Ritz, K. 1995. Growth responses of some fungi to spatially heterogeneous nutrients. *FEMS Microbiol. Ecol.* 16:269–280.

Ritz, K. and J. Crawford. 1990. Quantification of the fractal nature of colonies of *Trichoderma viride*. *Mycol. Res.* 94:1138–1152.

Ritz, K., J. Dighton, and K. E. Giller. 1994. *Beyond the Biomass: Compositional and Functional Analysis of Soil Microbial Communities*. Chichester, UK: John Wiley and Sons.

Robinson, C. H. 2001. Cold adaptation in Arctic and Antarctic fungi. *New Phytol.* 151:341–353.

Rodrigues, A., S. Gutierrez-Patricio, A. Miller et al. 2014. Fungal deterioration of stained-glass windows. *Int. Biodeter. Biodegr.* 90:152–160.

Rudgers, J. A. and K. Clay. 2005. Fungal endophytes in terrestrial communities and ecosystems. In: *The Fungal Community: Its Organization and Role in the Ecosystem*, 3rd Edition, ed. J. Dighton, J. F. White, and P. Oudemans, 423–442. Boca Raton, FL: Taylor & Francis.

Rutherford, J. 2011. Direct sensing of nutrient availability by fungi. *Fung. Biol. Rev.* 25:111–119.

Schimel, J. P. and J. Gulledge. 1998. Microbial community structure and global trace gases. *Global Change Biol.* 4:745–758.

Schmidt, O. 2007. Indoor wood-decay basidiomycetes: Damage, causal fungi, physiology, identification and characterization, prevention and control. *Mycol. Progress* 6:261–279.

Shen, M., L. Liu, D-W. Li et al. 2013. The effect of endophytic *Peyronellaea* from heavy metal-contaminated and uncontaminated sites on maize growth, heavy metal absorption and accumulation. *Fung. Ecol.* 6:539–545.

Sikes, B. A., K. Cottenie, and J. N. Klironomos. 2009. Plant and fungal identity determines pathogen protection of plant roots by arbuscular mycorrhizas. *J. Ecol.* 97:1274–1280.

Smith, M. L., J. N. Bruhn, and J. B. Anderson. 1992. The fungus *Armillaria bulbosa* is among the largest and oldest living organisms. *Nature*. 356:428–431.

Smith, S. E. and D. J. Read. 1997. *Mycorrhizal Symbiosis*. San Diego: Academic Press.

Sterflinger, K. 2010. Fungi: Their role in deterioration of cultural heritage. *Fung. Biol. Rev.* 24:47–55.

Suberkropp, K. and E. Chauvet. 1995. Regulation of leaf breakdown by fungi in streams: Influences of water chemistry. *Ecology* 76:1433–1445.

Swift M. J., O. W. Heal, and Anderson J. M. 1979. *Decomposition in Terrestrial Ecosystems*. London: Blackwell.

Termorshuizen, A. J. 2014. Root pathogens. In: *Interactions in Soil: Promoting Plant Growth*, ed. J. Dighton and J. A. Krumins, 119–137. Dordrecht: Springer.

Tilman, D., D. Wedin, and J. Knops. 1996. Productivity and sustainability influenced by biodiversity in grassland ecosystems. *Nature* 379:718–720.

Trappe, J. M. and D. L. Louma. 1992. The ties that bind: Fungi in ecosystems. In: *The Fungal Community: Its Organization and Role in the Ecosystem*, ed. G. C. Carroll and D. T. Wicklow, 17–27. New York: Marcel Dekker.

Tugay, T., N. N. Zhdanova, V. Zheltonozhsky, L. Sadovnikov, and J. Dighton. 2006. The influence of ionizing radiation on spore germination and emergent hyphal growth response reactions of microfungi. *Mycologia* 98:521–529.

Wainwright, M., K. Al-Wajeeh, and S. J. Grayston. 1997. Effect of silicic acid and other silicon compounds on fungal growth in oligotrophic and nutrient-rich media. *Mycol. Res.* 101:933–938.

Webster, J. and R. W. S. Webber. 2007. *Introduction to Fungi*. Cambridge, UK: Cambridge University Press, 841 pp.

Whipps, J. M. 2004. Prospects and limitations for mycorrhizas in biocontrol of root pathogens. *Can. J. Bot.* 82:1198–1227.

Xu, W., K.-L. Pang, and Z-H. Luo. 2014. High fungal diversity and abundance recovered in the deep-sea sediments of the Pacific Ocean. *Microb. Ecol.* 68:688–698.

Zak, J. C., R. Michael, R. Willig, D. L. Moorhead, and H. G. Wildman. 1994. Functional diversity of microbial communities: A quantitative approach. *Soil Biol. Biochem.* 26:1101–1108.

Zhdanova, N. N., V. A. Zakharchenko, V. V. Vember, and L. T. Nakonechnaya. 2000. Fungi from Chernobyl: Mycobiota of the inner regions of the containment structures of the damaged nuclear reactor. *Mycol. Res.* 104:1421–1426.

# CHAPTER 2

# Making Nutrients Available for Primary Production

Within ecosystems, primary production is carried out by autotrophic organisms. These organisms—plants—are able to fix carbon by the process of photosynthesis and build biomass by combining this fixed carbon with nutrient elements derived from the environment. The nutrients required for plant growth come from two main sources. The first source is the rock material underlying the soil. This rock may be of local origin, or—in areas that have been affected by glaciation—of remote geology. Rocks of the earth's crust contain a variety of the essential mineral nutrients that plants need, but the minerals are bound in complex chemical forms that make them poorly available for plant uptake. By the action of environmental factors (wind, water, and physical disturbance) along with the activities of bacteria, fungi, and plant roots, the surface of rocks can be weathered and degraded to finer particles, and the mineral nutrients released in a soluble form that can be accessed by plants. Some of these minerals will be carried in water to streams, rivers, and oceans, imparting fertility to these ecosystems. The second source of nutrients is by the breakdown or decomposition of dead plant and animal remains by microbes and animals. During decomposition, mineral nutrients are released in a soluble form as inorganic ions from the breakdown of the organic complexes within the plant and animal remains. This process is called mineralization, and provides fertility to the ecosystem. Decomposition and mineralization occur in terrestrial, freshwater, and marine ecosystems. In this chapter, we investigate the role that fungi play in these processes (Table 2.1). The bulk of the chapter deals with terrestrial ecosystems as this is where most of the information on these processes had been derived. The impact of decomposition activity within terrestrial ecosystems has a profound effect on the fertility of streams and rivers by the process of leaching. Here, nutrient elements in water percolate through the soil into water courses, carrying with them soluble nutrients derived in the terrestrial environment and that have not been immobilized into land plant tissue.

Table 2.1    Ecosystem Services Provided by Fungi in Relation to Soil Formation
and Making Nutrients Available for Plant Growth

| Ecosystem Service | | Fungal Functional Group |
|---|---|---|
| Soil formation | Rock dissolution | Lichens |
| | | Saprotrophs |
| | | Mycorrhizae |
| | Particle binding | Saprotrophs |
| | | Mycorrhizae |
| Providing fertility for primary production | Decomposition or organic residues | Saprotrophs (Ericoid and ectomycorrhizae) |
| | Nutrient mineralization | Saprotrophs (Ericoid and ectomycorrhizae) |
| | Soil stability (aggregates) | Saprotrophs Arbuscular mycorrhizae |

Note: Fungal groups in parentheses are regarded as of lesser importance in that function.

## 2.1 MAKING SOILS

Soils are a complex composition of weathered mineral rock, organic material derived from dead plant and animal remains together with the living biota of bacteria, actinomycetes, fungi, protozoa, nematodes, soil microarthropods, and other small fauna (Coleman et al., 2004; Wall, 2012). The holistic complement of abiotic and biotic components enables a functional soil to sustain plant growth (Dighton and Krumins, 2014). The importance of soil fertility has been known since the time of the development of agricultural practices in the Nile delta. The consequences of loss of stability of the tightly coupled interactions between the biotic and abiotic soil entities through mismanagement of agricultural soils and in combination with changes in climate result in the loss of soil fertility, soil erosion, and reduced crop yield.

Soils do not just occur—they are created by the breakdown of parent rock into mineral particles. It is the surface ionic exchange properties of these mineral particles that give soil its fertility. The greater the degree of dissociation of ions into soil solution, the more fertile the soil becomes, as plants are best able to access freely soluble nutrients. Weathering of parent rock material may be accomplished by a variety of abiotic factors. Brady and Weil (1999) describe the processes of mineral rock breakdown caused by weathering by wind and water, freeze–thaw cycles, and the effects of weak acids, formed by carbon dioxide combining with rainwater. However, there are a number of biotic factors that also influence the rate of parent rock breakdown, which influences the development of soils, of which fungi play a significant role as lichens, saprotrophs, and mycorrhizae.

### 2.1.1 Lichens and Biological Crusts

Lichens, symbiotic association of algae, bacteria, and ascomycete or, less frequently, basidiomycete fungi frequently colonize bare rock. Here, they are able to survive extreme environments of heat, cold, drought, and nutrient stress that other organisms are less able to tolerate. Approximately 8% of terrestrial ecosystems are lichen dominated, and in many of these systems, the ground cover by lichens is often very high, up to 100% (Honegger, 1991).

The definition of lichen is discussed by Hawksworth (1988). Definitions range from Berkely's 1869 suggestion that "it is quite impossible to distinguish some lichens from fungi" to Hawksworth's 1983 definition of "a stable self-supporting associ- ation of a fungus (mycobiont) and an alga or Cyanobacteria (photobiont)." Later, Hawksworth revised the definition to "A lichen is a stable self-supporting association of a mycobiont and a photobiont in which the mycobiont is the exhabitant," which suggests that the photobiont resides within the fungal tissue. Indeed, Sanders (2001) considers lichens to be "the interface between mycology and plant morphology." The algal symbiont is usually a green or yellow-green eucaryotic alga and, sometimes, blue-green prokaryotic Cyanobacteria. The algae are restricted to the upper zones of fungal tissue, where light is maximal for photosynthesis. The fungal associate is usually an Ascomycete or a Deuteromycete with occasional Basidiomyctes that are restricted to the genus *Omphalina* (Hawksworth, 1988). Fungi usually form the basal portion of the lichen, which may be differentiated into a stalk-like structure or podetia. Fungal tissues form the greater proportion of the biomass of lichens and are the supporting tissue for the algal symbiont. In addition to the combination of algae and fungi, other, nonphotosynthetic bacteria may also be present within the lichen (Banfield et al., 1999) and play a role in soil biogenesis.

Brady and Weil (1999) show that biogeochemical weathering of rock is a func- tion of water availability, the presence of organic acids, and complexation processes. Specifically, water is involved in hydration, hydrolysis, and dissolution. Hydration of oxides of iron and aluminum is an important process in rock degradation; for example, hematite ($Fe_2O_3$) is converted into ferrihydrate ($Fe_{10}O_{15} \cdot 9H_2O$). Hydrolysis is important in the release of essential nutrients for plant growth. For example, potassium is released from microcline, a feldspar ($KAlSi_3O_8$). Dissolution allows the dissociation of anions and cations from complex materials. For example, gypsum dissolves to release calcium and sulfate ions. In dry areas, the structure of lichens acts as point of condensation of water and a site on which atmospheric water can col- lect (Lange et al., 1994). Therefore, they are nuclei for water-related rock weathering processes. A review of rock weathering by lichens is given by Chen et al. (2000).

The presence of living organisms on rock surfaces increases the carbon dioxide concentration in the atmosphere, because of respiration, which when combined with rainfall forms carbonic acid. This weak organic acid is an important agent of dissolution of the calcite found in limestone and marble $\left( CaCO_3 + H_2CO_3 \Leftrightarrow Ca^{++} + 2HCO_3^- \right)$. Lichens and soil fungi and bacteria are organisms that produce organic acids, such as oxalic, citric, lichenic, and tartaric acids, which, in turn, contribute to the chemi- cal weathering of rocks by lowering pH and increasing the solubility of aluminum and silicon. They also form chelation products (complexes between inorganic ions and organic molecules) and release inorganic nutrient elements. For example, oxalic acid dissolves solid muscovite to produce soluble inorganic potassium and soluble, chelated aluminum ($K_2[Si_6Al_2]Al_4O_{20}(OH)_4 + 6C_2O_4H_2 + 8H_2O \Leftrightarrow 2K^+ + 8OH^- + 6C_2O_4Al^+ + 6Si(OH)_4^0$). Oxalic acid is known to be produced by fungal hyphae, whereas lichenic acid is specific to lichens. This action of lichens has also been reported to cause significant damage to both buildings and sculptures made of rock (Chen et al., 2000).

In terricolous lichens, the degree of interactions between the lichen and the underlying substrate can be characterized as Type 1 lichens (represented by the genus *Baeomyces*) with very intimate association between the lichen body and the underlying substrate; Type 2 lichens (e.g., *Peltgera*) with a leafy thallus and elaborate, but less intimate, system of attachment to the substrate; or Type 3 lichens (e.g., *Cladonia*), where the primary thallus is almost absent and the podetia have little contact with the substrate (Asta et al., 2001). Using thin sections for light and electron microscopy, Asta et al. (2001) showed that the lichen–rock interface is primarily associated with the fungal component of the lichen and consists of both individual hyphae and differentiated rhizomorphs. Although these rhizomorphs are thought to be important for translocation of water and nutrients, they do not have clearly differentiated internal structures for translocation (Sanders, 1997), as do the rhizomorphs of some ectomycorrhizal fungi (Duddridge et al., 1980). The interface between substratum and lichen in *Baeomyces* is more structured and results in reorientation of mineral particles, biodegradation of the walls of plant debris, and bonding between these two components. In contrast, *Cladonia* had a more diffuse association with the substrate and fungal hyphae escape from the lichen body and are incorporated into the soil (Asta et al., 2001). The production of polysaccharides by fungal hyphae is also important in the development of organomineral complexes, which bind mineral particles together and act as a carbon source promoting bacterial growth, which may enhance the rock degradation process (Asta et al., 2001).

Banfield et al. (1999) commented that, in addition to lichens possessing an upper layer of fungal hyphae containing photosynthetically active algae or Cyanobacteria, the fungal matrix is also a refuge for a community of nonphotosynthetic bacteria, the diversity and function of which are largely unknown. Their lichen/mineral rock weathering zone model (Barker and Banfield, 1998; Banfield et al., 1999) is shown in Figure 2.1. Zone 1 is the region of generation of lichen acids in the photosynthetic region of the lichen. Zone 2 consists of the area of biophysical disaggregation, where the fungal and nonphotosynthetic bacteria interact closely with weathered mineral rock particles, fungal hyphae, and rhizomorphs, which penetrate into fissures in the rock. Hyphal aggregations become narrower as the hyphae penetrate deeper into the underlying rock until only single hyphae exist (Ascaso and Wierzchos, 1995). Zone 2 is the area of most intense mineral weathering with maximal contact between cells, secreted polymers, and mineral surfaces. Here, clay particles and organic polymers are formed at the nanometer-scale and metal–lichen acid complexes occur, which do not occur deeper into the rock (Ascaso and Wierzchos, 1995), such as complexes of ferric oxide in *Acaroospora sinoptica*, aluminum in *Tremolecia atrata*, copper oxalate in *Acaraspora rugulosa* and *Lecidia theiodes*, and complexing of copper in the cortex of *Lecidia lacteal* and copper-psoromic acid in *Lecidella bullata*. In the underlying Zone 3, solutions containing lichen-derived organic acids effect chemical solubilization of the parent rock material. This is primarily a biogeochemical interaction and is not mediated by direct microbial contact. Finally, the bottom of

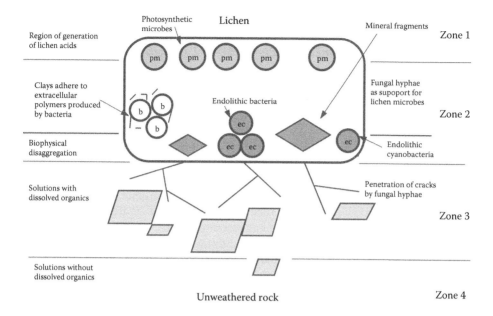

**Figure 2.1** Model indicating the four zones of activity within a mineral weathering lichen as depicted by Banfield et al. (1999). In Zone 1, photosynthetic members generate carbon and crystalline lichenic acids. In Zone 2, there is direct contact between microbes, organic products, and the mineral surface. In Zone 3, organic acids act to solubilize rock in the presence of direct rock/organism contact, particularly fungal hyphal penetration into cracks. Zone 4 is characterized by unweathered rock and inorganic chemical reactions.

Zone 3 represents the unweathered rock, where water can penetrate but does not carry organic acids. Ascaso and Wierzchos (1995) point out that there is a temporal component to the development of the lichen/soil interaction, in which microbial populations and diversity increase as the weathering continues and as a more diverse soil structure develops.

In contrast to this evidence of the role of lichens in the weathering of rock and the destruction of buildings and monuments made of rock, Mottershead and Lucas (2000) present evidence to suggest that the cover of *Aspicilia calcarea* and *Diploscistes diacapsis* lichens on calcareous stonework in Europe can protect against rock solubilization. They show evidence that lichen-protected areas of gypsum were 15 mm higher than the adjacent, uncovered areas, where the lichen layer increases the rate of shedding of rainwater containing acidic pollutants that would have eroded the rock surface.

In addition, the aerial parts of the lichen trap particles of dust, which together with dead parts of the lichen (Crittenden, 1991) contribute to the organic component of the protosoil produced. Thus, after a certain length of time (usually years), lichens contribute significantly to the formation of the mineral component of a new soil and, to some degree, to the organic component.

## 2.1.2 Role of Fungi in Rock Breakdown

Nonlichen fungi also produce organic acids that are capable of breaking down rock. Ascaso and Wierzchos (1995) cite studies showing that yeasts and filamentous fungi, such as *Aspergillus niger*, alone are involved in rock solubilization, releasing cations from amphibolite, biotite and orthoclase. *Penicillium* and yeasts were also found to be able to dissolve calcium-rich rocks, such as limestone, marble, and calcium phosphate (Chang and Li, 1998), and can be extended to include mollusk shells and corals (Golubic et al., 2007) and the surface of buildings (Bogomolova, 2007). Comprehensive reviews of rock weathering by fungi can be found in the work of Burford et al. (2003), Finlay et al. (2009), and Rosling et al. (2009). These reviews show that not only are lichen fungi important, but a range of other fungi are associated with the breakdown of specific rock minerals. Microfungi are highlighted by Burford et al. (2003) and saprotrophic and ectomycorrhizae by Finlay et al. (2009). The ability of fungi to penetrate into rock and through small fissures allows them to act both enzymatically and through acidic secretions to break down parent rock, and alter rock properties and secondary materials as the basis of a soil substrate (Fomina et al., 2010). Finlay et al. (2009) also suggest that fungal hyphae increase water holding capacity in rock fissures that may increase the likelihood of freezing effects on rock fracturing (Fomina et al., 2010)

Connolly et al. (1998) showed that the white rot wood decay fungus, *Resinicium bicolor*, could solubilize strontianite sand to release the strontium contained within. Fungi were found in small holes (3–10 μm diameter) in feldspars and horneblende (Jongmans et al., 1997), produced by micromolar concentrations of organic acids (succinic, citric, oxalic, formate, and malate) secreted by saprotrophic and ectomycorrhizal fungi associated with the overlying pine forest ecosystem. Thin sections of feldspars observed under the microscope have revealed fungal hyphae bearing cross walls in hyphal-generated tunnels in the rock (Hoffland et al., 2001). In addition to organic acid production, Martino and Perotto (2010) also suggest that the physical component of hyphal turgor pressure is also important in hyphal tunneling, allowing fungal hyphae to force their way through tiny fractures in rock. The physical structure and chemical composition of rock is probably more important than hardness *per se*, based on the number of fungal species isolated from rocks of different types after Burford et al. (2003), although the number of lichen species isolated from rock is significantly negatively related to rock hardness (number of lichen species = −11.1 × hardness + 7.57, $P$ = 0.016; calculated from their data presented in Table 1; Burford et al., 2003).

Rock weathering is significantly influenced by oxalates produced by fungal hyphae. Their actions can include the formation of cationic complexes in soil solution and at the soil–parent rock interface (Smits, 2009). Gadd (2004) postulates that organic acid production by fungal hyphae disassociates mineral nutrients from rock particles and makes them available to soil biota as well as forming metal complexes (e.g., metal oxalates) that alter mineral availability. However, the scale at which these molecules are produced probably do not have any effect on the bulk soil, but have been shown—via rock thin section and fluorescent microscopy—to be limited to a few microns of the hyphal surface (Smits, 2009; Sverdrup, 2009). Thus, it is assumed that there is importance in the tunneling abilities of fungal hyphae to provide intimate contact between the rock and the hyphal surface.

Mycorrhizal fungi in symbiotic association with plant roots play a role in the dissolution of parent rock material in more established soils. Mojallala and Weed (1978) showed that mycorrhizal soybeans used weathered potassium from the biotites, phlogopite, and muscovite. However, the potassium released was insufficient to sustain the enhanced growth of the mycorrhizal plants such that the tissue concentration of potassium was less in mycorrhizal than in nonmycorrhizal plants. Electron microprobe analysis of biotites showed that arbuscular mycorrhizal fungi of ryegrass roots increased the rock weathering with extensive potassium and some aluminum release from the edges of the phlogopite but not from muscovite (Hinsinger and Jaillard, 1993), where the rate of potassium release is related to the potassium demand by the plant.

April and Keller (1990) demonstrated changes in the mineral physical and chemical composition of soil in the rhizosphere of forest tree roots. They showed that in the presence of roots, phyllosilicate grains were fragmented and aligned with the long axis of the root, exposing a larger surface area for chemical attack. In addition, there was precipitation of amorphous aluminum oxides, opaline and amorphous silica, and calcium oxalate deposits in the roots. Also, kaolin in the rhizosphere had a higher thermal stability compared to kaolin in the bulk soil. These causative agents were potassium enrichment in the rhizosphere soil or preferential dissolution of biotite at the root–soil interface. Gobran et al. (1998) and Finlay et al. (2009) review the effects of rhizospheres on forest biogeochemistry. Gobran et al. (1998) showed that the rhizosphere/bulk soil ratio for bacterial populations is 10:50, and that for fungi is 5:10, showing that the rhizosphere is an important focus for the activities of microorganisms and that this can effect considerable physicochemical changes in the soil. They also showed that the abundance of weatherable minerals near the root surface were consistently less than in the bulk soil (Table 2.2), which

Table 2.2  Comparison between Rhizosphere and Bulk Soil Physicochemical Characteristics and Bulk Soil Content of Weatherable Minerals Expressed as Mineral Intensity as a Percentage of Quartz Peak at 100 in E Horizon of Forest Soils

| Characteristic | Rhizosphere Soil | Bulk Soil |
|---|---|---|
| Cation exchange capacity ($cmol_c$ $kg^{-1}$) | 4.41 | 12.16 |
| Exchangeable base cations ($cmol_c$ $kg^{-1}$) | 0.33 | 1.93 |
| Soluble base cations ($cmol_c$ $kg^{-1}$) | 0.10 | 0.46 |
| Titratable acidity ($cmol_c$ $kg^{-1}$) | 4.08 | 10.23 |
| Base saturation (%) | 7.47 | 16.13 |
| Organic matter (%) | 9.80 | 23.03 |
| Mineral (as% intensity of quartz peak) | | |
| Amphibole | 0.03 | 0.12 |
| Interstratified vermiculite | 0.54 | 1.14 |
| Plagioclase | 1.73 | 2.24 |
| K-feldspar | 1.28 | 1.29 |

Source: After Gobran, G.R. et al., Biogeochemistry, 42, 107–120, 1998.

they attribute to increased hydrogen ion and carbon dioxide content and the presence of complexing organic acids. Chang and Li (1998) investigated the ability of seven ectomycorrhizal fungal species to solubilize limestone, marble, and calcium phosphate. From plate clearing studies, only *Hysterangium setchellii, Rhizopogon vinicolor,* and *Suillus bovinus* formed halos around the colonies, indicating a degree of solubilization. In contrast, *Cenococcum geophilum, Hebeloma crustuliniforme, Laccaria laccata,* and *Piloderma croceum* did not clear the medium. Nonmycorrhizal fungi, including *Penicillium,* three species of *Azospirillum,* three isolates of *Pseudomonas fluorescens,* and a yeast also cleared the substrate. The effectiveness of these isolates in mineralizing calcium showed that the pseudomonads were most efficient, along with the yeast and *Penicillium.* Of the mycorrhizal fungi, *Laccaria laccata* showed no activity, but *Rhizopogon* and *Suillus* showed a slight increase in calcium release over the control, along with the *Azospirillum* isolates. In pure culture conditions, Paris et al. (1996) showed that the two ectomycorrhizal fungi, *Paxillus involutus* and *Pisolithus tinctorius,* produced oxalate in the presence of the mica, phlogopite, and that this process was not influenced by the availability of potassium or magnesium. *Pisolithus* produced oxalate in the presence of phlogopite with greatest oxalate production in the presence of ammonium nitrogen in the absence of phlogopite. On the other hand, *Paxillus* did not accumulate oxalate in the presence of ammonium or nitrate nitrogen in the absence of phlogopite. Similarly, *Paxillus involutus* and *Suillius variegatus* in mycorrhizal symbiosis with Scots pine seedlings were found to mobilize potassium from biotite and microcline by production of citric acid, which was produced in proportion to fungal hyphal biomass (Wallander and Wickman, 1999).

The role of ectomycorrhizal fungi in rock dissolution and ability to mobilize K, Mg, and Ca for tree nutrition has recently been reviewed by Landeweert et al. (2001) and Finlay et al. (2009) (Figure 2.2). Additionally, Thompson et al. (2001) showed that in certain circumstances, $NH_4$-N could be derived by ectomycorrhizal activity on feldspars in Miocene shales and, possibly, other rocks. In a comparison of mycorrhizal access to quartz, rock phosphate, and crystaline apatite, Koele et al. (2014) showed that both arbuscular- and ectomycorrhizal fungi gained access through tunneling to both solubilize phosphate and to extract rare earth elements for uptake into host plants. Gobran et al. (1998) suggest that, in addition to regulating nutrient fluxes and pools, the presence of abundant ectomycorrhizal hyphae in the rhizosphere of trees acts as a source of organic matter. This could act as a source and sink of available nutrients and, possibly, toxic elements. The ionic exchange sites on the surface of organic matter regulate the movement of these ions. Where ionic exchange forces are high, the elements are closely bound to the organic matter. Where ionic forces are less, the nutrients become more available in soil pore water for plant growth.

Fungi exist as part of a microbial community, so it is important to consider their actions along with other microorganisms, such as bacteria. An endolithic community in sandstone and granite was found to consist of a loose relationship between fungi, bacteria, and algal cells (Hirsch et al., 1995). Fungal species present included *Alternaria, Aspergillus, Aureobasidium, Candida, Cladosporium, Paecilomyces, Phoma, Penicillium,* and *Sporobolomyces.* The production of organic acids by this

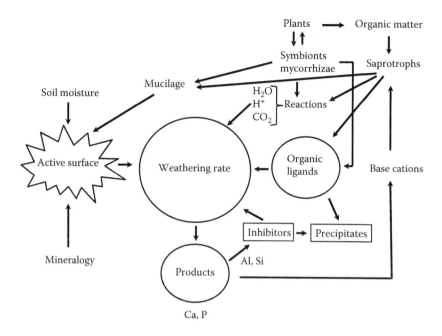

**Figure 2.2**  Chemical and biological interactions in soil formation. (After Finlay, R. et al., *Fung. Biol. Rev.*, 23, 101–106, 2009.)

assemblage of organisms was suggested to be responsible for the dissolution of rock, allowing the invasion by bacteria and other fungal species. Although *Aspergillus niger*, in culture, has been shown to effect solubilization of fluorapatite (Nahas et al., 1990), interactions between fungi and bacteria are likely to enhance rock dissolution and availability of nutrients from the rock. For example, Azcon et al. (1976) showed that enhanced acquisition of phosphorus from rock phosphate by lavender plants was achieved by the interaction of mycorrhizae and bacteria. They showed that there was a degree of synergism between the bacteria and mycorrhizal fungi and differences in behavior between the two mycorrhizal fungal species selected. However, in their study of maize root systems, Berthelin and Leyval (1982) compared the ability of arbuscular mycorrhizal root systems of maize to nonsymbiotic rhizospheric microflora and combinations of the two in the weathering of micas. In experimental systems, measures of maize growth (biomass), potassium, calcium, and magnesium uptake (derived from the breakdown of biotite) were similar in plants with nonsymbiotic rhizospheric microflora and arbuscular mycorrhizal root systems, but there was no synergistic effect of the combination of mycorrhizae and bacteria. Plant acquisition of nutrients from insoluble or poorly soluble sources is also enhanced by consortia of mycorrhizae, saprotrophic fungi, and bacteria. Singh and Kapoor (1998) showed that mung bean plants in association with a consortium of phosphate solubilizing organisms could better obtain phosphorus from rock phosphate than each organism alone. The consortium consisted of the arbuscular mycorrhizal fungus

*Glomus fasciculatum*, fungal saprotroph *Cladosporium herbarum*, and the bacterium *Bacillus circulans*. A field demonstration of the effect of rhizospheric microbial communities (including arbuscular mycorrhizae) on the release of phosphate from rock phosphate comes from the study of Vanlauwe et al. (2000) in Nigeria. The addition of rock phosphate to crops planted on low P soils showed an immediate response in terms of increased mycorrhizal colonization and enhanced growth. This increase in growth showed a combined effect of the mycorrhizae and associated rhizospheric bacteria in dissolving the rock phosphate to make it available for plant uptake. The pivotal role of mycorrhizae in this process was demonstrated by the fact that in the presence of high fungiverous nematode populations, the effect of added rock phosphate was significantly reduced, owing to grazing of the mycorrhizal mycelia.

The interactions between the mycorrhizal fungi and mycorrhizaspheric bacterial community appear to vary from a loose association to something near a symbiosis. However, the nature of the interactions of component organisms in the mycorrhizasphere is far from fully described nor are the physiological attributes and biogeochemical changes effected by these communities completely understood. It does seem, however, that the action of mycorrhizae and root surface bacterial communities may access a greater variety of nutrient elements from rock that had previously been thought (Thompson et al., 2001; Hoffland et al., 2001), and the relative contribution of nutrients derived in this fashion, compared to other sources, for plant growth in a variety of ecosystems has yet to be determined.

## 2.1.3 Fungal Contribution to Organic Matter Pool in Soil

In addition to the formation of soils, fungi in the form of lichens and cryptogamic soil crusts are important primary producers and contributors to the soil carbon pool (Lange et al., 1998), especially in arid ecosystems. Soil crusts can be formed by a diverse group of organisms including mosses, lichens, fungi, green and blue-green algae (Cyanobacteria), as well as other bacteria. States and Christensen (2001) identified 33 species of fungi associated with the lichens, bryophytes, and graminoids of the surface crusts of semidesert grassland ecosystems of Utah and Wyoming. These crust communities are an essential part of the soil formation and control the nutrient content and availability of soils for the invasion by other plant species. The mat-forming lichens (*Cladonia, Cetraria, Stereocaulon,* and *Alectora*), which grow in elevated microsites of boreal Arctic peatlands, grow upward and die off at the base. The dead bases input a considerable amount of organic matter into the developing soil (Crittenden, 1991, 2000).

In oligotrophic, sandy soils, vascular plants need to invest energy into root growth for water and nutrient acquisition. Mat forming lichens, however, trap both water and nutrients from the air, making them less dependent on roots and soil for their supply of water and nutrients; thus, the carbon cost of nutrient acquisition is reduced (Crittenden, 2000). This benefit allows lichens to be primary colonizers. In addition to adding organic matter in the form of lichen biomass, the nitrogen fixation by the Cyanobacteria photobiont of *Stererclaulon paschale* has been shown to be approximately 20 kg ha$^{-1}$ y$^{-1}$ N. This rate of fixation could provide a large proportion of the 10 to 40 kg ha$^{-1}$ y$^{-1}$ plant demand for nitrogen in upland spruce boreal forests of Canada and Russia (Crittenden, 2000).

The ability of fungal mycelia to form an extensive, long-lived network, which is more resilient to the vagaries of rapid changes in climatic conditions than bacteria, may allow components of the ecosystem that rely on fungal mycelia to respond more readily to systems that rely on the growth of populations of bacteria to carry out an ecophysiological function. A good example of this homeostasis can be seen in the study of Lange et al. (1994), who explored the role of soil crust lichens in carbon sequestration in the Namibian desert. In desert conditions, the main limitation to the survival of organisms is the scarcity of water. As was suggested earlier, the structure of lichens can act as a nucleus for water condensation, and lichens are able to survive periods of desiccation. Thus, lichens are well adapted as primary producers in the stressed conditions of deserts and periodically dry environments.

In the Namibian desert, the soil crust lichens *Acarospora schleicheri*, *Caloplaca volkii*, and *Lecidella crystalline* utilize available water from fog during early morning to photosynthesize before water becomes unavailable and, later light becomes limiting at night. Thus, much of the primary production of this ecosystem occurs during approximately a 3-h period of the day to provide an annual gross primary production of 32 g C m$^{-2}$; however, by factoring in respiration carbon loss the figure for net carbon gain by 100% lichen cover is in the order of 16 g C m$^{-2}$ y$^{-1}$ (Lange et al., 1994). Similarly, fast photosynthetic responses for cyanobacterial soil crust lichens have been found in the arid soils of Utah (Lange et al., 1998), in response to pulses of available water. The rapid response of lichens to short-term pulses of optimal environmental conditions to support photosynthesis, together with the ability of fungi to survive long periods of inhospitable conditions (S-strategists), provide lichens with a competitive advantage over vascular plants in extreme environments. The development of crust communities on soil also adds dead organic material to the soil (Crittenden, 1991), which is utilized as a food resource by saprotrophic fungi and bacteria, effecting mineralization of inorganic nutrients for plant uptake. This organic material is an essential component of soil and is combined with the mineral component of soil to form aggregates.

Once lichen communities have become established, their effect on the production of adjacent plant communities can be considerable. In a study of the mosaic of land cover in an Alaskan taiga ecosystem, Lamontagne (1998) demonstrated that net nitrogen mineralization was 7 times higher and nitrate nitrogen 40 times higher in lichen patches than adjacent forest islands. These higher figures are likely attributable to the lower nitrogen immobilization into lichen tissue than into tree tissue. Lichen-dominated areas of the upper slopes of the catchment are important producers of soluble mineral and dissolved organic nitrogen that runs off into the adjacent, lower lying forest islands. In these forest islands, the nutrients are utilized by immobilization into tree biomass in this landscape, consisting of a mosaic of forest islands covering 27% of the area, lichen patches covering 24%, and bedrock with crustose lichen cover occupying 49% of the catchment area, respectively.

## 2.2 BREAKING DOWN THE DEAD: ADDING FERTILITY

The nutrient component of soil comes partly from the dissolution of parent rock (see preceding discussion) and, secondarily, from the decomposition of dead plant and animal remains. Dead plant parts (above- and belowground) are returned to the soil where the activities of bacteria, saprotrophic fungi, and soil fauna degrade the complex organic components. They utilize the carbon skeletons for energy and, in the process, cleave off mineral elements, which they incorporate into their biomass and release into the soil water. These organisms rely on extracellular enzymes (Sinsabaugh and Liptak, 1997; Sinsabaugh, 2005) to degrade the complex organic molecules contained in the litters. By virtue of their high surface area/volume ratio (bacteria) or filamentous form (fungi), the degradative products of this enzyme activity can be absorbed into the cytoplasm of the organisms. The activity of these saprotrophic fungi is neatly summarized by Forsyth and Miyata (1984, p. 19), "under the silent, relentless chemical jaws of the fungi, the debris of the forest floor quickly disappears." However, the digestive process of animals, where enzyme activity occurs within a gut, is more efficient than secretion of enzymes outside the body, where some of the nutrients mineralized from the decomposing organic matter, escape absorption by the microorganisms and are released into the soil inorganic nutrient pool in soil water and is made available for plant growth. In addition, the rapid turnover of these organisms means that nutrients temporarily locked up or immobilized within their biomass soon become available as new, dead, organic resources for decomposition. Thus, nutrient elements are released into the soil solution as simple inorganic compounds through the process of mineralization providing soil fertility, and available nutrients are taken up by plants.

### 2.2.1 Input of Resources

The balance between rates of decomposition and mineralization and the rate of input of dead plant parts to soil determine the type of soil profile developed over time. Where decomposition is very slow, organic matter accumulates as peat. Where decomposition is rapid, as in agricultural soils and grasslands, a mineral soil profile is developed with low organic matter content that is often incorporated to a greater degree into the mineral component of the soil than in natural ecosystems. Soils that support poor rates of plant litter decomposition lead to an accumulation of partly decomposed, raw, humic material in which the nutrients are trapped in an organic form, with slow rates of release of nutrients in an inorganic form into pore water. The former soils are generally more fertile, being able to provide plant growth with a readily accessible source of nutrients. The latter soils are regarded as nutrient poor (although they may contain more total nutrient than the "fertile" soils) as most of the nutrients are in a plant unavailable form, or only available to plants with specific adaptations to these conditions (see ericoid and ectomycorrhizal association in the following chapter). The organization of soil structures in different ecosystems and the general description of soil structure are well described by Brady and Weil (1999) and summarized by Coleman et al. (2004). They show that the addition of plant and animal litter to the surface of soils is of greater importance in forest and woodland soils than in grassland soils, where the addition of dead root tissue to deeper soil layers is the most important organic input. This results in different soil profiles developing under different vegetation communities. Forest ecosystems have a clearly defined A horizon of progressively decomposing plant remains, underlying an $A_o$ horizon of relatively undecomposed and undecomposed plant remains. In contrast, in grassland communities, the input of aerial plant parts is less important than belowground plant parts, so the development of an $A_o$ layer is reduced and organic matter is more intimately mixed with the mineral soil. The role of fungi in the process of decomposition and mineralization is greatest in the $A_o$ and A horizons.

Data from Rodin and Bazilevich (1967) show that the rate of plant litter input into soil of different ecosystems is partly related to plant biomass (Table 2.3), but the amount of litter resident on the soil surface is related to the rate of decomposition (a combination of resource quality and environmental factors). Thus, the fertility and structure of soils is a function of climate, physicochemical composition (resource quality) of the plant litters, and the soil biotic (bacteria, fungi, and soil animals) composition and activity (Dighton, 1995; Dighton and Krumins, 2014).

In tropical forests, plant biomass is very high, and litter fall is high, but the litter on the soil is sparse, indicating a combination of climatic conditions conducive to decomposition and high resource quality. Hedger et al. (1993) showed that in tropical forest ecosystems, the input of litter is more important than in temperate forests as a nutrient return to the system. Tropical forests have continuous—rather than seasonal—litter input, and the quantity of litter can be triple that of temperate forests ($9 \text{ t ha}^{-1} \text{ y}^{-1}$ tropical; $3 \text{ t ha}^{-1} \text{ y}^{-1}$ temperate). The importance of litter in tropical regions is so significant that some 75% of phosphorus and 41% of potassium flux occur in the litter in tropical regions. One of the reasons for the reduced resource quality in forested ecosystems, compared with grassland or herbaceous ecosystems, is the diversity of litter types produced within a forest. Quantities cited by Dighton and Boddy (1989) show a forest litter composition of some 55% leaves; 10% fruits, buds, and flowers; 20% twigs; 10% branches; and 5% insect frass. The wood component may be underestimated and may be nearer to 40%. This woody component has a high lignin content and high C/nutrient ratio, making it much less degradable by fungi (Melillo et al., 1982). Ranges of carbon to nitrogen and carbon to phosphorus ratios in different plant residues are shown in Table 2.4. There is a temporal mosaic of litter input, with seasonally pulsed inputs of high resource quality litters

**Table 2.3 Trends in Primary Production and Litter Input to Decomposer System, in Relation to Latitude**

| Ecosystem | Plant Primary Production (t ha$^{-1}$) | Litter Biomass (t ha$^{-1}$) |
|---|---|---|
| Tundra | 30–130 | 1–4 |
| Temperate forest | 180–200 | 7.3–8.2 |
| Tropical forest | 400+ | 44 |

Source: After Rodin, L.E., Bazilevich, N.I., *Production and Mineral Cycling in Terrestrial Vegetation*, Oliver and Boyd, Edinburgh, 1967.

Table 2.4   Ranges of C/N and C/P Ratios in a Variety of Plant Residues

| Component | C/N | C/P |
|---|---|---|
| Herbaceous leaf litter | 15:1–160:1 | 25:1 |
| Tree leaf litter | 20:1–300:1 | |
| Woody litter | 300:1–500:1 | 1850:1 |
| Fungi | 6:1 | 15:1 |

Source: After Dighton, J., Can. J. Bot., 73, S1349–S1360, 1995; Swift, M.J. et al., Decomposition in Terrestrial Ecosystems, Blackwell Scientific, Oxford, 1979.

(flowers and buds), compared to the more continual input (twigs, branches), of lower resource quality. Thus, the continuous presence of fungal mycelia in the decomposer community provides a stable and constantly available mechanism for the decay of these resources whenever they become available. The rate at which the resources are utilized depends on the diversity of the fungal community available to colonize the resource, their enzymatic competence of that fungal assemblage, the nature of the available resources, and the climatic conditions. Taylor et al. (2014) used molecular profiling to show that fungal diversity *per se* was generally stable within a single forest ecosystem, but within that community there were numerous distinct ecological niche preferences shown by the fungal taxa. For example, they showed a significantly different fungal community in mineral soil than organic soil horizons. Taxa of both basidiomycete and ascomycete phyla and guild showed a range of niche specificity or broad niche use based on soil horizon, acidity, and soil moisture. Given that substrate diversity was not included in this study, our broad descriptions of the role of fungi in decomposition of dead organic matter and mineralization of nutrients has to be a very broad statement, as much more fine-scale detail is yet to be resolved.

## 2.2.2 Regulation of Decomposition by Chemical Stoichiometry

The rate at which a resource is decomposed is dependent on its chemical composition, edaphic factors (available moisture and temperature), and the colonization of the resource by appropriate saprotrophic organisms (Heal and Dighton, 1985; Heal et al., 1997). Many of these factors are discussed by Cooke and Rayner (1984). The input of different types (chemical composition and, hence, resource quality) of plant litter varies with ecosystem type (Dickinson and Pugh, 1974; Cadish and Giller, 1997). The general consensus is that the carbon/nitrogen and lignin/nitrogen ratios can be used as determinants of the resistance of resources to decomposition and ultimate mineralization of nutrients (Melillo et al., 1982; Thomas and Asakawa, 1993). Where the C/N or lignin/N ratios are high, there are reduced rates of decomposition compared to resources containing lower ratios. These ratios significantly impact rates of litter decomposition, mineralization, and degradation of carbon into dissolved organic matter (Silveira et al., 2011). Other secondary chemicals produced by plants, particularly polyphenols and tannins, also inhibit rates of decomposition of plant material by soil microorganisms (Harborne, 1997). However, in contrast, Vanlauwe et al. (1997) showed that there was often little negative relationship between the mineralization of nitrogen from leaf litter and either the polyphenol/N or lignin/N ratio of the resource. Despite this, it seems that in general both rates of decomposition (mass loss) and nitrogen mineralization are strongly correlated to the (lignin + polyphenol)/N ratio (Table 2.5).

In cold and wet climates, plant growth and biomass are low, being constrained by low temperatures and short photoperiods. Plant litter fall is comparatively high to the plant standing biomass, but often contains many secondary plant compounds, such as polyphenols and tannins as antiherbivore compounds. Consequently, the rate of decomposition is low because of the combination of poor litter quality for microbial decomposition and narrow windows of environmentally favorable conditions for microbial activity. This leads to accumulation of organic components of the soil and evolution of peaty soil profiles and soils with structured organic horizons. These high C/N ratio systems tend to be dominated by fungi (low bacterial/fungal ratio) as the main saprotrophic microorganism, with the greater fungal mass on low resource quality material providing bottom–up control of the faunal community (Heal and Dighton, 1986; Georgieva et al., 2005). In contrast, in agroecosystems bacteria dominate because of higher soil fertility, low C/N ratio of plant residues, and soil disturbance, which disrupts fungal hyphae. In a study of a chronosequence of 26 abandoned agricultural fields from 1 to 34 years from abandonment, van der Wal et al. (2006) showed that fungal biomass continued to increase in soil after abandonment as soil total P content declined (Figure 2.3). Kjøller and Struwe (1982) measured the abundance of fungal hyphae in different ecosystems, and the values suggest that hyphal length is a more important fungal investment of energy than biomass in cooler environments and where the available resources for decomposition are more recalcitrant (maximizing fungal surface area for enzyme activity), whereas biomass, with less hyphal extension, appears to be more important in warmer environments and where resources are of a higher quality. Schmidt (1999) shows the importance of fungal biomass in the high organic matter content soils of tundra ecosystems of Siberia. Hyphal lengths of 393 and 27 m $g^{-1}$ dry weight soil were found in the Levinson-Lessing valley, supporting typical tundra plant communities of dwarf willow communities and polygon soils, whereas low values of 9 m $g^{-1}$ dry weight soil can be found in the more fertile, lower organic content brown earth soils at Labaz.

In coniferous forests, plant biomass is high and litter fall is low, but accumulation occurs mainly because of the recalcitrance of the litter to decomposition (low

**Table 2.5   Regression Analysis (Regression Coefficients) of Decomposition (Mass Loss) and Nitrogen Release Rate and Determinants of Leaf Litter Resource Quality**

|  | Litter Mass Loss | N Mineralization |
| --- | --- | --- |
| C/N ratio | 0.74* | 0.61 |
| Lignin/N ratio | 0.68 | 0.42 |
| Polyphenol/N ratio | 0.54 | 0.76* |
| (Lignin + polyphenol)/N ratio | 0.77* | 0.68* |

*Source:* After Vanlauwe, B.J. et al., In: *Driven by Nature: Plant Litter Quality and Decomposition*, 157–166, ed. Cadish, G., Giller, K.E., CAB International, Wallingford, 1997.
*Significant regression ($\alpha$ = 0.05).

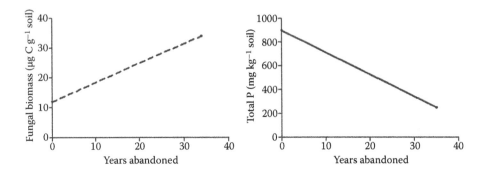

**Figure 2.3**   Increased dominance of fungi in abandoned agricultural soils in relation to declining P fertility. (Trend lines from data published by Van der Wal, A. et al., *Soil Biol. Biochem.* 38, 51–60, 2006.)

resource quality). Harborne (1997) describes the nature of plant phenolics, suggesting that external leaf phenolics are admixed with leaf waxes and have antifungal properties that reduce germination and growth of phyllosphere fungi (Table 2.6).

Although litter quality plays a major part in determining litter decomposition rates, the source and nature of the microbial inoculum initially colonizing that leaf litter also plays a part in determining initial mass loss (Strickland et al., 2009).

### 2.2.3  Resource Succession and Enzyme Competence

During seral succession of vegetation from herbaceous to forest ecosystems, there is a change in dominant plant species and plant form (Heal and Dighton, 1986). Along with this change in plant form, there is a general change in the diversity and complexity of resources entering the decomposer system. The initial seral stages are marked by an addition of high-quality resources to the decomposer community,

**Table 2.6   Antifungal Phenolics Obtained from Plant Surfaces**

| Phenolic | Source |
|---|---|
| Kaempferol 3-glucoside | Oak leaf |
| 4-Coumaric acid ester | Birch leaf |
| 6-Isopentenylnaringenin | Hops resin |
| 5-Pentadecylresorcinol | Mangifera fruit peel |
| Chrysin dimethyl ether | Heliochrysum leaf |
| Quercetin 7, 3′-dimethyl ether | Wedelia leaf |
| Sakuranetin | Ribes leaf gland |
| Lueone | Lupin leaf |
| Pinocembrin | Poplar leaf |

*Source:* After Harborne, J.B., In: *Driven by Nature: Plant Litter Quality and Decomposition*, 67–74, ed. Cadisch, G., Giller, K.E., Cambridge University Press, Cambridge, 1997.

consisting mainly of cellulose and a high C/N ratio and a low lignin content. Following forest canopy closure, woody resources and more recalcitrant leaf litters dominate (Attiwill and Adams, 1993). These litters have high lignin content and low C/N ratios and, therefore, decompose at a slower rate. As well as changes in the dominance of the fungal species or group with ecosystem succession, the degree of interaction between fungi and animals increases. There are more and more intimate associations between fungi and fauna in the exploitation of the more recalcitrant plant residues (Table 2.7).

Because of the variability in chemical composition of plant and animal remains, not all materials can be utilized by all fungal species. Differences exist in the ability of species to access simple or complex forms of carbohydrate and mineral nutrients. Decomposition is a product of enzyme activity, where the types of enzymes required are dependent on the substrates (chemical constituents) of the resource. Descriptions of the various ectoenzymes produced by fungi and their biochemical effects on organic resources in plant litters (Table 2.8) can be found in the work of Sinsabaugh and Liptak (1997) and Sinsabaugh (2005). The ability of different species of fungi to produce specific enzymes dictates, in part, the succession of fungi colonizing resources. For example, Osono et al. (2003) compared seven basidiomycete and four ascomycete fungi in terms of their ability to decompose larch leaf litter. Although there was considerable variation between fungi in each group, the basidiomycetes caused significantly greater lignin loss from litter compared with the ascomycetes (with 41.6 ± 3.0% and 50.2 ± 0.4% lignin remaining, respectively); however, this did not result in a significant difference in nitrogen loss from litter between basidiomycetes and ascomycetes (0.84 ± 0.01% and 0.81 ± 0.02% N remaining in litter, respectively).

The decomposition process is dynamic. The same suite of organisms is not present on the plant or animal remains (resource) for the duration of the process of decomposition as different fungi have different enzymatic capabilities, so their appearance on a resource will be dictated by (1) their ability to utilize the resource, (2) their rate of arrival at the resource either by growth or by transport as spores, etc., and (3) their ability to compete against other fungal species with similar physiological competence. In addition to enzymatic competency, there are other factors—such as relative growth rates, the production of antibiotic secondary metabolites, and environmental constraints—that influence the ability of specific fungi to colonize resources in the face of competition against other fungi (Dickinson and Pugh, 1974; Cook and Rayner, 1984; Frankland, 1992, 1998; Lockwood, 1992; Wicklow, 1992; Ponge, 2005). The colonization of resources by fungi is a function of the quality of the resource, rate of arrival of the fungal propagule (spore or hyphal fragment), and the competitive interaction between fungal species on the resource. The classic assumption is that the initial colonizers used soluble carbohydrate sources (sugars) and were later replaced with fungal species having greater enzymatic competence, which are able to break down organic sources of carbon such as cellulose, and lignin. However, there are few clear distinctions in the succession as many of the species overlap in time and space. The successional trends of fungi colonizing decomposing plant material have been described in more detail for the litter of the fern *Pteridium*

**Table 2.7**   Changes in Plant Forms, Their Residues, the Dominant Fungal Groups Effecting Plant Litter Decomposition, and the Interactions between Fungi and Animals during Plant Seral Succession from Herbaceous Ground Cover to High Forest

| | Lower Plants | Herbaceous Plants | Angiosperm Leaves | Coniferous Leaves | Wood |
|---|---|---|---|---|---|
| | Ecosystem Succession and Increasing Contribution of Component to Plant Residues | | | | |
| Cellulose (%) | 16–35 | 20–37 | 6–22 | 20–31 | 36–63 |
| Lignin (%) | 7–36 | 3–30 | 9–42 | 20–58 | 17–35 |
| C/N | 13–150 | 29–160 | 21–71 | 63–327 | 294–327 |
| | Changes in Dominant Fungal Groups | | | | |
| | 'Sugar fungi' Ascomycetes Mitosporic fungi | Yeasts 'Sugar fungi' Ascomycetes Mitosporic fungi Basidiomycetes | Ascomycetes Mitosporic fungi Basidiomycetes | Ascomycetes Mitosporic fungi Basidiomycetes | Basidiomycetes Ascomycetes Mitosporic fungi |
| | Fauna, Less Important | | Fauna, More Important | | |
| | Enchytraeids | Enchytraeids Oligochaetes Diptera | Oligochaetes Collembola Acari | Acari Collembola Oligochaetes | Insecta Arthropoda |

*Source:* After Heal, O.W., Dighton, J., In: *Microfloral and Faunal Interactions in Natural and Agro-Ecosystems*, ed. Mitchell, M.J., Nakas, J.P., 14–73, Martinus Nijhoff/Dr. W. Junk, Dordrecht, 1986; Dighton, J., In: *The Mycota IV: Environmental and Microbial Relationships*, 271–279, ed. Wicklow, D.T., Söderstrom, B., Springer Verlag, Berlin, 1997.

**Table 2.8  Fungal Enzyme Systems Associated with Degradation of Specific Plant Compounds**

| Plant Compound/Fungal Resource | Fungal Enzyme System |
|---|---|
| Lignin | Lignin peroxidase |
| | Manganese peroxidase |
| | Glucose oxidase |
| | Cellobiose oxidase |
| | Aryl alcohol oxidase |
| | Glyoxal oxidase |
| | Laccases |
| Cellulose | Exo-1,4-β glucanase |
| | Endo-1,4-β glucanase |
| | 1,4-β-Glucosidases |
| Hemicellulose | Endo-1,4-β xylanases |
| | Endo-1,4-β mannases |
| | 1,4-β Xylosidases |
| | 1,4-β-D Mannosidases |
| | 1,4-β-Glucosidases |
| | α-L-Arabinosidases |
| | α-Glucuronidases |
| | α-Galactosidases |
| | Acetyl xylan esterases |
| | Acetyl galactoglucomannan esterases |
| Pectin | Polygalacturonases |
| | Endo-1,4-α polygalacturonase |
| | Exo-1,4-α polygalacturonase |
| | Pectin lysases |
| | Pectin esterases |

*Source:* Information compiled from Sinsabaugh, R. L., and M. A. Liptak, Enzymatic conversion of plant biomass. In: *The Mycota IV, Environmental and Microbial Relationships*, 347–357, eds. D. T. Wicklow and B. Söderstrom. Berlin Heidelberg: Springer-Verlag, 1997.

*aquilinum* by Frankland (1992, 1998). She described changes from lesion forming *Rhizographus* and *Aureobasidium* on standing dead litter, through the colonization by basidiomycetes in relation to the rate of loss of cellulose and lignin and the consequential decrease in the C/N ratio from some 200:1 to 30:1. By microscopic observation of small samples of forest floor leaf litter, Ponge (1990, 1991, 2005) characterized the colonization of *Pinus sylvestris* needles into four stages (Figure 2.4). The first stage is characterized by the decomposition of freshly fallen leaves by fungal species (*Lophodermium, Ceuthospora,* and *Lophodermella*) that were probably present on and in the leaf at the time of abscission. These fungi cause browning of the leaf and decomposition of relatively available resources. This was followed by greater invasion of the leaf tissue by decomposing microfungi, such as *Verticicladium* (Stage 2), and basidiomycete fungi, such as *Marasmius* and *Collybia* (Stage 3). Finally, along with the entry of soil arthropods, was invasion by mycorrhizal fungi.

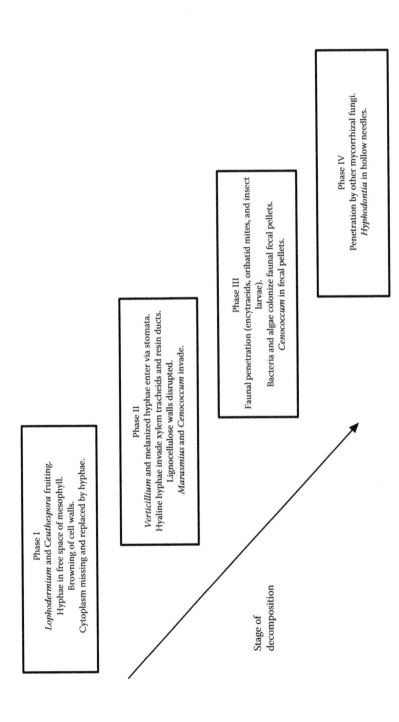

**Figure 2.4**   Schematic of the pattern of fungal colonization of individual pine needle leaf litter as observed by microscopy. (The information is compiled from Ponge, J.F., *Plant and Soil* 138, 99–113, 1991; and the diagram is modified from Dighton, J., Nutrient cycling by saprotrophic fungi in terrestrial habitats, In: *The Mycota IV: Environmental and Microbial Relationships*, 271–279, eds. D. T. Wicklow and B. Söderstrom, Berlin: Springer Verlag, 1997.)

In the early stages of leaf litter decomposition, the changes in the fungal community can be rapid. Over a 24-month incubation, a rapid change from phylloplane fungal communities to a changing community of saprotrophs was seen to develop on oak leaves (Voříšková and Baldrian, 2013) (Figure 2.5). Successional changes in fungal communities during the decomposition of plant litter show both niche restriction and niche overlap (Di Lonardo et al., 2013). Indeed, the production of enzymes changes not only as the resource chemistry changes, but also as fungi interact and compete with each other. Šnajdr et al. (2011) showed that at the point of interaction between two wood decomposing fungi, both *N*-acetylglucosaminidase, α-glucosidase and phosphomonoesterase are produced in relation to hyphal damage. They also show synergistic activities between fungi and bacteria in the production of enzymes. Using molecular methods, Peršoh et al. (2013) identified successions of fungi on European beech leaves, suggesting that fungal communities on living leaves and those on recently fallen or well-decomposed leaves were distinct from each other, but the endophytic fungal community at the start persisted throughout the decomposition process. Fungal successions have been reported in beech cupules (Fukasawa et al., 2012), which—despite being a woody resource similar in lignocellulose composition to coarse woody debris—have the characteristics of holocellulose decomposition similar to that of leaf litter, where *Xylaria*, *Geniculosporium*, and *Nigrospora* are responsible for most mass loss of the cupule. During the decomposition, *Xylaria* was constantly present, *Epicoccum*, *Ascochyta*, and *Phomopsis* were

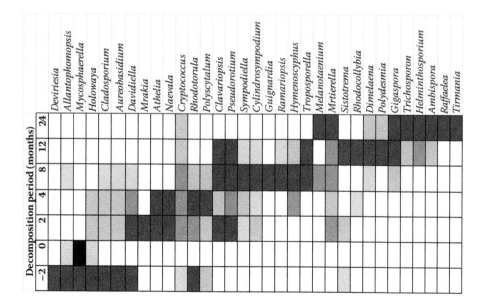

**Figure 2.5**    Heat map of abundance of fungal genera during the decomposition succession on oak leaves (darkest squares being most abundant). (Data from Voňšková, J., Baldrian, P. *ISME J.*, 7, 477–486, 2013.)

ephemeral early colonizers, and *Trichoderma, Geniculosporium,* and *Nigrospora* were later colonizers.

Phyllosphere fungi remain present at between 2% and 100% in the initial stages of leaf litter decomposition (Osono, 2006). Osono (2006) reports that of these fungi *Coccomyces, Genticulosporum,* and *Xylaria* participate in lignin decomposition, whereas others effect decomposition of less complex carbohydrates. The presence of fungal endophytes (e.g., *Neotyphodium coenophialum*) in grasses has been reported to retard decomposition rates (Omacini et al., 2004; Lemons et al., 2005) (Figure 2.6); however, in tall fescue grass and its associated endophyte, alkaloids did not seem to significantly affect grass leaf litter decomposition (Siegrist et al., 2010). Indeed, the presence of endophytes such as *Lophodermium* spp. can effect decomposition of pine needle litter to produce some 15% to 25% mass loss in 2 months by producing significant quantities of laccase, cellobiohydrolase, and β-glucosidase enzymes (Yuan and Chen, 2014).

In a model proposed by Swift et al. (1979), the changes from "sugar" fungi to basidiomycetes in relation to the changes in available resources and the influence of climatic stresses are presented. The model suggests that during initial decomposition, the carbohydrate component is used as an energy source until such time that the C/nutrient ratio approaches that of the decomposer organism (about 15:1 for P and 6:1 for N in fungi). Only then is there net conversion of organic nutrient to inorganic nutrient (net mineralization). In general, initial resource structure is chemically heterogenous, thus supporting a variety of fungal species. As decomposition proceeds, recalcitrant chemicals are left, which can be degraded only by a fungal flora that is capable of producing the necessary enzymes to degrade the complex resources. Hence, diversity is reduced. From these studies, the windows

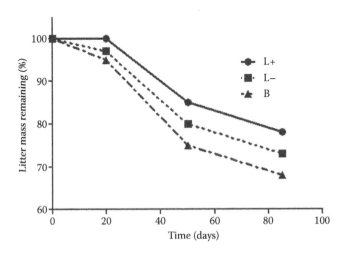

**Figure 2.6**    Leaf litter mass loss for *Lolium multiflorum* (L) without (–) or with (+) fugal endophytes compared to *Bromus unioloides* (B), which is naturally endophyte-free. (Data from Omacini, M. et al., *Oikos*, 104, 581–590, 2004.)

of opportunity for decomposition may be determined and the rate of substrate decomposition mapped, and feedbacks between changes in the substrate and environment and the fungal community dictate enzyme expression (Sinsabaugh, 2005) (Figure 2.7). Although climatic conditions dictate, to some degree, the mass and composition of plant material entering the decomposer system, climatic limitations do not necessarily relate to the lack of diversity in the fungal community, but rather to the lack of activity in that community. Indeed, Zak (1993) showed that there is great diversity in the fungal communities of desert ecosystems. This is attributable to the temporal heterogeneity imposed on the environment by pulses of availability of water and the rapid response to "windows of opportunity" by fungi. Fungi are better adapted to this periodic stress than bacteria because fungi are perennial, non-discrete organisms that are able to smooth out spatial heterogeneity. Fungal biomass may maintain a permanent presence, rather than having the peaks and troughs of populations seen in bacteria.

In general, the colonization of decomposing plant material in relation to resource quality has only been presented in reference to the chemical composition of the whole leaf. As fungal hyphae are of a small diameter (~5 μm), their pattern of growth, enzyme expression, and the subsequent changes in leaf litter chemistry occur at a sale of resolution much smaller than that of a whole leaf. Expression of enzymes and their effects on substrate chemistry and physical structure have recently been explored at the level of resolution of individual hyphae or areas of mycelium (Smart and Jackson, 2009; Šnajdr et al., 2011; Oberle-Kilic et al., 2013).

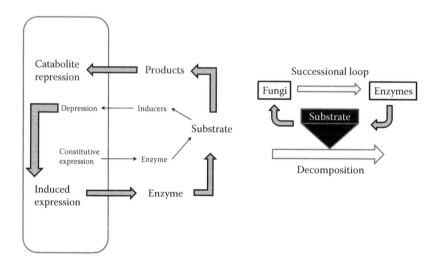

**Figure 2.7** Left: model for transcriptional regulation of enzyme production by fungi regulated by external signals; right: fungal community composition and enzyme production tied to resource decomposition through the successional loop. (After Sinsabaugh, R.L., *The Fungal Community: Its Composition and Role in the Ecosystem*, ed. Dighton, J. et al., 349–360, Taylor & Francis, Boca Raton, FL, 2005.)

## 2.2.4 Wood as an Example of a Recalcitrant Resource

Boddy and Watkinson (1995) show the importance of woody debris in the return of nutrient elements to terrestrial forest ecosystems (Table 2.9), where wood can represent 30–40% of the total biomass and 1–4 and 0.1–0.8 kg ha$^{-1}$ y$^{-1}$ of N and P, respectively. Decomposition of woody debris, and the mineralization of the nutrients contained within are primarily effected by basidiomycete fungi. Often, these fungi produce rhizomorphs or cords, which provide long-lived connections between islands of woody residues and allow reallocation of resources within an extensive fungal network. Movement of phosphorus was measured over distances of 1 m.

In forested systems, much dead wood remains in the canopy before recruitment to the forest floor. This standing dead material may have a different fungal community than wood on the forest floor. The work of Boddy and Rayner (1983) on oak wood in canopies showed that 12 basidiomycete fungal species dominated in the community. Of these, *Phellinus ferreus*, *Sterium gausapatum*, and *Vuilleminia comendens* were pioneer species of partially living branches, *Phlebia adiata* and *Coriolus versicolor* were secondary colonizers, and *Hyphoderma setigerum* and *Sterium hirsutum* related to insect activity. In wood, the interactions between fungi can be most clearly observed. The zones of interaction between adjacent, competing fungal colonies in wood are clearly demarked and have been mapped in three dimensions using wood as a resource (Rayner, 1978; Rayner and Boddy, 1988). The colonization pattern in wood is likely to be similar to that in straw, where colonization rate is correlated to relative growth rates of the fungi on agar (Robinson et al., 1993a) and modified by production of secondary metabolites modifying interactions. Where fungal interactions take place on straw, respiration was greater than where only one fungal species was present (Robinson et al., 1993b). This indicates that the maintenance of combative activities is energy-demanding and may affect the rate of decomposition.

**Table 2.9   Relative Contribution of Woody Debris and Nonwoody Litter to the Forest Floor in Temperate Woodland Ecosystems**

| | Nonwoody Litter Fall (kg ha$^{-1}$ y$^{-1}$) | | | Woody Litter Fall (kg ha$^{-1}$ y$^{-1}$) | | |
|---|---|---|---|---|---|---|
| | Biomass | N | P | Biomass | N | P |
| **Warm Temperate** | | | | | | |
| Broadleaf deciduous | 4236 | 36 | 3.8 | 891 | 2.6 | 0.8 |
| Broadleaf evergreen | 6484 | 55 | 3.7 | – | – | – |
| Needleleaf evergreen | 4432 | 28 | 2.7 | 1107 | 2.5 | 0.2 |
| **Cold Temperate** | | | | | | |
| Broadleaf deciduous | 3854 | 43 | 4.6 | 1046 | 3.7 | 0.2 |
| Broadleaf evergreen | 3590 | – | – | – | – | – |
| Needleleaf evergreen | 3144 | 26 | 3.2 | 602 | 1.1 | 0.1 |

*Source:* Boddy, L., Watkinson, S.C., *Can. J. Bot.*, 73, S1377–S1383, 1995.

Micro- and mesoinvertebrates can alter the growth rates of wood decomposing fungi by either consuming extra mycelial biomass or stimulating greater production of that biomass (A'Bear et al., 2014). In the presence of collembolan grazing, hyphal growth of *Hypholoma fasciculare* due to elevated temperatures was enhanced, but only minimal effects of faunal grazing were observed. In *Phallus impudicus*, growth was only reduced by grazing by *Protophorura*; in *Resiniculum bicolor*, grazing by both collembolans at both ambient and elevated temperatures reduced hyphal growth compared to an ungrazed system and negated the growth enhancement induced by elevated temperatures (Figure 2.8). This suggests that faunal and climate change interactions vary significantly between fungal species. Faunal grazing alters the competition between wood decomposing fungi. Rotheray et al. (2009) showed that collembola consistently preferred one fungal species over another where fungi were in competition on wood, likely affecting the rate of wood decomposition. In contrast, an isopod and nematode stimulated the growth of less competitive fungal species (Crowther et al., 2011a) (Figure 2.9), with the outcome of interactions

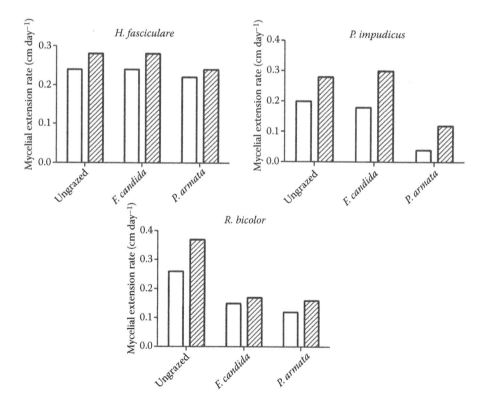

**Figure 2.8** Mycelial extension of three wood rotting fungi, *Hypholoma fasciculare*, *Phallus impudicus*, and *Resinicum bicolor*, when grazed by two collembola, *Folsomia candida* and *Protophorura armata*, at ambient (open bars) and elevated (hatched bars) temperatures. (Data from A'Bear, A.D. et al., *Fung. Ecol.*, 10, 34–43, 2014.)

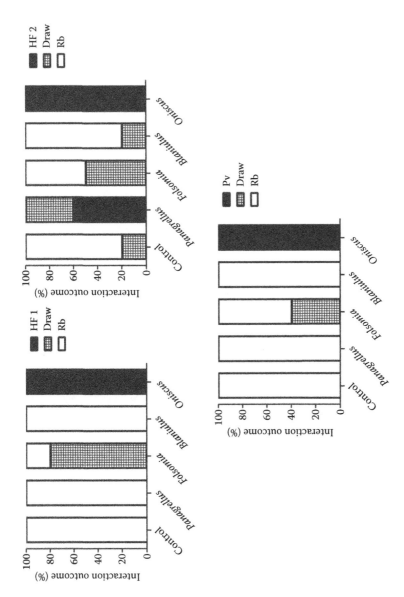

**Figure 2.9** Outcomes of competition between wood rotting fungi (*Hypholoma fasciculare*; Hf 1 and Hf 2) and *Resinicum bicolor* (Rb) as influenced by grazing from the nematode (*Panagrellus redivivus*), collembolan (*Folsomia candida*), myriapod (*Blanulus guttulans*), and isopod (*Oniscus asellus*). (Data from Crowther, T.W. et al., *Ecol. Lett.*, 14, 1134–1142, 2011.)

with the Isopod *Oniscus asellus* significantly altering wood decomposition rates by changing the fungal community, which in turn changes the community enzyme profile (Crowther et al., 2011b).

## 2.2.5 Nutrient Lockup and Translocation as an Ecological and Management Issue

During the course of decomposition, mineral nutrients are sequestered by decomposer soil organisms by being incorporated into the organism's biomass. The residence time of these elements is usually equivalent to the turnover time (life span) of that organism. During this period, the element is not in a soluble form in the soil solution, but is immobilized in microbial tissue. The amount of accumulation within the fungal component varies between ecosystems depending on the chemical composition of the plant parts available for decomposition and the main fungal groups involved in the process. Thus, shorter-lived, ephemeral molds, utilizing simple carbohydrates, have lower investment in biomass than longer-lived basidiomycetes, growing on woody resources; consequently, the potential accumulation in basidiomycetes is greater. Unlike bacteria, fungi are larger organisms and their rate of turnover is lower, particularly in the long-lived Basidiomycotina, which are important decomposers of recalcitrant organic compounds (Frankland et al., 1982).

Where the C/nutrient ratio of a resource is very high, as in wood, the model of Swift et al. (1979) proposes initial immobilization and import of free nutrient into organic form (fungal thallus) during the initial stages of decomposition until the fungal resource C/nutrient content is equivalent to that of the fungus. Fungal immobilization of nutrients can be considerable in that fungal hyphae have 193–272% greater N content and 104–223% greater P content than the pine needle litter on which they were found (Stark, 1972). Fungi are also important as temporary nutrient immobilizers. Fogel and Hunt (1983) demonstrated the importance of fungal biomass in a temperate Douglas fir forest ecosystem. For all nutrients except calcium, roots and mycorrhizae contained greater stocks than forest floor of fungi, but the amounts of Ca in fungi and forest floor were twice those for mycorrhizae and roots. Return of N, P, and K by mycorrhizae to soil was about 83–87% of the total tree return and 25–51% of the Ca and Mg return. The nutrient content of all elements, other than calcium, of fungal fruit bodies (mushrooms of both mycorrhizal and saprotrophic basidiomycetes) in a *Nothofagus* forest are more concentrated in fungal tissue than in forest floor material (Table 2.10) (Clinton et al., 1999).

Sinsabaugh et al. (1993) studied the mainly fungal derived, extracellular enzymes that are involved in wood decomposition, showing that the production of lignocellulase enzyme did not differ between different locations in a temperate forest ecosystem. However, the rate of immobilization (mainly fungal) of total nitrogen and total phosphorus into decomposing wood ranged from 2.2 to 4.4 µg g$^{-1}$ wood for P and from 43 to 139 µg g$^{-1}$ for P at the time when 80% mass loss was achieved. The spatial variability of this parameter was much greater than that for lignocellulase, but much less than that for acid phosphatase and *N*-acetylglucosaminase activity. Thus, where nutrient elements are less available, the fungi expend greater amounts of energy to produce enzymes to sequester the nutrients from organic sources. These

Table 2.10   Nutrient Concentrations (mg kg⁻¹) and C/N Ratio of Fungal Fruit Bodies
             and Underlying Forest Floor Substrates in a *Nothofagus* Forest Ecosystem

|       | Forest Floor Mushrooms | Forest Floor Substrate | Dead Wood Mushrooms | Dead Wood Substrate |
|-------|------------------------|------------------------|---------------------|---------------------|
| N     | 35                     | 8.7                    | 31                  | 1.5                 |
| P     | 4                      | 1.0                    | 5                   | 0.2                 |
| K     | 22                     | 3.5                    | 19                  | 0.1                 |
| Mg    | 0.8                    | 0.1                    | 1.6                 | 0.4                 |
| Ca    | 0.5                    | 7.0                    | 2.6                 | 2.9                 |
| Si    | 95                     |                        | 37                  |                     |
| Na    | 97                     |                        | 34                  |                     |
| Al    | 45                     |                        | 24                  |                     |
| C[a]  | 446                    |                        | 446                 |                     |
| C/N   | 13                     |                        | 19                  |                     |

*Source:* After Clinton, P.W. et al., *N. Z. J. Bot.*, 37, 149–153, 1999.
[a]  Carbon is expressed as mg g⁻¹.

results suggested a large degree of edaphic (soil condition) control over enzyme expression, which is closely related to the availability of inorganic N and P supplies in soil water. Similarly, the availability of N and P in surrounding soil influences the decomposition of leaf litter. In microcosms, Güsewell and Gessner (2009) incubated and decomposed cellulose filters on unsterilized sand containing contrasting N/P ratios or Carex leaf litter grown so as to obtain different N/P ratios. On cellulose, bacterial numbers peaked at the N/P ratio of 5, whereas fungal biomass (measured as ergosterol) peaked at N/P ratios between 15 and 45. The amount of N and P mineralized from leaf litter was dependent on the N/P ratio of the litter, depending on the most limiting of these nutrients. Thus, it can be seen that stoichiometric balance of nutrient availability can strongly affect fungal growth and the ability of these fungi to effect decomposition and nutrient mineralization. Thus, Sinsabaugh (2005) and Sinsabaugh et al. (1993) developed a model that contains both fungal (microbial) and soil nutrient controls over the expression of enzymes. Use of models like this can help us to better understand the complexities of decomposition and nutrient cycling processes by allowing hypothesis development, leading to the design of experiments that can logically alter single or multiple parameters to investigate the key processes and organisms that are responsible for driving ecosystem processes.

In practical forestry, particularly fast-growing trees, there is competition for essential nutrients between the fungi decomposing woody residues from logging and for growth by second rotation tree crop (Parfitt et al., 2001). To alleviate this competition, burning protocols have been established to rid the site of both woody debris and, incidentally, leaf litter and the nutrients they contain (Dighton, 1995); another option involves enhancing wood decomposition and N mineralization by chopping and incorporating woody remains into the soil (Jones et al., 1999; Shammas et al., 2003). A greater understanding of the interactions between nutrient availability, temporary nutrient immobilization, and alternative applications for postharvest residues, could lead to a more rational use of residues to provide sustainable forestry without the loss of nutrients from the ecosystem from burning and without the need for exogenous nutrients in the form of fertilizers (Jones et al., 1999).

Lodge (1993) discussed the role of fungi in nutrient cycling in tropical forest ecosystems (Table 2.11). These systems have large nutrient capital in plant biomass but are frequently limited by nutrient supply from soil. Much of this is attributable to phosphorus binding to aluminum and iron oxides and, thereby being less plant-available. Because of high rainfall, other plant essential nutrients, nitrogen, and potassium are likely to be leached from the rooting zone. Fungal biomass (5–15 mg $g^{-1}$ litter and 2.5–3 mg $g^{-1}$ soil; reports of 8–333 g $m^{-2}$) in these soils contains a large reservoir of nutrients that can be slowly released on death and decomposition. Using $^{14}C$ and $^{15}N$ labeling techniques for each of the cells, Marumoto et al. (1982) showed that the rate of carbon loss, as $CO_2$, was similar between microbe sources, but that the rate of mineralization of nitrogen as both $NH_4$ and $NO_3$-N was slower in the decomposition of fungal cells. Fungal retention of phosphorus in decaying leaf litter can increase P concentration 10-fold, where the concentration in fungal tissues can reach 5–36 mg $g^{-1}$ (Lodge, 1993). As fungal biomass is positively correlated to soil moisture and rainfall in these tropical systems, Lodge attributes maximal fungi to immobilization of nutrient elements at a time when there could otherwise be maximal leaching loss. Therefore, fungi are an important control on nutrient retention and release. Many of the fungi are basidiomycetes that form rhizomorphs and are associated with decomposing wood. Many of these have the ability to translocate nutrients between decomposed leaf litter to freshly fallen leaf litter to improve resource quality (lower the C/nutrient ratio) and enhance rates of decomposition. Owing to the activities of cord-forming fungi, Lodge (1993) demonstrated that the phosphorus content of recently fallen leaf litter could increase by 120–140% during the first 6 weeks of decomposition. Similarly, nitrogen could increase by 110–160%. Behera et al. (1991) found 36 species of fungi in the soils of a tropical forest site and showed that the composition and biomass of this community changed between seasons. The greatest biomass and species number occurred in January, after the rainy season, showing a positive correlation with both soil moisture and soil organic matter content.

Different nutrient elements and, in particular, metal ions may be immobilized for long periods in fungi (bioaccumulation). Fungi are nondiscrete organisms (having

Table 2.11 Proportion of Nutrient Elements Contained in Fungal Biomass in Wet Tropical Soil Systems, Demonstrating the Importance of Fungi as Nutrient Reservoirs

| Element | Fungal as % of Leaf Litter | Fungal as % of Soil Extractable |
|---------|---------------------------|--------------------------------|
| N | 1.6 | |
| P | 22.2 | 10.5 |
| K | 3.7 | 3.6 |
| Ca | 2.0 | 23.6 |
| Mg | | 3.2 |
| Na | | 3.1 |

Source: After Lodge, D.J., In: Aspects of Tropical Mycology, ed. Isaac, S. et al., 37–75, Cambridge University Press, Cambridge, 1993.

an extending hyphal network) and are able to translocate elements within the fungal thallus (Cairney, 1992). This could account for the spatial redistribution of elements. For example, if an element were always translocated away from dying regions, translocation would increase the length of time of immobilization into fungal components. Olsson and Jennings (1991) demonstrated that translocation of $^{14}$C and $^{32}$P through hyphal systems of *Rhizopus*, *Trichoderma*, and *Stemphylium* occurred by diffusion. The rate of translocation of carbon within the fungal thallus has been shown to react, in real time, to provide directional flow to the building phases of the hyphae (Olsson, 1995). In the face of high demand for nutrients and carbon at advancing hyphal fronts, nutrients and carbon are translocated acropetally through cytoplasmic flow and diffusion in the cytoplasm and apoplasm. In contrast to the diffusion of C and P, Gray et al. (1995) demonstrated that translocation of $^{137}$Cs through hyphae of *Schizophyllum commune* was slower than diffusion, suggesting incorporation of the element into structural components of the cytoplasm or hyphal wall, thereby reducing the rate of movement. This presents a plausible mechanism for the accumulation of radiocesium in basidiomycete fungi, as described by Dighton and Horrill (1988) and other researchers (data from Yoshida and Muramatsu, 1994). These studies suggest that accumulation levels could be high and long-lived.

The translocation of solutes through fungal tissues colonizing wood has been measured by tracer studies and has been shown to be of importance in allowing the colonization of low resource quality substrates. Wells and Boddy (1990) showed that 75% (*Phanerochaete velutina*) and 13% (*Phallus impudicus*) of the phosphorus added to a decomposed wood resource is translocated to newly colonized wood resources through mycelial cord systems. Maximum rates of P translocation are given as 7225 nmol P cm$^{-2}$ day$^{-1}$ through cords. Using the radioisotope $^{32}$P, Wells and Boddy (1990) demonstrated translocation of phosphorus through cords of the wood decomposing fungi *Phanerochaete velutina* and *Phallus impudicus*. They showed that cords were only formed in unsterile soil, suggesting that the trigger for cord formation is derived from other organisms and that the rate of translocation of phosphorus from decayed wood blocks to new wood blocks is of the order of 7 μmol cm$^{-2}$ day$^{-1}$ through cords. In field experimental manipulations, Wells and Boddy (1995a) showed that this translocation could be conducted over distances of up to 75 cm between decomposing resources on the forest floor and into living wild strawberry and moss plants. Translocation of phosphorus in mycelial cords is temperature-dependent, with greater rates of movement at higher temperatures (Wells and Boddy, 1995b). The effect of change from wet to dry soil conditions induces a thickening of the cord system of *Phanerochaete velutina* and a reduction in the translocation of phosphorus to a new wood resource; wetting appears to have no effect on cord structure or P movement (Wells et al., 2001).

## 2.2.6 Mycorrhizae as Decomposers

There is evidence that both ectomycorrhizal and ericaceous mycorrhizal fungi are able to access organic forms of nutrient (N and P), and thus may compete with saprotrophic fungi for resources in forested ecosystems (see Chapter 3). The importance of this interaction is not well understood, although negative interactions between saprotrophic and ectomycorrhizal fungi, in terms of mycorrhizal colonization of roots, have been found (Shaw et al., 1995). The interactions between fungi and bacteria in the decomposition of leaf litter may also not always be synergistic. In an incubation study of beech leaves, Møller et al. (1999) showed that the cellulolytic fungus *Humicola* sp. doubled the carbon utilization from leaves in combination with a mixed inoculum of soil bacteria. This increase in carbon utilization was positively related to greater β-*N*-acetylglucosaminidase and endoexocellulase activity of the fungus alone than in combination with bacteria.

Seasonal changes in ectomycorrhizal and saprotrophic fungi in northern hardwood forests alter the range of enzymes involved in C, N, and P cycling expressed by the fungal community. But the effects of deer exclusion and garlic mustard (an invasive plant) ground cover did not alter arbuscular mycorrhizal communities and their enzyme activity (Burke et al., 2011).

The close association between the presence of mycorrhizal fungi and decomposing organic matter has also been reported for arbuscular mycorrhizae (St. John et al., 1983), who's investigation allowed him to identify both fungal and faunal components and their interactions. Despite their limited enzyme capabilities, arbuscular mycorrhizal colonization of leaf litter appears to increase as litter decomposition advances. In a riparian tropical forests system, Posada et al. (2012) reported significantly greater AM fungal hyphae on well decomposed leaf litter at depth compared to less decomposed surficial litter. This correlated with increased saprotrophic fungal mass but was inversely correlated with root colonization. It is likely that these arbuscular mycorrhiza hyphae were accessing nutrients released from the litter decomposition (Figure 2.10).

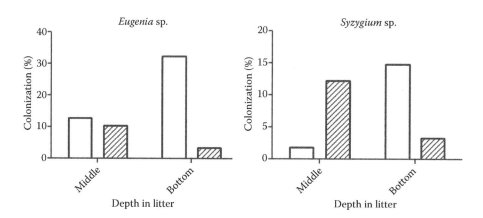

**Figure 2.10** Arbuscular mycorrhizal hyphal colonization of leaf litter (open bars) and roots (hatched bars) of two riparian tropical tree species, showing an inverse correlation between hyphae in leaf litter and colonization of the tree roots. (Data from Posada, R.H. et al., *Fung. Biol.*, 116, 747–755, 2012.)

## 2.2.7 Interactions between Decomposers and Other Organisms

It is important to remember at the outset that saprotrophic fungi involved with decomposition and nutrient cycling in soil do not perform that function in isolation. Plant and animal remains may be comminuted by soil fauna and subjected to enzyme attack by bacteria and actinomycetes. Interactions between these organisms are important in determining the rate of decomposition and diversity of soil biota. Micro- and mesoinvertebrates can alter the growth rates of wood decomposing fungi by either consuming extra mycelial biomass or stimulating greater production of that biomass (A'Bear et al., 2014), thus altering competition between wood decomposing fungi (Rotheray et al., 2009; Crowther et al., 2011a,b).

The high nutrient content of fungi makes them a good food source for animals. In addition to the direct, food chain, interaction, other soil faunal–fungal interactions also occur. Grazing of fungal hyphae by soil animals has been shown to affect both colonization of resources (litter) and rates of nutrient mineralization. Newell (1984a,b) showed that the effect of fungal grazing collembola altered the vertical distribution of competing mycelia of *Mycena galopus* and *Marasmius androcaceous* in a spruce forest floor. Preferential grazing of *Marasmius* restricted its growth to lower depths (*Mycena* dominated in the A soil horizon). Coleman et al. (1990) showed that reduction in microbial predators in ecosystems with high densities of soil fauna (forests) led to increased decomposition of litter (relief of grazing pressure). In contrast, in systems with low densities of soil fauna (agricultural soils) the effect of faunal reduction was to reduce decomposition (suggesting a synergistic interaction). In a comparison of tilled and nontilled agricultural soils, Beare et al. (1992) showed that the exclusion of fungiverous soil arthropods reduced litter dry mass loss by only 5%, but significantly altered nitrogen dynamics in surface litter of no-till soil. Saprotrophic fungi were responsible for as much as 86% of net nitrogen immobilization (1.8 g m$^{-2}$) into surface litters with the exclusion of fungiverous microarthropods.

In addition to plant litter, the remains of animals and their dung form specific organic resources that select for specific groups of fungi (Richardson, 2001). In a literature review, Richardson (2001) showed that there were highly significant differences in the fungal community structure between dung of six animal species. These fungi are often specific to their niche, but provide food for a variety of fungiverous animals (Hayashi and Tuno, 1998) and are, therefore, of significance in the ecosystem. We will not dwell on the autecology of these fungi here, but will return to their importance in fungal–faunal interactions in Chapter 5.

## 2.3 WHERE DOES SOIL BEGIN AND END?

In wet tropical forests, it may be difficult to determine where soil starts and ends. Because of the high rainfall, nutrients are leached from leaves and twigs in the canopy. These nutrients are added to rainfall percolating through the canopy and, thus, cause changes in the chemical composition of throughfall and stemflow. This modifies the nutrient availability in the forest floor. Additionally, some 7% of the total expected leaf litter fall never reaches the forest floor, but is trapped in the canopy of the tree or in the canopy of the understory shrub community (Hedger et al., 1993). Plant litter trapped in the canopies is held there by fungal hyphae and, particularly, rhizomorphs formed by species of the genera *Marasmius* and *Marasmiellus*, which effect the decomposition of the plant litter. These rhizomorphs and hyphae have become adapted to desiccating environments by their ability to produce copious amounts of mucilage and to be able to grow at low moisture potentials (−4 to −8 mPa). In an experimental litter manipulation in the canopy, Hedger et al. (1993) investigated the development of the contact zone between freshly added leaf litter and the fungal hyphae. Hedger et al. (1993) concluded that a large proportion of the hyphae invading new leaves originated from live leaves in the canopy. These fungi grow upward and away from previously trapped liter at a rate of 3–6 mm day$^{-1}$. The authors suggest that the fungi are endophytes within or saprotrophs on the surface of living leaves. The balance between the amounts of leaves trapped is a function of leaf weight, tensile strength of the retaining fungal structures, and weight loss due to decomposition. As the leaves decompose, mineralization will release nutrients that will wash to the forest floor in throughfall rain. Thus, the formation of "soil" in the tree canopy is a reality and probably has a significant impact on the fertility of the tropical forest ecosystem.

Lodge and Asbury (1988) demonstrated that the ability of fungi hyphae and cords to bind leaf litter together on the forest floor is important in preventing downslope loss of leaf litter in tropical forest ecosystems. The potential loss of organic matter, containing nutrients for plant growth, is prevented by the action of a number of, mainly, basidiomycete fungi that bind the leaf litter together. *Collybia*, *Marasmiellus*, *Marasmius*, and *Mycena* are the main fungi involved in forest floor litter trapping. The effect of litter binding by fungi increases with increasing ground slope. Lodge and Asbury (1988) concluded, from field manipulation experiments, that loss of litter was reduced by 35% from shallow slopes (<75% of angle) and 45% at greater slopes (75–90%). The reduction in leaf litter loss and subsequent incorporation of organic matter into the mineral soil is thought to prevent soil erosion during high rainfall periods.

The magnitude of effects of rainfall volume and leaching rates from canopy plant parts in wet tropical forests is much greater than that of other ecosystems. However, even in temperate forest ecosystems, the changes occurring in the stemflow water chemistry is enough to provide a suitable habitat for epiphytic lichen communities. Knops et al. (1996) demonstrated that the presence of the epiphytic lichen *Ramalina menziesii* on blue oak also altered throughfall chemistry, thus altering nutrient availability in the forest soil. They measured 590 kg ha$^{-1}$ of lichen biomass in the forest, in comparison with a standing crop of 958 kg ha$^{-1}$ of oak leaves. Trees with lichens had a higher deposition of total N, organic N, Ca, Mg, Na, and Cl in throughfall rain than trees without lichens. Trees with lichens had a lower throughfall of $SO_4$, and the concentration of $NO_3$, $NH_4$, K, and total P was not different. Lichen litter reduced the decomposition of oak leaf litter, such that release of N and P are reduced by 76% and 2%, respectively. However, the increased mineralization from lichen litter more than compensates for the reduction in oak litter decomposition, leading Knops et al. (1990) to conclude that the impact of both lichen leachates and effect on decomposition was unlikely to affect intrasystem or forest productivity. The evolution of a stable interaction between forest and epiphytic lichen community, however, has not been contemplated.

Hence, we see that fungi are major contributors to the fertility of soil by their action of decomposing organic residues derived from dead plant and animal remains. The activity of their exoenzymes removes the mass of dead remains and mineralizes the nutrients contained within, providing a source of nutrients for further primary production. During this process, however, fungi can perform an important function of regulating the release of nutrients in both space (translocation) and time (immobilization). These activities smooth out some of the heterogeneity seen in the distribution of resources to the decomposer community in the soil system. Patch accumulation of leaf litters on the forest floor of pine barrens ecosystems has been shown to be related to the density and distribution of stems of understory herbaceous vegetation (Dighton et al., 2000), who showed that the size of leaf litter patches was dependent on the density of leaf-trapping ericaceous stems. In addition, the quality of resources within patches depended on the litter patch size as the proportion of litter material composing the patches differed between patches of different size. The interpolation of the process rates occurring within different patches from the leaf litter patch scale to an ecosystem level scale of resolution, however, has yet to be made. The interactions between fungi and grazing fauna during the decomposition process may enhance the decomposition process and alter the outcomes of competition between fungi. So we cannot isolate the roles of fungi from other organisms in the soil.

## 2.4  KEEPING SOILS TOGETHER

Soils are prone to erosion by a number of factors; rainfall can wash soil away, especially on sloping ground, and dry soils can be displaced by wind erosion. Thus, in the development of a soil, the balance of inorganic, organic, and biotic components is of great importance to the physical stability of soils and their ability to support plant life.

An important component of stabilizing the mineral particles of a bare mineral soils are lichens, fungi, and bacteria as constituents of cryptogramic crusts (States and Christensen, 2001; Belnap and Lange, 2005), where they hold the mineral soil particles together. The fungal hyphae penetrate between the soil mineral particles and act as a web to physically retain soil particles. Carbohydrate secretions of both fungi and bacteria aid this process, acting as a glue to bind mineral particles together. The hydrophobic nature of some fungal hyphae alters the flow of water through soil colonized by cryptogramic crusts. Water tends to flow laterally across the surface soil in the presence of these crusts, rather than downward in uncolonized soil surfaces. This prevents the downward movement of soil particles and organic matter and reduces the risk of erosion of the developing soil structure.

Recently formed nutrients are also retained in the upper soil horizons by the hydophobicity of the lichen community, although the runoff of nutrients from crust communities may be of importance for the growth of surrounding plant communities (Eldridge et al., 2000).

Fungi involved in the mat-forming communities are subject to a variety of environmental stresses that occur on the soil surface, compared to the more buffered parts of the ecosystem. They are exposed to rapid and wide changes in water availability, temperature, and light. Certain adaptations of the fungal species having a lichen habit make them fit for existence in a stressful environment. In a recent study of melanins in fungal tissue, Gauslaa and Solhaug (2001) showed that the melanin content of fungal tissue within lichens reduced ultraviolet B (UVB) and UVA light penetration into the lichen, acting as a natural sunscreen. Removal of the orange pigment, parietin, in the lichen *Xanthoria parietina*, increased the damage caused by excessive light entering the lichen tissue (Solhaug and Gauslaa, 1996; Solhaug et al., 2003).

In more highly developed soils, soil aggregates are a formed by a physical combination of soil mineral particles, dead and living microbial components, and organic material derived from dead plant and animal remains (Tisdall and Oades, 1982). Microaggregates are classified as being less than 250 µm in diameter that themselves combine to form macroaggregates (>250 µm in diameter). An aggregation of aggregates is referred to as a soil crumb, the size and structure of which determine the texture and porosity of soil. In addition to the development of aggregates, soil organisms and their products are important in the stability of aggregates, their resistance to being physically disrupted.

The role of fungi and bacteria in the formation and stability of soil aggregates is of fundamental importance to both the fertility of soil and carbon storage and sequestration within soils. Coleman et al. (2004) summarize the development of soil aggregates by citing the work of Tisdall and Oades (1982). This process spans 5 orders of magnitude, from the cementation of clay particles, each of the order of 0.2 µm in diameter, through their interaction with microbial debris, and interactions with living bacteria and fungi (20 µm scale), making aggregates at 200 µm range to soil crumbs at 2000 µm in diameter (Table 2.12). The importance of bacteria and fungi in the development and stability of soil aggregates is further discussed in a review by Tisdall (1994). She concludes that bacteria play a major role in the formation and stabilization of microaggregates. The capsule surrounding many bacteria, especially Gram-negative bacteria, is composed of polysaccharides. This polysaccharide layer physically causes clay particles to adhere to the bacteria and, together with polyphenols attracted by ionic charges, protects the polysaccharide from microbial attack. This collection of clay particles and bacteria forms a microaggregate of about 20 µm in diameter. Additionally, saprotrophic fungal hyphae can grow between these microaggregates and continue the accumulation of material to produce larger aggregates, possibly by cation bridges between the hyphal polysaccharides and the clay particles. Saprotrophic fungi also influence soil properties. In a study on a sandy clay loam soil, Tisdall et al. (2012) noted that tensile strength of soil disks, abrasion resistance, and hot water

Table 2.12   Interaction of Soil Mineral Particles, Organic Matter, Bacteria, Fungi, and Root
             Material Spanning 5 Orders of Magnitude in Formation of Soil
             Microaggregates

| Scale of Interaction ($\mu m$) | Interaction between Components | Strength and Nature of Interaction |
|---|---|---|
| 0.2 | Amorphous aluminosilicates, oxides and organic polymers sorbed onto clay plates with electrostatic bonding and flocculation | Permanent, inorganic |
| 2 | Microbial and fungal debris (humic material) encrusted with inorganics | Persistent, organic |
| 20 | Plant, fungal and bacterial debris encrusted with inorganics | Persistent, organic |
| 200 | Roots and fungal hyphae aggregated with mineral particles | Medium term, organic |
| 2000 | Major binding of aggregate units to form a solid perforated with pores | Variable term, organic |

Source: After Coleman, D.C. et al., *Fundaments of Soil Ecology*, 2nd edn, Elsevier Academic
      Press, Amsterdam, 2004.

extractable carbohydrate, properties of aggregate stability, significantly increased
with increasing fungal hyphal length in soil when soils were colonized by saprotrophic fungi (Table 2.13).

Arbuscular mycorrhizal colonization of plant roots significantly increases soil
stability through the function of their extraradical hyphae. The presence of ryegrass
in an alfisol increased the content of stable macroaggregates from 36% to 78% in
6 months; in contrast, conversion of pasture (90% stable macroaggregates) into tilled
tomato crop reduced the aggregates to 58% because of the physical disruption of
the soil by agricultural practices. This increase in soil stability has been related to
the production of fungal carbohydrate secretions, which form cation bonds with
clay particles and other organic matter, especially the glycoprotein called glomalin (Wright and Upadhyaya, 1996, 1999). Using 16 soils from Mid-Atlantic states
of the United States and one from Scotland (UK), Wright and Upadhyaya (1998)

Table 2.13   Effect of Fungal Inoculation of Sterile Sand on Properties Associated
             with Aggregate Stability

| Treatment | Abrasion Resistance (g) | Tensile Strength (kPa) | HWEC (mg g$^{-1}$ soil) |
|---|---|---|---|
| Control | 9.4 a | 1.2 a | 0.53 a |
| *Chaetomium* sp. | 32.6 d | 6.7 b,c | 0.21 c,d |
| *Mucor* sp. | 35.6 e | 7.2 b,c | 0.16 e |
| *Curvularia inequalis* | 30.3 d | 10.2 d | 0.23 b,c |
| Unknown A | 30.4 d | 8.4 c,d | 0.27 a,b |
| Unknown B | 30.8 c | 5.2 b | 0.19 d |
| *Stemphylium* sp. | 18.9 b | 18.6 e | 0.23 b,c,d |

Source: Data from Tisdall, J.M. et al., *Soil Biol. Biochem.*, 50, 134–141, 2012.
Note:   Means followed by different letters in the same column are significantly different ($P < 0.05$);
        HWCE = hot water extractable carbohydrates.

used a monoclonal antibody technique to detect the glomalin content of soil aggregates. The glomalin content was then correlated with measures of aggregate stability. Across all sites, the relationship between the logarithm of glomalin content (expressed as mg g$^{-1}$ soil) and percent aggregate stability was positive, with a correlation coefficient of 0.86 (Table 2.14). Increased fungal and bacterial growth and the presence of organic matter act synergistically to develop a stable soil structure (St. John et al., 1983; Wright and Upadhyaya, 1998). For example, production of glomalin was significantly greater in the root systems of *Medicago sativa* when colonized by arbuscular mycorrhizal fungi, and there was a significant correlation between total hyphal length and hyphal density with the mean weight diameter of soil aggregates ($P = 0.048$ and 0.014, respectively; Bedini et al., 2009). In a meta-analysis of the effects of arbuscular mycorrhizal fungi on soil aggregation, Leifheit et al. (2014) found a positive effect of mycorrhizal inoculum with a mean effect of 0.2028, from a study of 175 studies in both greenhouse and field studies. They showed that short-term (2–5 months) pot studies at neutral pH in sandy soils produced the most positive mycorrhizal effects. However, interaction with soil fauna may influence the effect of mycorrhizae on soil aggregation. The presence of the collembolan *Proisotoma minuta* in pots containing *Plantago lanceolata* significantly increased mycorrhizal colonization of the roots and the abundance of water-stable soil aggregates of all sizes (Siddiky et al., 2012). It is suggested that fecal pellets from the collembolan act as resources for fungal hyphal growth, adding to the effect of the extraradical hyphae on soil aggregation.

The close juxtaposition of the mycorrhizal hyphae with the decomposing organic matter optimizes the ability of the mycorrhizae to capture mineralized nutrients for plant uptake. The interaction between mycorrhizal and nitrogen fixing, nodulating, bacteria on soil aggregate formation has been studied by Bethlenfalvay et al. (1999). Using soybeans inoculated with *Bradyrhizobium japonicum* and/or arbuscular mycorrhizae in a range of applied nitrogen fertilizers, they showed that maximal water-stable aggregate formation occurred in nodulated plants with ammonia fertilization. Nitrogen-deficient plants had sparse root development, and the integrity of soil aggregates was maintained by arbuscular mycorrhizal fungal hyphae. They suggested that water-stable soil aggregate formation occurred in three phases. During

**Table 2.14 Correlation Coefficients for Measures of Glomalin and Total Soil Carbon with Soil Aggregate Stability for a Range of Soils and Cropping History**

| | % Soil C | Easily Extractable Glomalin | Immunoreactive Easily Extractable Glomalin | Total Extractable |
|---|---|---|---|---|
| Easily extractable | 0.49 | | | |
| Immunoreactive easily extractable | 0.61 | 0.94 | | |
| Total extractable | 0.82 | 0.73 | 0.78 | |
| Aggregate stability (%) | 0.65 | 0.69 | 0.84 | 0.70 |

*Source:* After Wright, S.F., Upadhyaya, A., *Soil Biol. Biochem.*, 50, 134–141, 1998.
*Note:* All correlations are significant at less than 0.1%.

Phase 1, the mean aggregate size decreased because of an increase in bacterial numbers, whereas mycorrhizal fungi were in a lag phase of growth. This was suggested to be due to the cohesion of small aggregates being weakened by the metabolic activity of the increased bacterial population. During Phase 2, the rapidly developing network of mycorrhizal hyphae increased entanglement of small aggregates into larger aggregates owing to the production of glomalin. In Phase 3, compaction of soil by increased root growth was thought to possibly contribute to aggregate stabilization. In combination with other factors, it has thus been shown that glomalin from arbuscular mycorrhizal fungi is an important component in aggregate formation and stability. In their study of changes in soil characters under prairie restoration, Miller and Jastrow (1990) used path analysis to understand the interactive links between plant communities, their rooting characteristics, mycorrhizal associations, and the formation of stable aggregates in soil. The path of fine root length to root colonization and abundance of extraradical hyphal length gave the greatest correlation to aggregate size. It was of interest to note that this pathway was more strongly associated with fine roots, common to the native plant species, than the abundance of very fine roots, which are associated with nonindigenous plant species (Figure 2.11). This suggests that there could be some degree of coevolution of soil development and plant community type.

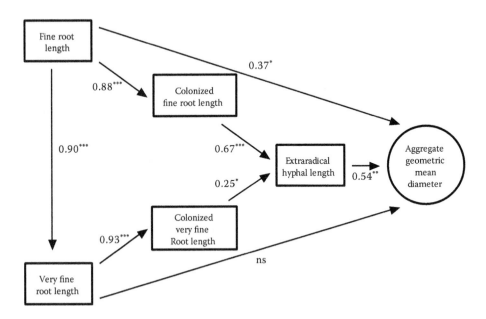

**Figure 2.11**  Path model relating the influence of root diameter class, arbuscular mycorrhizal infection and their relative contribution to the development of water-stable soil aggregates. Arrows indicate path coefficients, where *$P < 0.05$, **$P < 0.01$, and ***$P < 0.001$. (After Miller, R.M., Jastrow, J.D., *Soil Biol. Biochem.*, 22, 579–584, 1990.)

Although they do not produce glomalin, the presence of ectomycorrhizal fungi in the soil also assists in aggregate formation and soil stability. In a study of downslope soil erosion protection, Graf and Frei (2013) showed that inoculation of White alder with ectomycorrhizal inoculum of the hypogeous truffle *Melanogaster variegatus* significantly increased soil aggregate stability (Table 2.15) and tree root length. They also showed that there was a significant and positive correlation between root length per unit soil volume and abundance of stable aggregates, with a greater slope in mycorrhizal roots.

The changes in soil attributed to agricultural practices are manifold. One of the main effects is physical disturbance due to plowing and other mechanical disruption of soil. This disturbance can lead to direct damage to fungal hyphae, resulting in a shift of fungal community structure to favor those species that can withstand mycelial disruption. In a study of the effects of cultivation of steppe soils, Kurakov and Mirchink (1985) showed that of the mitosporic fungal community, *Penicillium* spp. was more abundant in cultivated soils than in virgin steppe. The change in species composition due to the initiation of agricultural practices caused a greater shift in fungal community structure than subsequent changes in crop species. Changes in fungal community composition following disturbance are largely attributable to changes in soil conditions and a result of the legacy of the former plant community. Invasion by fungal spore deposition appears to have a minimal effect on the fungal community; thus, the community is considered somewhat stable (Glinka and Hawkes, 2014). In Florida scrub vegetation, the strong plant–soil legacies (Lankau and Lankau, 2014) result in different fungal communities in soil under different plant species. This may be an important factor in the ability of invasive plant species to change soil conditions, such as nutrients (Chapman et al., 2006) and associated fungal (particularly mycorrhizal fungi) associations for their own benefit to improve competitiveness (Elgersma, 2014).

As a result of continual physical disturbance of soils under highly mechanized agricultural practices, soils become less stratified, with organic matter incorporated more evenly and to greater depths. Agricultural soils are primarily fertile because of the availability of inorganic nutrients in soil pore water provided by inorganic fertilizers, rather than being under fungal control of mineralization from organic material stocks (plant litters and their decomposition products). There has been heightened interest regarding the return to less intensive agricultural practices. Minimal or

**Table 2.15  Mean and Standard Deviation of Aggregate Stability [0,1] in Uncompacted soil (~15 kN m⁻³), Soil Planted with and without Mycorrhizal (*Melanogaster variegatus*) Alder (*Alnus incata*), or Compacted Soil (~19 kN m⁻³)**

| Treatment | Aggregate Stability |
|---|---|
| Uncompacted soil | 0.032 ± 0.041 |
| Soil plus alder | 0.059 ± 0.065 |
| Soil plus mycorrhizal alder | 0.380 ± 0.212 |
| Compacted soil | 0.476 ± 0.263 |

*Source:* Data from Graf, F., Frei, M., *Ecol. Eng.*, 57, 314–323, 2013.

no-till agricultural practices require a reduction in the physical disturbance of the soil structure, planting of crop species in and among native plant species, and a reduction in the use of herbicides and pesticides. The plant community of crop and native plants supports a greater diversity of natural predators for pests, and the return of dead plant parts to the soil surface is a source of endogenous, slow-release, fertilizer. The agricultural system, therefore, returns to a system that mimics natural ecosystems. The decomposer community shifts from a primarily bacterial dominated system toward one dominated by fungi. The impact of this fungal-driven decomposer community is that a wide variety of factors come into play in controlling the rates of release of mineral nutrients from an organic nutrient store. The temporary nutrient immobilization in fungal biomass, binding of organic material with mineral material in soil to increase aggregate formation and aggregate stability, as well as the decomposition of organic matter are integrated functions carried out by fungi in association with other soil organisms (Beare et al., 1994a,b).

## 2.5 NUTRIENT AVAILABILITY IN AQUATIC AND MARINE ECOSYSTEMS

### 2.5.1 Marine Ecosystems

Fungi are mainly known from the littoral zone, where they are important saprotrophs on algae (Zucarro and Mitchell, 2005), wood (Rama et al., 2014), and other organic debris. However, with the increased use of molecular tools, the presence of fungi as pelagic organisms in the marine water column (Richards et al., 2012) and in deep-sea sediments (Nagahama et al., 2001; Orsi et al., 2013) expands the niches in which fungi survive and function. A good review of the biology of marine fungi can be found in the book edited by Raghukumar (2012). The diverse taxonomic fungal groups found in marine ecosystems suggest that they should be regarded as an ecological—rather than a taxonomic—group (Hyde et al., 1998). The ascomycete order of *Halosphaeriales* is well represented in this ecosystem, by some 43 genera and 133 species. A lower diversity of basidiomycete and mitosporic fungi are found in marine ecosystems, but the distribution of species may be related to environmental conditions. Jones (1993) reports distinct fungal communities that are restricted to tropical and subtropical marine systems, which are distinct from those fungal species with cosmopolitan distribution. All marine fungi show physiological adaptations that allow them to survive under the stress of a high saline environment (Jennings, 1983; Clipson and Jennings, 1992), especially in their regulation of their osmotic potential and in deep-sea sediments to high hydrostatic pressure and low temperature (Nagango and Nagahama, 2012). Fungi even exist in the hypersaline conditions of the Dead Sea (at total salt concentrations of 340 g $l^{-1}$), where the number of species isolated has been reported to be 55 (Kis-Papo et al., 2001; Kis-Papo, 2005).

In marine water columns, fungi are most abundant in the top 30 cm (Wang et al., 2012), where a variety of filamentous fungi and Thraustochytrida are probably important in the decomposition of particulate organic matter and provide food for zooplankton. In their review of marine fungal diversity and function, Richards et al. (2012) show the considerable diversity of fungi in marine waters, but note that the diversity is less than that in other ecosystems because of the dilute carbon and nutrient resources in open ocean waters. Most pelagic organisms are small and largely have simple biochemical structures, so the need for a diverse fungal physiology to decompose these organisms is less important than in terrestrial environments (lower niche diversity). However, a number of pathogenic fungi closely related to *Cordyceps*, *Geomyces*, *Candida*, *Ustilago*, etc., have been found in marine ecosystems and are related to pathologies in marine invertebrates and mammals.

In benthic sediments, the role of Thrausochytrids in the decomposition of recalcitrant organic matter has probably been underestimated (Bongiorni, 2012). The role of a variety of other fungi revealed mainly by molecular methods from the deep sea is also largely unknown. In the Mariana Trench in the Pacific Ocean, sediments, clams, tubeworms, crabs, and mussels yielded some 90 yeast strains (Nagahama et al., 2001). Of these yeasts, some 40 isolates were red yeasts belonging to the genera *Rhodotortula* and *Sporobolomyces*. These red yeasts dominated the fungal community in sediments and animals collected at depths greater than 2000 m, whereas ascomycete yeasts dominated in sediments occurring at less than 2000 m depth. Adaptations of these yeasts to tolerate high hydrostatic pressures, high salinity, and low temperatures are discussed in the review by Nagango and Nagahama (2012). Using a variety of primers, Xu et al. (2014) obtained 48 molecular observational taxonomic units (OTUs) fungal taxa from deep-sea sediment samples taken from the Mariana Trench, of which 16 were for universal fungal primers, 17 from Ascomycota-specific primers, and 15 from Basidiomycota-specific primers. By RNA extraction and 454 pyrosequencing from subsurface sediment cores in deep ocean sites, Orsi et al. (2013) identified a wide diversity of fungi of Ascomycota, Basidiomycota, and Glomeromycota in sediments. There was a distinctly different fungal community between near-surface and deep samples, and the contribution of fungal RNA to other eukaryotic RNA appeared to be positively related to dissolved inorganic carbon and sulfide concentration of the sediment. The origin of these fungal propagules is largely unknown, but the abundance of most commonly aeolian dispersed fungi from arid lands is less than predicted, so it is assumed that other sources of fungal propagules play an important role.

Most marine fungi are known as saprotrophs on a variety of organic resources in the littoral zone. A range of ascomycete fungi appear to be true parasites of seaweeds, imparting a range of effects from minor discoloration of the thallus to the formation of galls and malformations that limit the photosynthetic function of the algae (Zuccaro and Mitchell, 2005). In addition, saprotrophs are present in degrading plant material derived from terrestrial sources as well as dead phytoplankton. However, it is interesting that the number of fungal species isolated from decaying seaweed (*Fucus serratus*) (84) was significantly lower than live material (252) (Zuccaro et al., 2008). Fungi associated with this seaweed showed phenological changes in community composition over the course of a year.

Newell (1996) suggests that these marine decomposers maximize their surface area and have a high substrate affinity to allow enzymes to easily diffuse into solid particles to effect mineralization. Penetration of the decomposing resource by tunneling or surface erosion into solid organic substrates is common in both bacteria and labyrinthulids and involves chains or series of single cells. Penetration of solid material by absorptive "ectoplasmic nets" or rhizoids is a strategy adopted by true chitrids, thrasustochitrids, and hyphochytrids. However, colonization of solid substrates by networks of "self-extending tubular reactors" (tubes of chitil-laminarin or cellulose-laminarin) signifies the activity of mycelial eumycotic fungi and oomycete protoctists.

Wood is a major resource for fungi in marine ecosystems (Rohrmann and Molitoris, 1992). Wooden breakwaters, jetties, and piers provide resources for the saprotrophic fungal community, whose effects cause large economic repercussions. Kohlmeyer and Kohlmeyer (1979) present a table of 107 fungal species isolated from decomposing wood in marine habitats. These include 73 Ascomycotina, two Basidiomycotina, and 29 mitosporic species. Rama et al. (2014) isolated 147 fungal OTUs from 50 logs in the intertidal and sea floor on the North Norwegian coast. Although ascomycetes dominated, members of Basidiomycota, Chytridiomycota, and Mucormycotina were also represented. Using wood baits of two tree species, Azevedo et al. (2011) showed that there was similar species richness between the fungal communities colonizing both beech (*Fagus sylvatica*) and pine (*Pinus pinaster*) wood in Portuguese waters, but the frequency of isolation of species differed significantly between the two substrates (Table 2.16).

As in terrestrial ecosystems, the low resource quality of wood appears to encourage tight linkages between fungi and fauna for its decomposition. Evidence suggests that wood boring marine mollusks preferentially settle and feed on wood that has previously been colonized by fungi and partially decomposed, rather than invade fresh wood. The associations have become so tight that, for example, the wood boring crustacean, the gribble (*Limnoria tripunctata*), has increased longevity when feeding on wood colonized by fungi. More importantly, it is incapable of reproduction on

Table 2.16   Frequency of Occurrence of Fungi from Wood Baits of Beech (*Fagus sylvatica*) and Pine (*Pinus pinaster*) from Two Marine Marinas in Portugal

|  | Beech | Pine |
|---|---|---|
| *Lulworthia* sp. | 97.9 | 45.1 |
| *Cirrenalia macrocephala* | 13.2 | 79.2 |
| *Corollospora maritima* | 27.8 | 45.8 |
| *Zalerion maritima* | 14.6 | 59.0 |
| *Cerisosporopsis halima* | 28.5 | 38.2 |
| *Halosphaeria appendiculata* | 46.5 | 12.5 |
| *Trichocladium achrasoprum* | 3.5 | 28.9 |
| *Periconia prolifica* | 20.8 | 1.4 |
| *Remispora quadriremis* | – | 19.4 |
| Richness (S) | 19 | 22 |
| Total number of specimens | 415 | 530 |

Source: Data from Azevedo, E. et al., *Anim. Biodivers. Conserv.*, 34, 205–215, 2001.

any substrate unless marine fungi are included as part of its diet. This may be attributable to the enhanced availability of proteins, essential amino acids, and vitamins, which are unavailable in the absence of fungi.

From samples of a variety of organic debris from intertidal sandy shores, the diversity of fungi was found to be dependent on temperature, rainfall, and salinity (Velez et al., 2013). Specific fungal species responded differently to rainfall and salinity, but overall diversity was positively correlated with increased temperature.

In salt marsh ecosystems, the decomposition of the salt marsh grass *Spartina* relies on a separate community of fungi to decompose leaves than those effecting decomposition of the roots and rhizomes, which make up to more than half of the plant biomass. Live fungal biomass is low (<20 mg g$^{-1}$ substrate), and their strategy is to grow rapidly within the substrate and immobilize nutrients into the fungal biomass. Fungi in this situation can contain some 75–100% of the total N of decaying cordgrass leaves. Leaves are mainly decomposed by ascomycetes, of which *Phaesosphaeria spartinicola* is dominant, and mitosporic fungi, but only a single basidiomycete species, *Nia vibrissa* (Kohlmeyer and Kohlmeyer 1979).

In salt marsh sediments, aerobic conditions quickly give way to anaerobic conditions with increasing depth. In the anaerobic environment, fungi cede to the abilities of bacteria to derive energy from chemoautotrophic processes. In the anaerobic zones, bacteria dominate over fungi. Mansfield and Bärlocher (1992) found that fungal biomass, measured as ergosterol content, was negatively related to redox potential in *Spartina* salt marshes, showing a rapid decline of fungal activity with increasing sediment depth, although fungi are a major agent of decomposition of *Spartina* plant parts (Meyers, 1974). However, when balsa wood panels were buried in an anaerobic salt marsh, they were colonized by fungi within 12 weeks. Many of the fungal species colonizing these wood blocks could not grow in entirely in anaerobic conditions, but were able to grow down from aerobic to anaerobic zones over distances of 5–10 mm in 15 days from resources in the aerobic zone. This evidence shows that fungi have the ability to conduct oxygen from aerobic regions, through hyphae, to advancing mycelial fronts, which are physiologically active in decomposition in the anaerobic zones (Padgett and Celio, 1990).

Mangrove swamps are the tropical equivalent of salt marsh habitats in the temperate world, but occur in more sheltered areas, away from the direct impact of wave action. The litter from these ecosystems is more diverse than that of salt marsh systems, and the high rate of primary production produces copious detritus supporting a large population and diversity of detritivore fungi, bacteria, and fauna (Kohlmeyer and Kohlmeyer, 1979). The mangrove fungi are almost exclusively saprotrophic, consisting of 23 species of ascomycete, 17 mitosporic species, and two basidiomycetes, and their community composition has been shown to be distinct from those associated in saltmarsh ecosystems (Newell, 1996). The fungal biomass in decaying mangrove leaves is much lower than that in decaying saltmarsh vegetation (<1 mg g$^{-1}$ compared to 60–85 mg g$^{-1}$, respectively; Newell and Fell, 1992; Newell, 1996), and bacteria represent a very small percentage of the decomposers (0.7 mg g$^{-1}$). Newell and Fell (1992) also showed that there were significant changes in fungal biomass during decomposition of red mangrove leaves. They suggest that the actual fungal biomass is greater than 1 mg g$^{-1}$ as many marine oomycetes (e.g., *Halophytopthora*) do not contain ergosterol. Mangrove leaves rapidly accumulate a large population of oomycete fungi (*Halophytophthora* spp.), but ascomycete fungal species, dominated by *Lulworthia grandispora*, may comprise 50% of the fungal community.

Export of plant detritus from mangrove ecosystems to the oceans is an important contribution to the nutrient loading of oceans, with between 60 and 260 t y$^{-1}$ of carbon, which is exported mainly as dissolved organic carbon (Lee, 1995). This may be an important component of nutrient additions to near-shore waters and a process in which fungi play a major role. However, Hyde and Lee (1995) point out that there are still many gaps in our knowledge of the role of fungi in nutrient cycling in mangrove ecosystems. They suggest that the rates of chemical transformations are dependent on the age and diversity of mangrove forest and contribution of terrestrial tree flora to the litter input of the various microhabitats within an area. They also suggest that the end product of fungal decomposition is likely to be dissolved organic matter, rather than particulate organic matter, of which there is scant understanding of its origins and movement in distribution in marine estuarine ecosystems.

## 2.5.2 Freshwater Ecosystems

According to Wong et al. (1998), there are more than 600 species of aquatic fungi, many of which have specific morphological and physiological adaptation to allow them to live in aquatic ecosystems, most of which are saprotrophs (Bärlocher, 2005). In a multivariate analysis of 146 samples from 92 aquatic systems, Gulis (2001) showed that the fungal species assemblage on wood and grass leaves have a different composition from those on tree leaf litter, showing that the diversity of allochthonous resources to aquatic systems determined fungal community composition.

Fungi have been isolated from spores suspended in the water column of streams, ponds, and lakes, growing on decaying vegetation and utilizing suspended organic matter in deep aquifers (Kuehn and Koehen, 1988). Aquatic hyphomycetes occur on almost all substrates in freshwater systems (Bärlocher, 1992, 2005). Fungal biomass is usually greater than bacterial biomass on decomposing leaf litter in aquatic ecosystems. Plant litter inputs into headwater streams in forested catchments can be of the order of 500 g dry mass m$^{-2}$ and reach peaks of more than 100 g m$^{-2}$ (Weigelhofer and Waringer, 1994). Measures of rates of decomposition of plant litters by aquatic fungi suggest that a range of carbohydrate resources can be utilized (Bergbauer et al., 1992), but that the rate of decomposition is reduced in mixed species fungal assemblages, compared to single species. Bergbauer et al. (1992) attribute this reduction in decomposition to the production of antimicrobial compounds that result in nonnutritional competition between fungal species. Wood is an important resource, representing up to 20% of the total plant litter input. Shearer (1992) estimates that some 250–800 t ha$^{-1}$ of woody debris enters stream systems of old-growth temperate coniferous forests and 40–130 t ha$^{-1}$ in mixed hardwood forests. The main addition of woody material to a stream ecosystem occurs in the winter as a result of windthrow and weather damage to branches and twigs. It is thus available for fungal colonization in the spring and throughout the year, whereas deciduous leaf litter enters as a pulse in the fall and is usually degraded before the next leaf fall or has been exported downstream (Shearer, 1992). This woody material is important in that its residence time is much greater than that of leaf litter and, therefore, forms a more stable environment for fungal community development. Despite the lack of white and brown rot fungi, which occur in terrestrial ecosystems, lignolytic aquatic hyphomycetes of the genera *Tricladium*, *Anguillospora*, and *Dendrospora* are dominant colonizers of woody material in streams. Shearer (1992) also shows that many species express a wide range of enzymes related to the decomposition of woody material. Of 20 species, seven species were shown to produce enzymes to degrade carboxymethyl cellulose, cellobiose, amylose, xylan, xylose, lignin, and pectin, whereas another four species could utilize five or more resources. The community of fungi colonizing woody resources alters along the length of a water course (habitat selection) and over time at the same location (resource succession). Gessner et al. (1997) provide a review of the role of fungi in plant litter decomposition in aquatic ecosystems. They provide a conceptual model of the interactions between the internal controls (litter quality), external controls (environmental variables), and the metabolic activity of fungi that determines the possible outcomes of decomposition in the aquatic system (Figure 2.12).

Fungi are not the only organisms that effect leaf litter decomposition in freshwater ecosystems. Gaur et al. (1992) identified 13 bacteria and 24 fungal species associated with the decomposition of water hyacinth leaves. They demonstrated that the bacteria were involved with the breakdown of polysaccharides and proteins and the fungi with the decomposition of cellulose and lignocellulose. Litter decomposition in the first few days, comprising some 30% of the litter mass loss, was predominantly nonmicrobial (faunal comminution). Bacteria were then the most important components of the microbial community, where they accessed dissolved organic matter. Fungi became important as the leached organic matter availability declined, and they were capable of entering the intact structural components of the leaf litter, which are more recalcitrant. Within the fungal community, Gaur et al. (1992) showed that there were successions of fungi, with fungi having greater enzymatic capabilities sharing a larger proportion of the community in later stages of decomposition and utilizing the more recalcitrant resources. Bacteria account for approximately 30% of the respiration from decomposing leaf litter of the dominant plant species in the Florida Everglades, and the remaining 70% comes from fungi (Hackney et al., 2000). Ergosterol measures of fungal biomass showed an increase on decaying leaves during the course of decomposition, but the fungal biomass was not related to nutrient levels in the water.

Leaf litter decomposition in freshwater habitats is related to both resource quality and environmental conditions. A review of the structure and function of aquatic fungi is given by Bärlocher (2005). The community composition of fungi is strongly influenced by the litter resource and, to a lesser extent, by the nature (chemistry and pH) of the aquatic system. Encalada et al. (2010) observed fungal communities on alder (*Alnus acuminata*) and Inga (*Inga spectabilis*) leaf litter decomposing in either a forest stream or a stream in a pasture. Fungal richness was significantly lower on Inga leaves (6 and 9 for forest and pasture, respectively) than alder (14 and 20,

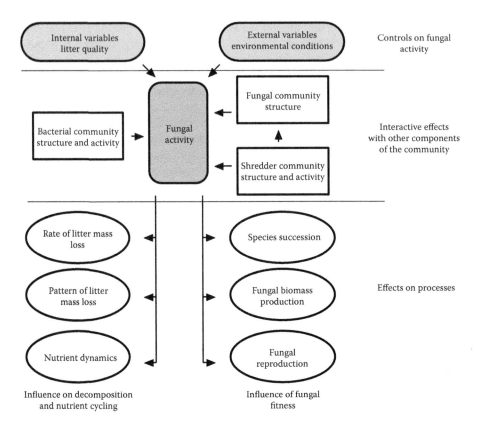

**Figure 2.12**  Schematic representation of decomposition in aquatic systems based on a fungal
dominated system. (After Gessner, M.O. et al., in *The Mycota IV: Environmental
and Microbial Relationships*, 303–322, ed. Wicklow, D.T., Soderstrom, B.,
Springer-Verlag, Berlin, 1997.)

respectively), although *Campylospora chaetocladia* and *Filosporella versimorpha*
dominated in both litter types and both streams. In a comparison of the decomposition and fungal colonization between two litter types at contrasting temperatures, higher rates of decomposition were found in alder leaves than in oak leaves,
with a significant increase in rate owing to the elevated temperature only in oak
(Gonçalves et al., 2013). Presumably because of niche restriction in the fungal community of oak litter, the fungal community on oak leaves did not change significantly
with increased temperature, but did so on alder leaves. In a similar study, Ferreira
and Chauvet (2011) showed that alder leaf litter decomposition was significantly
increased at elevated temperatures, particularly under high levels of available N and
P. The fungal community and biomass on decomposing leaves in streams is important for the shredder faunal community. Eutrophic streams significantly increased
fungal biomass and sporulation of aquatic hyphomycetes on oak and alder leaf litter,
and stimulated grater faunal biomass in eutrophic streams on oak litter (Gulis et

al., 2006). Suberkropp and Chauvet (1995) performed reciprocal exchange of yellow poplar leaves between streams differing in pH and nutrient composition. They showed that the rate of leaf decomposition varied by a factor of 9 between streams, the microbial activity (measured as ATP production) by a factor of 8, and fungal sporulation by a factor of 80.

A summary of the interactions between stream characteristics, leaf litter decomposition, and measures of fungal activity are given in Table 2.17. Litter decomposed faster in hardwater streams, with higher availabilities of nitrate-nitrogen and phosphorus and higher temperatures, stimulating greater fungal biomass and activity and a more diverse fungal species assemblage. About six species of fungi were common throughout all hardwater streams, whereas the softwater streams used in this study had only two species in common with the hardwater streams. Moreover, the species that occur between the streams were different, with *Anguillospora filiformis* and *Flagellospora curvula* being the two dominant species in softwater and not occurring in the hardwater streams. However, the occurrence of different species of fungi in aquatic systems is dependent on season. Both Gupta and Mehrota (1989) and Thomas et al. (1989) identified seasonal changes in fungal species composition of the community. Thomas et al. (1989) reported greatest conidial abundance in early autumn (500–600 conidia $l^{-1}$) and lowest during the winter (300–500 conidia $l^{-1}$). Using stepwise multiple regression analysis of the conidial abundance and a variety of environmental measures, they showed that spore abundance was significantly related to temperature ($R^2 = 0.62$) and somewhat related to a combination of both temperature and rainfall. Species composition, however, was not related to temperature, but to rainfall and conductivity.

These studies, however, do not relate to the impact of seasonal variations in plant litter composition and changes in the physicochemical composition of the litter that occur during succession. In a study of the comparative decomposition of yellow poplar (*Liriodendron tulipifera*), red maple (*Acer rubrum*), and white oak (*Quercus alba*) leaf material in streams, Griffith et al. (1995) demonstrated differences in decomposition constants between leaf species and between sites, where pH and temperature were the major variables between sites. In general, leaf litters decomposed more readily in streams of higher pH and higher temperature (Table 2.18), but some of the between-site differences were less obvious if the data were

Table 2.17 **Pearson Correlation Coefficients among Stream Variables, the Decomposition of Yellow Poplar Leaves and Fungal Activity in Contrasting Soft- and Hardwater Streams**

|  |  | $NO_3$-N | $PO_4$-P | Temperature |
|---|---|---|---|---|
| All streams | Decomposition constant (*k*) | 0.97 | 0.83 | 0.77 NS |
|  | Maximum ATP | 0.95 | 0.86 | 0.91 |
|  | Maximum sporulation | 0.90 | 0.46 NS | 0.72 NS |
| Hardwater streams | Decomposition constant (*k*) | 0.96 | 0.75 NS | 0.51 NS |
|  | Maximum ATP | 0.92 | 0.70 NS | 0.84 |
|  | Maximum sporulation | 0.97 | 0.66 NS | 0.85 NS |

*Source:* Data from Suberkropp, K., Chauvet, E., *Ecology*, 76, 1433–1445, 1995.
*Note:* NS indicates that correlation is not statistically significant.

Table 2.18    Decomposition Constants (–*k*) of Leaf Litters in Three Streams of Differing
              pH and Temperature

| Site | pH | Cumulative Degree Days | White Oak –*k* (Day⁻¹) | Red Maple –*k* (Day⁻¹) | Yellow Poplar –*k* (Day⁻¹) |
|------|-----|------------------------|------------------------|------------------------|----------------------------|
| SFR  | 4.3 | 233                    | 0.0020                 | 0.0037                 | 0.0058                     |
| WHR  | 6.2 | 424                    | 0.0059                 | 0.0106                 | 0.0068                     |
| HSR  | 7.7 | 393                    | 0.0038                 | 0.0091                 | 0.0081                     |

*Source:* Data from Griffiths, M.B. et al., *Oecologia*, 102, 460–466, 1995.

temperature-corrected, suggesting that this was a primary driving variable. The temporal pattern of production of pectinase enzyme was also different between leaf
litter types and may be related to differences in chemical composition as observed
via the physical changes between leaf litters during decomposition. Both white oak
and red maple exhibited skeletonization, whereas yellow poplar leaves just became
increasingly softer. Similar to terrestrial systems, temporal successions of aquatic
fungi occur as leaf litters decompose (Gessner et al., 1993). Fungal communities on
alder leaf litter were dominated by five to six species during early colonization at
2 weeks of incubation (*Flagellospora curvula, Tetracahetum elegans, Lemonniera
centrospharea, L. aquatica,* and *L. terrestris*). At 4 weeks, the species composition
was more equitable and consisted of a larger and persistent assemblage of 11 species at midsuccession fungal biomass (measured as ergosterol content) and conidial
production peak (Gessner et al., 1993).

    In a study of mixing leaf litter types, Jaibol and Chauvet (2012) noted signifi-
cant differences between fungal biomass, diversity, and rate of litter consumption
by the arthropod *Gammarus* between each litter type (oak, alder, walnut, and birch).
Mixing litters had little effect on any parameter, although fungal diversity in birch
mixtures was significantly greater than that in birch alone and consumption of walnut mixtures was significantly greater than consumption of walnut alone. Birch and
walnut litter in the birch, oak, and walnut litter mixtures induced significantly higher
grazing rates than these litters in any other mixture. Mixing alder and maple leaves
did not enhance litter decomposition rates or fungal diversity in a Nova Scotian
river, but did have a small effect of increasing the abundance of some members of
the shredder arthropod community (Taylor et al., 2007). Anthropogenic disturbance
(eutrophication) of streams was shown to significantly reduce shredded richness and
increase the fungal biomass, but not richness on decomposing leaf litter (Tolkkinen
et al., 2013). Dominance of fungal species appeared to be a major driver of litter decomposition rates, and fungal evenness was significantly lower in disturbed
streams than in neutral or acidic streams.

    Changes in resource quality of leaf litter may also result from prior occupancy
by fungi. For example, the presence of endophytes in leaves reduces the rate of
decomposition in aquatic systems. Maple leaves containing the endophyte/pathogen
*Rhytisma punctatum* decomposed about 37% slower than uninfected leaves (percent mass remaining at 77 days, 50.3 with *Rhytisma* and 31.4 without) (LeRoy et
al., 2011). The interaction between phylloplane fungal colonization of leaves and

their subsequent decomposition in both terrestrial and aquatic systems has been little explored.

Aquatic fungi serve as an important food source for invertebrates and alter the physical structure of the wood substrate to allow faunal penetration (Suberkropp, 1992). Graca et al. (1993) have demonstrated that some faunal shredders prefer to feed on leaves that are already colonized by fungi, whereas others consume fungal mycelia selectively. It is likely that the palatability of the resource is enhanced during fungal decomposition by the increase in nitrogen content during the initial stages of fungal attack (Gessner et al., 1997).

The role of fungi in aquatic ecosystems has not been studied as intensively as in terrestrial ecosystems. Wong et al. (1998) reiterate this in their review, saying that although fungi are the dominant decomposers in these systems, we know little about the mechanisms of decomposition or the interactions between fungi, and between fungi and other organisms. Although resource succession by saprotrophic fungi has been identified (Gessner et al., 1993), we know very little about the relationship between fungal biodiversity and ecological function. However, it would appear that resource quality of plant material dictates the community structure of the aquatic saprotrophic fungi colonizing it, and there may be many parallels between aquatic and terrestrial systems in the way in which materials are processed and utilized (Wagener et al., 1998).

## 2.6 CONCLUSIONS

Fungi are a diverse group of organisms. In the process of decomposition and mineralization of nutrients from organic sources, a range of taxonomic groups are involved. It has been shown in the preceding discussion that these fungi differ both in their enzymatic capabilities (their ability to decompose certain resources), their rate of growth (competitiveness), and their interactions with other organisms in the ecosystem. We have concentrated mainly on terrestrial ecosystems, as here we have gathered the greatest amount of information regarding the role of fungi in decomposition, measures of rates of nutrient mineralization from organic matter, and the effects of fungi on rock dissolution and soil development. We can see from the discussions on the activities of fungi in aquatic and marine ecosystems that they are involved in similar processes of decomposition of organic resources, and that the interactions between fungi and their changing environment during decomposition of these resources lead to similar patterns of resource succession, limitation of decomposition by resource quality, and impacts of environmental constraints on their activity in the same manner as in the terrestrial environment. However, information regarding measure of the amount of nutrient mineralization and the importance of this to plant primary production is less available than for terrestrial ecosystems. Soil development and stability is a unique feature of terrestrial ecosystems. Although lichens are abundant on rocky coastlines, there is little information regarding their input to the nutrient content of oceans by the process of rock dissolution. In salt marsh sediments, fungi probably play a role in binding particles together and reducing the possibility of erosion due to tidal and wave action. We know that approximately 50% of aquatic and salt marsh plants are mycorrhizal (Khan and Belik, 1995; Cooke and Lefor, 1998), but we have little information about the extent of mycorrhizal hyphal development into sediments in either aquatic or salt marsh habitats, or their roles in nutrient acquisition, sediment stabilization, or interactions with other biotic components.

Thus, although we can make generalized statements about the role of fungi in decomposition, nutrient cycling and nutrient accumulation, and the similarity of actions in different ecosystems, it must be remembered that there are other external influences on communities and function. For example, Hackney et al. (2000) showed that the external supply of nitrogen significantly influenced the ratio of fungi and bacteria colonizing plant leaves in the Everglades (Figure 2.13). In addition, Hawksworth (1991) has suggested that we may be able to identify some 5% of the possible total number of fungal species in the world. Of these, we may have isolated and investigated the physiology of a mere handful. We must therefore ask how confident we are in extrapolating these findings to fungi as a whole. Soil is an opaque medium where most of the fungal-mediated nutrient-cycling processes occur in terrestrial ecosystems. Bodies of water represent the aquatic and marine ecosystems that require specialized equipment to enter, in which to conduct experiments, and from which data may be collected. These are not easy habitats in which to study the activities and roles of fungi. Additionally, we are limited by our methodologies of study, especially in determining active fungal biomass and *in situ* measures of activity, enzyme production process rates at the scale at which they are performed by the fungal hyphae. Our level of knowledge is constantly growing, but it is far from complete.

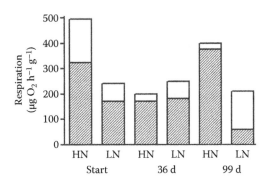

**Figure 2.13**  Relative contribution of bacterial (open bar) and fungal (hatched bar) respiration to total respiration of decomposing *Cladium* leaves in the Everglades (Florida, USA) at high (HN) and low (LN) levels of external nitrogen supply. (Data from Hackney, C.T. et al., *Mycol. Res.*, 104, 666–670, 2000.)

# REFERENCES

A'Bear, A. D., T. H. Jones, and L. Boddy. 2014. Potential impacts of climate change on inter-actions among saprotrophic cord-forming fungal mycelia and grazing soil invertebrates. *Fung. Ecol.* 10:34–43.

April, R. and D. Keller. 1990. Mineralogy of the rhizosphere in forest soils of the eastern United States. *Biogeochemistry* 9:1–18.

Ascaso, C. and J. Wierzchos. 1995. Study of the biodeterioration zone between the lichen thallus and the substrate. *Cryptogamic Bot.* 5:270–281.

Asta, J., F. Orry, F. Toutain, B. Souchier, and G. Villemin. 2001. Micromorphological and ultra-structural investigations of the lichen–soil interface. *Soil Biol. Biochem.* 33:323–338.

Attiwill, P. M. and M. A. Adams. 1993. Nutrient cycling in forests. *New Phytol.* 124:561–582.

Azcon, R., J. M. Barea, and D. S. Hayman. 1976. Utilization of rock phosphate in alkaline soils by plants inoculated with mycorrhizal fungi and phosphate solubilizing bacteria. *Soil Biol. Biochem.* 8:135–138.

Azevedo, E., M. F. Caeiro, R. Rebelo, and M. Barata. 2011. Biodiversity and characterization of marine mycota from Portuguese waters. *Anim. Biodivers. Conserv.* 34:205–215.

Banfield, J. F., W. W. Barker, S. A. Welch, and A. Taunton. 1999. Biological impact of mineral dissolution: Application of the lichen model to understanding mineral weathering in the rhizosphere. *Proc. Natl. Acad. Sci. U. S. A.* 96:3404–3411.

Barker, W. W. and J. F. Banfield. 1998. Zones of chemical and physical interaction at interfaces between microbial communities and minerals: A model. *Geomicrobiol. J.* 15: 223–244.

Bärlocher, F. 1992. Research on aquatic hyphomycetes: Historical background and over-view. In: *The Ecology of Aquatic Hyphomycetes.*, ed. F. Bärlocher, 1–15. Berlin: Springer-Verlag.

Bärlocher, F. 2005. Freshwater fungal communities. In: *The Fungal Community: Its Composition and Role in the Ecosystem*, ed. J. Dighton, J. F. White, and P. Oudemans, 39–59. Boca Raton, FL: Taylor & Francis.

Beare, M. H., M. L. Cabrera, P. F. Hendrix, and D. C. Coleman. 1994a. Aggregate-protected and unprotected pools of organic matter in conventional and no-tillage ultisols. *Soil Sci. Soc. Am. J.* 58:787–795.

Beare, M. H., P. F. Hendrix, and D. C. Coleman. 1994b. Water-stable aggregates and organic matter fractions in conventional and no-tillage soils. *Soil Sci. Soc. Am. J.* 58:777–786.

Beare M. H, R. W. Parmelee, P. F. Hendrix, W. Cheng, D. C. Coleman, and D. A. Crossley Jr. 1992. Microbial and faunal interactions and effects on litter nitrogen and decomposition in agroecosystems. *Ecol. Monogr.* 62:569–591.

Bedini, S., E. Pellegrino, L. Avio et al. 2009. Changes in soil aggregation and glomalin-related soil protein content as affected by the arbuscular mycorrhizal fungal species *Glomus mosseae* and *Glomus intraradices*. *Soil Biol. Biochem.* 41:1491–1496.

Behera, N., D. P. Pati, and S. Basu. 1991. Ecological studies of soil microfungi in tropical for-est soil of Orissa, India. *Tropical Ecology* 32:136–143.

Belnap, J. and O. L. Lange. 2005. Lichens and microfungi in biological soil crusts: Community structure, physiology, and ecological function. In: *The Fungal Community: Its Organization and Role in the Ecosystem*, ed. J. Dighton, J. F. White, and P. Oudemans, 117–138. Boca Raton, FL: Taylor & Francis.

Bergbauer, M., M. A. Moran, and R. E. Hodson. 1992. Lignocellulose decomposition by aero-aquatic fungi. *Microbial Ecol.* 23:159–167.

Berthelin, J. and C. Leyval. 1982. Ability of symbiotic and non-symbiotic rhizospheric microflora of maize (*Zea mays*) to weather micas and to promote plant growth and plant nutrition. *Plant Soil* 68:369–377.

Bethlenfalvay, G. J., I. C. Cantrell, K. L. Mihara, and R. P. Schreiner. 1999. Relationships
between soil aggregation and mycorrhizae as influenced by soil biota and nitrogen nutrition. *Biol. Fertil. Soils* 28:356–363.

Boddy, L. and A. D. M. Rayner. 1983. Ecological roles of basidiomycetes forming decay
communities in attached oak branches. *New Phytol.* 93:177–188.

Boddy, L. and S. C. Watkinson. 1995. Wood decomposition, higher fungi, and their role in
nutrient redistribution. *Can. J. Bot.* 73(Suppl. 1):S1377–S1383.

Bogomolova, E. V. 2007. Rock-inhabiting fungi: From biodeterioration to human disease.
In: *Fungi: Maultifaceted Microbes*, ed. B. N. Ganguli and S. K. Deshmukh. 49–6. New
Delhi: Anamya Publishers (Taylor & Francis).

Bongiorni, L. 2012. Thraustochytrids, a neglected component of organic matter decomposition
and food webs in marine sediments. In: *Biology of Marine Fungi*, ed. C. Raghukumar,
1–13. Berlin: Springer Verlag.

Brady, N. C. and R. R. Weil. 1999. *The Nature and Properties of Soils*. Upper Saddle River,
NJ: Prentice Hall.

Burford, E. P., M. Kierans, and G. M. Gadd. 2003. Geomycology: Fungi in mineral substrata.
*Mycologist* 17:98–107.

Burke, D. J., M. N. Weintrub, C. R. Hewins, and S. Kalisz. 2011. Relationship between soil
enzyme activities, nutrient cycling and soil fungal communities in a northern hardwood
forest. *Soil Biol. Biochem.* 43:795–803.

Cadish, G. and K. E. Giller. 1997. *Driven by Nature: Plant Litter Quality and Decomposition*.
Wallingford: CAB International.

Cairney, J. W. G. 1992. Translocation of solutes in ectomycorrhizal and saprotrophic rhizomorphs. *Mycol. Res.* 96:135–141.

Chang, T. T. and C. Y. Li. 1998. Weathering of limestone, marble, and calcium phosphate by ectomycorrhizal fungi and associated microorganisms. *Taiwan J. For. Sci.* 13:85–90.

Chapman, S. K., J. A. Langley, S. C. Hart, and G. W. Koch. 2006. Plants actively control nitrogen cycling: Uncorking the microbial bottleneck. *New Phytol.* 169:27–34.

Chen, J., H.-P. Blume, and L. Beyer. 2000. Weathering of rocks induced by lichen colonization—
A review. *Catena* 39:121–149.

Clinton, P. W., P. K. Buchanan, and R. B. Allen. 1999. Nutrient composition of epigeous fungal sporocarps growing on different substrates in a New Zealand mountain beech forest.
*N. Z. J. Bot.* 37:149–153.

Clipson, N. J. W. and D. H. Jennings. 1992. *Dendryphiella salina* and *Debaryomyces hansenii*:
Models for ecophysiological adaptation to salinity by fungi that grow in the sea. *Can.
J. Bot.* 70:2097–2105.

Coleman, D. C., D. A. Crossley, and P. Hendrix. 2004. *Fundaments of Soil Ecology*, 2nd
Edition. Amsterdam. Elsevier Academic Press.

Coleman, D. C., E. I. Ingham, H. W. Hunt, E. T. Elliott, C. P. P. Reid, and J. C. Moore. 1990.
Seasonal and faunal effects on decomposition in semiarid prairie meadows and lodgepole pine forest. *Pedobiologia* 34:207–219.

Connolly, J. H., W. C. Shortle, and J. Jellison. 1998. Translocation and incorporation of strontium carbonate derived strontium into calcium oxalate crystals by the wood decay fungus *Resinicium bicolor*. *Can. J. Bot.* 77:179–187.

Cooke, J. C. and M. W. Lefor. 1998. The mycorrhizal status of selected plant species from
Connecticut wetlands and transition zones. *Restoration Ecology* 6:214–222.

Cooke, R. C. and A. D. M. Rayner. 1984. *Ecology of Saprotrophic Fungi.* London: Longman.

Crittenden, P. D. 1991. Ecological significance of necromass production in mat-forming lichens. *Lichenologist* 23:323–331.

Crittenden, P. D. 2000. Aspects of the ecology of mat-forming lichens. *Rangefinder* 20: 127–139.

Crowther, T. W., L. Boddy, and T. Hefin Jones. 2011a. Outcomes of fungal interactions are determined by soil invertebrate grazers. *Ecol. Lett.* 14:1134–1142.

Crowther, T. W., T. Hefin Jones, L. Boddy, and P. Baldrian. 2011b. Invertebrate grazing determines enzyme production by basidiomycete fungi. *Soil Biol. Biochem.* 43:2060–2068.

Di Lonardo, D. P., F. Pinzari, D. Lunghini, O. Maggi, V. M. Granio, and A. M. Persiani. 2013. Metabolic profiling reveals a functional succession of active fungi during the decay of Mediterranean plant litter. *Soil Biol. Biochem.* 60:210–219.

Dickinson, C. H. and G. J. F. Pugh. 1974. *Biology of Plant Litter Decomposition.* London: Academic Press.

Dighton, J. 1995. Nutrient cycling in different terrestrial ecosystems in relation to fungi. *Canadian Journal of Botany* 73(Supp. 1):S1349–S1360.

Dighton, J. 1997. Nutrient cycling by saprotrophic fungi in terrestrial habitats. In: *The Mycota IV: Environmental and Microbial Relationships,* ed. D. T. Wicklow and B. Söderstrom, 271–279. Berlin: Springer Verlag.

Dighton, J. and L. Boddy. 1989. Role of fungi in nitrogen, phosphorus and sulphur cycling in temperate forest ecosystems. In: *Nitrogen, Phosphorus and Sulphur Cycling in Temperate Forest Ecosystems,* ed. L. Boddy, R. Marchant, and D. J. Read, 269–298. Cambridge: Cambridge University Press.

Dighton, J. and A. D. Horrill. 1988. Radiocaesium accumulation in the mycorrhizal fungi *Lactarius rufus* and *Inocybe longicystis,* in upland Britain. *Trans. Br. Mycol. Soc.* 91:335–337.

Dighton, J. and J. A. Krumins (eds.). 2014. *Interaction in Soil: Promoting Plant Growth.* Dordrecht: Springer Science and Business Media, 231 pp.

Dighton, J., A. S. Morale-Bonilla, R. A. Jimînez-Nûñez, and N. Martînez. 2000. Determinants of leaf litter patchiness in mixed species New Jersey pine barrens forest and its possible influence on soil and soil biota. *Biol. Fertil. Soils.* 31:288–293.

Duddridge, J. A., A. Malibari, and D. J. Read. 1980. Structure and function of mycorrhizal rhizomorphs with special reference to their role in water transport. *Nature* 287:834–836.

Eldridge, D. L., E. Zaady, and M. Shachk. 2000. Infiltration through three contrasting biological soil crusts in patterned landscapes in the Negev, Israel. *Catena* 40:323–336.

Elgersma, K. J. 2014. Soils suppressing and promoting non-native plant invasions. In: *Interactions in Soil: Promoting Plant Growth,* ed. J. Dighton and J. A. Krumins, 181–202. Dordrecht: Springer Science and Business.

Encalada, A. C., J. Calles, V. Ferriera, C. M. Canhoto, and A. S. Graça. 2010. Riparian land use and the relationship between benthos and litter decomposition in tropical montane streams. *Freshwater Biol.* 55:1719–1733.

Ferreira, V. and E. Cahuvet. 2011. Synergistic effects of water temperature and dissolved nutrients on litter decomposition and associated fungi. *Freshwater Biol.* 17:551–564.

Finlay, R., H. Wallander, M. Smits et al. 2009. The role of fungi in biogenic weathering in boreal forest soils. *Fung. Biol. Rev.* 23:101–106.

Fogel, R. and G. Hunt. 1983. Contribution of mycorrhizae and soil fungi to nutrient cycling in a Douglas-fir ecosystem. *Can. J. For. Res.* 13:219–232.

Fomina, M., E. P. Burford, S. Hillier, M. Klerans, and G. M. Gadd. 2010. Rock-building fungi. *Geomicrobiol. J.* 27:624–629.

Forsyth, A. and K. Miyata. 1984. *Tropical Nature.* New York: Charles Scribner's Sons.

Frankland, J. C. 1992. Mechanisms in fungal succession. In: *The Fungal Community: Its Organization and Role in the Ecosystem*, ed. G. C. Carroll and D. T. Wicklow, 393–401. New York: Marcel Dekker.

Frankland, J. C. 1998. Fungal succession—Unravelling the unpredictable. *Mycol. Res.* 102: 1–15.

Frankland, J. C., J. N. Hedger, and M. J. Swift. 1982. *Decomposer Basidiomycetes: Their Biology and Ecology.* Cambridge: Cambridge University Press.

Fukasawa, Y., O. Tateno, Y. Hagiwara, D. Hirose, and T. Osono. 2012. Fungal succession and decomposition of beech cupule litter. *Ecol. Res.* 27:735–743.

Gadd, G. M. 2004. Mycotransformation of organic and inorganic substrates. *Mycologist* 18:60–70.

Gaur, S, P. K. Singhal, and S. K. Hasiji. 1992. Relative contributions of bacteria and fungi to water hyacinth decomposition. *Aquat. Bot.* 43:1–15.

Gauslaa, Y. and K. A. Solhaug. 2001. Fungal melanins as a sun screen for symbiotic green algae in the lichen *Lobaria pulmonaria. Oecologia* 126:462–471.

Georgieva, S., S. Christensen, H. Petersen, P. Gjelstrup, and K. Thorup-Kristensen. 2005. Early decomposer assemblages of soil organisms in litterbags with vetch and rye roots. *Soil Biol. Biochem.* 37:1145–1155.

Gessner, M. O., K. Suberkropp, and E. Chauvet. 1997. Decomposition of plant litter by fungi in marine and freshwater ecosystems. In: *The Mycota IV: Environmental and Microbial Relationships*, ed. D. T. Wicklow and B. Soderstrom, 303–322. Berlin: Springer-Verlag.

Gessner, M. O., M. Thomas, A.-M. Jean-Louis, and E. Chauvet. 1993. Stable successional patterns of aquatic hyphomycetes on leaves decaying in a summer cool stream. *Mycol. Res.* 97:163–172.

Glinka, C. and C. V. Hawkes. 2014. Environmental controls on fungal community composition and abundance over 3 years in native and degraded shrublands. *Microb. Ecol.* 68:807–817.

Gobran, G. R., S. Clegg, and F. Courchesne. 1998. Rhizospheric processes influencing the biogeochemistry of forest ecosystems. *Biogeochemistry* 42:107–120.

Golubic, S., G. Radtke, and T. Le Campion-Alsumard. 2007. Endolithic fungi. In: *Fungi: Mutifaceted Microbes*, ed. B. N. Ganguli and S. K. Deshmukh, 38–48. New Delhi: Anamaya Publishers (Taylor & Francis).

Gonçalves, A. L., M. A. S. Graça, and C. Canhoto. 2013. The effect of temperature on leaf decomposition and diversity of associated aquatic hyphomycetes depends on the substrate. *Fung. Ecol.* 6:546–553.

Graca, M. A. S., L. Maltby, and P. Calow. 1993. Importance of fungi in the diet of *Gammarus pulex* and *Asellus aquaticus*: II. Effects on growth, reproduction and physiology. *Oecologia* 96:304–309.

Graf, F. and M. Frei. 2013. Soil aggregate stability related to soil density, root length, and mycorrhiza using site-specific *Alnus incata* and *Melanogaster variegatus s. l. Ecol. Eng.* 57:314–323.

Gray, S. N., J. Dighton, S. Olsson, and D. H. Jennings. 1995. Real-time measurement of uptake and translocation of 137Cs within mycelium of *Schizophyllum commune* Fr. by autoradiography followed by quantitative image analysis. *New Phytol.* 129:449–465.

Griffith, M. B., S. A. Perry, and W. B. Perry. 1995. Leaf litter processing and exoenzyme production on leaves in streams of differing pH. *Oecologia* 102:460–466.

Gulis, V. 2001. Are there any substrate preferences in aquatic hyphomycetes? *Mycol. Res.* 105:1088–1093.

Gulis, V., V. Ferriera, and A. S. Graça. 2006. Stimulation of leaf litter decomposition and associated fungi and invertebrates by moderate eutrophication: Implications for stream assessment. *Freshwater Biol.* 51:1655–1669.

Gupta, R. K. and R. S. Mehrota. 1989. Seasonal periodicity of aquatic fungi in tanks at Kurukshetra, India. *Hydrobiologica* 173:219–229.

Güsewell, S. and M. O. Gessner. 2009. N:P ratios influence litter decomposition and colonization by fungi and bacteria in microcosms. *Funct. Ecol.* 23:211–219.

Hackney, C. T., D. E. Padgett, and M. H. Posey. 2000. Fungal and bacterial contributions to the decomposition of *Cladium* and *Typha* leaves in nutrient enriched and nutrient poor areas of the Everglades, with a note on ergosterol concentrations in Everglades soils. *Mycol. Res.* 104:666–670.

Harborne, J. B. 1997. Role of phenolic secondary metabolites in plants and their degradation in nature. In: *Driven by Nature: Plant Litter Quality and Decomposition*, ed. G. Cadisch and K. E. Giller, 67–74. Cambridge: Cambridge University Press.

Hawksworth, D. L. 1988. The variety of fungal–algal symbioses, their evolutionary significance, and the nature of lichens. *Bot. J. Linn. Soc.* 96:3–20.

Hawksworth, D. L. 1991. The fungal dimension of biodiversity: Magnitude, significance, and conservation. *Mycol. Res.* 95:641–655.

Hayashi, T. and N. Tuno. 1998. Notes on the lesser dung flies emerged from fungi in Japan (Diptera, Sphaeroceridae). *Med. Entomol. Zool.* 49:357–359.

Heal, O. W., J. M. Anderson, and M. J. Swift. 1997. Plant litter quality and decomposition: An historical overview. In: *Driven by Nature: Plant Litter Quality and Decomposition*, ed. G. Cadish and K. E. Giller, 3–30. Wallingford: CAB International.

Heal, O. W. and J. Dighton. 1985. Resource quality and trophic structure in the soil system. In: *Ecological Interactions in Soil*, ed. A. H. Fitter, D. Atkinson, D. J. Read, and M. B. Usher, 339–354. Oxford: Blackwell.

Heal, O. W. and J. Dighton. 1986. Nutrient cycling and decomposition of natural terrestrial ecosystems. In: *Microfloral and Faunal Interactions in Natural and Agro-Ecosystems*, ed. M. J. Mitchell and J. P. Nakas, 14–73. Martinus Nijhoff/Dr. W. Junk.

Hedger, J., P. Lewis, and H. Gitay. 1993. Litter trapping by fungi in moist tropical forest. In: *Aspects of Tropical Mycology*, ed. S. Isaac, J. C. Frankland, R. Watling, and A. J. S. Whalley. Cambridge, UK: Cambridge University Press.

Hinsinger, P. and B. Jaillard. 1993. Root-induced release of interlayer potassium and vermiculitization of phlogopite as related to potassium depletion in the rhizosphere or ryegrass. *J. Soil. Sci.* 44:525–534.

Hirsch, P., F. E. W. Eckhardt, and R. J. Palmer. 1995. Fungi active in weathering of rock and stone monuments. *Can. J. Bot.* 73(Suppl. 1):1384–1390.

Hoffland, E., R. Landeweert, T. W. Kuyper, and N. van Breemen. 2001. (Further) links from rocks to plants. *Trends Ecol. Evol.* 16:544.

Honegger, R. 1991. Functional aspects of the lichen symbiosis. *Annu. Rev. Plant Physiol. Plant Mol. Biol.* 42:553–578.

Hyde, K. D., E. B. Gareth Jones, E. Leano, S. B. Pointing, A. D. Poonyth, and L. L. P. Vrijmoed. 1998. Role of fungi in marine ecosystems. *Biodivers. Conserv.* 7:1147–1161.

Hyde, K. D. and Y. Lee. 1995. Ecology of mangrove fungi and their role in nutrient cycling: What gaps occur in our knowledge? *Hydrobiologia* 295:107–118.

Jaibol, J. and E. Chauvet. 2012. Fungi involved in the effects of litter mixtures on consumption by shredders. *Freshwater Biol.* 57:1667–1677.

Jennings, D. H. 1983. Some aspects of the physiology and biochemistry of marine fungi. *Biol. Rev.* 58:423–459.

Jones, E. B. G. 1993. Tropical marine fungi. In: *Aspects of Tropical Mycology*, ed. S. Isaac, J. C. Frankland, R. Watling, and A. J. S. Whalley, 73–89. Cambridge: Cambridge University Press.

Jones, H. E., M. Madeira, L. Herraez et al. 1999. The effect of organic-matter management on the productivity of *Eucalyptus globulus* stands in Spain and Portugal: Tree growth and harvest residue decomposition in relation to site and treatment. *For. Ecol. Manage.* 122:73–86.

Jongmans, A. G., N. van Breemen, U. Lundstrom et al. 1997. Rock-eating fungi. *Nature* 389:682–683.

Khan, A. G. and M. Belik. 1995. Occurrence and ecological significance of mycorrhizal symbiosis in aquatic plants. In: *Mycorrhiza: Structure, Function, Molecular Biology and Biotechnology*, ed. A. Varma and B. Hock, 627–666. Berlin: Springer-Verlag.

Kis-Papo, T. 2005. Marine fungal communities. In: *The Fungal Community: Its Composition and Role in the Ecosystem*, ed. J. Dighton, P. Oudemans, and J. F. White, 61–92. Boca Raton, FL: Taylor & Francis.

Kis-Papo, T., I. Grishkan, A. Oren, S. P. Wasser, and E. Nevo. 2001. Spatiotemporal diversity of filamentous fungi in the hypersaline Dead Sea. *Mycol. Res.* 105:749–756.

Kjøller, A. and S. Struwe. 1982. Microfungi in ecosystems: Fungal occurrence an activity in litter and soil. *Oikos* 39:391–418.

Knops, J. M. H., T. H. Nash, and W. H. Schlesinger. 1996. The influence of epiphytic lichens on the nutrient cycling of an oak woodland. *Ecol. Monographs* 66:159–179.

Koele, N., I. A. Dickie, J. D. Blum, J. D. Gleason, and L. de Graaf. 2014. Ecological significance of mineral weathering in ectomycorrhizal and arbuscular mycorrhizal ecosystems from a field-based comparison. *Soil Biol. Biochem.* 69:63–70.

Kohlmeyer, J. and E. Kohlmeyer. 1979. *Marine Mycology: The Higher Fungi*. New York: Academic Press.

Kuehn, K. A. and R. D. Koehen. 1988. A mycofloral survey of an artesian community within the Edwards aquifer of central Texas. *Mycologia* 80:646–652.

Kurakov, A. V. and T. G. Mirchink. 1985. The structure of complexes of saprotrophic microscopic fungi in the cultivation of typical sierozem. *Mikol. Fitopatol.* 19:207–211.

Lamontagne, S. 1998. Nitrogen mineralization in upland Precambrian Shield catchments: Contrasting the role of lichen-covered bedrock and forested areas. *Biogeochemistry* 41:53–69.

Landeweert, R., E. Hoffland, R. D. Finlay, T. W. Kuyper, and N. van Breemen. 2001. Linking plants to rocks: Ectomycorrhizal fungi mobilize nutrients from minerals. *Trends Ecol. Evol.* 16:248–254.

Lange, O. L., J. Belnap, and H. Reichenberger. 1998. Photosynthesis of the cyanobacterial soil-crust lichen *Collema tenax* from arid lands in southern Utah, USA: Role of water content on light and temperature responses of $CO_2$ exchange. *Funct. Ecol.* 12:195–202.

Lange, O. L., A. Meyer, H. Zellner, and U. Heber. 1994. Photosynthesis and water relations of lichen soil crusts: Field measurements in the coastal fog zone of the Namib Desert. *Funct. Ecol.* 8:253–264.

Lankau, E. W. and R. A. Lankau. 2014. Plant species capacity to drive soil fungal communities contributes to differential impacts of plant–soil legacies. *Ecology* 95:3221–3228.

Lee, S. Y. 1995. Mangrove outwelling: A review. *Hydrobiologia* 295:203–212.

Leifheit, E. F., S. D. Veresoglou, A. Lehmann, E. K. Morris, and M. C. Rillig. 2014. Multiple factors influence the role of arbuscular mycorrhizal fungi in soil aggregation—A meta-analysis. *Plant Soil* 374:523–537.

Lemons, A., K. Clay, and J. A. Rudgers. 2005. Connecting plant–microbial interactions above and belowground: A fungal endophyte affects decomposition. *Oecologia* 145:595–604.

LeRoy, C. J., D. G. Fischer, K. Halstead, M. Pryor, J. K. Bailey, and J. A. Schweitzer. 2011. A fungal endophyte slows litter decomposition in streams. *Freshwater Biol.* 56:1426–1433.

Lockwood, J. L. 1992. Exploitation Competition. In: *The Fungal Community: Its Organization and Role in the Ecosystem*, ed. G. C. Carrol and D. T. Wicklow, 243–263. New York: Marcel Dekker.

Lodge, D. J. 1993. Nutrient cycling by fungi in wet tropical forests. In: *Aspects of Tropical Mycology*, ed. S. Isaac, J. C. Frankland, R. Watling, and A. J. S. Whalley, 37–75. Cambridge: Cambridge University Press.

Lodge, D. J. and C. E. Asbury. 1988. Basidiomycetes reduce export of organic matter from forest slopes. *Mycologia* 80:888–890.

Mansfield, S. D. and F. Bärlocher. 1993. Seasonal variation of fungal biomass in the sediment of a salt marsh in New Brunswick. *Microb. Ecol.* 26:37–45.

Martino, E. and S. Perotto. 2010. Mineral transformations by mycorrhizal fungi. *Geomicrobiol. J.* 27:609–623.

Marumoto, T., J. P. E. Anderson, and K. H. Domsch. 1982. Decomposition of $^{14}C$ - and $^{15}N$- labeled microbial cells in soil. *Soil Biol. Biochem.* 14:461–467.

Melillo, J. M., J. D. Aber, and J. F. Muratore. 1982. Nitrogen and lignin control of hardwood leaf litter decomposition dynamics. *Ecology* 63:621–626.

Meyers, S. P. 1974. Contribution of fungi to biodegradation of Spartina and other brackish marshland vegetation. *Veroff. Inst. Meeresforsch. Bremerh. Suppl.* 5:357–375.

Miller, R. M. and J. D. Jastrow. 1990. Hierarchy of root and mycorrhizal fungal interactions with soil aggregates. *Soil Biol. Biochem.* 22:579–584.

Mojallala, H. and S. B. Weed. 1978. Weathering of micas by mycorhizal soybean plants. *Soil Biol. Biochem.* 42:367–372.

Møller, J., M. Miller, and A. Kjøller. 1999. Fungal–bacterial interaction on beech leaves: Influence on decomposition and dissolved organic carbon quality. *Soil Biol. Biochem* 31:367–374.

Mottershead, D. and G. Lucas. 2000. The role of lichens in inhibiting erosion of soluble rock. *Lichenologist* 32:601–609.

Nagahama, T., M. Hammamoto, T. Nakase, H. Takami, and K. Horikoshi. 2001. Distribution and identification of red yeasts in deep-sea environments around the northwest Pacific Ocean. *Antonie van Leeuwenhoek* 80:101–110.

Nagango, Y. and T. Nagahama. 2012. Fungal diversity in deep-sea environments. *Fung. Ecol.* 5:463–471.

Nahas, E., D. A. Banzatto, and L. C. Assis. 1990. Fluorapatite solubilzation by *Aspergillus niger* in vinasse medium. *Soil Biol. Biochem.* 22:1097–1101.

Newell, K. 1984a. Interaction between two decomposer basidiomycetes and a collembolan under Sitka spruce: Distribution, abundance and selective grazing. *Soil Biol. Biochem.* 16:227–233.

Newell, K. 1984b. Interactions between two decomposer basidiomycetes and a collembolan under Sitka spruce: Grazing and its potential effects on fungal distribution and litter decomposition. *Soil Biol. Biochem.* 16:235–239.

Newell, S. Y. 1996. Established and potential impacts of eukaryotic mycelial decomposers in marine/terrestrial ecotones. *J. Exp. Mar. Biol. Ecol.* 200:187–206.

Newell, S. Y. and J. W. Fell. 1992. Ergosterol content of living and submerged, decaying leaves and twigs of red mangrove. *Can. J. Microbiol.* 38:979–982.

Oberle-Kilic, J., J. Dighton, and G. Arbuckle-Keil. 2013. Atomic force microscopy and micro-ATR–FT-IR imaging reveals fungal enzyme activity at the hyphal scale of resolution. *Mycology* 4:44–53.

Olsson, S. 1995. Mycelial density profiles of fungi on heterogenous media and their interpretation in terms of nutrient reallocation patterns. *Mycol. Res.* 99:143–153.

Olsson, S. and D. H. Jennings. 1991. Evidence for diffusion being the mechanism of translocation in the hyphae of three moulds. *Exp. Mycol.* 15:302–309.

Omacini, M., E. J. Chaneton, C. M. Ghersa, and P. Otero. 2004. Do foliar endophytes affect grass litter decomposition? A microcosm approach using *Lolium multflorum*. *Oikos* 104:581–590.

Orsi, W., J. F. Biddle, and V. Edgecomb. 2013. Deep sequencing of subseafloor eukaryotic rRNA reveals active fungi across marine and subsurface provinces. *PLoS ONE* 8(2):e56335. doi.10.1371/journal.pone.0056335.

Osono, T. 2006. Role of phyllosphere fungi of forest trees in the development of decomposer fungal communities and decomposition process of leaf litter. *Can. J. Microbiol.* 52:701–716.

Osono, T., Y. Fukasawa, and H. Takeda. 2003. Roles of diverse fungi in larch needle-litter decomposition. *Mycologia* 95:820–826.

Padgett, D. E. and D. A. Celio. 1990. A newly discovered role for aerobic fungi in anaerobic salt marsh soils. *Mycologia* 82:791–794.

Parfitt, R. L., G. J. Salt, and S. Saggar. 2001. Post-harvest residue decomposition and nitrogen dynamics in *Pinus radiata* plantations of different N status. *For. Ecol. Manage.* 154:55–67.

Paris, F., B. Botton, and F. Laperyrie. 1996. In vitro weathering of phlogopite by ectomycorrhizal fungi. *Plant Soil* 179:141–150.

Peršoh, D., J. Segert, A. Zigan, and G. Rambold. 2013. Fungal community composition shifts along a leaf degradation gradient in European beech forest. *Plant Soil* 362:175–186.

Ponge, J. F. 1990. Ecological study of a forest humus by observing a small volume. I. Penetration of pine litter by mycorrhizal fungi. *Eur. J. For. Pathol.* 20:290–303.

Ponge, J. F. 1991. Succession of fungi and fauna during decomposition of needles in a small area of Scots pine litter. *Plant and Soil* 138:99–113.

Ponge, J. F. 2005. Fungal communities: Relation to resource succession. In: *The Fungal Community: Its Composition and Role in the Ecosystem*, ed. J. Dighton. J. F. White, and P. Oudemans, 169–180. Boca Raton, FL: Taylor & Francis.

Posada, R. H., S. Madriñan, and E.-L. Rivera. 2012. Relationships between the litter colonization by saprotrophic and arbuscular mycorrhizal fungi with depth in a tropical forest. *Fung. Biol.* 116:747–755.

Raghukumar, C. (ed.). 2012. *Biology of Marine Fungi*. Berlin: Springer-Verlag.

Rama, T., J. Norden, M. L. Davey, G. H. Mathiassen, J. W. Spatafora, and H. Kauserud. 2014. Fungi ahoy! Diversity on marine wooden substrata in the high North. *Fung. Ecol.* 8:46–58.

Rayner, A. D. M. 1978. Interactions between fungi colonizing hardwood stumps and their possible role in determining patterns of colonization and succession. *Ann. Apppl. Biol.* 89:505–517.

Rayner, A. D. M. and L. Boddy. 1988. *Fungal Decomposition of Wood*. Chichester: John Wiley.

Richards, T. A., Jones, M. D. M., G. Leonard, and D. Bass. 2012. Marine fungi: Theory ecology and molecular diversity. *Annu. Rev. Mar. Sci.* 4:495–552.

Richardson, M. J. 2001. Diversity and occurrence of coprophilous fungi. *Mycol. Res.* 105:387–402.

Robinson, C. H., J. Dighton, and J. C. Frankland. 1993a. Resource capture by interaction by fungal colonizers of straw. *Mycol. Res.* 97:547–558.

Robinson, C. H., J. Dighton, J. C. Frankland, and P. A. Coward. 1993b. Nutrient and carbon dioxide release by interacting species of straw-decomposing fungi. *Plant Soil* 151:139–142.

Rodin, L. E. and N. I. Bazilevich. 1967. *Production and Mineral Cycling in Terrestrial Vegetation.* Edinburgh: Oliver and Boyd.

Rohrmann, S. and H. P. Molitoris. 1992. Screening for wood-degrading enzymes in marine fungi. *Can. J. Bot.* 70:2116–2123.

Rosling, A., T. Roose, A. M. Herrmann, F. A. Davidison, R. D. Finlay, and G. M. Gadd. 2009. Approaches to modelling mineral weathering by fungi. *Fung. Biol. Rev.* 23:138–144.

Rotheray, T. D., L. Boddy, and T. Hefin Jones. 2009. Collembola foraging responses to interacting fungi. *Ecol. Entomol.* 34:125–132.

Sanders, W. B. 1997. Fine structural features of rhizomorphs (*sensu lato*) produced by four species of lichen fungi. *Mycol. Res.* 101:319–328.

Sanders, W. B. 2001. Lichens: The interface between mycology and plant morphology. *BioScience* 51:1025–1035.

Schmidt, N. 1999. Microbial properties and habitats of permafrost soils on Taimyr Peninsula, central Siberia. *Ber. Polarforschung.* 340:1–183.

Shammas, K., A. M. O'Connell, T. S. Grove, R. McMurtrie, P. Damon, and S. J. Rance. 2003. Contribution of decomposing harvest residues to nutrient cycling in a second rotation *Eucalyptus globus* plantation in south western Australia. *Biol. Fertil. Soils* 38:228–235.

Shaw, T. M., J. Dighton, and F. E. Sanders. 1995. Interactions between ectomycorrhizal and saprotrophic fungi on agar and in association with seedlings of lodgepole pine (*Pinus contorta*). *Mycol. Res.* 99:159–165.

Shearer, C. A. 1992. The role of woody debris. In: *The Ecology of Aquatic Hyphomycetes*, ed. F. Barlocher. Berlin: Springer-Verlag.

Siddiky, M. R. K., J. Kohler, M. Cosme, and M. C. Rillig. 2012. Soil biota effects on soil structure: Interactions between arbuscular mycorrhizal fungal mycelium and collembolan. *Soil Biol. Biochem.* 50:33–39.

Siegrist, J. A., R. L. McCulley, L. P. Bush, and T. D. Phillips. 2010. Alkaloids may not be responsible for endophyte-associated reductions in tall fescue decomposition rates. *Funct. Ecol.* 24:460–468.

Silveira, M. L., K. R. Reddy, and N. B. Comerford. 2011. Litter decomposition and soluble carbon, nitrogen, and phosphorus release in a forest ecosystem. *Eur. J. Soil Sci.* 1:86–96.

Singh, S. and K. K. Kapoor. 1998. Effects of inoculation of phosphate-solubilizing microorganisms and arbuscular mycorrhizal fungus on mungbean grown under natural soil conditions. *Mycorrhiza* 7:149–153.

Sinsabaugh, R. L. 2005. Fungal enzymes at the community scale. In: *The Fungal Community: Its Composition and Role in the Ecosystem*, ed. J. Dighton, P. Oudemans, and J. F. White, 349–360. Boca Raton, FL: Taylor & Francis.

Sinsabaugh, R. L., R. K. Antibus, A. E. Linkins, and C. A. McClaughtery. 1993. Wood decomposition: Nitrogen and phosphorus dynamics in relation to extracellular enzyme activity. *Ecology.* 74:1586–1593.

Sinsabaugh, R. L. and M. A. Liptak. 1997. Enzymatic conversion of plant biomass. In: *The Mycota IV, Environmental and Microbial Relationships*, ed. D. T. Wicklow and B. Söderstrom, 347–357. Berlin: Springer-Verlag.

Smart, K. A. and C. R. Jackson. 2009. Fine scale patterns of microbial extracellular enzyme activity during leaf litter decomposition in a stream and its floodplain. *Microb. Ecol.* 58:591–598.

Smits, M. M. 2009. Scale matters? Exploring the effects of scale on fungal–mineral interactions. *Fung. Biol. Rev.* 23:132–137.

Šnajdr, J., P. Dobiášová, T. Větrovský et al. 2011. Saprotrophic basidiomycete mycelia and their interspecific interactions affect the spatial distribution of extracellular enzymes in soil. *FEMS Microb. Ecol.* 78:80–90.

Solhaug, K. A. and Y. Gauslaa. 1996. Parietin, a photoprotective secondary product of the lichen *Xanthoria parietina*. *Oecologia* 108:412–418.

Solhaug, K. A., Y. Gauslaa, L. Nybakken, and W. Bilger. 2003. UV-induction of sun-screenng pigments in lichens. *New Phytol.* 158:91–100.

St. John, T. V., D. C. Coleman, and C. P. P. Reid. 1983. Association of vesicular–arbuscular mycorrhizal hyphae with soil organic particles. *Ecology* 64:957–959.

Stark, N. 1972. Nutrient cycling pathways and liter fungi. *Bioscience* 22:355–360.

States, J. S. and M. Christensen. 2001. Fungi associated with biological soil crusts in desert grasslands of Utah and Wyoming. *Mycologia* 93:432–439.

Strickland, M. S., E. Osburn, C. Lauber, N. Fierer, and M. A. Bradford. 2009. Litter quality is in the eye of the beholder: Initial decomposition rates as a function of inoculum characteristics. *Funct. Ecol.* 23:627–636.

Suberkropp, K. 1992. Interactions with invertebrates. In: *The Ecology of Aquatic Hyphomycetes*, ed. F. Barlocher. Berlin: Springer-Verlag.

Suberkropp, K. and E. Chauvet. 1995. Regulation of leaf breakdown by fungi in streams: Influences of water chemistry. *Ecology* 76:1433–1445.

Sverdrup, H. 2009. Chemical weathering of soil minerals and the role of biological processes. *Fung. Biol. Rev.* 23:94–100.

Swift, M. J., O. W. Heal, and J. M. Anderson. 1979. *Decomposition in Terrestrial Ecosystems*. Oxford: Blackwell Scientific.

Taylor, D. L., T. N. Hollingsworth, J. W. McFarland, N. J. Lennon, C. Nusbaum, and R. W. Ruess. 2014. A first comprehensive census of fungi in soil reveals both hyperdiversity and fine-scale niche partitioning. *Ecol. Monogr.* 84:3–20.

Taylor, B. R., C. Mallaley, and J. F. Cairns. 2007. Limited evidence that mixing leaf litter accelerates decomposition or increases diversity of decomposers in streams of Eastern Canada. *Hydrobiologia* 592:405–422.

Thomas, R. J. and N. M. Asakawa. 1993. Decomposition of leaf litter from tropical forage grasses and legumes. *Soil Biol. Biochem.* 25:1351–1361.

Thomas, K., G. A. Chilvers, and R. H. Norris. 1989. Seasonal occurrence of conidia of aquatic hyphomycetes (Fungi) in Lees Creek, Australian Capital Territory. *Aust. J. Mar. Freshwater Res.* 40:11–23.

Thompson, R. M., C. R. Townsend, D. Craw, R. Frew, and R. Riley. 2001. (Further) links from rocks to plants. *Trends Ecol. Evol.* 16:543.

Tisdall, J. M. 1994. Possible role of soil microorganisms in aggregation in soils. *Plant Soil* 159:115–121.

Tisdall, J. M., S. E. Nelson, K. G. Wilkinson, S. E. Smith, and B. M. McKenzie. 2012. Stabilization of soil against wind erosion by six saprotrophic fungi. *Soil Biol. Biochem.* 50:134–141.

Tisdall, J. M. and J. M. Oades. 1982. Organic matter and water-stable aggregates in soil. *J. Soil Sci.* 33:141–163.

Tolkkinen, M., H. Mykrä, A.-M. Markkola et al. 2013. Decomposer communities in human-impacted streams: Species dominance rather than richness affects leaf decomposition. *J. Appl. Ecol.* 50:1142–1151.

Van der Wal, A., J. A. van Veen, W. Smant et al. 2006. Fungal biomass development in a chronosequence of land abandonment. *Soil Biol. Biochem.* 38:51–60.

Vanlauwe, B. J. Diels, N. Sangina, and R. Merckx 1997. Residue quality and decomposition: An unsteady relationship? In: *Driven by Nature: Plant Litter Quality and Decomposition*, ed. G. Cadish and K. E. Giller, 157–166. Wallingford: CAB International.

Vanlauwe, B., O. C. Nwoke, J. Diels, N. Sanginga, R. J. Carsky, J. Dekers, and R. Merckz. 2000. Utilization of rock phosphate by crops on a representative toposequence in the Northern Guinea savanna zone of Nigeria: Response by *Mucuna pruriens, Lallab purpureus* and maize. *Soil Biol. Biochem.* 32:2063–2077.

Velez, P., M. C. González, E. Rosique-Gil et al. 2013. Community structure and diversity of marine ascomycetes from coastal beaches of the southern Gulf of Mexico. *Fung. Ecol.* 6:513–521.

Voříšková, J. and P. Baldrian. 2013. Fungal community on decomposing leaf litter undergoes rapid successional changes. *ISME J.* 7:477–486.

Wagener, S. M., M. W. Oswood, and J. P. Schimel. 1998. Rivers and soils: Parallels in carbon and nutrient processing. *BioScience* 48:104–108.

Wall, D. H. (ed.) 2012. *Soil Ecology and Ecosystem Services.* Oxford, UK: Oxford University Press, 406 pp.

Wallander, H. and T. Wickman. 1999. Biotite and microcline as potassium sources in ectomycorrhizal and non-mycorrhizal *Pinus sylvestris* seedlings. *Mycorrhiza* 9:25–32.

Wang, G., X. Wang, X. Liu, and Q. Li. 2012. Diversity and biogeochemical function of planktonic fungi in the ocean. In: *Biology of Marine Fungi*, ed. C. Raghukumar, 71–88. Berlin: Springer Verlag.

Weigelhofer, G. and J. A. Waringer. 1994. Allochthonous input of coarse particulate matter (CPOM) in a first to fourth order Austrian forest stream. *Int. Revue Ges. Hydrobiol.* 79:461–471.

Wells, J. M. and L. Boddy. 1990. Wood decay, and phosphorus and fungal biomass allocation, in mycelial cord systems. *New Phytol.* 116:285–295.

Wells, J. M. and L. Boddy. 1995a. Phosphorus translocation by saprotrophic basidiomycete mycelial cord systems on the floor of a mixed deciduous woodland. *Mycol. Res.* 99:977–999.

Wells, J. M. and L. Boddy. 1995b. Effect of temperature on wood decay and translocation of soil-derived phosphorus in mycelial cord systems. *New Phytol.* 129:289–297.

Wells, J. M., J. Thomas, and L. Boddy. 2001. Soil water potential shifts: Developmental responses and dependence on phosphorus translocation by the saprotrophic, cord-forming basidiomycete *Phanerochaete velutina. Mycol. Res.* 105:859–867.

Wicklow, D. T. 1992. Interference Competition. In: *The Fungal Community: Its Organization and Role in the Ecosystem*, ed. G. C. Carrol and D. T. Wicklow, 265–274. New York: Marcel Dekker.

Wong, M. K. M., T.-K. Goh, I. J. Hodgkiss, K. D. Hyde, V. M. Ranghoo, C. K. M. Tsui, W.-H. Ho, W. S. W. Wong, and T.-K. Yuen. 1998. Role of fungi in freshwater ecosystems. *Biodivers. Conserv.* 7:1187–1206.

Wright, S. F. and A. Upadhyaya. 1996. Extraction of an abundant and unusual protein from soil and comparison with hyphal protein from arbuscular mycorrhizal fungi. *Soil Sci.* 161:575–586.

Wright, S. F. and A. Upadhyaya. 1998. A survey of soils for aggregate stability and glomalin, a glycoprotein produced by hyphae of arbuscular mycorrhizal fungi. *Plant Soil* 198:97–107.

Wright, S. F. and A. Upadhyaya. 1999. Quantification of arbuscular mycorrhizal fungi activity by the glomalin concentration on hyphal traps. *Mycorrhiza* 8:283–285.

Xu, W., K.-L. Pang, and Z.-H. Luo. 2014. High fungal diversity and abundance recovered in the deep-sea sediments of the Pacific Ocean. *Microb. Ecol.* 68:388–398.

Yoshida, S. and Y. Muramatsu. 1994. Accumulation of radiocesium in basidiomycetes collected from Japanese forests. *Sci. Total Environ.* 157:197–205.

Yuan, Z. and L. Chen. 2014. The role of endophytic fungi individuals and communities in the decomposition of *Pinus massoniana* needle litter. *PLoS ONE* 9(8): e105911. doi10.1371 /journalpone.0105911.

Zak, J. C. 1993. The enigma of desert ecosystems: The importance of interactions among the soil biota to fungi. In: *Aspects of Tropical Mycology*, ed. S. Isaac, J. C. Frankland, R. Watling, and A. J. S. Whalley, 59–71. Cambridge: Cambridge University Press.

Zuccaro, A. and J. I. Mitchell. 2005. Fungal communities of seaweeds. In: *The Fungal Community: Its Composition and Role in the Ecosystem*, ed. J. Dighton, P. Oudemans, and J. F. White, 533–579. Boca Raton, FL: Taylor & Francis.

Zucarro, A., C. L. Schoch, J. W. Spatafora, J. Kohlmeyer, S. Draeger, and J. I. Mitchell. 2008. Detection and identification of fungi intimately associated with the brown seaweed *Fucus serratus*. *Appl. Environ. Microbiol.* 74:931–941.

# CHAPTER 3

# Role of Fungi in Promoting Primary Production

The role of fungi in primary production goes beyond making nutrients available to plants. In the form of lichens, the whole symbiotic association between fungi, algae, and bacteria is involved in primary production, where the fungal partner acts as a supportive network for photosynthetically active algae and bacteria. There are intimate associations between primary producers and fungi, of which many are symbiotic. These include mycorrhizae associated with plant roots and endophytes, which can occur in multiple plant parts. Such interactions enhance nutrient availability for primary production; confer plant tolerance to drought, salt, and heavy metals; and provide a degree of protection from pathogenic fungi and bacteria and herbivory.

## 3.1 LICHENS AS PRIMARY PRODUCERS

The role of lichens in primary production can be important in ecosystems where lichens compose a large proportion of the plant biomass. An important function of the fungal component of lichens is to support and protect the photosynthetic apparatus contained in the prokaryotic symbiont. Solhaug and Gauslaa (1996) showed that by extracting the lichen *Xanthoria parietina* with 100% acetone, they were able to extract the compound parietin, without damage to the lichen. However, at high light intensities, it was found that extracted lichens showed a reduction in photosynthetic oxygen production, evidencing damage to the photosynthetic apparatus in the absence of the blue light filtering chemical produced by the fungus, despite the fact that parietin and melanin provide protection against ultraviolet radiation under which these chemicals are formed in the lichen thallus (Solhaug et al., 2003). Thus, both the physical support provided by the fungus and its ability to produce beneficial chemicals aid the process of primary production in lichens.

Crustose and foliose soil lichens are major components of the plant biomass in many cold, wet environments, where vascular plants are less able to survive. Beymer and Kolpatek (1991) showed that approximately 28 kg C ha$^{-1}$ was fixed by the lichen crust community in a pinyon pine and juniper forest in a semiarid environment in the Grand Canyon, USA. They estimated that approximately 34%

to 36% of this fixed carbon becomes incorporated into soil organic matter. In mat-forming lichens, Crittenden et al. (1994) showed that lichen growth was limited by the availability of nitrogen in oligotrophic environments, which is essential for chitin production in the mycobiont. Crittenden (1989) reported that there is very little nitrogen available in the substratum on which these lichens grow and that they are very dependent on intercepting nitrogen in precipitation. The efficiency of nitrogen interception can often be close to 100%, but, at certain times, lichens can be a source of leached nitrogen and potassium for other plants. Indeed, this form of nitrogen capture can be equivalent to the N fixation capacity of those lichens containing nitrogen fixing bacterial phycobionts (Table 3.1) and scavenging nutrient from rainfall (Table 3.2). In some cases, as in the soil crust communities, bacteria, in association with lichens and fungi, may fix significant quantities of nitrogen. Belnap (2002) showed that between 1 and 13 kg N ha$^{-1}$ y$^{-1}$ could be fixed by crust communities in the deserts of Utah.

Growth of the fruiticose lichens, *Cetraria* spp., *Cladonia* spp., and *Alectoria nigricans*, in the Norwegian high arctic can be between 2.4 and 10.6 mg g$^{-1}$ per week or between 2.5% and 11.2% of the original lichen biomass in one 10-week season (Cooper and Wookey, 2001). Similarly, biomass of arctic tumbleweed lichen, *Masonhalea richardsonii*, increases by about 10% per year in Alaska (Peck et al., 2000). These rates of growth are similar to those reported by Kärenlampi (1971). These lichens provide a large amount of the winter feed of reindeer and, in the island of Svalbard, may become severely depleted in biomass because of the intense graz-ing pressure, low rates of growth, and the indirect effect of reindeer trampling on

**Table 3.1  Accumulation and Loss of N in Two Mat-Forming Lichen Species during 82 Days of Growth**

|                                    | *Stereocaulon Paschale* | *Cladonia Stellaris* |
|------------------------------------|:-----------------------:|:--------------------:|
| Increment in total biomass N       | 758                     | 95                   |
| Inorganic N in rainfall deposited  | 31                      | 31                   |
| Inorganic N in rainfall retained   | 27                      | 25                   |
| N lost as organic N                | 19                      | 11                   |
| N fixation                         | 669                     | 0                    |

*Source:* Crittenden, P.D., in: *Nitrogen, Phosphorus and Sulphur Utilization by Fungi*, ed. Boddy, L. et al., 243–268, Cambridge University Press, Cambridge, 1989.
*Note:*  Values are expressed as mg N m$^{-2}$ of pure lichen cover.

**Table 3.2  Range of Nutrient Retention by Mat-Forming Lichens from Rainfall**

|                        | % Nutrient Retention | | |
|------------------------|:----------:|:---------:|:------------:|
| **Mat Lichen Species** | **NO$_3$-N** | **NH$_4$-N** | **K** |
| *Stereocaulon paschale* | 86 to 100 | 40 to 99 | −37 to +90 |
| *Cladonia stellaris*    | 62 to 99  | 50 to 97 | −978 to +65 |

*Source:* Crittenden, P.D., in: *Nitrogen, Phosphorus and Sulphur Utilization by Fungi*, ed. Boddy, L. et al., 243–268, Cambridge University Press, Cambridge, 1989.

lichen survival. In soil crust lichens, photosynthetic rates can be between 5 and 7 $\mu$mol $CO_2$ m$^{-2}$ s$^{-1}$, which is approaching the maximum rates achieved by plant leaves in full sunlight (10–20 $\mu$mol $CO_2$ m$^{-2}$ s$^{-1}$) (Belnap and Lange, 2005). In these mixed lichen crust communities, annual carbon fixation can be of the order of 120–137 kg C ha$^{-1}$, and photosynthesis terrestrial lichens are occurring up to 40°C and small but measurable amounts at –12°C to –22°C for *Cladonia* spp., especially by altering their dark respiration rates depending on the temperature (Lange and Green, 2003; Belnap and Lange, 2005).

In temperate forest ecosystems, epiphytic lichens can form a significant proportion of the net primary production of the ecosystem. Using tethered arboreal lichens, Sillett et al. (2000) showed that the colonization of experimental branches was highest in clearcut and old growth Douglas fir forests and lowest in young (10 years old, 1.5 m tall) forests. In general, there was improved lichen colonization and growth on rough branches, compared to smooth branches, but this preference was forest-dependent. For the lichen *Lobaria oregana*, there was greater colonization of smooth bark in the clearcuts, no difference between barks in the young forest, and a significant preference for rough bark in old-growth stands. Terricolous lichen species in Sweden have been shown to have growth rates of 0.2 to 0.4 g g$^{-1}$ dry weight (30–70 g m$^{-2}$) in Sweden, but epiphytes in the same locality only produce 0.01 to 0.02 g g$^{-1}$ (1–4 g m$^{-2}$) (Palmqvist and Sundberg, 2000). The greater biomass accumulation of ground-inhabiting species is attributed to their better water-holding capacity and greater light levels, compared to arboreal habitats. As epiphytes, lichens are able to successfully utilize the mineral nutrients intercepted by or leached from tree canopies and which run down the branches and trunks as stem flow. Differences in growth rate and colonization potential may be related to light levels. In a study of light use efficiency of five macrolichen species, Palmqvist and Sundberg (2000) showed that there was a significant positive correlation between intercepted irradiance and growth, when lichens were wet. They demonstrated that there was a range of between 0.5% and 2% of the light use efficiency per dry weight at a standard energy equivalent of light between lichens grown in low and high light regimes. Energy conversion efficiency into biomass of *Lobaria pulmonaria* in British Columbia range between 0.6% and 1.3% (Coxson and Stevenson, 2007) compared to a maximum of 4.6% and 6% for C3 and C4 plants, respectively (Zhu et al., 2008). Lichens were only able to fix carbon during brief exposure to light fleck penetrating the forest canopy of a maximum of about 160 min per day in old-growth forest and 70 min per day in even aged stands of cedar–hemlock forest.

In tropical ecosystems, the production of lichen biomass is limited by the high rates of dark respiration, leading to a low net rate of carbon accumulation. Lange et al. (2000) determined that within the genus *Leptogium*, between 47% and 88% of the carbon gained during photosynthesis was lost as respiration, thus limiting productivity. Epiphytic fruticose lichens and *Tillandisa* in desert ecosystems reduced water retention in soil under vegetation with epiphytic by intercepting precipitation and atmospheric water (Stanton et al., 2014). However, the interception of water from fog and rainfall by lichens increases nitrogen in leachate water, resulting in higher

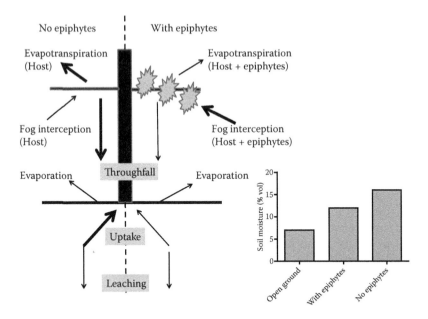

**Figure 3.1**    Diagrammatic representation of the influence of arboreal epiphytic lichens on water relations of *Eulychnia* and *Caesalpinia*. Arrows of different thicknesses represent proportional mass of water flow. Inset graph shows the influence of epiphytes on the soil moisture content. (After Stanton, D.E. et al., *Funct. Ecol.*, doi:10.1111/1365-1435.12249, 2014.)

nitrogen concentrations in soil. These changes in edaphic factors have impacts on the growth potential of host plant species in the high-altitude deserts of South America (Figure 3.1).

## 3.2 MYCORRHIZAE HELP PRIMARY PRODUCTION

The mycorrhizal habit is an old symbiotic association between plant roots and fungi that arose some 450 million years ago in the evolution of terrestrial plants (Humphreys et al., 2010). The association is mostly related to enhancing nutrient acquisition by the host plant. Although the mycorrhizal fungi contributes only about 1% of total ecosystem biomass, the percentage of net primary production represented by mycorrhizal fungi can be 14–15% or 45% in young forest stands and 75% in mature stands, when combined with the fine root biomass supporting the mycorrhizal fungal tissue (Vogt et al., 1982). However, the main role of mycorrhizal symbioses may not be during the early, productive stages of plant succession in ecosystems, but rather in the protective stage where most resources are entrained in plant biomass, where mycorrhizae regulate the cycling of nutrients from decomposing organic matter back into plants and by reducing nutrient loss from the ecosystem (Pankow et al., 1991).

In these associations, the fungus provides increased acquisition of nutrients and water from soil in return for carbohydrates from host plant photosynthesis.

### 3.2.1 Mycorrhizal Habit

Mycorrhizae are symbiotic associations between fungi and plant roots; their description and function have been detailed in many texts (Harley, 1969; Harley and Smith, 1983; Smith and Read, 1997; Peterson et al., 2004; Brundrett, 2009; Bonfante and Genre, 2010; Dighton, 2011). The ecology and role of mycorrhizae in ecosystems has also been explored in a variety of texts (Allen, 1991; Read et al., 1992; Varma and Hock, 1995; Mukerji, 1996; Itoo and Reshi, 2013). Approximately 95% of all vascular plants have a mycorrhizal association (Brundrett, 1991). Traditionally, mycorrhizal associations have been divided into a range of categories, based on the taxonomy of the fungal associate and the physical form of the interactions between the root and the fungus in the mycorrhizal structures produced in the symbiosis. A list of mycorrhizal forms, their plant associates, and the key features of the mycorrhizae is given in Table 3.3.

Among the most common types of mycorrhizal association are arbuscular mycorrhizae (Figure 3.2), which are formed mainly by zygomycete fungal species (Glomeromycota). As many as 25,000 plant species are recorded as being associated with arbuscular mycorrhizae (Brundrett, 2009). These fungi are mainly associated with herbaceous vegetation, grasses, and tropical and some temperate trees. The association is characterized by fungal penetration within the host root cortical cells and the development of a variously developed, treelike branching of the hyphae between the host cell wall and plasmolemma, called an arbuscule. It is here that the surface area of the interface between plant host and fungus is optimized

**Table 3.3   Outline of Some of the Features of Different Types of Mycorrhizal Associations**

| Mycorrhizal Type | Host Plant Group | Characteristics | Fungal Associate |
|---|---|---|---|
| Arbuscular mycorrhizae | Herbaceous plants, grasses, some trees | Formation of arbuscules within cortical cells of host root | Glomeromycota |
| Ectomycorrhizae | Coniferous and deciduous trees | Formation of a sheath or mantle of fungal tissue around the root surface and a Hartig net of fungal penetration between the cortical cells to the endodermis | Basidiomycota Ascomycota |
| Ericoid mycorrhizae | Ericaceaea | Hyphal coils within the host root cortical cells | *Rhizoscyphus Oidiodendron*, etc. |
| Arbutoid mycorrhizae | Arbutus | Hyphal coils within the host root cortical cells | Basidiomycota Ascomycota |
| Orchidaceous mycorrhizae | Orchids | Fungal propagule carried in the seed of the plant | Basidiomycota |

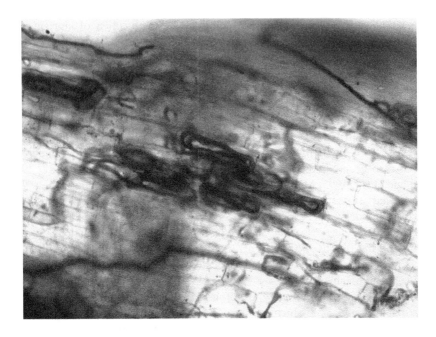

**Figure 3.2    (See color insert.)** Arbuscular mycorrhizal hyphal coils in Knieskern's beaked
sedge (*Rhynchospora knieskernii*) from the New Jersey pine barrens. (From
Dighton, J. et al., *Bartonia*, 66, 24–27, 2013.)

for nutrient and carbohydrate exchange. In some instances, vesicles are formed in
some cortical cells that contain storage materials (lipids and phosphates), giving
rise to the vesicular–arbuscular mycorrhizal type. This name is now reserved for a
limited number of associations, mainly with the fungal genus *Glomus* (Smith and
Read, 1997). The arbuscular mycorrhizal association is formed with a large number
of plant species and a relative small diversity of fungal species. Because these fungi
do not produce large fruiting structures as in the Basidiomycotina, the identification
of the fungal partner is by the anatomy of spores, which may be produced within or
outside of the host root.

   The ectomycorrhizal habit consists of an association between, mainly, tree
species and a range of fungal taxa consisting mainly of Basidiomycotina and
Ascomycotina. There are estimates of about 8000 plant species associated with
between 7 and 10,000 fungal species (Taylor and Alexander, 2005). The fungus does
not penetrate into the host cortical cells, but only between them, forming a Hartig
net. On the surface of the root, a sheath or covering of fungal material develops.
This surface structure may be varying degrees of complexity from a loose weft of
hyphae to highly organized pseudoparenchymatous structures. It is the structure
of the sheath, degree of branching, induced by change in cytokinins, and nature of
emanating hyphae or hyphal strands that allow morphological identification of these
mycorrhizae (Agerer, 1987, p. 99; Ingleby et al., 1990: Goodman et al., 1996–2000).
Ectomycorrhizal associations are formed between a limited number of plant species

and a huge number of fungal species. In addition to ectomycorrhizae, ectendomy-corrhizal associations also occur with tree species. These associations have both ectomycorrhizal and arbuscular mycorrhizal structural characteristics (Laiho and Mikola, 1964).

Ericoid mycorrhizae are similar in structure to arbuscular mycorrhizae, but are associated solely with members of the Ericales (Ericaceae, Empetraceae, Epicaridaceae, Diapensiaceae, and Prionotocaceae). All of these groups are sclero-phyllous evergreens and reside in habitats where both nitrogen and phosphorus are sparsely available. The root systems of these plants consist of very fine roots con-taining a single layer of cortical cells, which the mycorrhizal fungi penetrate to form hyphal coils, rather than arbuscules (Read, 1996). The fungi associated with this type of symbiosis are still not completely identified, but consist of a relative few genera, including *Hymenoscyphus* and *Oidiodendron*. Closely associated with these mycorrhizae are the arbutoid mycorrhizae.

Orchidaceous mycorrhizae are unique in terms of the obligate nature of the asso-ciation. The importance of the mycorrhizal association for seed germination and the initial establishment of the plant has been reviewed by Zettler and McInnes (1992) and Rasmussen and Wigham (1994). The fungal partner is usually ascribed to the genus *Rhizoctonia*, and there has been such evolution of the obligateness of the asso-ciation that the fungus is transported in the seed of the plant.

Further details of the structure of all mycorrhizal associations can be found in the work of Peterson and Farquhar (1994) and Smith and Read (1997). For the pur-poses of demonstrating the role of mycorrhizae in ecosystem process, the following discussions will be limited mainly to the role of arbuscular mycorrhizae, ericoid mycorrhizae, and ectomycorrhizae.

### 3.2.2 Nutrient Uptake

The major ecosystem function of mycorrhizae is to assist host plants in the acqui-sition of mineral nutrients from soil. Approximately 95% of plants are mycorrhizal, where the presence of the fungal partner suppresses root hair development, and the function of the root hair is largely replaced by fungal hyphae. These hyphae have two major benefits for sequestering nutrients. They are of smaller diameter than root hairs and can penetrate more easily and to a greater distance from the root into the soil, thus exploring a greater volume of soil and presenting a greater surface area for nutrient absorption than could the root–root hair system alone (Nye and Tinker, 1977; Owusu-Bennoha and Wild, 1979; Clarkson, 1985; Hetrick, 1991; Marschner and Dell, 1994), extending the depletion zone around the root up to 110 mm in mycor-rhizal clover (Clark and Zeto, 2000). The actual effect of the mycorrhizal association depends on the rate of growth of the extraradical hyphae of the fungal species, with *Acaulospora laevis* having hyphal extension rates of approximately 20 mm week$^{-1}$, but that of *Glomus* spp. less than 10 mm week$^{-1}$. In some cases, in the ectomycorrhi-zal condition, the fungal partner has evolved not only individual extraradical hyphae, but may also develop mycelial structures called strands or rhizomorphs that have a distinct structure with conductive elements analogous to the vascular tissue of plants.

These strands have been shown to be important in long-distance transport of nutrients and water (Duddridge et al., 1980).

The energetic costs per unit of nutrient absorbed are thought to be of significant benefit to the plant (Vogt et al., 1982; Harley and Smith, 1983; Fitter, 1991), because it is reported that the extraradical mycelium for ectomycorrhizal pine seedlings accounted for only 5% of the potential nutrient absorbing system dry weight (fungi and roots) and the mycelium accounted for 75% of the potential absorbing area and more than 99% of the absorbing length, representing a small investment in structural carbohydrate (Rousseau et al., 1994) (Table 3.4). In fact, arbuscular mycorrhizae colonizing roots of corn (*Zea mays*) and barley (*Hordeum vulgaris*) may account for more than 83% of the soil fungal hyphae (Kabir et al., 1996). The underground network of mycorrhizal hyphae has been estimated to be equivalent to 10% of net photosynthesis of the host plant and may be some 20–30% of soil microbial biomass, so the return to the plants accessing this network needs to compensate for the carbohydrate investment into its maintenance (Leake et al., 2004). However, the analysis of this cost–benefit equation for arbuscular mycorrhizae in natural conditions may only be realized at specific times where nutrient (P) demand is greater than readily available supplies in soil, else the cost of maintenance of the mycorrhizal symbiont is equivalent to the cost of root maintenance (Table 3.5) (Fitter, 1991). As each plant in an ecosystem may be incurring nutrient stresses at different times, it is likely that the shared cost of maintaining a mycorrhizal network may be less per plant than that estimated by Leake et al. (2004) if plants contribute only in times of plenty. Soil biodiversity is essential for soil stability as it has been estimated that agricultural practices, urbanization, and soil salinity have resulted in an estimated loss of between 8 and 70+ Gt $y^{-1}$ of carbon and a potential sink for between 2 and 4 Gt $y^{-1}$ (Fitter, 2005). Fitter also suggests that the evolution of Glomeromycota (arbuscular mycorrhizae) as obligate symbionts resulted in great diversity of species, in which each species has selected efficiency for one function (P acquisition, water acquisition,

Table 3.4   **Plant and Fungal Parameters for Pine Tree Seedlings Colonized by Ectomycorrhizal Fungi** *Pisolithus Tinctorius* **and** *Cenococcum Geophilum* **Showing Enhanced Nutrient Uptake Capacity of Mycorrhizal Plants due to Extraradical Hyphal Development**

| Plant/Fungal Parameter | *Pisolithus* | *Cenococcum* | Nonmycorrhizal Plant |
|---|---|---|---|
| Mycorrhizal infection (%) | 69.5 | 66.5 | 0 |
| Fine root diameter (µm) | 477 | 573 | 299 |
| Root tip ratio | 3.72 | 1.39 | 1.55 |
| Fine root area (mm²) | 4.02 | 1.49 | 1.30 |
| Hyphal area (mm² g⁻¹ soil) | 33.8 | 28.1 | 1.5 |
| Rhizomorph area (mm² g⁻¹ soil) | 13.6 | 0 | 0 |
| Total fungal area (mm² g⁻¹ soil) | 47.4 | 28.1 | 1.5 |
| Hyphal length (m g⁻¹ soil) | 6.42 | 2.8 | 0.28 |
| Rhizomorph length (m g⁻¹ soil) | 0.36 | 0 | 0 |
| Total fungal length (m g⁻¹ soil) | 6.78 | 2.8 | 0.28 |

*Source:* Rousseau, J.V.D. et al., *New Phytol.*, 128, 639–644, 1994.

**Table 3.5  Cost to Plants of Maintenance of Arbuscular Mycorrhizal Infection**

| | |
|---|---|
| Biomass of mycorrhizal fungus | 10–20% of root biomass |
| Cost of growth and maintenance of the fungus | 1–10% of fungal biomass day$^{-1}$; i.e., 0.1–1% of root biomass day$^{-1}$ |
| Root maintenance cost | About 1.5% of root biomass day$^{-1}$ |

Source: Fitter, A.H., *Experientia*, 47, 350–355, 1991.
Note:  Cost of maintaining mycorrhizae ≡ root maintenance cost.

pathogen defense, protection from salinity, etc.) as each of these functions requires contrasting physical structures of the association. This specificity of function may play a major role in the cycling of C, N, and P, and control plant community composition and invasiveness of plant species into a community (Fitter, 2005).

Overall, the association between a plant and its mycorrhizal partner increases plant growth (Karst et al., 2008), although effects vary with plant growth metric, and tree and mycorrhizal fungal genus (Figure 3.3). For example, pine seedlings inoculated with the ectomycorrhizal fungi *Suillus*, *Laccaria*, *Lactarius*, *Hebeloma*, *Pisolithus*, and *Rhizopogon* in the nursery grew significantly better in the field after a wildfire in Northern Portugal (Franco et al., 2014). The seedling trees maintained

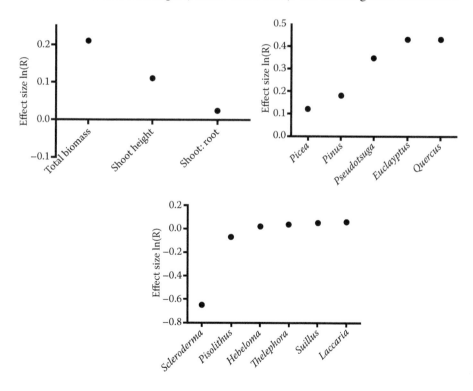

**Figure 3.3**  Effect of size of mycorrhizae on plant growth. (From Karst, J. et al., *Ecology*, 89, 1032–1042, 2008.)

a mycorrhizal community consisting primarily of the inoculated genera even after 5 years of growth.

### 3.2.3 Distribution of Mycorrhizal Types in Relation to Nutrient Availability

The distribution of mycorrhizal types is dependent on the geographic distribution of the host plant species and the nature of the soil. Read (1991a) focused on the geographical distribution of the main mycorrhizal types around the world, and his results showed that the arbuscular mycorrhizal habit was dominant in temperate and tropical grasslands, tropical forests, and desert communities, ectomycorrhizae were dominant in temperate and arctic, forested ecosystems, and ericoid mycorrhizae were most common in the boreal heathland ecosystems. He equated this distribution with that occurring along an altitudinal cline (Read, 1991a,b). Where soil development is constrained by climatic conditions (extreme North and South latitudes or high altitudes), plant communities develop a high number and concentration of secondary metabolites (lignin, polyphenols, etc.) that make their litter recalcitrant to decomposition. Here, organic matter accretes on the soil surface at a faster rate than it can be decomposed, leading to the accumulation of raw, undecomposed humic material in which ericoid plants and mycorrhizae dominate and the production of enzymes to access organic forms of nutrients is beneficial. At midlatitudes and at the midrange of altitude, coniferous and deciduous forest ecosystems dominate with their predominantly ectomycorrhizal fungal symbionts. Here, a mixed range of plant litter resources provides a mixture of easily decomposed and recalcitrant resources, providing nutrients in both an inorganic and organic form. The ectomycorrhizae would be expected to have a range of physiological functions from efficient inorganic nutrient uptake to a high degree of enzyme activity for acquisition of nutrients that are poorly plant available. At low and equatorial latitudes, low altitudes, and in certain ecosystems at midlatitudes (grasslands), arbuscular mycorrhizae dominate as plant litter material is usually readily decomposed providing soils with higher concentrations of inorganic nutrients. The arbuscular mycorrhizae are therefore probably more adapted for efficiency of inorganic nutrient uptake and have lower abilities to access organic or poorly soluble forms of nutrients (Figures 3.4 and 3.5). A summary of the functional roles of different mycorrhizal types can be found in the work of Leake and Read (1997).

The mycorrhizal fungi forming associations with ericaceous plant communities are capable of producing enzymes (protease and phosphatase) enabling the host plant to access organic forms of nutrients directly as a response to the low availability of inorganic nutrients, which is caused by low rates of decomposition by the saprotrophic microbial community. This concept of a direct cycling system was proposed by Went and Stark (1968). Ericoid-dominated ecosystems tend to be nitrogen-limited, and the production of mycorrhizal-generated enzymes affords the plant community with a greater access to organic forms of nitrogen (Stribley and Read, 1980; Bajwa and Read, 1985; Leake and Read, 1989, 1990a,b). Indeed, Read and Kerley (1995) show that ericoid mycorrhizal plants

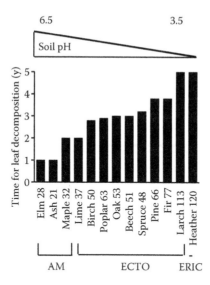

**Figure 3.4** Ecological distribution of mycorrhizal types in relation to plant leaf litter resource quality (presented as C/N ratios adjacent to tree name) of a selection of tree species and heather (*Calluna vulgaris*), the rate of decomposition of that plant litter, and the pH of soil. (Redrawn from Read, D.J., *Experientia*, 47, 376–391, 1991a.)

derive some 96% of their nitrogen from organic sources (hydrolyzable organic N, humin, and other recalcitrant N) in highly organic soils. Evidence for the use of organic nitrogen and phosphorus by ericoid mycorrhizae comes from a number of studies. Mitchell and Read (1981), Myers and Leake (1996), and Leake and Miles (1996) showed that *Vaccinum macrocarpon* could access phosphate from inositol hexaphosphate (a commonly occurring phosphorus compound in organic soils) and both P and N from phosphodiesters from nuclei (Figure 3.6). The ericoid mycorrhizal fungus, *Hymenoscyphus ericae*, can decompose chitin and transfer some 40% of the nitrogen contained in *N*-acetylglucosamine to its host plants, *Vaccinium macrocarpon* and *Calluna vulgaris* (Kerley and Read, 1995). Similarly, the ericoid mycorrhizae, *Oidiodendron maius* and *Acremonium strictum*, of salal (*Gautheria shallon*) can utilize the amino acid, glutamine, the peptide, glutathione and the protein, and bovine serum albumin (BSA) as nitrogen sources (Xiao and Berch, 1999), and the mycorrhizal endophytes of *Woollsia pungens* are able to degrade glutamine, argenine, and BSA (Chen et al., 2000).

Complex organic forms of phosphorus can be accessed by ericoid mycorrhizae, which are capable of producing phosphatase enzymes (Pearson and Read, 1975; Mitchell and Read, 1981; Straker and Mitchell, 1985), although in low pH soils, where concentrations of iron and aluminum are greater than 100 mg l⁻¹, these metals are inhibitory to phosphatase production by the ericoid mycorrhizal fungus, *Hymenoscyphus ericae* (Shaw and Read, 1989). However, in low pH soils, ericoid mycorrhizal associations have been said to "detoxify" the ecosystem by assimilation

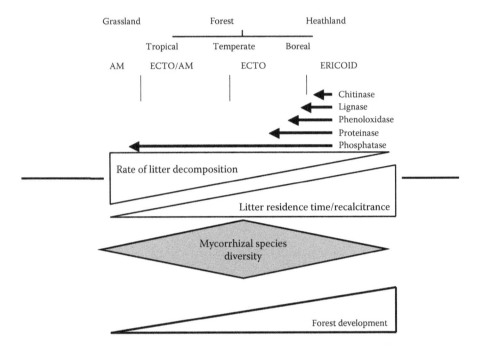

**Figure 3.5**    Relationship between the dominance of mycorrhizal type in ecosystems (above
the line) and to forest development (below the line) to the changes in plant litter resource quality, its rate of decomposition, and the enzyme competence of
the mycorrhizal community. (Modified from Read, D.J., *Experientia*, 47, 376–391,
1991a; Dighton, J., Mason, P.A., in: *Developmental Biology of Higher Fungi*, ed.
Moore, D. et al., 163–171, Cambridge University Press, Cambridge, 1985.)

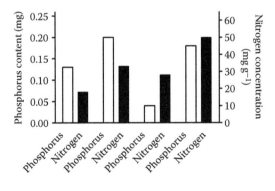

**Figure 3.6**    Shoot phosphorus content (open bars) and nitrogen concentration (solid bars) of
the ericaceous plant *Vaccimium macrocarpon* in the presence (M) or absence
(NM) of ericoid mycorrhizal inoculum when provided with orthophosphate
(Ortho-P) or nutrients supplied in the form of nuclei. (Data from Myers, M.D.,
Leake, J.R., *New Phytol.*, 132, 445–452, 1996.)

of phenolic and aliphatic acids (Leake and Read, 1991) and complexing toxic metal ions (Bradley et al., 1982). There is little documented evidence of the role of ericoid mycorrhizae in these cultivated forms, where the extent of root colonization is much higher than expected, based on their survey of native and cultivated blueberry (*Vaccinium corymbosum*) in the United States (Goulart et al., 1993).

In coniferous, deciduous or mixed forest biomes, the array of plant litter chemistry is diverse, with a mixture of readily degradable and recalcitrant materials. In these ecosystems, ectomycorrhizae dominate as soils develop a "mor" and "moder" type of humus. In these ecosystems, phosphorus as well as nitrogen can be limiting to plant growth. Again, the ability of mycorrhizal fungi to behave as saprotrophs to effect a "direct cycling" of nutrients from partially decomposed organic residues is a benefit to the plant community. In this context, the ability of ectomycorrhizae to produce a range of enzymes is a benefit, which allows the host plant to obtain both nitrogen and phosphate from organic resources and to compete against immobilization by the saprotrophic soil microbial community (Dighton, 1991).

Ectomycorrhizae have been shown to produce nitrogen-degrading protease enzymes (Abuzinada and Read, 1986a,b, 1989; Read et al., 1989; Leake and Read, 1990a,b; Zhu et al., 1994; Tibbett et al., 1999; Anderson et al., 2001), phosphate solubilizing acid phosphatase enzymes (Bartlett and Lewis, 1973; Dighton, 1983; Antibus et al., 1992, 1997; Leake and Miles, 1996; Joner and Johansen, 2000), and other enzymes (Giltrap, 1982; Durall et al., 1994)—enabling them to utilize forest floor carbon. Ectomycorrhizae in culture and in symbiosis can use peptides and proteins as nitrogen. Four tree species in mycorrhizal association with the fungus *Hebeloma crustuliniforme* were shown to be able to incorporate up to 53% of the total N contained in proteins or peptides, whereas nonmycorrhizal tree seedlings could not access nitrogen from these organic sources (Abuzinada and Read, 1986a,b). In ectomycorrhizal associations, the uptake of nitrogen from alanine or ammonium was 10 times higher than that from nitrate sources (Wallander et al., 1997). In arctic regions, where decomposition and nutrient mineralization is constrained by low temperatures, Tibbett et al. (1998a) suggest that there has been a preadaptation of *Hebeloma* species to utilize nitrogen in the form of proteins and glutamic acid, which are often released from organic matter during freezing (Tibbett et al., 1998b,c). Here, cold active phosphomonoesterase enzyme is only produced by *Hebeloma* when grown at 6°C. Thus, there is competition between the saprotrophic and mycorrhizal fungi for readily available nutrients (Kaye and Hart, 1997). For example, nitrifying bacteria can consume up to 70% of the total available $NH_4$ as an energy source, and be in direct competition with plant roots and their mycorrhizae (Norton and Firestone, 1996; Kaye and Hart, 1997). The ectomycorrhizal fungi *Laccaria bicolor* can utilize ammonium, nitrate, and urea as sources of nitrogen, and *Hebeloma* spp. can also use BSA. None of the mycorrhizal fungi could utilize nitrogen in the form of ethylenediamine or putrescine, suggesting that the ectomycorrhizal fungi could not compete with saprotrophic fungi for resources in decaying animal carcasses (Yamanaka, 1999).

The role of ectomycorrhizae in tree canopy "soils" may be considerable for some tropical and subtropical trees with high epiphyte loading. For example, in

*Nothofagus menziesii* in New Zealand, "soil" created in the canopy by dead epiphytes has a very high organic content (86%) compared to soil on the ground (10%); the adventitious roots of *Nothofagus* develop in the canopy in this "soil" and are colonized by a variety of ectomycorrhizal fungi (*Clavulina, Cortinarius, Cenococcum, Leotia, Inocybe, Laccaria, Lactarius, Russula,* and *Thelephora*) that are likely using organic nutrient sources to supplement nutrients gained from soil on the ground (Orlovich et al., 2013).

Approximately 50% of the phosphorus in a Norway spruce forest is in the form of organic P; thus, the benefit of the ability of ectomycorrhizal-associated forest trees to produce phosphatase enzymes is evident (Häussling and Marschner, 1989). In a laboratory study, Repáč (1996a,b) showed that ectomycorrhizal colonization of tree roots increased in the presence of organic matter. It is the juxtaposition of roots, fungal hyphae, and the nutrient-rich organic material that provides the best option for the mineralization and direct uptake of nutrients by the roots, minimizing the chance for leakage loss to drainage water. In organic-rich soil horizons, there is a 2- to 2.5-fold increase in acid phosphatase activity in the rhizosphere as compared to the bulk soil (Häussling and Marschner, 1989). The ability of ectomycorrhizal fungi to access and incorporate phosphorus from complex organic forms of P, such as inositol hexaphosphate, has been demonstrated a number of times (Dighton, 1983; Mousain and Salsac, 1986; Antibus et al., 1992, 1997), and the regulation of the expression of this enzyme by external concentrations of orthophosphate has been reported by MacFall et al. (1991). Indeed, Antibus et al. (1992) showed that in some ectomycorrhizal fungi there was a greater uptake of phosphorus from organic than inorganic sources, because of the action of acid phosphatase and phytase enzymes (Table 3.6), and access is greater

**Table 3.6** Incorporation of $^{32}$P-Labeled Phosphorus (CPM mg dm$^{-1}$ h$^{-1}$) into Ectomycorrhizal Fungal Mycelia from either Inorganic (P$_i$) or Organic (P$_o$) Sources due to the Activity of Mycelial Surface or Soluble Acid Phosphatase (pNPPase – pNPP Release mg dm$^{-1}$ h$^{-1}$) or Phytase (nmol P Release mg/protein h) Enzyme Activity

| Fungal Species | P source | pNPPase | | Phytase | | $^{32}$P Uptake |
|---|---|---|---|---|---|---|
| | | Mycelium | Soluble | Mycelium | Soluble | |
| Amanita rubescens | P$_i$ | 1.8 | 60.0 | 55.2 | 0.7 | 17.0 |
| | P$_o$ | 1.1 | 45.0 | 36.3 | 0.4 | 5.2 |
| Entoloma sericeum | P$_i$ | 15.6 | 77.4 | 83.8 | 0.3 | 136.5 |
| | P$_o$ | 18.3 | 41.4 | 60.0 | 0.3 | 4000 |
| Hebeloma crustuliniforme | P$_i$ | 0.6 | 4.8 | 2.0 | 0.0 | – |
| | P$_o$ | 0.6 | 11.0 | 2.7 | nd | – |
| Lactarius sp. | P$_i$ | 8.2 | nd | 265.1 | nd | 56.1 |
| | P$_o$ | 6.3 | nd | 198.4 | nd | 148.4 |
| Scleroderma citrinum | P$_i$ | 3.7 | 1.0 | 74.3 | nd | 68.5 |
| | P$_o$ | 4.7 | 1.0 | 2.3 | nd | 93.3 |
| Cenococcum geophilum | P$_i$ | 0.1 | nd | 13.1 | nd | 2.6 |
| | P$_o$ | 0.7 | nd | 23.2 | nd | 457.9 |

*Source:* Antibus, R.K. et al., *Can. J. Bot.*, 70, 794–801, 1992.

in ectomycorrhizal trees than in arbuscular mycorrhizal tree species (Antibus et al., 1997; Figure 3.7). In addition, there is evidence showing that ectomycorrhizae (e.g., *Paxillus involutus*) can access phosphorus from the complex inorganic forms of phosphate, such as calcium phosphate, but only in the presence of available ammonium or nitrate nitrogen (Lapeyrie et al., 1991; Chang and Li, 1998). The ectomycorrhizal fungal communities in the roots of some tropical legume tree species allow exploitation of phosphorus in deeper soil layers than are being colonized by surface feeder roots. Newberry et al. (1997) suggest that mycorrhizae are able to keep phosphorus cycling in the biotic components of the forest through a strong interaction between the phosphate acquisition capacity of the mycorrhizae, the environmental controls of phosphate release, and the seasonal demands from P by the trees, especially during mast years—which they call a phenological and climatic ectomycorrhizal response (PACER), a process that optimizes phosphate utilization and minimizes phosphate leaching loss. The role of ectomycorrhizae in cycling N and P from organic sources is discussed by Buée et al. (2007, 2009).

There is, however, variability between fungal species in their ability to produce enzymes (Dighton, 1983, 1991; Lapeyrie et al., 1991), and Read (1991b) suggests that certain species, such as *Laccaria laccata* and *Pisolithus tinctorius*, are poor enzyme producers, relying on enhancing nutrient uptake of mineral nutrients derived from the breakdown of organic residues by the saprotrophic microbial community, whereas other species (*Paxillus involutus*, *Lactarius* spp., *Amanita* spp., and *Suillus* spp.) have a greater degree of enzyme competency. The idea is supported by the observations that *Lactarius controversus*, *Paxillius involutus*, *Piloderma croceum*, and *Pisolithus tinctorus* mycelia accumulated no more nitrogen from BSA as a nitrogen source than they did from a basal medium, whereas *Suillus bovinus* had greater access to the nitrogen in BSA (Bending and Read, 1996). A word of caution in the interpretation of enzyme studies in pure culture comes from the work of Anderson et al. (2001), who showed that some of the variation in the ability of different isolates

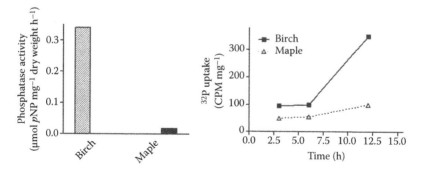

**Figure 3.7** Acid phosphatase activity of birch (ectomycorrhizal) and maple (arbuscular mycorrhizal) roots (left) and the uptake of radioactively labeled phosphorus from an organic P source (inositol polyphosphate) (right), demonstrating the benefit of ectomycorrhizal associations in a mixed forest ecosystem for access to poorly available nutrient sources. (Data from Antibus, R.K. et al., *Mycorrhiza*, 7, 39–46, 1997.)

of *Pisolithus tinctorius* to utilize organic sources of nitrogen is attributable to the length of maintenance of the isolate on agar culture. Longer storage times appear to enhance organic nitrogen utilization potential.

Ectomycorrhizal fungi produce mycorrhizae with contrasting exploration types, defined by the extent and nature of extraradical hyphae and rhizomorphs (Agerer, 2001). It has been suggested that mycorrhizae with exploration types with greater reach may be more effective in nutrient acquisition that exploration types with less reach. Although ectomycorrhizae with greater amounts of extraradical hyphae tended to show more degradative enzyme activity than those with fewer hyphae, it was shown in an African lowland forest that phylogeny of the mycorrhizal fungus was a better predictor of nutrient acquisition (Tedersoo et al., 2012).

Some plant species alter their dominant mycorrhizal symbiont depending on age or edaphic conditions. In the case of willow (*Salix repens*), there appears to be different functional significance between the arbuscular mycorrhizal and ectomycorrhizal associate (van der Heijden, 2001). The arbuscular mycorrhizal fungus *Glomus mosseae* had a low rate of root colonization, but showed large short-term effects on shoot growth and root length. However, the ectomycorrhizal fungus *Hebeloma leucosarx* had high levels of root colonization and improved host plant growth over a longer term. Arbuscular mycorrhizal colonization resulted in higher shoot P uptake, shoot growth, root growth, and response duration in plants collected in December compared with those collected in March, whereas the ectomycorrhizal and nonmycorrhizal treatment showed no differences between cuttings collected on different dates. The differential effects of the two mycorrhizal types could be related to the availability of nutrients at different times of the year and the differences in function of the two types of mycorrhiza.

Varying plant growth responses can be seen when plants are associated with different mycorrhizal associates (Villeneuve et al., 1991a). However, most of these studies are conducted in the laboratory or in a greenhouse, in somewhat artificial conditions. Similarly, field observation of the advantage of inoculation of tree seedlings with a variety of mycorrhizal species frequently shows that there are differences in growth rates of the host plant with different mycorrhizal fungal symbionts, that mycorrhizal plants perform better than nonmycorrhizal plants (especially in disturbed situations), and that the inoculated mycorrhizal species are frequently replaced by native mycorrhizal flora (Villeneuve et al., 1991b).

Demonstrations of the effect of different ectomycorrhizae in the field are more difficult to obtain (Miller, 1995). Jones et al. (1990) showed that soil type influenced the performance of ectomycorrhizae but demonstrated that, in general, *Laccaria proxima* induced a higher level of tissue phosphorus content in willow (*Salix viminalis*) compared with *Thelephora terrestris*. In field-grown birch, Dighton et al. (1990) injected radioactive inorganic phosphorus into soil in zones around birch trees whose mycorrhizal community was known to be dominated by different ectomycorrhizal species, based on the appearance of fruit bodies. They measured the incorporation of $^{32}$P into the leaves of trees and found that the influx of phosphorus into leaves was higher when influenced by mycorrhizal communities dominated by *Hebeloma* spp. than by communities dominated by either *Laccaria* spp. or *Lactarius* spp.

(Figure 3.8). Evidence from the evaluation of enzyme production my mycorrhizal fungi also suggests that there are significant differences in the ability of different fungal species to produce the enzyme (Dighton, 1983; Leake and Read, 1990a,b; Antibus et al., 1992, 1997) and that the availability of the inorganic form of the nutrient in soil has a negative feedback on the enzyme production (Sinsabaugh and Liptak, 1997). Given these facts and the information that the root system of individual forest trees can maintain a community of many ectomycorrhizal fungal species at the same time (Zak and Marx, 1964; Gibson and Deacon, 1988; Palmer et al., 1994; Allen et al., 1995; Shaw et al., 1995), it is therefore possible that the ectomycorrhizal community on root systems is functionally plastic and able to be changed at spatial and temporal scales to optimize resource utilization, as the local environmental conditions change. Tibbett (2000) indicates that in both ericoid and ectomycorrhizal symbioses, the extraradical hyphae exhibit significant morphological and physiological plasticity (Bending and Read, 1995a,b; Cairney and Burke, 1996) that makes them ideally suited for the exploitation of patchily distributed nutrient resources.

The relationship between biotic diversity, or biodiversity, in ecosystems, the function of that diversity and the stability of ecosystems has been a matter of debate in the ecological world for many years (McCann, 2000; Schwartz et al., 2000). Attention has been focused on the role of diversity in function, and a number of manipulative experiments have been performed where the outcome of altered plant species diversity on ecosystem functions has been measured (Tilman 1993, 1997, 1999; Tilman et al., 1996). We know that there is considerable diversity in soil biota, and questions have been asked regarding the soil microorganisms' function (Ritz et al., 1994; Dighton and Krumins, 2014) as much of the previous research had concentrated on measuring biomass, rather than function. Radiata pine seedlings performed better in the presence of three ectomycorrhizal symbionts than with one (Chu-Chou and Grace, 1985), and two ectomycorrhizal species were found to produce larger host plants compared with just one (Parladé and Alvarez, 1993; Reddy and Natarajan,

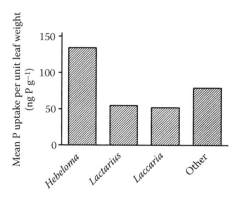

**Figure 3.8**  Uptake of inorganic phosphorus supplied to the upper 5 cm of soil in areas dominated by different ectomycorrhizal species under birch trees in the field. (Data from Dighton, J. et al., *New Phytol.*, 116, 655–661, 1990.)

1997). The yield of competing Douglas fir seedlings was enhanced as the number of ectomycorrhizal partners was increased (Perry et al., 1989). Under low fertility situations in Sweden, Jonsson et al. (2001) showed that the growth of birch trees was higher when they were associated with eight ectomycorrhizal species than comparable plants associated with single fungal species.

By manipulating both the actual diversity and species composition of ectomycorrhizae on birch seedlings, Baxter and Dighton (2001) were able to demonstrate that changes in plant performance were related to the diversity *per se* of the ectomycorrhizal community they supported, rather than the actual species composition of that mycorrhizal community. As the community of mycorrhizae increased, the proportional representation of each species declined, but the total number of mycorrhizal root tips per plant increased. In response to the increased mycorrhizal diversity, plant shoot biomass declined, but root biomass increased (Figure 3.9), and phosphorus

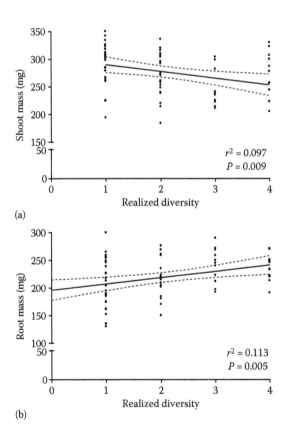

(a)

(b)

**Figure 3.9** Effect of ectomycorrhizal species diversity on shoot and root growth of birch seedlings when inoculated with a random mixture of mycorrhizal fungal species, selected from a species pool. (Data from Baxter, J.W., Dighton, J., *New Phytol.*, 152, 139–149, 2001.)

uptake from organic P sources was enhanced by increased mycorrhizal diversity (Baxter and Dighton, 2005a,b). Using stepwise multiple regression analysis, they showed, however, that changes in plant biomass (root and shoot) and plant phosphate content were significantly correlated with ectomycorrhizal diversity, rather than the level of root colonization. This suggests that there is some function of diversity and the interaction of ectomycorrhizal fungi on the same root system that has an influence on plant performance, rather than either the total level of mycorrhizal infection or the species composition of the mycorrhizal community. However, the study by Jonsson et al. (2001) suggests that the effects of mycorrhizal fungal diversity may depend on the context in which they exist. The community structure may be beneficial, detrimental, or neutral, depending on the nutritional conditions of the soil, plant age, or other factors. Hence, Fitter (1985, 1991) says that it is not always obvious what the function of mycorrhizae is in a natural ecosystem, but that a clear nutritional benefit accrues in an agricultural context. In his commentary on this work, Leake (2001) correctly points out that the study of Baxter and Dighton (2001) was conducted under laboratory conditions and that plant and fungal responses may be different in a more realistic field situation. However, these results show how little we really understand the complexities of fungal interactions and the role of diversity in the mycorrhizal community.

Cairney (1999) discussed the range of ectomycorrhizal species and their varied physiological functions, suggesting that we know relatively little about variation in the physiology of the few fungi that we have studied extensively in the laboratory, let alone the myriad other species about which we know very little and especially those fungi that we have yet to encourage to grow in culture. In addition, in their review article, Cairney and Burke (1996) cite examples of heterogeneity of function of mycelia of the same ectomycorrhizal fungus as it exploits pockets of different resources in the soil. They suggest that this heterogeneity of function drives the ability of ectomycorrhizae to exploit resources, and it drives enzyme expression, nutrient uptake, and translocation within the mycorrhizal system (Cairney, 1992).

In warmer, moist environments, decomposition is rapid and organic matter rapidly becomes incorporated into the soil mineral matrix; nutrient cycling occurs at a more rapid pace, and there is a higher concentration of inorganic nutrients than in more organic soils. Under these conditions in temperate grasslands and tropical forests and grasslands, arbuscular mycorrhizae dominate, where phosphorus tends to be a limiting nutrient (Read, 1991b) and the symbiosis appears to confer a greater efficiency in mineral nutrient acquisition (Hetrick, 1989; Jeffries and Barea, 1994). Bolan (1991) suggests that the arbuscular mycorrhizal benefit for phosphate uptake into plants is attributable to three factors: (1) exploitation of a larger soil volume, (2) faster movement of phosphate into the root via fungal hyphae, and (3) the ability to solubilize complex inorganic forms of phosphate. He suggests that mycorrhizal fungi may help to overcome the three rate-limiting steps of phosphate uptake by increasing the rate of diffusion into plant roots, increasing the phosphate concentration at the root surface, and increasing the rate of phosphate dissociation from the surface of soil particles (Figure 3.10).

Arbuscular mycorrhizae are of particular importance in agriculture (Gianinazzi and Schüepp, 1994), with significant influence on biogeochemical cycling and

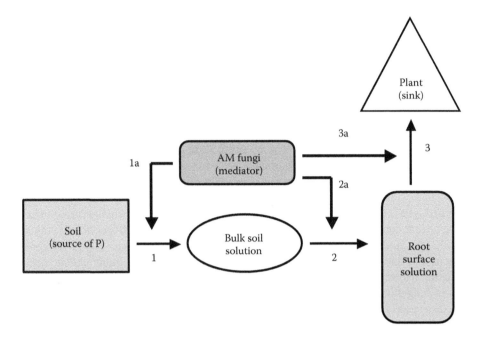

**Figure 3.10**  Rate-limiting processes in the uptake of phosphorus by plants and the role of
arbuscular mycorrhizae in overcoming these limitations. Thin arrows represent
flows in the nonmycorrhizal condition; 1, release of P from soil particles; 2, diffusion to the root surface; 3, uptake by the plant. Thick arrows indicate the influence of the mycorrhizae; 1a, chemical modification of the P release mechanism
by enzyme or organic acid production; 2a, decreasing the diffusion distance by
the exploitation of soil by extraradical hyphae; 3a, reducing the threshold concentration of P required to permit transfer of P across the plant cell membrane.
(Adapted from Bolan, N.S., *Plant Soil*, 134, 189–207, 1991.)

sustainability by improving plant nutrition, preventing root pathogens, and improving soil structure by binding soil particles together with mycelia (Jeffries and Barea,
1994). As a consequence of the relatively high availability of inorganic to organic
sources of nutrient in these soils, these mycorrhizal types have only limited enzyme
expression, although they are capable of producing phosphatase enzymes to solubilize poorly available phosphates in soil (Azcon et al., 1976; Singh and Kapoor, 1998).
For example, nonmycorrhizal big bluestem grass (*Andropogon gerardii*) could
access phosphorus from glycerophosphate and adenosine monophosphate, but not
from phytic acid, ribonucleaic acid (RNA), adenosine triphosphate (ATP), or CMP
(cytidine 2'- and 3'-monophosphate), whereas in symbiosis with *Glomus etunicatum*,
plants could access all forms of organic phosphorus and plant uptake was 500- to
600-fold higher in the mycorrhizal plants (Jayachandran et al., 1992) (Figure 3.11).
The ability of arbuscular mycorrhizae to solubilize phosphate may be an important
factor in permitting plants to grow in calcareous soils in which phosphate is limited
because of complexing with heavy metal ions.

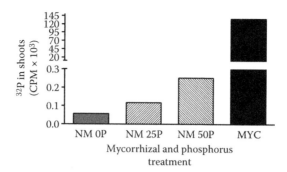

**Figure 3.11** Incorporation of radioactive phosphorus into shoots of big bluestem (*Agropyron gerardii*) from an organic phosphorous source (cytidine diphosphate) when in symbiotic association with the arbuscular mycorrhizal fungus (*Glomus etunicatum*) (MYC) in comparison with phosphorus incorporation in the presence of 0, 25, or 50 mg P kg$^{-1}$ of the organic phosphate in the absence of mycorrhizae. (Data from Jayachandran, K. et al., *Soil Biol. Biochem.*, 24, 897–903, 1992.)

The inability of the calcifuge plant species—*Carex pilulifera, Deschampsia flexuosa, Holcus mollis, Luzula pilosa, Nardus stricta*, and *Veronica officinalis*—to grow on limestone is related to their inability to decouple the iron–phosphate complexes to derive both elements, which are essential to their growth. Calcicole species, however, appear to have developed mechanisms of acquiring both P and Fe from these soils via the production of organic acids in the rhizosphere (Ström, 1997; Lee, 1999) (Table 3.7). Part of this ability may be linked to the arbuscular mycorrhizal associations of the calcicoles, although Lee (1999) points out that we know little of the role of mycorrhizae in the process of adaptations of calcicolous plants. However, there is evidence suggesting that fungi can produce organic acids (Azcon et al., 1976; Bolan, 1991; Singh and Kapoor, 1998). Arbuscular mycorrhizae are not only limited to enhancing phosphorus uptake into the host plant as enhanced nitrogen uptake has also been observed (Clark and Zeto, 2000). However, this may be related to the induced demand by achieving greater plant size because of the mycorrhizal effect of overcoming phosphate limitations. In particular, the interaction between mycorrhizae and nitrogen-fixing leguminous plants is of importance in assisting the delivery of phosphate to plants to maximize nitrogen fixation in root nodules (Azcón-Aguilar et al., 1979; Peoples and Craswell, 1992; Herrera et al., 1993). Additionally, there is some evidence of arbuscular mycorrhizae being able to utilize organic forms of nitrogen (Ames et al., 1983).

Although there is a large body of information showing that different species of trees associate with different communities of ectomycorrhizae, there is less information on the host plant/arbuscular mycorrhizal species specificity. It is generally thought that there is low specificity; however, Eom et al. (2000) showed that after 4 months of growth there were significantly different arbuscular mycorrhizal communities developing under different component plant species from a tallgrass prairie (Figure 3.12).

Table 3.7 Production of Organic Acids in Rhizosphere of Calcifuge and Calcicole Plant Species, Showing Adaptation of Calcicoles in Order to Solubilize Phosphate and Essential Heavy Metals

| Plant strategy | Species | Organic Acid Production (m mol m⁻³ Soil Solution), Where Root Weights Are Equivalent | | | | | | | | | | |
|---|---|---|---|---|---|---|---|---|---|---|---|---|
| | | Monocarboxylic | | | | Dicarboxylic | | | Tricarboxylic | | | Sum |
| | | Lactic + Acetic | Proprionic | Formic | Pyruvic | Malic + Succinic | Tartaric | Oxalic | Citric | Isocitric | Aconitic | |
| Calcifuge | *Deschampsia* | 8.9 | 0.4 | 4.8 | 0.4 | 0.3 | 0.2 | 1.7 | 1.0 | 0.2 | 0.2 | 17.3 |
| | *Viscaria* | 7.9 | 0.3 | 5.1 | 0.3 | 1.4 | 0.2 | 3.1 | 0.5 | 0.2 | 0.2 | 11.7 |
| Calcicole | *Gypsophila* | 8.0 | 0.3 | 4.3 | 0.4 | 3.7 | 13.4 | 6.2 | 4.5 | 0.2 | 0.2 | 40.8 |
| | *Sanguisorba* | 11.9 | 0.3 | 5.9 | 0.4 | 1.3 | 0.2 | 7.9 | 3.8 | 0.2 | 0.9 | 32.6 |

*Source:* Ström, L., *Oikos*, 80, 459–466, 1997.
*Note:* The role of mycorrhizal fungi is not implied, but it is probable that they could be involved in increasing the function of organic acid production.

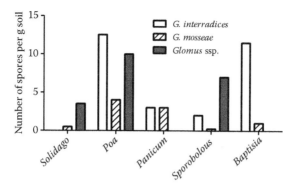

**Figure 3.12** Mean spore densities of arbuscular mycorrhizal fungi *Glomus interradices*, *G. mosseae*, and *Glomus* spp., indicating differences in the mycorrhizal community structure on roots of five plant species (*Solidago missouriensis*, *Poa pratense*, *Panicum virginiatum*, *Sporobolus heterolepis*, and *Baptisia bracteata*). (Data from Eom, A.-H. et al., *Oecologia*, 122, 435–444, 2000.)

Koide et al. (2000) showed that the role of arbuscular mycorrhizae was probably not directly an effect on plant growth, but a change in the rate of uptake of phosphorus and phosphorus use efficiency by plants. Mycorrhizal colonization of roots of *Lactuca* and *Abutilon* spp. increased the rate of phosphorus uptake (phosphorus efficiency index) by 23% and 32%, respectively, but it had no effect on the nonmycorrhizal plant *Beta* sp. However, the mycorrhizal association significantly reduced the phosphorus use efficiency of *Lactuca*, although it did not alter that of *Abutilon* or *Beta*, leading to a slight increase in growth of *Lactuca*, a significant increase in growth of *Abutilon*, and no effect on *Beta*. This indicates that mycorrhizal colonization of roots of different plant species has different effects and that the resulting outcome may influence more than the growth of the host plant, including its relative fitness within the plant community. Results from a meta-analysis of published reports show that the benefit of adding mycorrhizal inoculum in agroecosystems appears to increase plant growth and yield only in cases where soil P availability or natural mycorrhizal inoculum is low (Lekberg and Koide, 2005).

Fitter (1985) suggested that we knew relatively little about the ecological significance and ecosystem functioning of arbuscular mycorrhizae in field conditions. Most information regarding mycorrhizal function and physiology comes from laboratory/greenhouse studies and studies of mycorrhizae in an agricultural context. Plant responses in natural ecosystems often show different responses. For example, mycorrhizal inoculation of clover in acid grassland ecosystems where plant growth was severely limited in the absence of added phosphorus had no effect on plant growth; by contrast, clover responded positively to the addition of one of the two arbuscular mycorrhizae in the second year, showing an improvement in yield on brown earth soil of higher pH and fertility (Rangeley et al., 1982). Fitter suggested that, in comparison with laboratory experiments, the difference in response in the field could be attributable to the interconnectedness of plants via mycorrhizae, the effects of faunal grazing reducing the function of mycorrhizae, or differences in the longevity of roots compared to artificial systems. Some of these concepts have been explored more recently. Very little influence of mycorrhizal association of natural grasses could be seen in the uptake of phosphorus and a variety of heavy metals (Sanders and Fitter, 1992a,b), although the authors suggest that the benefit of the association occurs seasonally, during times when phosphorus availability is low and plant demand is high. This is possibly a reason for the maintenance of the mycorrhizal association in the community.

## 3.3 SUCCESSION AND PLANT COMMUNITY COMPOSITION

The existence of successions of ectomycorrhizal species during primary succession is supported by the findings of Jumpponen et al. (1999, 2002) on the Lyman glacier forefront. In the different plant successional stages they identified, they found 68 ectomycorrhizal species belonging to 25 genera, with no single ectomycorrhizal species occurring in all three successional sites. The authors also found that ectomycorrhizal species diversity increased to a maximum where tree canopies started to overlap. The appearance of spores of a variety of fungal genera in the feces of pika, voles, chipmunks, marmots, mountain goats, and mule deer on the forefront of Lyman Glacier, forms an inoculum source allowing colonization of the newly developing soils by early successional and slow-growing tree species (*Abies lasiocarpa*, *Larix lyalii*, *Tsuga mertensiana*, and *Salix* spp.). "Safe sites" for seed germination and mycorrhizal establishment by animal dispersed propagules consisted of concave surfaces of course rocky particles, which were ideal for trapping tree seeds and protecting them from desiccation (Jumpponen et al., 1999). Under these harsh environmental conditions, the dark-septate mycorrhizal fungus, *Phialocephalia fortinii*, significantly enhanced the growth of lodgepole pine (*Pinus contorta*), which is an early colonizer of the glacier forefront, but only in the presence of available nitrogen (Jumpponen et al., 1998). During the succession of plants in this recent glacial till, microbial communities change from bacterial domination to fungal-dominated communities. During this change, carbon use efficiency changes from a high rate of carbon respiration to an accumulating phase, thus indicating that fungi are a stabilizing force in the developing ecosystem and facilitating net carbon fixation into biomass (Ohtonen et al., 1999).

In a study of primary succession on sand dunes, Sikes et al. (2012) saw greater diversity in arbuscular mycorrhizal fungi in early succession and mycorrhizae producing greater amounts of extraradical hyphae and arbuscules later in succession, a probable response to ecosystem filters.

During primary succession on lava fields in Iceland, mycorrhizal fungi were absent from young lava flows but dominated by the ectomycorrhizal fungus *Russula aeruginea* and ericoid fungus *Melinionmyces bicolor* on older larva fields at later stages of succession (Cutler et al., 2014). The actual community composition of mycorrhizae during changes in plant communities undergoing primary or secondary succession may not be well characterized as discrete and changing communities—rather, the community will be dominated by specific species depending on the plant community assembly, climate, and edaphic conditions at that point and the specific traits exhibited by the dominant fungal species (Dickie et al., 2013) (Figure 3.13). The increase of organic matter in soil is a filter for mycorrhizal type—for example, dark septate hyphal fungal associations of roots are more frequent than arbuscular mycorrhizae in soils with a more defined humus layer (Kauppinen et al., 2014). The relative efficiencies of these mycorrhizae for nutrient acquisition have not been fully determined.

In forest succession and forest growth, the function of mycorrhizae appears to have a more dramatic effect on plant growth in oligotrophic systems than in fertile systems. In localized areas of nutrient-poor soils, such as volcanic fields and glacial outwash (Gehring and Whitham, 1994; Jumpponen et al., 1998, respectively), growth of pinyon pine on cinder soils was doubled by the addition of ectomycorrhizae compared to the effect of mycorrhizae on adjacent loam soil. This fact was attributed to the role of mycorrhizae in the cinder soil to overcome multiple stresses of cinder soil having half the moisture content, one-third of the available phosphorus of the loam, and no mineralizable nitrogen. Growth of lodgepole pine on glacier outwash soil was enhanced by the dark septate mycorrhizal fungus *Phialocephala fortinii*, rather than ectomycorrhizae, because of the ability of the fungus to enhance phosphate acquisition in this nutrient-poor ecosystem. During ecosystem succession to a forested community, the quality and quantity of organic resources available to the decomposer community change (Heal and

**Figure 3.13** Proportion of sites in earliest, early, and mature stages of ecosystem development containing ectomycorrhizal species (top) and arbuscular mycorrhizal species (bottom) showing successional changes in mycorrhizal community composition. Gray, mycorrhizal absent; black, mycorrhiza present; *n*, number of sites examined. (Data from Dickie, I.A. et al., *Plant Soil*, 367, 11–39, 2013.)

Dighton, 1986). Similar changes occur during a forest rotation (Cromack, 1981; Polglase et al., 1992; Hughes and Fahey, 1994). A change from high-quality resources (low lignin C/N) to low-quality resources (high lignin C/N) over time would imply that fungi occurring during the later stages of forest development or in more mature forests would benefit from greater enzyme competency than in early stages of forest development. Fleming et al. (1986) proposed the concept of mycorrhizal succession, which by observations of fruit body production of ectomycorrhizal species appear to occur (Dighton and Mason, 1985; Dighton et al., 1986; Last et al., 1987). However, there is some debate over the suitability of using fruit bodies as an index of mycorrhizal abundance and dominance, compared to actual measures of mycorrhizal root tip abundance (Termorshuizen and Schaffers, 1989; Egli et al., 1993; Yamada and Katsuya, 2001).

The nature and degree of mycorrhizal and plant establishment is subject to a series of ecological filters (*sensu* Jumpponen and Egerton-Warburton, 2005; Figure 3.14) where there is initial selection for compatibility between host and fungus, environmental tolerance to edaphic and climatic conditions, and biotic interactions of competition, synergism, and trophic interactions (Fitzsimons et al., 2008). A change from largely arbuscular mycorrhizal associations in early succession to ectomycorrhizae in later stages leads to different interactions with root herbivores,

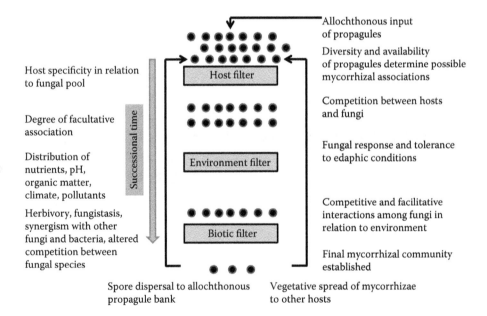

**Figure 3.14**    Conceptual development of fungal–plant community assemblage during succession in relation to available inoculum potential and biological and environmental filters modifying the final community composition. (From Jumpponen, A., Egerton-Warburton, L.M., in: *The Fungal Community: Its Organization and Role in the Ecosystem*, 3rd edn, ed. Dighton, J. et al., 139–168, Taylor & Francis, Boca Raton, FL, 2005.)

where arbuscular mycorrhizae increase plant defenses against grazers, and ECM form a protective barrier of hyphae around the host root that increases root tough- ness and longevity (up to 45% longer than arbuscular mycorrhizal roots). These are part of the biotic filters that are part of the feedback between plants and mycor- rhizae during succession (Rasmann et al., 2011). Two hypotheses have been posed to explain successional changes in arbuscular mycorrhizal fungal species. One of these hypotheses suggests that mycorrhizal fungi are the driving force (drivers); the second suggests that changes in mycorrhizal species is dependent on the plant and environmental conditions and the mycorrhizae are considered "passengers" (Hart et al., 2001; Figure 3.15).

During secondary succession, mycorrhizal dependency of plants is influenced by edaphic conditions and the stage of development of a plant community. By influenc- ing host plant growth, fecundity, and tolerance to stress, mycorrhizae can contribute to the success of different plants in a community to alter plant community composi- tion. These effects can often be seen in vegetation successions during both primary and secondary seral succession. For example, 3 years after agricultural disturbance and as the abandonment of agricultural practices progresses, vegetation changes from a ruderal community to grassland, shrubland, and early successional and late successional woodlands. The ruderal vegetation, composed primarily of annual spe- cies, is largely nonmycorrhizal. After 2–3 years, perennials are recruited into the community and most plant species have arbuscular mycorrhizal symbionts, and only in later stages of woodland establishment are arbuscular mycorrhizae displaced by

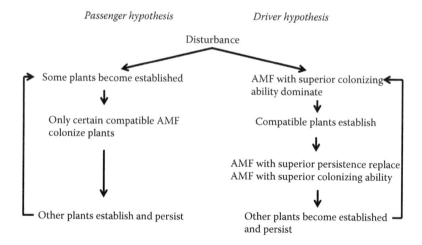

**Figure 3.15** A model proposing two alternate mechanisms for changes in community struc- ture of arbuscular mycorrhizal fungi (AMF) communities through time. The pas- senger hypothesis proposes that mycorrhizal communities are determined by the plant community, whereas in the driver hypothesis the mycorrhizae deter- mine the plant species by interspecific differences in colonization and persis- tence potential of the fungi. (From Hart, M.M. et al., *Mycologia*, 93, 1186–1194, 2001.)

ectomycorrhizae (Barni and Siniscalco, 2000). Changes in mycorrhizal status were related to the effect of disturbance (tillage) in the agricultural setting, reducing the ability of mycorrhizal establishment. Later, soil physical stability increased along with plant species diversity, leading to widespread arbuscular mycorrhizal colonization of roots as both water relations and nutrient availability were altered (drought became more of a problem and available nutrients in soil were reduced). A change in plant host species and accumulation of organic plant residues in the woodland phase lead to a dominance of ectomycorrhizae, which have the ability to utilize organic forms of nutrients. During secondary succession, late successional plant species seem to be more dependent on their mycorrhizal associations than early successional species. During primary colonization of sand dunes, mycorrhizal colonization of root systems of plants are more important later in the dune formation, as initial colonizers (ruderal species) tend to be less mycorrhizal (Çakan and Karataş, 2006). Close ties between plants and their mycorrhizae result from positive feedback between plants and soil compared to negative feedback in early successional stages (Kardol et al., 2006). During primary succession, the role of mycorrhizal fungi in supporting plant establishment and primary production depends on the dispersal of fungal propagules to root systems as there is no legacy of their presence from a previous plant community.

The arbuscular mycorrhizal community assembly on herbaceous annual plants, herbaceous perennial plants, and woody plant species shows contrasting traits with plant feedback altering fungal communities from season to season and with different fungal communities associated with different plant life forms (Lopez-Garcia et al., 2014). Lopez-Garcia et al. (2014) found that fungal communities were sensitive to disturbance in June but were resilient in September. Mycorrhizal communities were more similar over time on herbaceous annuals than semiwoody plants, wherein the more ruderal annual plants were more reliant on mycorrhizal establishment via spore populations as inoculum and the semiwoody plants were more reliant on established mycelial networks.

Superimposed on the successional story is the degree of heterogeneity within the ecosystem at any one time point in succession. These small-scale changes in environmental conditions can also significantly affect the community and, therefore, probably the functions of mycorrhizal communities. For example, the patchy distribution leaf litter resources on the forest floor may influence mycorrhizal community structure and function as different resources are available. Repeated harvesting of forest floor leaf litter in a Swedish spruce forest has been shown to reduce the abundance of ectomycorrhizae on roots, but not the number of species (Mahmood et al., 1989), although Baar and De-Vries (1995) showed that complete removal of the leaf litter on a Scots pine forest floor in the Netherlands increased the diversity of mycorrhizal fungal species, whereas doubling the leaf litter reduced the diversity below that of control plots where leaf litter was left unmanipulated. In experiments investigating the effects of leaf litter extracts on the growth of ectomycorrhizae in culture, Baar et al. (1994) showed that extracts of pine leaf litter reduced the growth of *Laccaria proxima* and *Rhizopogon luteolus* and only affected the growth of *Paxillus involutus* and *Xercomus badius* at high concentrations. However, extracts of the grass *Deschampsia flexuosa* inhibited the growth of *L. proxima, P. involutus,* and *R. luteolus,* but enhanced the growth of *Laccaria bicolor.* Koide et al. (1998b,c) showed that the polyphenols catechin and epicatechin gallate act similar to pine leaf litter water extracts in stimulating the growth of *Suillus intermedius* and reducing the growth of *Amanita rubescens,* but also noted that volatile compounds α- and β-pinene had differential effects on a range of ectomycorrhizal fungi. This study suggests that the phenolic content and composition of leaf litter can exert a significant control on the ectomycorrhizal communities developing within the vicinity of the litter.

In a mixed forest ecosystem in the New Jersey pine barrens, Dighton et al. (2000) showed that there were localized patches of leaf litter occupying the forest floor. Liter patch size (large, small, or nonexistent) was dictated by the density of stems of the ericaceous understory vegetation stem density acting as a litter dam. Large litter patches had different leaf species composition than the small patches, which altered both soil chemistry and physical conditions such that different ectomycorrhizal communities developed on the pine and oak roots invading those leaf litter patches, with significantly higher ectomycorrhizal diversity in larger patches ($r^2 = 0.734$). Pitch pine seedlings favored a mixed leaf liter community of oak and pine, over either of the leaf litter species alone, with different ectomycorrhizal fungal community developing on roots in each of the leaf litter types. During the decomposition of oak leaves, phosphorus was immobilized into the decomposing leaf litter, whereas nitrogen was immobilized in pine. By determining the phosphatase activity of each of the mycorrhizal types found on the roots and relating the enzyme production with the percentage contribution of the mycorrhizal type to the whole community, Conn and Dighton (2000) showed that the mycorrhizal communities on roots exploiting oak and oak/ pine mixed leaf litters had a higher proportion of phosphatase-producing mycorrhizae than on pine (Figure 3.16). They attributed this to the lack of available phosphorus in oak-containing litters, where phosphorus is being immobilized during the initial stages of decomposition. Thus, there appears to be some positive interaction between the local environmental conditions and the development of ectomycorrhizal communities in relation to the ability of mycorrhizae to utilize the resources available.

Another example of the influence of plants and plant litter on the community structure of ectomycorrhizae comes from evidence showing that *Rhododendron maximum* (an ericaceous shrub) severely reduces regeneration of hardwood and coniferous seedlings in the southern Appalachians. Walker et al. (1999) showed that litter manipulations within these forests did not affect the total mycorrhizal colonization of tree roots, but altered the distribution of *Cenococcum geophilum* mycorrhizae. However, it was noticed that after the first year, hemlock seedlings regenerating in rhododendron thickets had significantly less ectomycorrhizal colonization of their roots (19%) than trees outside of the thickets (62%). Within the ectomycorrhizal community, root colonization of 1-year-old hemlocks by *C. geophilum* was significantly higher in the presence of rhododendron (10.4%) than without (4.6%), although this difference was lost after 2 years of growth. The effect of the difference in mycorrhizal colonization of roots within and outside rhododendron thickets resulted in a 50% reduction in seedling shoot biomass in the second year.

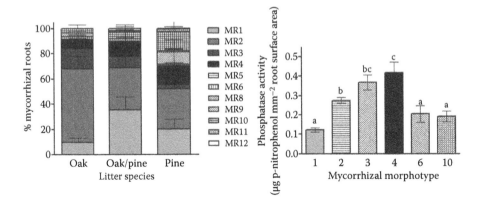

**Figure 3.16**   Effect of pine and oak leaf litter and their mixture on the ectomycorrhizal species development and phosphatase activity of pine roots invading the litter. The
graph on the left shows the proportional contribution of mycorrhizal types to the
community in each leaf litter type. The graph on the right shows that mycorrhizal
types found more frequently in the oak litter produce more phosphatase enzyme
as this litter type immobilizes P in the initial stages of decomposition. (Data from
Conn, C., Dighton, J., *Soil Biol. Biochem.*, 32, 489–496, 2000.)

Patches of nutrients arising from localized additions to the soil from urine, feces,
and dead animal bodies also affect mycorrhizal communities. Clear successions of
mycorrhizal fungi fruit bodies develop over a time course of animal decomposition,
where later successions favor the appearance of *Laccaria bicolor* and *Hebeloma*
spp., which have an affinity for high ammonium content in soil (Table 3.8), and
exploitation of subterranean mole middens by *Hebeloma radicosum*, which—under
these conditions—is able to "defend" its site of occupancy against the more common
*H. spoliatum* (Sagara, 1995).

## 3.4  PLANT COMMUNITIES

In addition to improving plant growth, the effect of mycorrhizal associations can lead
to improvements in overall plant fitness. This improved fitness, if asymmetric, can
be a method of providing competitive advantage to those plant species or individuals that respond the most to the effects of mycorrhizal colonization. These highly
responsive plants will therefore become more dominant in the community. However,
there is a need to look at the whole system of interactions between mycorrhizae and
other soil (micro) organisms, where the interactions and feedbacks between these
soil organisms and the plant community can generate a greater understanding of the
development of plant communities in terms of competition, synergisms, and imbalances caused by the introduction of invasive species. This study encompasses both
theoretical and practical ecology (Bever et al., 2010).

The effects of mycorrhizae on the increase in reproductive potential of plants
have been noted by Koide et al. (1988a), Stanley et al. (1993), Lewis and Koide (1990),

**Table 3.8 Fungal Fruit Body Appearance and Successions on Two Localized Substrates and an Ammonium-Treated Control Site (500 g m$^{-2}$ N) in a Pine Forest in Japan**

| Species | Time (days) | | | | | | | |
|---|---|---|---|---|---|---|---|---|
| | 0 | 100 | 200 | 300 | 400 | 500 | 600 | 1700 |
| **Human Feces** | | | | | | | | |
| *Ascobolus hansenii* | + | | | | | | | |
| *Peziza* sp. | + | | | | | | | |
| *Laccaria laccata* | | + | | | | + | | |
| **Dead Cat** | | | | | | | | |
| *Ascobolous denudatus* | + | | | | | | | |
| *Tephrocybe tesquorum* | + | | | | | | | |
| *Hebeloma spoliatum* | | + | + | | | | | |
| *Lactarius chrysorrheus* | | | | | | | + | + |
| *Mitrulia* sp. | | | | | | | + | |
| **Aqueous Ammonia** | | | | | | | | |
| *Ascobolous denudatus* | + | + | | | | | | |
| *Amblyosporium botrytis* | + | + | | | | | | |
| *Pseudombryophila deerata* | | + | | | | | | |
| *Tephrocybe tesquorum* | | + | + | | | | | |
| *Coprinus echinosporum* | | + | + | | | | | |
| *Peziza* spp. | | | + | | | | | |
| *Tephrocybe ambusta* | | | + | | | | | |
| *Laccaria bicolor* | | | | | | + | + | + |
| *Hebeloma* sp. | | | | | | + | + | |

*Source:* Sagara, N., *Can. J. Bot.*, 73, S1423–S1433, 1995.

Bryla and Koide (1990), and Koide and Lu (1992), where the increased reproductive potential has led to improvement in offspring vigor in terms of increased seedling germination, leaf area, root/shoot ratio, and root enzyme production. Heppell et al. (1998) showed that offspring of arbuscular mycorrhizal-infected *Abutilon theophrasti* were significantly larger than offspring of nonmycorrhizal parents, and under high-density conditions, improved even more because of the effects of early self-thinning in the mycorrhizal condition. This advantage was also transferred to the next generation in terms of total seed production. The influence of mycorrhizae can, however, differ significantly between plant species and, according to Janos (1980), can be a significant factor in determining plant species composition in the tropics.

In studies of the obligate mycorrhizal association of bluebell, the temporal allocation of phosphorus and carbon between the soil and plant is regulated through the mycorrhizae (Merryweather and Fitter 1995a), and the role of arbuscular mycorrhizal association of roots of bluebells increases with age as, during the aging process, bluebell bulbs descend further into the soil to zones where phosphorus becomes increasingly depleted and mycorrhizal colonization increases (Merryweather and

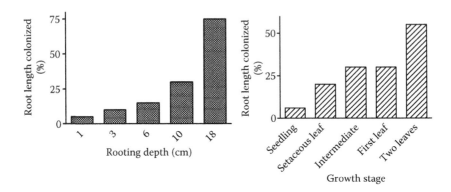

**Figure 3.17**  Effect of rooting depth (left) and stage of plant development (right) on arbuscular mycorrhizal development of roots of the bluebell (*Hyacinthoides non-scripta*) collected in the wild. (Data from Merryweather, J., Fitter, A.H., *New Phytol.*, 129, 629–636, 1995a.)

Fitter, 1995b) (Figure 3.17). Reducing the arbuscular mycorrhizal association of bluebells, through the application of benomyl, reduced plant phosphorus concentration in the vegetative parts of the plant, but preferentially allocated phosphorus to flowers and seeds, suggesting some control of the plant's fitness by the presence of mycorrhizal symbiosis (Merryweather and Fitter, 1996).

The distribution of fungal species in a mixed community of arbuscular mycorrhizal plant species is not homogenous and differs between plant species of a grassland community (Johnson et al., 1992; Eom et al., 2000). In a similar manner, van der Heijden et al. (1998) showed that the arbuscular mycorrhizal fungal community strongly influenced the plant species composition of members of a European calcareous grassland ecosystem, which was constructed in mesocosms. At low mycorrhizal species diversity, the plant species diversity varied widely as the arbuscular mycorrhizal species in the community are altered. At high mycorrhizal species diversity, altering the species composition of the mycorrhizal fungi did not cause such large changes in the plant species composition. At these high mycorrhizal diversities, nutrient acquisition by the host plant community increased, leading to greater biomass accumulation. This information reinforces the idea that there are functional feedbacks between the mycorrhizal fungal associate and the plant, such that the plant species can dictate the fungal species assemblage and *vice versa*, where a resultant patchy mycorrhizal influence on growth between plant species would have considerable effects on the structure of the plant community if the growth of some species is enhanced more than others. At high mycorrhizal diversity, however, each plant and each plant species has a greater chance of associating with an efficient mycorrhizal species. In this case, the asymmetry in benefit is lost, an even more beneficial effect of the mycorrhizae is seen throughout the plant community, and a shift in plant species community structure is unlikely.

The number of ramets produced by the clonal plant *Prunella vulgaris* was influenced by different arbuscular mycorrhizal fungal species (Streitwolf-Engel et al.,

2001), but stolon length and spacing between daughter plantlets were determined by host genotype, not directly under the influence of the mycorrhizal partner. These effects may partially explain the interactions between the grasses *Holcus lanatus* and *Dactylis glomerata* in the presence and absence of mycorrhizal colonization, where the degree of mycorrhizal colonization altered the proportion of each plant in the community, although *Holcus* always dominated over *Dactylis* (Watkinson and Freckleton, 1997). The effect of mycorrhizal colonization of roots slightly altered the competition/plant density response surface, suggesting that the increase in plant performance conferred on the plant by the mycorrhizal association was compensated for by changes in the intra- and interspecific competition strengths.

Taking Simpson's paradox as the basic model, where the response of the whole may not be based on the response of the individual parts, Allison and Goldberg (2002) explored the responses of individual plant species in communities to both arbuscular mycorrhizal association and availability of phosphorus in soil. Their conclusion from the metadata could not predict an overall community response that was the sum of consistent trends in the response of the component plant species. Thus, they suggested that the direction of response of each individual plant species to the degree of mycorrhizal infection in relation to P supply was different. As a consequence, there was no net community response (Figure 3.18). The varied responses of the individual plants species to both mycorrhizal colonization and environmental variables would lead to changes in community structure as conditions changed.

Fire within forested ecosystems is often a natural event that maintains both plant and fungal diversity, pushing the ecosystem back to a previous point in successional history. There are many examples of changes in ectomycorrhizal species' composition of the fungal community resulting from forest fire (Visser, 1995; Jonsson

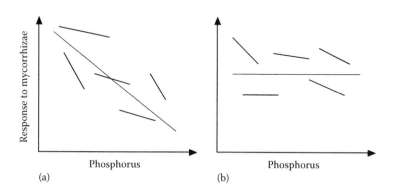

(a)                                    (b)

**Figure 3.18**  Models of the response between arbuscular mycorrhizal plants to mycorrhizal infection and soil phosphorus availability. (a) Graph depicts each plant species in the community responding in the same way, with a reduction of mycorrhizal colonization of roots with increasing P supply. In this situation, the net ecosystem effect is for a general reduction in mycorrhizal associations. (b) Graph depicts a variable response of each plant species in the community, resulting in a net lack of mycorrhizal response throughout the ecosystem. (From Allison, V.J., Goldberg, D.E., *Funct. Ecol.*, 16, 346–352, 2002.)

et al., 1999a) during the reestablishment of a mature forest (Frankland, 1992, 1998; Boerner et al., 1996). However, the nature of the ectomycorrhizal community establishing on the next rotation of forest trees is dependent on the degree of damage to the former mycorrhizal community. Where the effects of fire on the soil surface organic matter and soil is minimal, there will be a residual ectomycorrhizal community on the dying roots of the former forest trees. If reestablishment of the forest is rapid, these dying roots will act as a source of mycorrhizal inoculum, thus maintaining similar species diversity in the new forest similar to what existed in the old (Baar et al., 1999; Jonsson et al., 1999b). Thus, the forest can maintain some degree of continuity and stability. However, as nutrient conditions (influenced by the degree of nutrient mineralization from the fire and/or loss of organic matter) together with changes in physical characteristics of the soil (increased heating due to solar radiation absorbance by a dark soil surface) may affect the relative survival of the mycorrhizal species and their physiological function. In dry sclerophyllous shrub communities in Australia, the effect of fire on arbuscular mycorrhizal colonization of roots appears to be more closely related to the density of host plants than a direct influence of fire on the mycorrhizae (Torpy et al., 1999). In both arbuscular- and ectomycorrhizal-dominated natural ecosystems, where disturbance is minimal (cf. agroecosystems), the persistence of a mycelial network of mycorrhizal fungal hyphae increases the stability of the ecosystem as all plant species can tap into this common symbiotic system.

## 3.5 PLUGGING INTO THE HYPHAL NETWORK

In the 1960s, Bjorkman (1960) discovered that radioactive carbon injected into the stem of a pine tree could be detected in an adjacent achlorophyllous *Monotropa* plant. He also noted that there were ectomycorrhizal associations of the roots of both the *Monotropa* plant and the tree. He therefore suggested that there could be a carbohydrate exchange between plants by the mycorrhizal connection between them. This interaction between autotrophic plant and mycoheterotrophic plants is important in maintaining plant diversity and evolution (Kennedy et al., 2011). The concept of interconnectedness of plants within the ecosystem by mycorrhizal bridges has been a subject of research that has been dotted with examples of inconclusive results and skepticism. However, it is now believed that in many ecosystems there is a common network of mycorrhizal hyphae in the soil into which roots of many plant species develop their mycorrhizae and likely share resources between themselves depending on environmental stressors that differentially influence the growth of individual species (Simard et al., 2012).

The importance of this belowground hyphal network is indicated in that about 25–63% of gross primary production of a forest ecosystem is allocated belowground to roots and their mycorrhizae. In a meta-analysis of extraradical hyphal growth of ectomycorrhizal in forests, Ekblad et al. (2013) showed that a considerable portion of tree belowground C allocation goes into mycorrhizal structures that have a relatively long residence time (rhizomorphs with a longevity of 7–22 months, which is

2–9 times longer than that of mycorrhizal root tips) and may be poorly consumed by grazers (Table 3.9).

Suggestions of transfer of nutrients from a "donor" plant of one species to another came from the movement of phosphorus between *Lolium* and *Plantago* through arbuscular mycorrhizal connections, wherein one plant was stressed by clipping aboveground parts (Heap and Newman 1980a,b), but not between either of these mycorrhizal plants and the nonmycorrhizal cabbage (*Brassica oleracea*) (Newman and Eason, 1989), or between arbuscular and ectomycorrhizal plants (Eason et al., 1991). In the same manner, $^{14}$C label applied to pine tree seedlings was preferentially transferred to neighboring, unlabeled pine tree seedlings, rather than to neighboring plant species associated with arbuscular mycorrhizae (Read et al., 1985). Carleton and Read (1990) showed that there could be transfer of phosphorus and carbon between pine trees and feather moss (*Pleurozium schreberi*) communities in the understory over distances of several centimeters, because of the interconnecting ectomycorrhizal fungi associated with the roots of both plant species.

Transfer of carbon between plants in the field was demonstrated by Simard et al. (1997a,b,c), wherein carbon was transferred from paper birch (*Betula papyrifera*) to Douglas fir (*Pseudotsuga menziesii*) seedlings growing in partial or deep shade (Figure 3.19). They showed that the amount of carbon transferred between plants could form a significant proportion of the carbon contained in the shoots (13% for *P. menziesii* and 45% for *B. papyrifera*), which they suggest could considerably supplement the limited photosythetically derived carbon in the shaded plant. In trenching experiments, Douglas fir seedlings planted into trenched areas in a birch-dominated community had approximately half the diversity of mycorrhizal fungi associated with their roots than counterparts planted into untrenched plots. The increased mycorrhizal diversity in untrenched plots significantly increased the photosynthetic capacity of the Douglas fir seedlings (Simard et al., 1997a), showing the importance of a common mycorrhizal network. Indeed, Wu et al. (2002) have shown that some 24% of $^{14}$C label occurring in the underground parts of pine seedlings was allocated to the extraradical hyphal component of their ectomycorrhizal association.

**Table 3.9    Production of Extrametrical Hyphae (EMM) from Ectomycorrhizal Trees in Contrasting Biomes**

| Biome | Dominant Tree | Average Age (y) | EMM Production (kg ha⁻¹ per growing season) |
|---|---|---|---|
| Boreal | *Picea abies* | 125 | 151 |
|  | *Pinus sylvestris* |  |  |
| Boreonemoral | *Picea abies* | 50 | 188 |
|  | *Pinus sylvestris* |  |  |
| Nemoral | *Picea abies* | 60 | 138 |
|  | *Quercus robur* |  |  |
| Warm temperate | *Pinus palustris* | 20 | 611 |
|  | *Pinus taeda* |  |  |

*Source:* Ekblad, A. et al., *Plant Soil*, 366, 1–27, 2013.

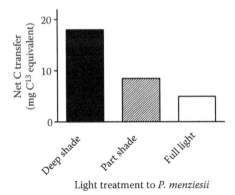

Light treatment to *P. menziesii*

**Figure 3.19**  Carbon transfer from birch to Douglas fir seedlings through ectomycorrhizal connections in the field where the fir trees were subjected to various degrees of shading. (Data from Simard et al., 1997a.)

Interplant transfer of phosphorus in arbuscular mycorrhizal tallgrass prairie communities showed greater transfer within forbs and cool season $C_3$ grasses than in $C_4$ grasses (Figure 3.20), suggesting that there is some selectivity in the process of interplant transfer (Walter et al., 1996). This difference between plant groups may confer an advantage to plants that are capable of greater interplant transfers that plants with lesser abilities.

The existence of a common mycorrhizal network could account for the ease by which some plants can associate with a variety of mycorrhizal types. For example, the

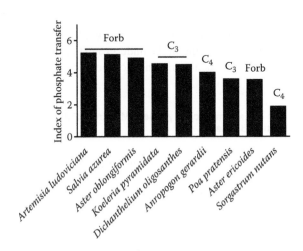

**Figure 3.20**  Index of phosphorus transfer from radioisotopically labeled donor plant (*Andropogon gerardii*) to the shoots of neighboring plant species in a tallgrass prairie community. (Data from Walter, L.E.F. et al., *Am. J. Bot.*, 83, 180–184, 1996.)

displacement of arbuscular mycorrhizae by ectomycorrhizae in *Alnus, Heliathemum,* and *Eucalyptus* with increasing tree age (Read et al., 1977; Chilvers et al., 1987) is reported by Lodge and Wentworth (1990). Ectomycorrhizae displaced arbuscular mycorrhizae on *Populus* and *Salix* with increasing soil moisture (Lodge and Wentworth, 1990), and multiple mycorrhizal types may occur within the same plant species (Largent et al., 1980; Dighton and Coleman, 1992; Smith et al., 1995).

The importance of the mycorrhizal network in ecosystem dynamics is summarized by Simard et al. (2012). Growth of individual plants is enhanced in the presence of available and compatible mycorrhizae in the network (Dickie et al., 2004), the transfer of carbon and nutrients from established to newly colonizing plants (Teste et al., 2009), and the ability of the network to confer disease resistance from one plant to another (Song et al., 2010). In approximately 48% of studies, mycorrhizal networks have been shown to stimulate seedling plant growth (van der Heijden and Horton, 2009).

Where plants are under stress, access to a common mycorrhizal network may be of importance. For example, an experimental 50% defoliation of lodgepole pine in a mixed lodgepole pine (*Pinus contorta*)–Engelmann spruce (*Picea engelmannii*) forest in Yellowstone National Park resulted in no change in mycorrhizal colonization (142 mycorrhizal tips/core) or species richness (~5.0 species/core), compared to control plots (Cullings et al., 2001). However, the ecosystem-dominant ectomycorrhizal species, *Inocybe* sp., became rare in defoliation plots, whereas both Agaricoid and Suilloid species became dominant in both the defoliated and control plots (Figure 3.21). Ectomycorrhizal fungal species associating with both lodgepole pine and Engelmann spruce were affected by defoliation, which suggests that changing the photosynthetic capacity of one species can affect mycorrhizal associations of

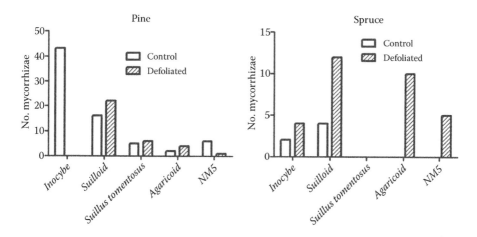

**Figure 3.21**   Effect of faunal defoliation (right column of each pair of bars) on the ectomycorrhizal community structure of lodgepole pine (*Pinus contorta*) and Englemann spruce (*Picea engelmannii*). (Data from Cullings, K.W. et al., *Oecologia*, 127, 533–539, 2001.)

neighboring trees of a different species. A similar response was found in conspecifics where the mycorrhizal benefit to defoliation was greatest closest to adjacent intact plants and decreased with distance from the intact plant (Pietkäinen and Kytöviita, 2007). The competitive dominance of dwarf birch (*Betula nana*) in the arctic may be enhanced by warming climate trends as conspecific exchange of carbon by mycorrhizal networks and root grafting increase with increased soil warming (Deslippe and Simard, 2011).

Within an ectomycorrhizal network, hyphal growth may be rapid and large mycelial networks, many of which are aggregated into cords or rhizomorphs, develop (Read, 1991a). These rhizomorphs allow translocation of nutrients from distal parts of the extraradical mycelial network to the root (Finlay and Read, 1986a,b) in an analogous way to mycelial cord systems of wood rotting fungi (Rayner et al., 1985; Wells and Boddy, 1990, 1995; Cairney, 1992; Boddy, 1999). The mycorrhizal mycelium can, in fact, be so dense in these humic soil horizons that they have been termed "mats" (Griffiths et al., 1990), where they may form almost 10% to 20% of the top 10 cm of soil in a temperate forest ecosystem (Cromack et al., 1988) and account for 45% to 55% of the total soil biomass (Cromack et al., 1979). Aguilera et al. (1993) showed that these mat-forming ectomycorrhizal communities in Douglas fir forests are important in increasingly removing organic nitrogen from the soil pool and immobilizing it into high C/N ratio fungal tissue as forest growth progresses. Although the forest soil thus becomes enriched with organic nitrogen as the forest matures, this N becomes increasingly less available to plant growth. The patchy existence of nutrients or accessible resources for mycorrhizal utilization in soil would indicate that these fungi would be adaptable to be able to exploit a variety of resources as and when they become available. Indeed, Tibbett (2000) indicates that in both ericoid and ectomycorrhizal symbioses, the extraradical hyphae exhibit significant morphological and physiological plasticity (Bending and Read, 1995a,b; Cairney and Burke, 1996) that makes them ideally suited for the exploitation of patchily distributed nutrient resources. The density of hyphae of ectomycorrhizal has also been shown to alter in response to both the concentration and nature of nitrogen resources offered. Dickie and Koide (1998) showed that hyphal foraging was increased by the production of less dense hyphal growth at low concentrations of nitrogen in either an inorganic or an organic form. It is suggested that this response, which is similar to that seen for saprotrophs (Ritz, 1995; Rayner, 1991; Rayner et al., 1994), affords the mycelia greater abilities to exploit patchily distributed resources. In the same way, contrasting arbuscular mycorrhizal communities associate with different plants (*Trifolium repens* and *Lolium perenne*), but the same community exists on plants of each species over a wide geographical range and is influenced only by local edaphic conditions (Hazard et al., 2013).

## 3.6 INTERACTIONS WITH OTHER MICROBES

In addition to interactions between ectomycorrhizal fungi and saprotrophic fungi in soil, close associations of these mycorrhizal fungi and bacteria within the rhizosphere have been reported. This tripartite arrangement may be regarded as essential to the functioning of mycorrhizae, where mycorrhizae can be dependent on nutrients released from bacterial action and bacteria feeding on fungal hyphae. The details of many of these interactions are still unclear, especially the endobacteria that occur within fungal hyphae (Bonfante and Anca, 2009; Buée et al., 2009). In his reviews, Garbaye (1991, 1994) discusses the importance of these so-called "helper bacteria" (Garbaye and Bowen, 1987, 1989; Duponnois and Garbaye, 1990, 1991; Fitter and Garbaye, 1994). It is suggested that these bacteria assist in the fungal/plant recognition system (Garbaye and Duponnois, 1992), receptivity of the host root to the mycorrhizal fungus, enhancement of the mycorrhizal fungal mycelial growth (Duponnois and Garbaye, 1990; Garbaye and Duponnois, 1992) before root colonization, alteration of the rhizospheric soil by altering pH, production of ion complexing compounds (siderphores) and the nutrient balance (Wallander and Nylund, 1991), and germination of fungal propagules (Ali and Jackson, 1989), which is probably most important in arbuscular mycorrhizae (Von Alten et al., 1993). However, the main ecological implication of these helper bacteria is to enhance the rate of root colonization, assist in the acquisition of nutrients by production of enzymes, and to complex ions that help to detoxify soil for plant growth. In addition, there are synergistic interactions between mycorrhizae and other bacteria in soil, particularly root nodulating and free-living nitrogen-fixing bacteria (Barea et al., 1997). These associations are particularly important in agroecosystems where the sustainability of soil fertility is of prime importance. The actual variation in the species composition of bacteria around ectomycorrhizal roots in the field, however, is largely unknown. New methods for the identification of these communities (Mogge et al., 2000) together with more information about the physiological attributes of these bacteria, such as phosphate solubilization (Berthelin and Leyval, 1982; Leyval and Berthelin, 1983; Singh and Kapoor, 1998) will allow us a greater understanding of the ecological and ecosystem processes that accrue from these close associations between mycorrhizae and bacteria.

Fungi do not exist alone in the environment. In soil, in particular, mycorrhizal fungi are in close juxtaposition with a range of other fungi, bacteria, and fauna. Thus, the effect of fungi may be considerably affected by their interactions with these other organisms. The interaction between mycorrhizae and saprotrophs in litter decomposition is one example of an interaction that may alter the rates of decomposition, nutrient mineralization, and the ultimate fate of the nutrients released during the decomposition process. We have seen that many ectomycorrhizae have the capacity to act as decomposers. Gadgil and Gadgil (1971, 1975) first suggested that there could be a strong interaction between mycorrhizal tree roots and the saprotrophic community in soil, where the presence of roots can suppress the rate of decomposition of leaf litter. Berg and Lindeberg (1980) repeated Gadgil and Gadgil's experiment in a northern coniferous forest and found the opposite effect—that the presence of tree roots did not influence the rate of leaf litter decomposition. In a laboratory study of controlled mycorrhizal inoculation of trees in the presence and absence of saprotrophic fungi, Dighton et al. (1987) showed that the presence of a saprotrophic fungus, *Mycena galopus*, reduced the decomposition potential of the ectomycorrhizal fungi, *Suillus luteus* and *Hebeloma crustuliniforme*, associated with seedling pine roots. Zhu and Ehrenfeld (1996) revisited this argument in an oligotrophic forest system to show that the presence of roots increased the activities of saprotrophic fungi and soil fauna to increase the rate of leaf litter decomposition and nutrient mineralization. In particular, the nematode faunal population of rooted chambers was significantly enhanced. The difference in results between studies is probably related to the overall fertility of the soil and the relative dependence of the system on readily available or unavailable sources of nutrients. The balance between the relative abilities of the saprotrophic fungal community and the ectomycorrhizal community to effect leaf litter decomposition is also dependent on the species composition of the two groups of fungi. Colpaert and van Tichelen (1996) showed that the decomposition of beech leaf litter is much less in the presence of Scots pine tree seedlings colonized by the ectomycorrhizal species *Thelephora terrestris*, *Suillus bovinus*, or *Paxillus involutus* than in the presence of the saprotroph *Lepista nuda* (Table 3.10). Indeed, nitrogen mineralization only occurred in the presence of *Lepista*. Durall et al. (1994) suggest that ectomycorrhizal fungi are capable of effecting leaf litter decomposition in the absence of a competing saprotroph, but that saprotrophic fungi are superior competitors for the organic resources and suppress the decomposing abilities of the ectomycorrhizal fungi. Lindahl et al. (1999) showed that the interaction between the ectomycorrhizal fungi *Suillius luteus* and *Paxillus involutus* and the wood decomposing saprotroph *Hypholoma fasciculare* resulted in a net transfer of phosphorus from the saprotroph to the mycorrhizal fungi. This suggests a positive synergistic activity of the mycorrhizae, which can more readily absorb and translocate away mineral nutrients derived from the activity of a saprotroph than the saprotroph itself can (Table 3.11). The authors show that up to 25% of the P present in the mycelium of the saprotroph is captured by the mycorrhizal fungi and translocated to the host tree within 30 days, whereas the reciprocal transfer of phosphorus was 3 orders of magnitude lower. However, the interaction between these different functional groups of fungi is still far from clear. As Singer and da Silva Araujo (1979) state, the differences in the dependence of tropical forest trees on ectomycorrhizae, rather than arbuscular mycorrhizae, may be linked to the ability of the

**Table 3.10  Mass of Beech Leaf Litter Remaining and Respiration per Unit Leaf Litter Weight after the Decomposition of 2 g Leaf Material in 26 Days in the Presence of Scots Pine Seedlings in the Absence (Control) or Presence of Ectomycorrhizal Associates *Thelephora terrestris* or *Suillius bovinus* or in the Presence of the Saprotrophic Basidiomycete *Lepista nuda***

| Fungus | Litter Dry Weight (g) | Decomposition Constant ($k$) | Litter Respiration (mg $CO_2$ day$^{-1}$ g$^{-1}$) |
|---|---|---|---|
| Control | 1.51 | 0.12 | 2.0 |
| *Thelephora terrestris* | 1.48 | 0.15 | 4.5 |
| *Suillius bovinus* | 1.39 | 0.22 | 11.0 |
| *Lepista nuda* | 0.92 | 1.00 | 14.0 |

*Source:* Colpaert, J.V., van Tichelen, K.K., in: *Fungi and Environmatal Change*, ed. Frankland, J.C., 109–128, Cambridge University Press, Cambridge, 1996.

Table 3.11 Partitioning of Radiophosphorus Label Applied Either to Wood that Was Being Decomposed by *Hypholoma fasciculare* or to Scots Pine Seedlings Colonized by Either *Paxillus involutus* or *Suillus variegates* mycorrhizae

| Application of Label | Site of Label Measured | Partitioning of $^{32}$P | |
| --- | --- | --- | --- |
| | | *Paxillus* | *Suillus* |
| *Hypholoma* labeled | Fraction outside wood block | 8 | 7 |
| | Fraction in plant | 12 | 14 |
| | Fraction in plant shoot | 29 | 15 |
| Mycorrhiza labeled | Fraction outside plant | 24 | 22 |
| | Fraction in wood | 0.09 | 0.15 |

Source: Lindahl, B. et al., *New Phytol.*, 144, 183–193, 1999.
Note: The data show greater transfer of phosphorus from the decomposing wood to the plant than from the plant to the wood.

ectomycorrhizal associates to effect leaf litter decomposition. They showed that in a white podsol campinarana soil, trees were obligatorily ectomycorrhizal where there were large accumulations of raw humus and low rates of leaf litter decomposition. In contrast, trees in a latisol-terra-firma soil, in which the rate of leaf litter breakdown was much higher, the fungal community is dominated by saprotrophs, the rate of mineral nutrient availability is higher, and the trees are primarily associated with arbuscular mycorrhizae.

Evidence from natural abundance isotopic ratios, however, suggests that there may not be much direct competition between saprotrophic and ectomycorrhizal fungi for either nitrogen or carbohydrates (Hobbie et al., 1999). By analyzing the $\delta^{15}$N and $\delta^{13}$C values of fruit bodies of mycorrhizal and saprotrophic fungi, vegetation, and soils from young, deciduous-dominated and older, coniferous-dominated sites forest sites, they showed that mycorrhizal fungi had consistently higher N and lower C values than saprotrophic fungi. Foliar $\delta^{13}$C values were always isotopically depleted relative to both fungal types (Table 3.12). It is suggested that isotopic

Table 3.12 Variation in Nitrogen and Carbon Stable Isotope Signatures of Various Component Parts of a Mixed Forest Ecosystem, Showing Differences in Signatures between Ectomycorrhizal and Saprotrophic Fungi

| | | $\delta^{15}$N ($\lambda$) | $\delta^{13}$C ($\lambda$) |
| --- | --- | --- | --- |
| Fungi | Mycorrhizae | 4.2 | −25.6 |
| | Saprotrophs | −1.3 | −22.6 |
| Plant parts | Alder leaves | −1.3 | −29.7 |
| | Spruce leaves | −3.7 | −28.8 |
| | Fine roots | −2.0 | −27.5 |
| Soil an soil nitrogen | Mineral soil | 6.0 | −25.6 |
| | Organic soil | 0.6 | −27.5 |
| | Ammonium-N | −0.7 | |
| | Nitrate-N | −0.1 | |

Source: Hobbie, E.A. et al., *Oecologia*, 118, 353–360, 1999.

Table 3.13    Influence of Inoculation of Maize with Wither Arbuscular Mycorrhizae (AM),
Beneficial Bacteria (*Pseudomonas fluorescens* Pf4) or in Combination

|                      | Control | Pf4  | AM   | AM + Pf4 |
|----------------------|---------|------|------|----------|
| Shoot mass (g)       | 334[a]  | 401[b] | 399[b] | 407[b]   |
| Root mass (g)        | 68[a]   | 115[b] | 58[a]  | 88[ab]   |
| Spike mass (g)       | 141[a]  | 190[b] | 214[b] | 217[b]   |
| Grains/spike         | 497[a]  | 619[b] | 652[b] | 634[b]   |
| Grain mass/spike (g) | 120[a]  | 163[b] | 183[b] | 185[b]   |

Source: Berta, G. et al., *Mycorrhiza*, 24, 161–170, 2014.
Note: Significant differences between treatments (columns) are indicated with superscript
letters.

fractionation by mycorrhizal fungi during the transfer of nitrogen to plants may be attributable to enzymatic reactions in the fungi that produces isotopically depleted amino acids, which are passed on to the host plant. Thus, the authors maintain, the mycorrhizal association of the trees maintains a higher level of $\delta^{15}N$ enrichment in the plant, owing to the changes exerted by the mycorrhizal fungus. Enriched carbon signatures of mycorrhizal fungi compared to those of foliage may be attributable to the fungal use of isotopically enriched photosynthate, such as simple sugars, in contrast to the mixture of compounds present in decomposing leaves. These methods provide interesting information on the rates of transfers of nutrients and carbon within the ecosystem, but the details of the transfers (sources and sinks) and transfer rates (source and sink strengths) are currently a major focus of research.

However, not all plant growth benefits come from the mycorrhizae as plant growth-promoting rhizobacteria are an important component in the rhizosphere, which may work synergistically with mycorrhizae or provide other services. For example, in conjunction with arbuscular mycorrhizae in field-grown maize, plant growth promoting rhizobacteria were equally good or better in promoting vegetative growth but were only equally good in promoting yield over uninoculated plants, and there was no observable synergistic interaction (Berta et al., 2014; Table 3.13). The ectomycorrhizal status of a root can significantly change the community of root-associated bacteria. In addition, mycorrhizal helper bacteria not only assist in hyphal growth and the establishment of mycorrhizal associations. With enhanced gene expression for protein kinase and transcription factors (Schrey et al., 2005), these streptomycetes and fluorescent pseudomonads also assist in making phosphate available from complex inorganic sources and produce siderophores to enhance root acquisition to, for example, iron, and increase plant protection from pathogens (Frey-Klett et al., 2005).

## 3.7  MYCORRHIZAE AND STRESS TOLERANCE

The role of mycorrhizal symbioses in defense of host plants against stress in summarized by Dighton (2009). The geographical distribution of mycorrhizae in relation to climatic zones was alluded to in the preceding discussion. At extreme latitudes and elevations, the mycorrhizal habit is largely lost with limited survival of mycorrhizal fungi above 76–78°N and increased altitude on Mount Fuji (Wu et al., 2004). In these stressed environments, dark-septate hyphae—rather than mycorrhizae—dominate as fungal endophytes of roots (Ruotsalainen and Kytöviita, 2004). Where mycorrhizae occur in cold environments, adaptations such as the production of mannitol and trehalose assist in cryoprotection (Tibbett et al., 2002).

### 3.7.1 Drought Tolerance

Drought imposes an osmotic stress on plants, which can be overcome in arbuscular mycorrhizal plants by reductions in proline, soluble sugars, and oxidative damage, all of which are stress-induced chemicals (Porcel and Ruiz-Lozano, 2004; Figure 3.22), as seen in ectomycorrhizal trees (Shi et al., 2002). The presence of arbuscular mycorrhizal associates of papaya roots significantly increased leaf water potential and plant biomass under water stress conditions, where mycorrization also reduced the levels of ethylene and 1-aminocyclopropane-1-carboxylic acid, stress-induced plant metabolites (Cruz et al., 2000; Table 3.14).

Mycorrhizal associations have a significant impact on plant water relations and can help to alleviate drought stress (Sánches-Diaz and Honrubia, 1994; Cheplick

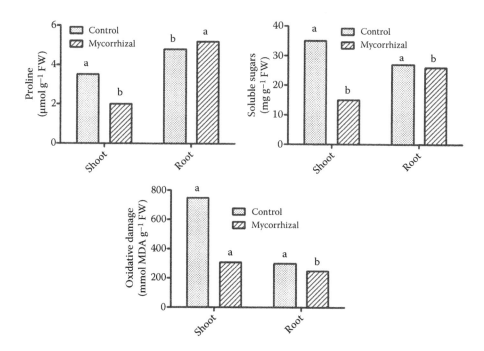

**Figure 3.22**  Comparison of control and mycorrhizal soybean under drought stress for production of proline (left), soluble sugars (center), and oxidative damage to lipids (right). (Data from Porcel, R., Ruiz-Lozano, J.M., *J. Exp. Bot.*, 55, 1743–1750, 2004.)

Table 3.14   Influence of Inoculation of Papaya with Arbuscular Mycorrhizal Fungus
(*Gigaspora marginata*) on Root Mass and Stress Indicators
1-Aminocyclopropane-1-Carboxylic Acid (ACC) and Ethylene

|  | Root Fresh Weight (g) | | ACC (nmol g⁻¹ Fresh Weight) | | Ethylene (ppm) | |
|---|---|---|---|---|---|---|
|  | Irrigated | Water Stress | Irrigated | Water Stress | Irrigated | Water Stress |
| Non-AM | 55 | 99 | 0.14 | 0.62 | 0.93 | 1.41 |
| AM | 86 | 141 | 0.06 | 0.41 | 1.35 | 1.23 |

*Source:* Cruz, A.F. et al., *Mycorrhiza*, 10, 121–123, 2000.
*Note:* Tree with mycorrhiza had significantly higher leaf water potential than control trees.

et al., 2000). This benefit can arise from direct water flow through fungal hyphae, improvement in the plant phosphate nutrition, and altered hormonal balance. The effect of arbuscular mycorrhizal infection of the tropical trees *Acacia nilotica* and *Leucaena leucocephala* benefited *Leucaena* most in the presence of droughty conditions. The addition of phosphorus to soil improved the growth of both plant species, and the addition of mycorrhizae mirrored the effect of adding P, but the effect of mycorrhizae was greater than the effect of P addition in *Leucaena* under drought stress (Michelsen and Rosendahl, 1990). Fruit production of pepper plants was similar between mycorrhizal plants under drought to that of nonmycorrhizal plants supplied with adequate water, suggesting that mycorrhizae could help acquire adequate soil water (Mena-Violante et al., 2006). The presence of arbuscular mycorrhizal colonization of a "bald" mutant of barley (a mutant with suppressed root hair development) showed that mycorrhizae provide an equivalent benefit to plant growth in drought conditions to the possession of root hairs, but significantly increased P acquisition (Li et al., 2014). These results are in concurrence with the meta-analysis of Augé (2001), who came to the conclusion that arbuscular mycorrhizae helped in drought conditions, but that the effect on host plants was much less P acquisition and plant growth enhancement.

Plant and mycorrhizal adaptations in desert ecosystems are essential for effective drought tolerance. Plants in the Chihuahuan desert can be divided into two groups: (1) annuals with thin roots (118–375 µm diameter) and low arbuscular mycorrhizal root colonization (6.25%) and (2) perennials with thicker roots (202–818 µm diameter) and high levels of colonization (72%). Ephemeral annuals may not have time to benefit from mycorrhizal associations, but longer-lived perennials benefit from the water acquisition properties of their arbuscular mycorrhizal associations and therefore establish more obligate mycorrhizal associations (Collier et al., 2003). *Helianthemum almeriense* mycorrhizal with the desert truffle, *Terfezia claveryi* showed increased survival, 26%greater water potential, 92% higher transpiration rate, 45% higher stomatal conductance, and 88% higher photosynthesis under drought stress of a matric potential of −0.5 MPa than nonmycorrhizal plants (Morte et al., 2000). In contrast, drought reduced the mycorrhization of *Zygophyllum dumosum* and other plant species in the Negev desert, where the degree of root colonization by arbuscular mycorrhizae was positively correlated to seasonal increases in soil moisture (Jacobson, 1997; He et al., 2002). Similarly, pinyon pine in severe drought years have 50% less ectomycorrhizal colonization and altered community composition compared to trees under less stress (Swaty et al., 2004), suggesting both a quantitative and qualitative response of the mycorrhizal community to drought. In semiarid ecosystems, the retention of spores and propagules during adverse conditions imparts a degree of resilience to the ecosystem as they can be stimulated to form mycorrhizae when plant growth resumes (Barea et al., 2011).

In ectomycorrhizae, drought tolerance has been shown to be increased more by *Cenococcum geophilum* than by *Lactarius* spp. (Jany et al., 2003). This is possibly associated with the melanin deposition in the fungal cell walls of *Cenococcum*, as—by comparing a melanin inhibited with the control strain of *Cenococcum geophilum*—Fernandez and Koide (2013) showed reduced fungal growth in the melanin-inhibited strain under water stress and related this to the significantly reduced hyphal diameter of melanin-inhibited mycelium. However, the relationship of this finding to the tolerance of mycorrhizal plants was not tested.

### 3.7.2 Halotolerance

Where irrigation plays a large part in the management of agroecosystems, the evaporation of water often leaves localized increases in soil salinity. Increased soil salinity can reduce the germination of arbuscular mycorrhizal spores and reduce extraradical hyphal growth (Juniper and Abbott, 1993; Semones and Young, 1995; Baker et al., 1995; Johnson-Green et al., 2001). Thus, plants growing in these saline soils have a reduced mycorrhizal component, which is probably detrimental to their growth and survival. However, there is some degree of tolerance to salinity in arbuscular mycorrhizae (Sengupta and Chaudhuri, 1990), which—despite the fact that mycorrhizal function is reduced in these highly saline soils—could still be of benefit in the revegetation of salt-degraded soils (Johnson-Green et al., 2001). Using a split root technique, the mycorrhizal colonization of clover roots exposed to saline soil was significantly reduced to 45%, compared to 65% in no-saline soil. However, if the whole root system were in salty soil, the degree of colonization was significantly increased (76%), suggesting that when the whole plant is stressed, there is greater investment in the mycorrhizal component to overcome the stress but reliance on the unstressed part of the root system in a heterogenous environment (differences between the two chamber halves) (Füzy et al., 2007).

A meta-analysis of the role of arbuscular mycorrhizal fungi in halotolerance has shown that mycorrhizal inoculation significantly enhances shoot and root growth as well as increased uptake of N, P, and K, while suppressing Na uptake (Chandrasekaran et al., 2014). Chandrasekaran et al. (2014) also report that mycorrhizal plants produce more antioxidant enzymes compared with nonmycorrhizal plants, which also provides protective effects for the host plant (Figure 3.23). Arbuscular mycorrhizae have been shown to increase the levels of antioxidants super dismutase, catalase, peroxidase, ascorbic peroxidase, and glutathione reductase in host plants under salt stress (Evelin and Kapoor, 2014).

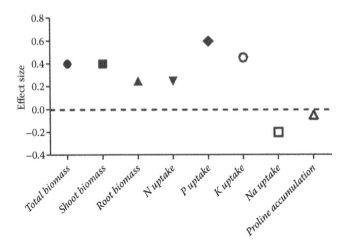

**Figure 3.23** Statistically significant arbuscular mycorrhizal plant responses to salt stress. (Data from a metadata analysis by Chandrasekaran, M. et al., *Mycorrhiza*, 24, 611–625, 2014.)

## 3.8 MYCORRHIZAE IN AQUATIC AND ESTUARINE SYSTEMS

Fungi are essentially aerobic organisms, and their physiological functions are inhibited by inundation with water. Waterlogged soils are generally anoxic except at the water–soil interface at the soil surface. In these conditions, iron, sulfur, and manganese compounds may be produced by the anaerobic bacterial communities and, together with the reduction in oxygen availability, reduce root and mycorrhizal growth (Khan and Belik, 1995). Gadgil (1972) demonstrated that a short period of inundation (7 weeks) reduced the phosphate uptake by radiata pine and Douglas fir, but that both P uptake and succinic dehydrogenase activity of the mycorrhizae was negligible after 14–16 weeks of waterlogging conditions. Khan (1993) showed that as one moved from plants growing in water up the stream bank, roots of the tree *Casuarina cunninhamiana* became increasingly colonized by arbuscular and ecto-mycorrhizae. Thus, it can be inferred that the physiological function of mycorrhizae on trees is reduced under waterlogged conditions as the mycorrhizal colonization of the root system is impaired.

However, mycorrhizal fungi are able to survive and successfully colonize plants growing in aquatic and salt marsh ecosystems. Plants found in aquatic or developing shorelines have mycorrhizal colonization rates of 50% to 100% (respectively) (Khan and Belik, 1995; Cooke and Lefor, 1998), but arbuscular mycorrhizal colonization of root systems of semiaquatic grasses is reduced as water depth increased (Miller, 2000). Arbuscular mycorrhizal colonization of the roots of the emergent aquatic, *Lythrum salicaria*, increased significantly with increasing phosphate levels in water, until a threshold of 1000 μg PO$_4$ l$^{-1}$ was exceeded, after which mycorrhizae did not form (White and Charvat, 1999). It is interesting that mycorrhizal infection did not have a positive effect on total plant biomass at low levels of P availability, but significantly reduced plant growth at the highest level of phosphate addition (47.5 mg $PO_4$ $l^{-1}$), where no mycorrhizae were seen to develop. At intermediate levels of P availability, however, mycorrhizal colonization increased the rate of root growth over that of shoot biomass, thus increasing the root/shoot ratio (Figure 3.24).

Mycorrhizae are known to exist in salt marshes around the globe (Rozema et al., 1986; van Duin et al., 1989; Sengupta and Chaudhuri, 1990; Hildebrandt et al., 2001), including marshes along the East Coast of the United States (Cooke et al., 1993; Hoefnagels et al., 1993). Within salt marsh plant communities, the colonization of roots by arbuscular mycorrhizae appears to be species-dependent. Consistent reports indicate that species such as *Spartina alterniflora* is nonmycorrhizal, whereas *S. cynosuroides* is frequently mycorrhizal (Hoefnagels et al., 1993; Cooke and Lefor, 1998). *Spartina alterniflora* and *S. cynosuroides* are similar in growth form and both grow in intertidal marsh elevations in East Coast of the United States. *S. alterniflora* is a dominant grass in high-salinity marshes, whereas *S. cynosuroides* is a dominant grass in brackish marshes (Smith and Read, 1997), but they do overlap in their toler- ance of brackish water (Parrondo et al., 1978; Stribling, 1998). A variety of factors affect the degree of root colonization by these fungi. Using ergosterol as an indicator of fungal colonization in roots, fungi in the living roots of *Spartina* spp. were con- firmed in North Carolina (Padgett and Celio, 1990) and in New Brunswick, Canada (Mansfield and Bärlocher, 1993). Both studies found that the greatest fungal biomass coincided with periods of active root growth during the summer months (Van Duin et al., 1989). This variation in degree of mycorrhizal colonization could be the cause of reports of nonmycotrophy in *S. alterniflora* and other species, depending on when samples were taken. This increase in root colonization during times of maximal plant growth also suggests a link between the depletion of readily available nutri- ents and the development of mycorrhizae in an effort to increase the efficiency of

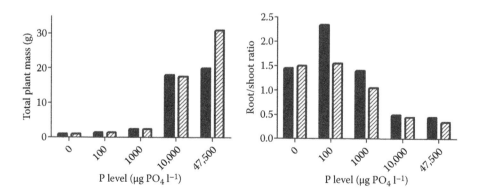

**Figure 3.24** Total plant mass (left) and root/shoot ratio (right) of the emergent aquatic plant *Lythrum salicarina* with (solid bars) and without (open bars) arbuscular mycor- rhizal inoculum grown in hydroponic sand culture at different levels of P supply. (Data from White, J.A., Charvat, I., *Mycorrhiza*, 9, 191–197, 1999.)

scavenging for scarce resources. In a waterlogged pioneer zone, the degree of root colonization of *Jaumea carnosa* was significantly reduced at higher elevations than in the channels or marine sites. Colonization was not related to redox potential, but to the higher levels of nitrogen in the sediments of the channels and creeks, where faunal activity increased sediment aeration and plant litter decomposition (Brown and Bledsoe, 1996).

In many salt marsh ecosystems, it is phosphorus—rather than nitrogen—that is generally recognized as the primary limiting nutrient (Valiela and Teal, 1974). This provides a possible explanation for the evolution of a close association of salt marsh plants and arbuscular mycorrhizae for the purpose of enhancing P acquisition. In conducting a nitrogen and phosphorus enrichment experiment in Louisiana, Buresh et al. (1980) found that plants fertilized with phosphorus had increased phosphorus content as a result of luxury uptake, as no growth increase was observed. In the same study, it was determined that only a limited amount of phosphorus was apparently available to *S. alterniflora*, because in nitrogen-enriched areas increased growth successively led to decreased phosphorus content in plants. Other studies have reported initial increases in mycorrhizal colonization as phosphorus availability increased and then subsequent decreases in colonization as phosphorus concentrations continued to increase (Johnson, 1998; White and Charvat, 1999). In a study of artificially manipulating periods of tidal inundation, high and low salinity, the addition of phosphorus, and the presence and absence of mycorrhizal inoculum, McHugh and Dighton (2004) showed that mycorrhizal colonization of *Spartina alterniflora* was much less than in *S. cynosuroides*. The hyphal colonization of roots of *S. alterniflora* was reduced in higher slaine conditions, but a similar decrease was not found in the abundance of arbuscules of *S. cynosuroides* in this study or in another study involving a mycorrhizal halophyte (Allen and Cunningham, 1983). There were little differences in the level of mycorrhizal colonization of roots with effective depth (duration of inundation) for either species, but although Cooke et al. (1993) found vesicles in roots of salt marsh species *Spartina patens* and *Distichlis spicata* down to 42 cm in depth, arbuscules were only found to a depth of 37 cm, with a significant reduction in abundance below 25 cm. McHugh and Dighton (2004) also found no effect of mycorrhizal inoculum on the shoot biomass or P content of either plant species; however, mycorrhizal colonization of the roots of *S. cynosuroides* increased the total amount of nitrogen assimilated by the plant. In both species, however, inoculation resulted in more shoots per pot, or increased tillering. Increased tillering resulting from mycorrhizal colonization has been observed in dune grass (Gemma and Koske, 1997) and wetland rice (Solaiman and Hirata, 1998) and could serve a useful function of significantly enhancing the rates of lateral spread in field plantings and subsequently affect rates of soil stabilization in restoration projects. Jasper (1994) suggests that mycorrhizae are important in revegetation projects by (1) enhancing plant establishment through improved nutrition, (2) maintaining diversity and altering plant competitive fitness, (3) contributing to the recycling of resources and increasing ecosystem stability, and (4) stabilizing soil.

## 3.9 FUNGAL ENDOPHYTES AND PRIMARY PRODUCTION

Mycorrhizae are not alone in helping plant growth. Fungal endophytes also promote plant growth and increase defense against herbivores. The interaction between the two fungal functional groups can influence the development of plant communities. In an assemblage of five plant species, the final plant assemblage differed between inoculation with endophytes, mycorrhizae, or a mixture of both (Rillig et al., 2014), based on the proportional plant mass of each plant species. The outcome was mainly a result of effects of different inocula on the growth of *Leucanthemum vulgare* in relation to *Trifolium pretense* (Figure 3.25). The presence of the endophytic fungus *Alternaria alernata* in the invasive forb *Centaurea stoebe* significantly enhanced the allelopathic effect of the invader on native North American grass *Koeleria macran-tha* (Aschehoug et al., 2014).

Endophytes are mainly ascomycete fungi that live within the host plant that do not cause any disease symptoms. They can be beneficial to host plant growth, provide defense against herbivory, and increase tolerance to environmental stressors. These fungi may represent only one stage in the life cycle of the fungus, and for some they become insect pathogens (Suryanaraynan, 2013). These fungi may lie along a mutualist–pathogen continuum and are only maintained where the bene-fit imparted by the fungus is balanced by the energetic cost of maintenance of the

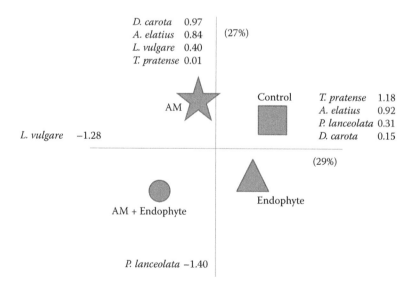

**Figure 3.25**  Effects of inoculation with arbuscular mycorrhizae (AM) fungal endophytes or combination of both on plant community composition of a contrived plant assemblage of five species: *Leucathemium vulgare, Trifolium pratense, Daucus carota, Arrhenathrum elatius,* and *Plantago lanceolata.* (Data from Rillig, M.C. et al., *Oecologia,* 174, 263–270, 2014.)

symbiont within the host tissue (Arnolds, 2007). Some tree endophytes express both pathogenic and mutualistic traits, and needle colonization by the fungus increases with needle age, suggesting that they may initiate leaf fall in low light conditions (Sieber, 2007), thus reducing the energy cost of supporting less photosynthetically productive needles (Figure 3.26). Indeed, Slippers and Wingfield (2007) show that fungi of the Botryosphaeriaceae, as endophytes of woody plants, are also potent pathogens with a latent phase. The potential change from mutualist to pathogen with environmental stress could be an important influence on plant communities in a changing climate.

The hypothesis of "defensive mutualism" has been ascribed to fungal endophytes in general as they have been reported to protect their host plant from insect and vertebrate herbivores, disease organisms, and environmental stressors (Belesky and Malinowski, 2000; White and Bacon, 2012). Much of this defense comes about by the production of secondary compounds produced by the endophyte, such as alkaloids, tannins, terpenoids, and glucosides, which are enhanced by the presence of grazing pressure or chemical stress (Clay, 2009; Popay, 2009). Additionally, endophytes enhance the production of both reactive oxygen species and auxins in host plant cells, produced by the fungus to stimulate the host plant to produce antioxidants to reduce cell death (Rodriguez et al., 2009; Torres et al., 2012). The results of this can be seen in, for example, increased phenolics in endophyte-colonized plants as antiherbivore compounds and decreased levels of proline, a stress *indicator* (Figure 3.27). Defense against herbivory has been shown to be important in grasses (Tadych et al., 2009; Bultman et al., 2009), but it has also been suggested, although less well proven, to be true for endophytes of coniferous trees (Pirttilä and Wäli, 2009).

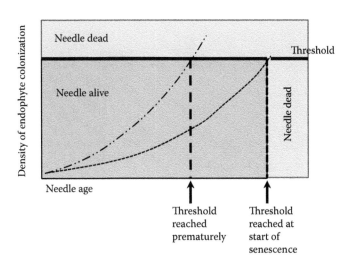

**Figure 3.26**  Relationship between needle age and endophyte density in the threshold model, where density increases to the threshold prematurely under conditions of stress leading to premature needle death. (After Sieber, T.N., *Fung. Biol. Rev.*, 21, 75–89, 2007.)

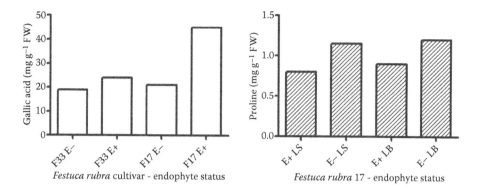

**Figure 3.27** Effect of fungal endophyte (E+) on phenolic content (left) and proline content (right) compared to endophyte-free (E–) *Festuca rubra* plants to demonstrate the stress protective effect of increased phenolics and reduction of stress induced proline content of plants by endophyte colonization. (Data from Torres, M.S. et al., *Fung. Ecol.*, 5, 322–330, 2012.)

Colonization of tall fescue by the endophytes *Epichloë* and *Neotyphodium* has been shown to reduce aphid populations on plants because of the fungal production of alkaloids (Bultman et al., 2012). The effect of *Beauvaria, Hypocrea, Gibberella, Metarhizium, Trichoderma*, and *Fusarium* in bean (*Vicia faba* and *Phaseolus vulgaris*) on the leaf miner *Liriomyza huidobrensis* caused insect mortality in 13–15 days. *Hypocrea*, in particular, reduced the longevity of offspring from 18 days to 11 days in control, decreased the number of pupae 4-fold, and reduced adult longevity 3-fold (Akutse et al., 2013). The endophytes *Epichloë* and *Neotyphodium* not only reduced the population of aphids on tall fescue, but also exerted their effects through the production of alkaloids in aphid parasitoids. These multitrophic effects may have a greater impact on insect communities than earlier thought (Bultman et al., 2012; Figure 3.28). However, Faeth and Saari (2012) argue that this defense is not always

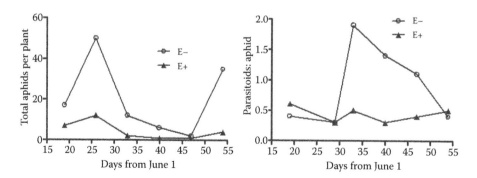

**Figure 3.28** Effect of an endophytic fungus on the population of aphids per plant (left) and proportion of aphids parasitized (right). (Data from Bultman, T.L. et al., *Fung. Ecol.*, 5, 372–378, 2012.)

effective because of the evolution of specialist herbivores that can detoxify defense chemicals and the effect of alkaloids on predators and parasitoids of herbiverous arthropods, releasing their populations from grazing pressure.

In contrast to the negative effects of endophytes on grazers, some endophytes actively attract insects to disseminate spores. The evolutionary link between the Clavicipitalean fungus *Epichloë* and flies has been suggested to have arisen because the fungus is an obligate outcrosser, and the encouragement of fly larval grazing on ascospores in the perithecia of the fungus allows a greater chance of spores being vectored to another fungus by the adult fly. Flies are attracted to the fungal stroma by volatiles produced by the fungus (Bultman and Leuchtmann, 2008).

The role of endophytes in providing host plant tolerance to environmental stressors (temperature, drought, and salinity) is well outlined by Rodriguez et al. (2009). In Petri plate studies, fungal endophytes of wheat significantly increased the seed germination rate, wherein the seed was placed adjacent to fungal hyphae under conditions of drought or heat stress (Hubbard et al., 2012). Increased seedling growth was only induced under heat stress for one fungal isolate. In comparison with fungicide-treated plants, tall fescue—naturally infected with *Neotyphodium*—recovered significantly faster from drought stress, and contained higher levels of plant sugars, alcohols, and amino acids as well as increased fungal metabolites including loline alkaloids (Nagabhyru et al., 2013). Barley plants were shown to be more tolerant to waterlogging when colonized by the endophytic fungus *Epichloë* with lower malondialdehyde content and reduced leakage of electrolytes (Song et al., 2015).

Reports of the effect of fungal endophytes in increasing host plant resistance to disease range from positive, through neutral, to negative, depending on the plant species and fungal pathogen (Wiewióra et al., 2015). The study conducted by Wiewióra et al. (2015) on perennial ryegrass endophyte protection against the fungal pathogens *Dreshslera siccans* and *Fusarium* spp., showed almost complete protection (95% reduction in disease frequency) against the pathogens in one ecoregion of Poland but only an average of 76% reduction in another region. The dominant endophytes in roots of Norway spruce, *Phialocephala fortini* and *Acephala applanata*, appear to have protective effects against the tree pathogen *Hetreobasidion parviporum* in drained peatlands in South Finland (Terhonen et al., 2014).

Endophytes may also protect plants from heavy metal toxicity. More than 50% of leaf or stem samples of six plant species growing in a lead–zinc mining area in China yielded fungal endophytes (Li et al., 2012). Of these fungal isolates, many showed enhanced growth in culture in the presence of heavy metal, compared to the control (Figure 3.29), although the effect of the endophyte on tolerance and growth of the host plants was not tested.

Bringing all these factors together, Rudgers and Clay (2007) demonstrate, from both the literature and long-term experimental manipulations of grass endophytes, that these fungi have a significant impact on ecosystems. Because of the plant fitness benefits of endophytes, endophytic tall fescue dominated the plant community by outcompeting other perennial grasses and forbs. This was largely attributable to the deterrence of vole herbivory in endophyte-colonized grass, where voles concentrated on nonendophyte-colonized plant species in the community. Endophyte-reduced

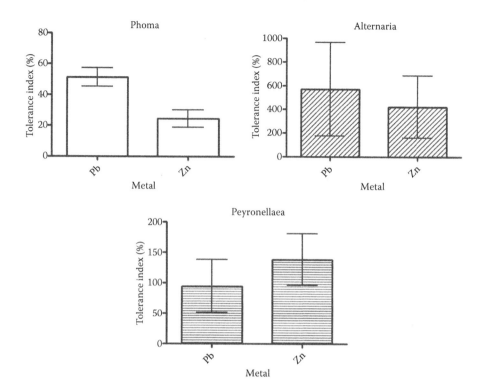

**Figure 3.29** Average tolerance index (% of mycelial growth compared to control) of three fungal endophytes under stress from lead or zinc. (Data from Li, H.-Y. et al., *Fung. Ecol.*, 5, 309–313, 2012.)

insect herbivory reduced abundances of leaf-chewing invertebrates more than sap-sucking invertebrates and, consequently, the population of their predators. Foliar endophytes even reduced belowground herbivory and the populations of root knot nematodes and Japanese beetle larvae. Changes in leaf litter chemistry between endophyte-colonized and uncolonized plants altered the rates of decomposition and nutrient mineralization and the abundance of some soil arthropods. These changes influence plant performance and contribute to changes in plant community composition. Thus, endophytes can have significant influences at the community and landscape levels.

## 3.10 CONCLUSION

Fungi interact with plants in a positive manner by being either mycorrhizal or endophytic symbionts. In both cases, they are able to enhance plant growth via the greater acquisition of nutrients and water. This enhancement of plant growth increases C fixation in plant biomass, thus having a potential global influence on C cycling. It is believed that the evolution of these associations by a large diversity of fungal species is maintained as each fungal species imparts slightly different functional qualities (enzyme competency, efficiency of nutrient transfer, degree of plant protection, etc.) and as such, have contrasting effects on different plant hosts and in differing environments. These interactions not only increase plant growth and fitness, but do so in an asymmetric manner, leading to alteration of plant competitive or synergistic abilities. This results in a strong influence of the fungal partner on plant community composition. Both mycorrhizae and endophytes alter the susceptibility of their host plant to disease, herbivory, and environmental stressors, thus maintaining greater plant biomass in their presence.

## REFERENCES

Abuzinadah, R. A. and D. J. Read. 1986a. The role of proteins in the nitrogen nutrition of ectomycorrhizal plants: I. Utilization of peptides and proteins by ectomycorrhizal fungi. *New Phytol.* 103:481–493.

Abuzinadah, R. A. and D. J. Read. 1986b. The role of proteins in the nitrogen nutrition of ectomycorrhizal plants: III. Protein utilization by *Betula, Picea* and *Pinus* in mycorrhizal association with *Hebeloma crustuliniforme. New Phytol.* 103:507–514.

Abuzinadah, R. A. and D. J. Read. 1989. The role of proteins in the nitrogen nutrition of ectomycorrhizal plants: V. The utilization of peptides by birch (*Betula pendula* L.) infected with different mycorrhizal fungi. *New Phytol.* 112:55–60.

Agerer, R. 1987. *Colour Atlas of Ectomycorrhizae.* Munich: Einhorn-Verlag.

Agerer, R. 2001. Exploration types of ectomycorrhizae. *Mycorrhiza* 11:107–114.

Aguilera, L. M., R. P. Griffiths, and B. A. Caldwell. 1993. Nitrogen in ectomycorrhizal mat and non-mat soils of different-age Douglas-fir forests. *Soil Biol. Biochem.* 25(8): 1015–1019.

Akutse, K. S., N. K. Maniania, K. K. M. Fiaboe, J. van den Berg, and S. Ekesi. 2013. Endophytic colonization of *Vicia faba* and *Phaseolus vulgaris* (Fabaceae) by fungal pathogens and their effects on the life-history parameters of *Liriomyza huidobrensis* (Diptera: Agromyzidae). *Fung. Ecol.* 6:293–301.

Ali, A. N. and R. M. Jackson. 1989. Stimulation of germination of spores of some ectomycorrhizal fungi by other organisms. *Mycol. Res.* 93:182–186.

Allen, M. F. 1991. *The Ecology of Mycorrhizae.* Cambridge, UK: Cambridge University Press.

Allen, E. B., M. F. Allen, D. J. Helm, J. M. Trappe, R. Molina, and E. Rincon. 1995. Patterns and regulation of mycorrhizal plant and fungal diversity. *Plant Soil* 170:47–62.

Allen, E. B. and G. L. Cunningham. 1983. Effects of vesicular–arbuscular mycorrhizae on *Distichlis spicata* under three salinity levels. *New Phytol.* 93:227–236.

Allison, V. J. and D. E. Goldberg. 2002. Species-level versus community-level patterns of mycorrhizal dependence on phosphorus: An example of Simpson's paradox. *Funct. Ecol.* 16:346–352.

Ames, R. N., C. P. P. Reid, L. K. Porter, and R. B. Clark. 1983. Hyphal uptake and transport of nitrogen from two $^{15}$N-labeled sources by *Glomus mosseae*, a vesicular–arbuscular mycorrhizal fungus. *New Phytol.* 95:381–396.

Anderson, I. C., S. M. Chambers, and J. W. G. Cairney. 2001. Variation in nitrogen source utilization by *Pisolithus* isolates maintained in axenic culture. *Mycorriza* 11:53–56.

Antibus, R. K., D. Bower, and J. Dighton. 1997. Root surface phosphatase activities and uptake of 32P-labelled inositol phosphate in field-collected gray birch and red maple roots. *Mycorrhiza* 7:39–46.

Antibus, R. K, R. L. Sinsabaugh, and A. E. Linkins. 1992. Phosphatase activities and phosphorus uptake from inositol phosphate by ectomycorrhizal fungi. *Can. J. Bot.* 70:794–801.

Arnolds, A. E. 2007. Understanding the diversity of foliar endophytic fungi: Progress, challenges, and frontiers. *Fung. Biol. Rev.* 21:51–66.

Aschehoug, E. T., R. M. Callaway, G. Newcombe, N. Tharayil, and S. Chen. 2014. Fungal endophyte increases the allelopathic effects of an invasive forb. *Oecologia* 175:285–291.

Augé, R. M. 2001. Water relations, drought and vesicular–arbuscular mycorrhizal symbiosis. *Mycorrhiza* 11:3–42.

Azcon, R., J. M. Barea, and D. S. Hayman. 1976. Utilization of rock phosphate in alkaline soils by plants inoculated with mycorrhizal fungi and phosphate solubilizing bacteria. *Soil Biol. Biochem.* 8:135–138.

Azcón-Aguilar, C., R. Azcón, and J. M. Barea. 1979. Endomycorrhizal fungi and *Rhizobium* as biological fertilizers for *Medicago sativa* in normal cultivation. *Nature* 279:325–327.

Baar, J. and F. W. de Vries. 1995. Effects of manipulation of litter and humus layers on ectomycorrhizal colonization potential in Scots pine stands of different age. *Mycorrhiza* 5:267–272.

Baar, J., T. R. Horton, A. M. Kretzer, and T. D. Bruns. 1999. Mycorrhizal colonization of *Pinus muricata* from resistant propagules after a stand-replacing wildfire. *New Phytol.* 143:409–418.

Baar J., W. A., Ozinga, I. L. Sweers, and T. W. Kuyper. 1994. Stimulatory and inhibitory effects of needle litter and grass extracts on the growth of some ectomycorrhizal fungi. *Soil Biol. Biochem.* 26:1073–1079.

Bajwa, R. and D. J. Read. 1985. The biology of mycorrhiza in the Ericaceae. IX. Peptides as nitrogen sources for the ericoid endophyte and for mycorrhizal and non-mycorrhizal plants. *New Phytol.* 101:459–467.

Baker, A., J. I. Sprent, and J. Wilson. 1995. Effects of sodium chloride and mycorrhizal infection on the growth and nitrogen fixation of *Prosopis juliflora. Symbiosis* 19:39–51.

Barea, J. M., C. Azcon-Aguilar, and R. Azcon. 1997. Interactions between mycorrhizal fungi and rhizosphere micro-organisms within the context of sustainable soil–plant systems. In: *Multitrophic Interactions in Terrestrial Systems*, ed. A. C. Gange and V. K. Brown, 65–77. Oxford: Blackwell Science.

Barea, J. M., J. Palenzuela, P. Corneju et al. 2011. Ecological and functional roles of mycorrhizas in semi-arid ecosystems of Southeast Spain. *J. Arid Environ.* 75:1292–1301.

Barni, E. and C. Siniscalco. 2000. Vegetation dynamics and arbuscular mycorrhiza in old-field successions of the western Italian Alps. *Mycorrhiza* 10:63–72.

Bartlett, E. M. and D. H. Lewis. 1973. Surface phosphatase activity of mycorrhizal roots of beech. *Soil Biol. Biochem.* 5:249–257.

Baxter, J. W. and J. Dighton. 2001. Ectomycorrhizal diversity alters growth and nutrient acquisition of gray birch (*Betula populifolia* Marshall) seedlings in host–symbiont culture conditions. *New Phytol* 152:139–149.

Baxter, J. W. and J. Dighton. 2005a. Phosphorus source alters host plant response to ectomycorrhizal diversity. *Mycorrhiza* 15:513–523.

Baxter, J. W. and J. Dighton. 2005b. Diversity-functioning relationships in ectomycorrhizal fungal communities. In: *The Fungal Community: Its Organization and Role in the Ecosystem* (3rd Edition), ed. J. Dighton, J. F. White, and P. Oudemans, 383–398. Boca Raton, FL: Taylor & Francis.

Belesky, D. P. and D. P. Malinowski. 2000. Abiotic stresses and morphological plasticity and chemical adaptations of *Neotyphodium*-infected tall fescue plants. In: *Microbial Endophytes*, ed. C. W. Bacon and J. F. White, 455–484. New York: Marcel Dekker.

Belnap, J. 2002. Nitrogen fixation in biological soil crusts from southeast Utah. *Biol. Fertil. Soils* 35:128–135.

Belnap. J. and O. L. Lange. 2005. Lichens and microfungi in biological soil crusts: Community structure, physiology, and ecological functions. In: *The Fungal Community: Its Organization and Role in the Ecosystem* (3rd Edition), ed. J. Dighton, J. F. White, and P. Oudemans, 117–138. Boca Raton, FL: Taylor & Francis.

Bending, G. D. and D. J. Read. 1995a. The structure and function of the vegetative mycelium of ectomycorrhizal plants: V. Foraging behaviour and translocation of nutrients from exploited litter. *New Phytol.* 130:401–409.

Bending, G. D. and D. J. Read. 1995b. The structure and function of the vegetative mycelium of ectomycorrhizal plants: VI. Activities of nutrient mobilizing enzymes in birch litter colonized by *Paxillus involutus* (Fr.) Fr. *New Phytol.* 130:411–417.

Bending, G. D. and D. J. Read. 1996. Nitrogen mobilization from protein–polyphenol complex by ericoid and ectomycorrhizal fungi. *Soil Biol. Biochem.* 28(12):1603–1612.

Berg, B. and T. Lindberg. 1980. Is litter decomposition retarded in the presence of mycorrhizal roots in forest soil? *Swedish Coniferous Project Internal Report* 95.

Berta, G., A. Copetta, E. Gamalero et al. 2014. Maize development and grain quality differentially affected by mycorrhizal fungi and growth-promoting pseudomonad in the field. *Mycorrhiza* 24:161–170.

Berthelin, J. and C. Leyval. 1982. Ability of symbiotic and non-symbiotic rhizospheric microflora of maize (*Zea mays*) to weather micas and to promote plant growth and plant nutrition. *Plant Soil* 68:369–377.

Bever, J. D., I. A. Dickie, E. Facelli et al. 2010. Rooting theories of plant community ecology in microbial interactions. *Trends Ecol. Evol.* 25:468–478.

Beymer, R. J. and J. M. Klopatek. 1991. Potential contribution of carbon by microphytic crusts in Pinyon–Juniper woodlands. *Arid Soil Res. Rehabilitation* 5:187–198.

Bjorkman, E. 1960. *Monotropa hypopitys* L. an epiparasite on tree roots. *Physiol. Plant.* 13:308.

Boddy, L. 1999. Saprotrophic cord-forming fungi: Meeting the challenge of heterogenous environments. *Mycologia.* 91:13–32.

Boerner, R. E., B. G. DeMars, and P. N. Leicht. 1996. Spatial patterns of mycorrhizal infectiveness of soils long a successional chronosequence. *Mycorrhiza* 6, 79–90.

Bolan, N. S. 1991. A critical review on the role of mycorrhizal fungi in the uptake of phosphorus by plants. *Plant Soil* 134:189–207.

Bonfante, P. and I-A. Anca. 2009. Plants, mycorrhizal fungi, and bacteria: A network of interactions. *Annu. Rev. Microbiol.* 63:363–383.

Bonfante, P. and A. Genre. 2010. Mechanisms underlying beneficial plant–fungus interactions in mycorrhizal symbiosis. *Nat. Commun.* 1:48 doi:10.1038/ncomms1046.

Bradley, R., A. J. Burt, and D. J. Read. 1982. The biology of the mycorrhiza in the Ericaceae: VIII. The role of mycorrhizal infection in heavy metal resistance. *New Phytol.* 91:197–209.

Brown, A. M. and C. Bledsoe. 1996. Spatial and temporal dynamics of mycorrhizas in *Jaumea carnosa*, a tidal saltmarsh halophyte. *J. Ecol.* 84:703–715.

Brundrett, M. C. 1991. Mycorrhizas in natural ecosystems. *Adv. Ecol. Res.* 21, 171–313.

Brundrett, M. C. 2009. Mycorrhizal associations and other means of nutrition of vascular plants: Understanding the global diversity of host plants by resolving conflicting information and developing reliable means of diagnosis. *Plant Soil* 320:37–77.

Bryla, D. R. and R. T. Koide. 1990. Regulation of reproduction in wild and cultivated *Lycopersicum esculentum* Mill. *Oecologia* 84:82–92.

Buée, M., P. E. Courtney, D. Mignot, and J. Garbaye. 2007. Soil niche effect on species diversity and catabolic activities in an ectomycorrhizal fungal community. *Soil Biol. Biochem.* 39:1947–1955.

Buée, M., W. De Boer, F. Martin, L. Van Overbeek, and E. Jurkevitch. 2009. The rhizosphere zoo: An overview of plant-associated communities of microorganisms, including phages, bacteria, archaea, and fungi, and some of their structuring factors. *Plant Soil* 321:189–212.

Bultman, T. L., A. Aguilera, and T. J. Sullivan. 2012. Influence of fungal isolates infecting tall fescue on multitrophic interactions. *Fung. Ecol.* 5:372–378.

Bultman, T. L. and A. Leuchtmann. 2008. Biology of the *Epichloë – Botanophilia* interaction: An intriguing association between fungi and insects. *Fung. Rev.* 22:131–138.

Bultman, T. L., T. J. Sullivan, M. H. Cortez, T. J. Pennings, and J. L. Andersen. 2009. Extension to and modulation of defensive mutualism in grass endophytes. In: *Defensive Mutualism in Microbial Symbiosis,* ed. J. F. White and M. Torres, 301–317. Boca Raton, FL: Taylor & Francis.

Buresh, R. J., R. D. DeLaune, and W. H. Patrick. 1980. Nitrogen and phosphorus distribution and utilization by *Spartina alterniflora* in a Louisiana gulf coast marsh. *Estuaries* 3:111–112.

Cairney, J. W. G. 1992. Translocation of solutes in ectomycorrhizal and saprotrophic rhizomorphs. *Mycol. Res.* 96, 135–141.

Cairney, J. W. G. 1999. Intraspecific physiological variation: Implications for understanding functional diversity in ectomycorrhizal fungi. *Mycorrhiza.* 9:125–135.

Cairney, J. W. G. and R. M. Burke. 1996. Physiological heterogeneity within fungal mycelia: An important concept for a functional understanding of the ectomycorrhizal symbiosis. *New Phytol.* 134:685–695.

Çakan, H. and Ç. Karataş. 2006. Interactions between mycorrhizal colonization and plant life forms along the successional gradient of coastal sand dunes in the eastern Mediterranean, Turkey. *Ecol. Res.* 21:301–310.

Carleton, T. J. and D. J. Read. 1990. Ectomycorrhizas and nutrient transfer in conifer–feathermoss ecosystems. *Can. J. Bot.* 69:778–784.

Chandrasekaran, M., S. Boughattas, S. Hu, S.-H. Oh, and T. Sa. 2014. A meta-analysis of arbuscular mycorrhizal effects on plants grown under salt stress. *Mycorrhiza* 24:611–625.

Chang, T. T. and C. Y. Li. 1998. Weathering of limestone, marble, and calcium phosphate by ectomycorrhizal fungi and associated microorganisms. *Taiwan J. For. Sci.* 13:85–90.

Chen, J., H.-P. Blume, and L. Beyer. 2000. Weathering of rocks induced by lichen colonization—A review. *Catena* 39:121–149.

Cheplick, G. P., A. Perera, and K. Koulouris. 2000. Effect of drought on the growth of *Lolium perenne* genotypes with and without fungal endophytes. *Funct. Ecol.* 14:657–667.

Chilvers, G. A., F. F. Lapeyrie, and D. P. Horan. 1987. Ectomycorrhizal vs endomycorrhizal fungi within the same root system. *New Phytol.* 107:441–448.

Chu-Chou, M. and L. J. Grace. 1985. Comparative efficiency of the mycorrhizal fungi *Laccaria laccata, Hebeloma crustuliniforme* and *Rhizopogon* spp. on the growth of radiata pine seedlings. *N. Z. J. Bot.* 23:417–424.

Clark, R. B. and S. K. Zeto. 2000. Mineral acquisition by arbuscular mycorrhizal plants. *J. Plant Nutrition* 23:867–902.

Clarkson, D. T. 1985. Factors affecting mineral nutrient acquisition by plants. *Annu Rev. Plant Physiol.* 36:77–115.

Clay, K. 2009. Defensive mutualism and grass endophytes: Still valid after all these years. In: *Defensive Mutualism in Microbial Symbiosis* ed. J. F. White and M. Torres, 9–20. Boca Raton, FL: Taylor & Francis.

Collier, S. C., C. T. Yarnes, and R. P. Herman. 2003. Mycorrhizal dependency of Chihuahuan desert plants is influenced by life history strategy and root morphology. *J. Arid Environ.* 55:223–229.

Colpaert, J. V. and K. K. van Tichelen. 1996. Mycorrhizas and environmental stress. In: *Fungi and Environmental Change*, ed. J. C. Frankland, N. Magan, and G. M. Gadd, 109–128. Cambridge: Cambridge University Press.

Conn, C. and J. Dighton. 2000. Litter quality influences on decomposition, ectomycorrhizal community structure and mycorrhizal root surface acid phosphatase activity. *Soil Biol. Biochem.* 32:489–496.

Cooke, J. C., R. H. Butler, and G. Madone. 1993. Some observations on the vertical distribution of vesicular arbuscular mycorrhizae in roots of salt marsh grasses growing in saturated soils. *Mycologia* 85:547–550.

Cooke, J. C. and M. W. Lefor. 1998. The mycorrhizal status of selected plant species from Connecticut wetlands and transition zones. *Restor. Ecol.* 6:214–222.

Cooper, E. J. and P. A. Wookey. 2001. Field measurements of the growth rates of forage lichens and the implications by grazing by Svalbard reindeer. *Symbiosis* 31:173–186.

Coxson, D. S. and S. K. Stevenson. 2007, Growth rates of *Lobaria pulmonaria* to canopy structure in even-aged and old-growth cedar-hemlock forests on central-interior British Columbia. *For. Ecol. Manage.* 242:5–16.

Crittenden, P. D. 1989. Nitrogen relations of mat-forming lichens. In *Nitrogen, Phosphorus and Sulphur Utilization by Fungi*, ed. L. Boddy, R. Marchant, and D. J. Read, 243–268. Cambridge: Cambridge University Press.

Crittenden, P. D., I. Katucka, and E. Oliver. 1994. Does nitrogen supply limit growth of lichens? *Crypt. Bot.* 4:143–155.

Cromack, K. 1981. Below-ground processes in forest succession. In *Forest Succession: Concepts and Applications*, ed. D. A. West, H. H. Shugart, and D. B. Botkin, 361–373. New York: Springer-Verlag.

Cromack, K., B. L. Fichter, A. M. Moldenke, and E. I. Ingham. 1988. Interactions between soil animals and ectomycorrhizal fungal mats. *Agric. Ecosyst. Environ.* 24:161–168.

Cromack, K., P. Sollins, W. C. Granstein, T. Speidel, A. W. Todd, G. Spycher, and Y-Li Ching. 1979. Calcium oxalate accumulation and soil weathering in mats of the hypogeous fungus *Hysterangium crassum. Soil Biol. Biochem.* 11:463–487.

Cruz, A. F., T. Ishii, and K. Kadoya. 2000. Effects of arbuscular mycorrhizal fungi on tree growth, leaf water potential, and levels of 1-aminocyclopropane-1-carboxylic acid and ethylene in the roots of papaya under water-stress conditions. *Mycorrhiza* 10:121–123.

Cullings, K. W., D. R. Vogler, V. T. Parker, and S. Makhija. 2001. Defoliation effects on the ectomycorrhizal community of a mixed *Pinus contorta/Picea engelmannii* stand in Yellowstone Park. *Oecologia* 127:533–539.

Cutler, N. A., D. CL. Chaput, and C. J. van der Gast. 2014. Long-term changes in soil microbial communities during primary succession. *Soil Biol. Biochem.* 69:359–370.

Deslippe, J. R. and S. W. Simard. 2011. Below-ground carbon transfer among *Betula nana* may increase wit warming in Arctic tundra. *New Phytol.* 192:689–698.

Dickie, I. A., R. C. Guza, S. E. Krazewski, and P. B. Reich. 2004. Shared ectomycorrhizal fungi between a herbaceous perennial (*Helianthemum bicknellii*) and oak (*Quercus*) seedlings. *New Phytol.* 164:375–382.

Dickie, I. A. and R. T. Koide. 1998. Tissue density and growth response of ectomycorrhizal fungi to nitrogen source and concentration. *Mycorrhiza* 8:145–148.

Dickie, I. A., L. B. Martinez-Garcia, N. Cole et al. 2013. Mycorrhizas and mycorrhizal fungal communities throughout ecosystem development. *Plant Soil* 367:11–39.

Dighton, J. 1983. Phosphatase production by mycorrhizal fungi. *Plant Soil* 71:455–462.

Dighton, J. 1991. Acquisition of nutrients from organic resources by mycorrhizal autotrophic plants. *Experientia* 47:362–369.

Dighton, J. 2009. Evaluation of mycorrhizal symbioses as defense in extreme environments. In: *Defensive Mutualism in Microbial Symbiosis*, ed. J. F. White and M. Torres, 199–216. Boca Raton, FL: Taylor & Francis.

Dighton, J. 2011. Mycorrhiza. In: *Eukaryotic Microbes*, ed. M. Schaechter, 73–83. Amsterdam: Elsevier, Academic Press.

Dighton, J. and D. C. Coleman. 1992. Phosphorus relations of roots and mycorrhizas of *Rhododendron maximum* L. in the southern Appalachians, N. Carolina. *Mycorrhiza* 1:175–184.

Dighton, J., T. Gordon, R. Mejia, and M. Sobel. 2013. Mycorrhizal status of Knieskern's beaked sedge (*Rhynchospora knieskernii*) in the New Jersey pine barrens. *Bartonia* 66:24–27.

Dighton, J. and J. A. Krumins. 2014. *Interactions in Soil: Promoting Plant Growth*, Dordrecht: Springer Science + Business Media.

Dighton, J. and P. A. Mason. 1985. Mycorrhizal dynamics during forest tree development. In: *Developmental Biology of Higher Fungi*, ed. D. Moore, L. A. Casselton, D. A. Wood, and J. C. Frankland, 163–171. Cambridge: Cambridge University Press.

Dighton, J., P. A. Mason, and J. M. Poskitt. 1990. Field use of 32P tracer to measure phosphate uptake by birch mycorrhizas. *New Phytologist* 116:655–661.

Dighton, J., A. S. Morale-Bonilla, R. A. Jimînez-Nûñez, and N. Martînez. 2000. Determinants of leaf litter patchiness in mixed species New Jersey pine barrens forest and its possible influence on soil and soil biota. *Biol. Fertil. Soils* 31:288–293.

Dighton, J., J. M. Poskitt, and D. M. Howard. 1986. Changes in occurrence of basidiomycete fruit bodies during forest stand development: With specific reference to mycorrhizal species. *Transactions of the British Mycological Society* 87:163–171.

Dighton, J., E. D. Thomas, and P. M. Latter. 1987. Interactions between tree roots, mycorrhizas, a saprotrophic fungus and the decomposition of organic substrates in a microcosm. *Biol. Fertil. Soils* 4:145–150.

Duddridge, J. A., A. Malibari, and J. D. Read. 1980. Structure and function of mycorrhizal rhizomorphs with special reference to their role in water transport. *Nature* 287:834–836.

Duponnois, R. and J. Garbaye. 1990. Some mechanisms involved in growth stimulation of ectomycorrhizal fungi by bacteria. *Can. J. Bot.* 68:2148–2152.

Duponnois, R. and J. Garbaye. 1991. Mycorrhization helper bacteria associated with the Doulas fir–*Laccaria laccata* symbiosis: Effects in aseptic an in glasshouse conditions. *Ann. Sci. For.* 48:239–251.

Durall, D. M., A. W. Todd, and J. M. Trappe. 1994. Decomposition of $^{14}$C-labelled substrates by ectomycorrhizal fungi in association with Douglas fir. *New Phytol.* 127:725–729.

Eason, W. R., E. I. Newman, and P. N. Chuba. 1991. Specificity of interplant cycling of phosphorus: The role of mycorrhizas. *Plant Soil* 137:267–274.

Egli, S., R. Amiet, M. Zollinger, and B. Schneider. 1993. Characterization of *Picea abies* (L) Karst. ectomycorrhizas: Discrepancy between classification according to macroscopic versus microscopic features. *TREE.* 7:123–129.

Ekblad, A., H. Wallander, D. L. Godbold et al. 2013. The production and turnover of extrametrical mycelium of ectomycorrhizal fungi in forest soils: Role in carbon cycling. *Plant Soil* 366:1–27, doi 10.1007/s11104-013-1630-3.

Eom, A-H., D. C. Hartnett, and G. W. T. Wilson. 2000. Host plant species effects on arbuscular mycorrhizal fungal communities in tallgrass prairie. *Oecologia* 122:435–444.

Evelin, H. and R. Kapoor. 2014. Arbuscular mycorrhizal symbiosis modulates antioxidant response in salt stressed *Trigonella foenum-graecum* plants. *Mycorrhiza* 24:197–208.

Faeth, S. H. and S. Saari. 2012. Fungal grass endophytes and arthropod communities: Lessons from plant defence theory and multitrophic interactions. *Fung. Ecol.* 5:364–371.

Fernandez, C. W. and R. T. Koide. 2013. The function of melanin in the ectomycorrhizal fungus *Cenococcum geophilum* under water stress. *Fung. Biol.* 6:479–486.

Finlay, R. D. and D. J. Read. 1986a. The structure and function of the vegetative mycelium of ectomycorrhizal plants: I. Translocation of $^{14}$C-labellled carbon between plants interconnected by a common mycelium. *New Phytol.* 103:143–156.

Finlay, R. D. and D. J. Read. 1986b. The structure and function of the vegetative mycelium of ectomycorrhizal plants: II. The uptake and distribution of phosphorus by mycelial strands interconnecting host plants. *New Phytol.* 103:157–165.

Fitter, A. 1985. Functioning of vesicular–arbuscular mycorrhizas under field conditions. *New Phytol.* 99:257–265.

Fitter, A. H. 1991. Cost benefits of mycorrhizas: Implications for functioning under natural conditions. *Experientia* 47:350–355.

Fitter, A. H. 2005. Darkness visible: Reflections on underground ecology. *J. Ecol.* 93:231–243.

Fitter, A. H. and J. Garbaye. 1994. Interactions between mycorrhizal fungi and other soil organisms. *Plant Soil* 159:123–132.

Fitzsimons, M. S., R. M. Miller, and J. D. Jastrow. 2008. Scale-dependent niche axes of arbuscular mycorrhizal fungi. *Oecologia* 158:117–127.

Fleming, L. V., J. W. Last, and F. T. Deacon. 1986. Ectomycorrhizal succession in a Scottish birch wood. In: *Physiological and General Aspects of Mycorrhizae*, ed. V. Gianinazzi-Pearson and S. Gianinazzi, 259–264. Paris: INRA.

Frankland, J. C. 1992. Mechanisms in fungal succession. In *The Fungal Community: Its Organization and Role in the Ecosystem*, ed. G. C. Carroll and D. T. Wicklow, 383–401. New York: Marcel Dekker.

Frankland, J. C. 1998. Fungal succession—Unravelling the unpredictable. *Mycol. Res.* 102:1–15.

Franco, A. R., N. R. Sousa, M. A. Ramos, R. S. Oliveira, and P. M. L. Castro. 2014. Diversity and persistence of ectomycorrhizal fungi and their effect on nursery-grown *Pinus pinaster* in a post-fire plantation in Northern Portugal. *Microb. Ecol.* 68:761–772.

Frey-Klett, P., M. Chavatte, M-L. Clausse et al. 2005. Ectomycorrhizal symbiosis affects functional diversity of rhizosphere fluorescent pseudomonads. *New Phytol.* 165:317–328.

Füzy, A., T. Tóth, and B. Biró. 2007. Mycorrhizal colonization can be altered by the direct and indirect effect of drought and salt in a split root experiment. *Cereal Res. Comm.* 35:401–404.

Gadgil, P. D. 1972. Effect of waterlogging on mycorrhizas of radiata pine and Douglas fir. *N. Z. J. For. Sci.* 2:222–226.

Gadgil, R. L. and P. D. Gadgil. 1971. Mycorrhiza and litter decomposition. *Nature* 233:133.

Gadgil, R. L. and P. D. Gadgil. 1975. Suppression of litter decomposition by mycorrhizal roots of *Pinus radiata*. *N. Z. Jl. For. Res.* 5:33–41.

Garbaye, J. 1991. Biological interactions in the mycorrhizosphere. *Experientia* 47:370–375.

Garbaye, J. 1994. Helper bacteria: A new dimension to the mycorrhizal symbiosis. *New Phytol.* 128:197–210.

Garbaye, J. and G. D. Bowen. 1987. Effect of different microflora on the success of ectomy-corrhizal inoculation of *Pinus radiata. Can. J. For. Res.* 17:941–943.

Garbaye, J. and G. D. Bowen. 1989. Stimulation of mycorrhizal infection of *Pinus radiata* by some microorganisms associated with the mantle of ectomycorrhizas. *New Phytol.* 112:383–388.

Garbaye, J. and R. Duponnois. 1992. Specificity and function of mycorrhization helper bac-teria (MHB) associated with the *Pseudotsuga menziesii–Laccaria laccata* symbiosis. *Symbiosis* 14:335–344.

Gehring, C. A. and T. G. Whitham. 1994. Comparisons of ectomycorrhizae on Pinyon pines (*Pinus edulis*; Pinaceae) across extremes of soil type and herbivory. *Am. J. Bot.* 81:1509–1516.

Gemma, J. N. and R. E. Koske. 1997. Arbuscular mycorrhizae in sand dune plants of the north Atlantic coast of the U.S.: Field and greenhouse inoculation and presence on mycorrhi-zae in planting stock. *J. Environ. Manage.* 50:251–264.

Gianinazzi, S. and H. Schüepp. 1994. *Impact of Arbuscular Mycorrhizas on Sustainable Agriculture and Natural Ecosystems.* Basel, Switzerland: Birkhäuser Verlag.

Gibson, F. and J. W. Deacon. 1988. Experimental study of establishment of ectomycorrhizas in different regions of birch root systems. *Trans. Br. Mycol. Soc.* 91:239–251.

Giltrap, N. J. 1982. Production of polyphenol oxidases by ectomycorrhizal fungi with special reference to *Lactarius* spp. *Trans. Br. Mycol. Soc.* 78:75–81.

Goodman, D., D. M. Durall, J. A. Trofymow, and S. M. Berch. 1996–2000. *A Manual of Concise Descriptions of North American Ectomycorrhizae.* Victoria, B.C: Mycologue Publications.

Goulart, B. L., M. L. Schroeder, K. Demchak, J. P. Lynch, J. R. Clark, R. L. Darnell, and W. F. Wilcox. 1993. Blueberry mycorrhizae: Current knowledge and future directions. *Acta Horticulturae* 346:230–239.

Griffiths, R. P., B. A. Caldwell, K. Cromack, and R. Y. Morita. 1990. Microbial dynamics and chemistry in Douglas fir forest soils colonised by ectomycorrhizal mats: I. Seasonal variation in nitrogen chemistry and nitrogen cycle transformation rates. *Can. J. For. Res.* 20:211–218.

Harley, J. L. 1969. *The Biology of Mycorrhiza.* London: Leonard Hill.

Harley, J. L. and S. E. Smith. 1983. *Mycorrhizal Symbiosis.* London: Academic Press.

Hart, M. M., R. J. Reader, and J. N. Klironomos. 2001. Life-history strategies of arbuscular mycorrhizal fungi in relation to their successional dynamics. *Mycologia* 93:1186–1194.

Häussling, M. and H. Marschner. 1989. Organic and inorganic soil phosphates and acid phos-phatase activity in the rhizosphere of 80-year old Norway spruce [*Picea abies* (L.) Karst.] trees. *Biol. Fertil. Soils* 8:128–133.

Hazard, C., P. Gosling, C. J. van der Gast, D. T. Mitchell, F. M. Doohan, and G. D. Bending. 2013. The role of local environment and geographical distance in determining com-munity composition of arbuscular mycorrhizal fungi at the landscape scale. *ISME J.* 7:498–508.

He, X., S. Mouratov, and Y. Steinberger. 2002. Temporal and spatial dynamics of vesicular–arbuscular mycorrhizal fungi under the canopy of *Zygophyllum dumosum* Boiss. in the Negev Desert. *J. Arid Environ.* 52:379–387.

Heal, O. W. and J. Dighton. 1986. Nutrient cycling and decomposition of natural terrestrial ecosystems. In: *Microfloral and Faunal Interactions in Natural and Agro-Ecosystems,* ed. M. J. Mitchell and J. P. Nakas, 14–73. Martinus Nijhoff/Dr. W. Junk.

Heap, A. J. and E. I. Newman. 1980a. Links between roots by hyphae of vesicular arbuscular mycorrhizas. *New Phytol.* 85:169–171.

Heap, A. J. and E. I. Newman. 1980b. The influence of vesicular arbuscular mycorrhizas on phosphorus transfere between plants. *New Phytol.* 85:173–179.

Heppell, K. B., D. L. Shumway, and R. T. Koide. 1998. The effect of mycorrhizal infection of *Abutilon theophrasti* on competitiveness of offspring. *Funct. Ecol.* 12:171–175.

Herrera, M. A., C. P. Salamanca, and J. M. Barea. 1993. Inoculation of woody legumes with selected arbuscular mycorrhizal fungi and rhizobia to recover desertified mediterranean ecosystems. *Appl. Environ. Microbiol.* 59:129–133.

Hetrick, B. A. D. 1989. Acquisition of phosphorus by VA mycorrhizal fungi and the growth responses of their host plants. In: *Nitrogen, Phosphorus and Sulphur Cycling in Temperate Forest Ecosystems*, ed. L. Boddy, R. Marchant, and D. J. Read, 205–226. Cambridge: Cambridge University Press.

Hetrick, B. A. D. 1991. Mycorrhizas and root architecture. *Experientia* 47:355–362.

Hildebrandt, U., K. Janetta, F. Ouziad, B. Renne, K. Nawrath, and H. Bothe. 2001. Arbuscular mycorrhizal colonization of halophytes in central European salt marshes. *Mycorrhiza* 10:175–183.

Hobbie, E. A., S. A. Macko, and H. H. Shugart. 1999. Insights into nitrogen and carbon dynamics of ectomycorrhizal and saprotrophic fungi from isotopic evidence. *Oecologia* 118:353–360.

Hoefnagels, M. H., S. W. Broom, and S. R. Shafer. 1993. Vesicular–arbuscular mycorrhizae in salt marshes in north Carolina. *Estuaries* 16:851–858.

Hubbard, M., J. Germida, and V. Vujanovic. 2012. Fungal endophytes improve wheat seed germination under heat and drought stress. *Botany* 90:137–149.

Hughes, J. W. and T. J. Fahey. 1994. Litterfall dynamics and ecosystem recovery during forest development. *For. Ecol. Manage.* 63:181–198.

Humphreys, C. P., P. J. Franks, M. Rees, M. I. Bidartondo, J. R. Leake, and D. J. Beerling. 2010. Mutualistic mycorrhiza-like symbiosis in the most ancient group of land plants. *Nat. Commun.* doi:10.1038/ncomms1105.

Ingleby, K., P. A. Mason, F. T. Last, and L. V. Fleming. 1990. *Identification of Ectomycorrhizas*. London: Institute of Terrestrial Ecology Research Pub. No. 5.

Itoo, Z. A. and Z. A. Reshi. 2013. The multifunctional role of ectomycorrhizal associations in forest ecosystem processes. *Bot. Rev.* 79:371–400.

Jacobson, K. M. 1997. Moisture and substrate determine VA-mycorrhizal fungal community distribution and structure in an arid grassland. *J. Arid Environ.* 35:59–75.

Janos, M. P. 1980. Mycorrhizae influence tropical succession. *Biotropica* 12 (suppl):56–64.

Jany, J-L., F. Martin, and J. Garbaye. 2003. Respiration activity of ectomycorrhizas from *Cenococcum geophilum* and *Lactarius* sp. in relation to soil water potential in five beech forests. *Plant Soil.* 255:487–494.

Jasper, D. A. 1994. Management of mycorrhizas in revegetation. *Plant Soil* 159:211–219.

Jayachandran, K., A. P. Schwab, and B. A. D. Hetrick. 1992. Mineralization of organic phosphorus by vesicular–arbuscular mycorrhizal fungi. *Soil Biol. Biochem.* 24:897–903.

Jeffries, P. and J. M. Barea. 1994. Biogeochemical cycling and arbuscular mycorrhizas in the sustainability of plant–soil systems. In: *Impact of Arbuscular Mycorrhizas on Sustainable Agriculture and Natural Ecosystems*, ed. S. Gianinazzi and H. Schüepp, 101–115. Basel, Switzerland: Birkhäuser Verlag.

Johnson, N. C. 1998. Responses of *Salsola kali* and *Panicum virgatum* to mycorrhizal fungi, phosphorus and soil organic matter: Implications for reclamation. *J. Appl. Ecol.* 35:86–94.

Johnson, N. C., D. Tilman, and D. Wedin. 1992. Plant and soil controls on mycorrhizal communities. *Ecology* 73:2034–2042.

Johnson-Green, P., N. C. Kenkel, and T. Booth. 2001. Soil salinity and arbuscular mycorrhizal colonization of *Puccinella nuttallana*. *Mycol. Res.* 105:1094–1110.

Joner, E. J. and A. Johansen. 2000. Phosphatase activity of external hyphae of two arbuscular mycorrhizal fungi. *Mycol. Res.* 104:81–86.

Jones, M. D., D. M. Durall, and P. B. Tinker. 1990. Phosphorus relationships and production of extramatrical hyphae by two types of willow ectomycorrhizal at different soil phosphorus levels. *New Phytol.* 115:259–267.

Jonsson, L., A. Dahlberg, M.-C. Nilsson, O. Zackrisson, and O. Karen. 1999a. Ectomycorrhizal fungal communities in late-successional Swedish boreal forests, and their composition following wildfire. *Mol. Ecol.* 8:205–215.

Jonsson, L., A. Dahlberg, M.-C. Nilsson, O. Karen, and O. Zackrisson. 1999b. Continuity of ectomycorrhizal fungi in self-regulating boreal *Pinus sylvestris* forests studied by comparing mycobiont diversity on seedlings and mature trees. *New Phytol.* 142:151–162.

Jonsson, L. M., M.-C. Nilsson, D. A. Wardle, and O. Zackrisson. 2001. Context dependent effects of ectomycorrhizal species richness on tree seedling productivity. *Oikos* 93:353–364.

Jumpponen, A. and L. M. Egerton-Warburton. 2005. Mycorrhizal fungi in successional environments: A community assembly model incorporating host plant, environmental, and biotic filters. In: *The Fungal Community: Its Organization and Role in the Ecosystem* (3rd Edition) ed. J. Dighton, J. F. White, and P. Oudemans, 139–168. Boca Raton, FL: Taylor & Francis.

Jumpponen, A., K. G. Mattson, and J. M. Trappe. 1998. Mycorrhizal functioning of *Phialocephala fortinii* with *Pinus contorta* on glacier forefront soil: Interactions with soil nitrogen and organic matter. *Mycorrhiza* 7:261–265.

Jumpponen, A., J. M. Trappe, and E. Cazares. 1999. Ectomycorrhizal fungi in Lyman Lake Basin: A comparison between primary and secondary successional sites. *Mycologia* 91:575–582.

Jumpponen, A., J. M. Trappe, and E. Cázares. 2002. Occurrence of ectomycorrhizal fungi on the forefront of retreating Lyman Glacier (Washington, USA) in relation to time since deglaciation. *Mycorrhiza* 12:43–49.

Juniper, S. and L. Abbott. 1993. Vesicular–arbuscular mycorrhizas and soil salinity. *Mycorrhiza* 4:45–57.

Kabir, Z., I. P. O'Halloran, and C. Hamel. 1996. The proliferation of fungal hyphae in soils supporting mycorrhizal and non-mycorrhizal plants. *Mycorrhiza* 6:477–480.

Kardol, P., T. M. Bezemer, and W. H. van der Heijden. 2006. Temporal variation in plant–soil feedback controls succession. *Ecol. Lett.* 9:1080–1088.

Kärenlampi, L. 1971. Studies on the relative growth rate of some fruiticose lichens. *Reports from the Kevo Subarctic Research Station* 7:33–39.

Karst, J., L. Marzcak, M. D. Jones, and R. Turkington. 2008. The mutualism–parasitism continuum in ectomycorrhizas: A quantitative assessment using meta-analysis. *Ecology* 89:1032–1042.

Kauppinen, M., K. Raveala, P. R. Wäli, and A. L. Routsalainen. 2014. Contrasting preferences of arbuscular mycorrhizal and dark septate fungi colonizing boreal and subarctic *Avenella flexuosa*. *Mycorrhiza* 24:171–177.

Kaye, J. P. and S. C. Hart. 1997. Competition for nitrogen between plants and soil microorganisms. *Trends Ecol. Evol.* 12:139–143.

Kennedy, A. H., D. Lee Taylor, and L. E. Watson. 2011. Mycorrhizal specificity in the fully mycoheterotrphic *Hexalectris* Raf. (Orchidaceae: Epidendroideae*). *Mol. Ecol.* 20:1303–1316.

Kerley, S. J. and D. J. Read. 1995. The biology of mycorrhizas in the Ericaceae: XVIII. Chitin degradation by *Hymenoscyphus ericae* and transfer of chitin-nitrogen to the host plant. *New Phytol.* 131:369–375.

Khan, A. G. 1993. Occurrence and importance of Mycorrhizae in aquatic trees of New South Wales, Australia. *Mycorrhiza* 3.:31.

Khan, A. G. and M. Belik. 1995. Occurrence and ecological significance of mycorrhizal symbiosis in aquatic plants. In: *Mycorrhiza: Structure, Function, Molecular Biology and Biotechnology*, ed. A. Varma and B. Hock, 537–543. Berlin: Springer-Verlag.

Koide, R. T., M. Li, J. Lewis, and C. Irby. 1988a. Role of mycorrhizal infection in the growth and reproduction of wild versus cultivated plants: I. Wild vs. cultivated oats. *Oecologia* 77:537–543.

Koide, R. T. and X. Lu. 1992. Mycorrhizal infection of wild oats: Nutritional effects on offspring growth and reproduction. *Oecologia* 90:218–226.

Koide, R., D. L. Shumway, and C. M. Stevens. 2000. Soluble carbohydrates of red pine (*Pinus resinosa*) mycorrhizas and mycorrhizal fungi. *Mycol. Res.* 104:834–840.

Koide, R. T., L. Suomi, and R. Berghage. 1998b. Tree-fungus interactions in ectomycorrhizal symbiosis. In *Phytochemical Signals and Plant–Microbe Interactions*, ed. Romeo, J. T., K. R. Downum, and R. Verpoorte Vol. 32, New York: Plenum Press.

Koide, R. T., L. Suomi, C. M. Stevens, and L. McCormick. 1998c. Interactions between needles of *Pinus resinosa* and ectomycorrhizal fungi. *New Phytol.* 140:539–547.

Kroehler, C. J., R. K. Antibus, and A. E. Linkins. 1988. The effects of organic and inorganic phosphorus concentration on the acid phosphatase activity of ectomycorrhizal fungi. *Can. J. Bot.* 66:750–756.

Laiho, O. and P. Mikola. 1964. Studies of the effects of some eradicants on mycorrhizal development in forest nurseries. *Acata For. Fenn.* 77:1–34.

Lange, O. L., B. Budel, A. Meyer, H. Zellner, and G. Zotz. 2000. Lichen carbon gain under tropical conditions: Water relations and $CO_2$ exchange of three *Leptogium* species of a lower montane rainforest in Panama. *Flora Morphol. Geobot. Oekophysiol.* 195:172–190.

Lange, O. L. and T. G. A. Green. 2003. Lichens show that fungi can acclimate their respiration to seasonal changes in temperature. *Oecologia* 142:11–19.

Lapeyrie, F., J. Ranger, and D. Vairelles. 1991. Phosphate-solubilizing activity of ectomycorrhizal fungi in vitro. *Can. J. Bot.* 69:342–346.

Largent, D. L., N. Sugihara, and C. Wishner. 1980. Occurrence of mycorrhizae on ericaceous and pyrolaceous plants in northern California. *Can. J. Bot.* 59:2274–2279.

Last, F. T., J. Dighton, and P. A. Mason. 1987. Successions of sheathing mycorrhizal fungi. *Trends Ecol. Evol.* 2:157–161.

Leake, J. R. 2001. Is diversity of ectomycorrhizal fungi important for ecosystem function? *New Phytol.* 152:1–8.

Leake, J., D. Johnson, D. Donnelly, G. Muckle, L. Boddy, and D. J. Read. 2004. Networks of power and influence: The role of mycorrhizal mycelium in controlling plant communities and agroecosystem functioning. *Can. J. Bot.* 82:1016–1045.

Leake, J. R. and W. Miles. 1996. Phosphodiesters as mycorrhizal P sources: I. Phosphodiesterase production and utilization of DNA as a phosphorus source by the ericoid mycorrhizal fungus *Hymenoscyphus ericae*. *New Phytol.* 132:435–443.

Leake, J. R. and D. J. Read. 1989. The biology of mycorrhiza in the Ericaceae: XIII. Some characteristics of the extracellular proteinase activity of the ericoid endophyte *Hymenoscyphus ericae*. *New Phytol.* 112:69–76.

Leake, J. R. and D. J. Read. 1990a. Chitin as a nitrogen source for mycorrhizal fungi. *Mycol. Res.* 94:993–995.

Leake, J. R. and D. J. Read. 1990b. Proteinase activity in mycorrhizal fungi: I. The effect of extracellular pH on the production and activity of proteinase by the ericoid endophytes of soils of contrasted pH. *New Phytol.* 115:243–250.

Leake, J. R. and D. J. Read. 1991. Experiments with ericoid mycorrhizae. In *Methods in Microbiology 23*, ed. J. R. Norris, D. J. Read, and A. K. Varma, 435–459. London: Academic Press.

Leake, J. R. and D. J. Read. 1997. Mycorrhizal fungi in terrestrial habitats. In: *The Mycota IV*, ed. D. T. Wicklow and B. Soderstrom, 281–301. Berlin: Springer-Verlag.

Lee, J. A. 1999. The calcicole–calcifuge problem revisited. *Adv. Bot. Res.* 29:1–30.

Lekberg, Y. and R. T. Koide. 2005. Is plant performance limited by abundance of arbuscular mycorrhizal fungi? A meta-analysis of studies published between 1988 and 2003. *New Phytol.* 168:189–204.

Lewis, J. D. and R. T. Koide. 1990. Phosphorus supply, mycorrhizal infection and plant offspring vigor. *Funct. Ecol.* 4:695–705.

Leyval, C. and J. Berthelin. 1983. Effets rhizopheriques de plantes indicatrices de grands types de pedogenese sur quelques groupes bacteriens modifiant l`etat de mineraux. *Rev. Ecol. Sol.* 20:191–206.

Li, H.-Y., D.-W. Li, C.-M. He, Z.-P. Zhou, T. Mei, and H.-M. Xu. 2012. Diversity and heavy metal tolearance of endophytic fungi from six dominant plant species in a Pb–Zn mine wasteland in China. *Fung. Ecol.* 5:309–313.

Li, T., G. Lin, X. Zhang, Y. Chen, S. Zhang, and B. Chen. 2014. Relative importance of an arbuscular mycorrhizal fungus (*Rhizophagus intrardices*) and root hairs in plant drought tolerance. *Mycorrhiza* 24:595–602.

Lindahl, B., J. Stenlid, S. Olsson, and R. Finlay. 1999. Translocation of $^{32}$P between interacting mycelia of a wood-decomposing fungus and ectomycorrhizal fungi in microcosm systems. *New Phytol.* 144:183–193.

Lodge, D. J. and T. R. Wentworth. Negative associations among VA-mycorrhizal fungi and some ectomycorrhizal fungi inhabiting the same root system. *Oikos* 57, 347–356. 1990.

Lopez-García, Á., C. Azcón-Aguilar, and J. M. Barea. 2014. The interactions between plant life form and fungal traits of arbuscular mycorrhizal fungi determine the symbiotic community. *Oecologia* 176:1075–1086.

MacFall, J., S. A. Slack, and J. Iyer. 1991. Effects of *Hebeloma arenosa* and phosphorus fertility on root acid phosphatase activity of red pine (*Pinus resinosus*) seedlings. *Can. J. Bot.* 69:380–385.

Mahmood, K., K. A. Malik, K. H. Sheikh, and M. A. K. Lodhi. 1989. Allelopathy in saline agricultural land: Vegetation successional changes and patch dynamics. *J. Chem. Ecol.* 15:565–579.

Mansfield, S. D. and F. Bärlocher. 1993. Seasonal variation of fungal biomass in the sediment of a salt marsh in New Brunswick. *Microb. Ecol.* 26:37–45.

Marschner, H. and B. Dell. 1994. Nutrient uptake in mycorrhizal symbiosis. *Plant Soil* 159:89–102.

McCann, K. S. 2000. The diversity–stability debate. *Nature* 405:228–233.

McHugh, J. M. and J. Dighton. 2004. Influence of mycorrhizal inoculation, inundation period, salinity and phosphorus availability on the growth of two salt marsh grasses, *Spartina alterniflora* Lois. and *Spartina cynosuroides* (L.) Roth. In nursery systems. *Rest. Ecol.* 12:533–545.

Mena-Violante, H., O. Ocampo-Jiménez, L. Dendooven, G. Martinéz-Soto, J. González-Castañeda, F. T. Davies Jr., and V. Olalde-Portuga. 2006. Arbuscular mycorrhizal fungi enhance fruit growth and quality of chile ancho (*Capsicum annum* L cv San Luis) plants exposed to drought. *Mycorrhiza* 16:261–267.

Merryweather, J. and A. Fitter. 1996. Phosphorus nutrition of an obligately mycorrhizal plant treated with the fungicide benomyl in the field. *New Phytol.* 132:307–311.

Merryweather, J. and A. H. Fitter. 1995a. Arbuscular mycorrhiza and phosphorus as controlling factors in the life history of *Hyacinthoides non-scripta* (L.) Chouard ex Rothm. *New Phytol.* 129:629–636.

Merryweather, J. and A. H. Fitter. 1995b. Phosphorus and carbon budgets: Mycorrhizal contribution in *Hyacinthoides non-scripta* (L.) Chouard ex Rothm. under natural conditions. *New Phytol.* 129:619–627.

Michelsen, A. and S. Rosendahl. 1990. The effect of VA mycorrhizal fungi, phosphorus and drought stress on the growth of *Acacia nilotica* and *Leucana leucocephala* seedlings. *Plant Soil* 124, 7–13.

Miller, S. L. 1995. Functional diversity in fungi. *Can. J. Bot.* 73(Suppl 1):S50–S57.

Miller, S. P. 2000. Arbuscular mycorrhizal colonization of semi-aquatic grasses along a wide hydrologic gradient. *New Phytol.* 145:145–155.

Mitchell, D. T. and D. J. Read. 1981. Utilization of inorganic and organic phosphate by the mycorrhizal endophytes of *Vaccinium macrocarpon* and *Rhododendron ponticum*. *Trans. Br. Mycol. Soc.* 76(2):255–260.

Mogge, B., C. Loferer, R. Agerer, and P. Hutzler. 2000. Bacterial community structure and colonization patterns of *Fagus sylvatica* L. ectomycorrhizospheres as determined by fluorescence *in situ* hybridization and confocal laser scanning microscopy. *Mycorrhiza* 9:271–278.

Morte, A., C. Lovisolo, and A. Schubert. 2000. Effect of drought stress on growth and water relations of the mycorrhizal association *Helianthemum almeriense–Terfezia claveryi*. *Mycorrhiza* 10:115–119.

Mousain, D. and L. Salsac. 1986. Utilisation du phytate et activites phosphatases acides chez *Pisolithus tinctorius,* basidiomycete mycorhizien. *Physiol. Veg.* 24:193–200.

Mukerji, K. G. 1996. *Concepts in Mycorrhizal Research.* Dordrecht: Kluwer Academic Publishers.

Myers, M. D. and J. R. Leake. 1996. Phosphodiesters as mycorrhizal P sources: II. Ericoid mycorrhiza and the utilization of nuclei as phosphorus and nitrogen source by *Vaccinium macrocarpon*. *New Phytol.* 132:445–452.

Nagabhyru, P., R. D. Dinkins, C. L. Wood, C. W. Bacon, and C. L. Schardl. 2013. Tall fescue endophyte effects on tolerance to water-deficit stress. *BMC Plant Biol.* 13:127–144.

Newberry, D. McC., I. J. Alexander, and J. A. Rother. 1997. Phosphorus dynamics in a lowland African rain forest: The influence of the ectomycorrhizal trees. *Ecol. Monogr.* 67:367–409.

Newman, E. I. and W. R. Eason. 1989. Cycling of nutrients from dying roots to living plants, including the role of mycorrhizas. *Plant Soil* 115:211–215.

Norton, J. M. and M. K. Firestone. 1996. N dynamics in the rhizosphere of *Pinus ponderosa* seedlings. *Soil Biol. Biochem.* 28:351–362.

Nye, P. H. and P. B. Tinker. 1977. *Solute Movement in the Soil–Root System.* University of California Press.

Ohtonen, R., H. Fritze, T. Pennanen, A. Jumpponen, and J. M. Trappe. 1999. Ecosystem properties and microbial community changes in primary succession on a glacial forefront. *Oecologia* 119:239–246.

Orlovich, D. A., S. J. Draffin, R. L. Daly, and S. L. Stephenson. 2013. Piracy in the high trees: Ectomycorrhizal fungi form an aerial 'canopy soil' microhabitat. *Mycologia* 105:52–60.

Owusu-Bennoah, E. and A. Wild. 1979. Autoradiography of the depletion zone of phosphate around onion roots in the presence of vesicular arbuscular mycorrhiza. *New Phytol.* 82:133–140.

Padgett, D. E. and D. A. Celio. 1990. A newly discovered role for aerobic fungi in anaerobic salt marsh soils. *Mycologia* 82:791–794.

Palmer, J. G., O. K. Miller, and C. Gruhn. 1994. Fruiting of ectomycorrhizal basidiomycetes on unburned and prescribed burned hard-pine/hardwood plots after drought-breaking rainfalls on the Allegheny Mountains of southwestern Virginia. *Mycorrhiza* 4:93–104.

Palmqvist, K. and B. Sundberg. 2000. Light use efficiency of dry matter gain in five macrolichens: Relative impact of microclimate conditions and species-specific traits. *Plant Cell Environ.* 23:1–14.

Pankow, W., T. Boller, and A. Wimken. 1991. The significance of mycorrhizas for protective ecosystems. *Experientia* 47:391–394.

Parladé, J. and I. F. Alvarez. 1993. Coinoculation of aseptically grown Douglas fir with pairs of ectomycorrhizal fungi. *Mycorrhiza* 3:93.

Parrondo, R. T., J. G. Gosselink, and C. S. Hopkinson. 1978. Effects of salinity and drainage on the growth of three salt marsh grasses. *Bot. Gazette* 139:102–107.

Pearson, V. and D. J. Read. 1975. The physiology of the mycorrhizal endophyte of *Calluna vulgaris. Trans. Br. Mycol. Soc.* 64:1–7.

Peck, J.-L. E., J. Ford, B. McCune, and B. Daly. 2000. Tethered transplants for estimating biomass growth rates of the arctic lichen *Masonhalea richardsonii. Bryologist* 103:499–454.

Peoples, M. B. and E. T. Craswell. 1992. Biological nitrogen fixation: Investments, expectations and actual contribution to agriculture. *Plant Soil* 141:13–39.

Perry, D. A., H. Margolis, C. Choquette, R. Molina, H. Marschner, and J. M. Trappe. 1989. Ectomycorrhizal mediation of competition between coniferous tree species. *New Phytol.* 112:501–511.

Peterson, R. L. and M. L. Farquhar. 1994. Mycorrhizas—Integrated development between roots and fungi. *Mycologia* 86:311–326.

Peterson, R. L., H. B. Massicott, and L. H. Melville. 2004. *Mycorrhizas: Anatomy and Cell Biology*. Ottawa: NRC–CNRC Research Press, CABI. pp. 173.

Pietkäinen, A. and M–M. Kytöviita. 2007. Defoliation changes mycorrhizal benefit and competitive interactions between seedlings and adult plants. *J. Ecol.* 95:639–647.

Pirttilä, A. M. and P. R. Wäli. 2009. Conifer endophytes. In: *Defensive Mutualism in Microbial Symbiosis*, ed. J. F. White and M. Torres, 235–246. Boca Raton, FL: Taylor & Francis.

Polglase, P. J., P. M. Attiwill, and M. A. Adams. 1992. Nitrogen and phosphorus cycling in relation to stand age of *Eucalyptus regnans* F. Muell: III. Phosphatase activity and pools of labile soil P. *Plant Soil* 142:177–185.

Popay, P. 2009. Insect herbivory and defensive mutualisms between plants and fungi. In: *Defensive Mutualism in Microbial Symbiosis*, ed. J. F. White and M. Torres, 347–366. Boca Raton: CRC Taylor & Francis.

Porcel, R. and J. M. Ruiz-Lozano 2004. Arbuscular mycorrhizal influence on leaf water potential, solute accumulation, and oxidative stress in soybean plants subjected to drought stress. *J. Exp. Bot.* 55:1743–1750.

Rangeley, A., M. J. Daft, and P. Newbold. 1982. The inoculation of white clover with mycorrhizal fungi in unsterile hill soil. *New Phytol.* 92:89–102.

Rasmann, S., T. L. Bauerle, K. Poveda, and R. Vannette. 2011. Evolutionary ecology of plant defenses: Predicting root defense against herbivores during succession. *Funct. Ecol.* 25:368–379.

Rasmussen, H. N. and D. F. Wigham. 1994. Seed ecology of dust seeds *in situ*: A new study technique and its application in terrestrial orchids. *Am. J. Bot.* 80:1374–1378.

Rayner, A. D. M. 1991. The challenge of the individualistic mycelium. *Mycologia* 83:48–71.

Rayner, A. D. M., G. S. Griffith, and H. G. Wildman. 1994. Induction of metabolic and morphogenetic changes during mycelial interactions among species of higher fungi. *Trans. Biochem. Soc.* 22:389–394.

Rayner, A. D. M., K. A. Powell, W. Thompson, and D. H. Jennings. 1985. Morphogenesis of vegetative organs. In: *Developmental Biology of Higher Fungi*, ed. D. Moore, L. A. Casselton, D. A. Wood, and J. C. Frankland, 249–279. Cambridge: Cambridge University Press.

Read, D. J. 1991a. Mycorrhizas in ecosystems. *Experientia* 47:376–391.

Read, D. J. 1991b. Mycorrhizas in ecosystems—Nature's response to the "Law of the Minimum." In *Frontiers in Mycology*, ed. D. L. Hawksworth. Wallingford, UK: CAB International.

Read, D. J. 1996. The structure and function of the ericoid mycorrhizal root. *Ann. Bot.* 77:365–376.

Read, D. J., R. Francis, and R. D. Finlay. 1985. Mycorrhizal mycelia and nutrient cycling in plant communities. In: *Ecological Interactions in Soil, Plants, Microbes and Animals*, ed. A. H. Fitter, 193–213. Oxford UK: Blackwell Scientific.

Read, D. J. and S. Kerley. 1995. The status and function of ericoid mycorrhizal systems. In: *Mycorrhiza: Structure, Function, Molecular Biology and Biochemistry*, ed. A. Varma and B. Hock, 499–520. Berlin: Springer Verlag.

Read, D. J., H. Kianmehr, and A. Malibari. 1977. The biology of mycorrhiza of *Helianthemum* Mill. *New Phytol.* 78:305–312.

Read, D. J., J. R. Leake, and A. R. Langdale. 1989. The nitrogen nutrition of mycorrhizal fungi and their host plants. In: *Nitrogen, Phosphorus and Sulphur Cycling in Temperate Forest Ecosystems*, ed. L. Boddy, R. Marchant, and D. J. Read, 181–204. Cambridge: Cambridge University Press.

Read, D. J., D. H. Lewis, A. Fitter, and I. J. Alexander. 1992. *Mycorrhizas in Ecosystems*. Wallingford, U.K.: CAB International.

Reddy, M. S. and K. Natarajan. 1997. Coinoculation efficiency of ectomycorrhizal fungi on *Pinus patula* seedlings in a nursery. *Mycorrhiza* 7:133–138.

Repáč, I. 1996a. Effects of forest litter on mycorrhiza formation and growth of container-grown Norway spruce (*Picea abies* (L.) Karst.) seedlings. *Lesnictvi Forestry* 42:317–324.

Repáč, I. 1996b. Inoculation of *Picea abies* (L.) Karst., seedlings with vegetative inocula of ectomycorrhizal fungi *Suillus bovinus* (L.: Fr.) O. Kuntze and *Inocybe lacera* (Fr.) Kumm. *New For.* 12:41–54.

Rillig, M. C., S. Wendt, J. Antonovics et al. 2014. Interactive effects of root endophytes and arbuscular mycorrhizal fungi on an experimental plant community. *Oecologia* 174:263–270.

Ritz, K. 1995. Growth responses of some fungi to spatially heterogeneous nutrients. *FEMS Microbiol. Ecol.* 16:269–280.

Ritz, K., J. Dighton, and K. E. Giller. 1994. *Beyond the Biomass: Compositional and Functional Analysis of Soil Microbial Communities*. Chichester, U.K.: John Wiley & Sons.

Rodriguez, R. J., C. Woodward, Y-O. Kim, and R. S. Redman. 2009. Habitat-adapted symbiosis as a defense against abiotic and biotic stress. In: *Defensive Mutualism in Microbial Symbiosis*, ed. J. F. White and M. Torres, 335–346. Boca Raton, FL: Taylor & Francis.

Rousseau, J. V. D., D. M. Sylvia, and A. J. Fox. 1994. Contribution of ectomycorrhiza to the potential nutrient-absorbing surface of pine. *New Phytol.* 128:639–644.

Rozema, J., W. Arp, J. van Diggelen, M. van Esbroek, R. Broekman, and H. Punte. 1986. Occurrence and ecological significance of vesicular arbuscular mycorrhiza in the salt marsh environment. *Acta Bot. Neer.* 35:457–467.

Rudgers, J. A. and K. Clay. 2007. Endophyte symbiosis with tall fescue: How strong are the impacts on communities and ecosystems? *Fung. Biol. Rev.* 21:107–124.

Ruotsalainen, A. L. and M. M. Kytöviita. 2004. Mycorrhiza does not alter low temperature impact on *Gnaphalium norvegicum*. *Oecologia* 140:226–233.

Sagara, N. 1995. Association of ectomycorrhizal fungi with decomposed animal wastes in forest habitats: A cleaning symbiosis? *Can. J. Bot.* 73 (Suppl. 1): S1423–S1433.

Sánches-Díaz, M. and M. Honrubia. 1994. Water relations and alleviation of drought stress in mycorrhzal plants. In: *Impact of Arbuscular Mycorrhizas on Sustainable Agriculture and Natural Ecosystems*, ed. S. Gianinazzi and H. Schüepp, 167–178. Basel, Switzerland: Birkhäuser Verlag.

Sanders, I. J. and A. H. Fitter. 1992a. The ecology and functioning of vesicular–arbuscular mycorrhizas in co-existing grassland species: I. Seasonal patterns of mycorrhizal occurrence and morphology. *New Phytol.* 120:517–524.

Sanders, I. J. and A. H. Fitter. 1992b. The ecology and functioning of vesicular–arbuscular mycorrhizas in co-existing grassland species: II. Nutrient uptake and growth of vesicular–arbuscular mycorrhizal plants in a semi-natural grassland. *New Phytol.* 120:525–533.

Schrey, S. D., M. Schellhammer, M. Ecke, R. Hampp, and M. T. Tarkka. 2005. Mycorrhizal helper bacterium *Streptomyces* AcH 505 induces differential gene expression in the ectomycorrhizal fungus *Amanita muscaria*. *New Phytol.* 168:205–216.

Schwartz, M. W., C. A. Brigham, J. D. Hoeksema, K. G. Lyons, M. H. Mills, and P. J. van Mantgem. 2000. Linking biodiversity to ecosystem function: Implications for conservation ecology. *Oecologia* 122:297–305.

Semones, S. W. and D. R. Young. 1995. VAM association in the shrub *Myrica cerifera* on a Virginia, USA barrier island. *Mycorrhiza* 5:423–429.

Sengupta, A. and S. Chaudhuri. 1990. Vesicular arbuscular mycorrhiza (VAM) in pioneer salt marsh plants of the Ganges river delta in West Bengal (India). *Plant Soil* 122:111–113.

Shaw, G. and D. J. Read. 1989. The biology of mycorrhiza in the Ericaceae XIV. Effects of iron and aluminum on the activity of acid phosphatase in the ericoid endophyte *Hymenoscyphus ericae* (Read) Korf and Kernan. *New Phytol.* 113:529–533.

Shaw, T. M., J. Dighton, and F. E. Sanders. 1995. Interactions between ectomycorrhizal and saprotrophic fungi on agar and in association with seedlings of lodgepole pine (*Pinus contorta*). *Mycol. Res.* 99:159–165.

Shi, L. M. Guttenberger, I. Kottke, and R. Hampp. 2002. The effect of drought on mycorrhizas of beech (*Fagus sylvatica* L.): changes in community structure, and the content of carbohydrates and nitrogen storage bodies of the fungus. *Mycorrhiza* 12:303–311.

Sieber, T. N. 2007. Endophytic fungi in forest trees: Are they mutualists. *Fung. Biol. Rev.* 21:75–89.

Sikes, B. A., H. Maherali, and J. N. Klironomos. 2012. Arbuscular mycorrhizal fungal communities change among three stages of primary sand dune succession but do not alter plant growth. *Oikos* 212:1791–1800.

Sillett, S. C., B. McCune, J-L. E. Peck, and T. R. Rambo. 2000. Four years of epiphyte colonization in Douglas-fir forest canopies. *Bryologist* 103:661–669.

Simard, S. W., K. J. Beiler, M. A. Bingham, J. R. Deslippe, L. J. Philip, and F. P. Teste. 2012. Mycorrhizal networks: Mechanisms, ecology and modelling. *Fung. Biol. Rev.* 26:39–60.

Simard, S. W., M. D. Jones, D. M. Durall, D. A. Perry, D. D. Myrold, and R. Molina. 1997a. Reciprocal transfer of carbon isotopes between ectomycorrhizal *Betula payrifrea* and *Pseudotsuga menziesii*. *New Phytol.* 137:529–542.

Simard, S. W., D. A. Perry, M. D. Jones, D. D. Myrold, D. M. Durall, and R. Molina. 1997b. Net transfer of carbon between ectomycorrhizal tree species in the field. *Nature* 338:579–582.

Simard, S. W., D. A. Perry, J. E. Smith, and R. Molina. 1997c. Effects of soil trenching on occurrence of ectomycorrhizas of *Pseudotsuga menziesii* seedlings grown in mature forests of *Betula papyrifera* and *Pseudotsuga menziesii*. *New Phytol.* 136:327–340.

Singer, R. and I. de J. Da Silva Araujo. 1979. Litter decomposition and ectomycorrhiza in Amazonian forests: I. A comparison of litter decomposition and ectomycorrhizal Basidiomycetes in latosol-terra-firme rain forest and white podsol campinarana. *Acta Amaz.* 9:25–41.

Singh, S. and K. K. Kapoor. 1998. Effects of inoculation of phosphate-solubilizing micro-organisms and arbuscular mycorrhizal fungus on mungbean grown under natural soil conditions. *Mycorrhiza* 7:149–153.

Sinsabaugh, R. L. and M. A. Liptak. 1997. Enzymatic conversion of plant biomass. In: *The Mycota IV*, ed. D. Wicklow and B. Soderstrom, 347–357. Berlin: Springer-Verlag.

Slippers, B. and M. J. Wingfield. 2007. Botrysphaeriaceae as endophytes and latent pathogens of woody plants: Diversity, ecology and impact. *Fung. Biol. Rev.* 21:90–106.

Smith, J. E., R. Molina, and D. A. Perry. 1995. Occurrence of ectomycorrhizas on ericaceous and coniferous seedlings grown in soils from the Oregon Coast Range. *New Phytol.* 129:73–81.

Smith, S. E. and D. J. Read. 1997. *Mycorrhizal Symbiosis*. San Diego: Academic Press.

Solaiman, M. Z. and H. Hirata. 1998. *Glomus*-wetland rice mycorrhizas influenced by nursery inoculation techniques under high fertility soil conditions. *Biol. Fertil. Soils* 27:92–96.

Solhaug, K. A. and Y. Gauslaa. 1996. Parietin, a photoprotective secondary product of the lichen *Xanthoria parietina*. *Oecologia* 108:412–418.

Solhaug, K. A., Y. Gauslaa, L. Nybakken, and W. Bilger. 2003. UV-induction of sun-screening pigments in lichens. *New Phytol.* 158:81–100.

Song, M., X. Li., K. Saikkonen, C. Li, and Z. Nan. 2015. An asexual *Epichloë* endophyte enhances waterlogging tolerance of *Hordeum brevisubulatum*. *Fung. Ecol.* 13:44–52.

Song, Y. Y., R. S. Zeng, J. F. Xu, J. Li, X. Shen, and W. G. Yihdego. 2010. Interplant communication of tomato plants through underground common mycorrhizal networks. *PLoS One* 5(10): e13324.

Stanley, M. R., R. T. Koide, and D. L. Schumway. 1993. Mycorrhizal symbiosis increases growth, reproduction and recruitment of *Abutidon theophrasti* medic. in the field. *Oecologia* 94:30–35.

Stanton, D. E., J. H. Chávez, L. Villegas et al. 2014. Epiphytes improve host plant water use by microenvironment modification. *Funct. Ecol.* 28:1274–1283.

Straker, C. J. and D. T. Mitchell. 1985. The characterization and estimation of polyphosphates in endomycorrhizal of the Ericeae. *New Phytol.* 99:431–440.

Streitwolf-Engel, R., M. G. A. van der Heijden, A. Wiemken, and I. R. Sanders. 2001. The ecological significance of arbuscular mycorrhizal fungal effects on clonal reproduction in plants. *Ecology* 82:2846–2859.

Stribley, D. P. and D. J. Read. 1980. The biology of mycorrhiza in the Ericaceae: VII. The relationship between mycorrhizal infection and the capacity to utilize simple and complex organic nitrogen sources. *New Phytol.* 86:365–371.

Ström, L. 1997. Root exudation of organic acids: Importance to nutrient availability and the calcifuge and calcicole behaviour of plants. *Oikos* 80:459–466.

Suryanaraynan, T. S. 2013. Endophyte research: Going beyond isolation and metabolite documentation. *Fung. Ecol.* 6:561–568.

Swaty, R. L., R. J. Deckert, T. G. Whitham, and C. A. Ghering. 2004. Ectomycorrhizal abundance and community composition shifts with drought: Predictions from tree rings. *Ecology* 85:1072–1084.

Tadych, M., M. S. Torres, and J. F. White. 2009. Diversity and ecological roles of Clavicipitaceous endophytes of grasses. In: *Defensive Mutualism in Microbial Symbiosis*, ed. J. F. White and M. Torres, 247–256. Boca Raton, FL: Taylor & Francis.

Taylor, A. F. S. and I. Alexander. 2005. The ectomycorrhizal symbiosis: Life in the real world. *Mycologist* 19:102–112.

Tedersoo, L., T. Naadel, M. Bahram et al. 2012. Enzymatic activities and stable isotope patterns of ectomycorrhizal fungi in relation to phylogeny and exploration types in an afrotropical forest. *New Phytol.* 195:832–843.

Terhonen, E., S. Keriö, H. Sun, and F. O. Asiegbu. 2014. Endophytic fungi of Norway spruce roots in boreal pristine mire, drained peatland and mineral soil and their inhibitory effect on *Heterobasidion parviporum in vitro*. *Fung. Ecol.* 9:17–26.

Termorshuizen, A. J. and A. P. Schaffers. 1989. The relation in the field between fruitbodies of mycorrhizal fungi and their mycorrhizas. *Agric. Ecosyst. Environ.* 28:509–512.

Teste, F. P., S. W. Simard, D. M. Durall, R. D. Guy, M. D. Jones, and A. L. Schoomaker. 2009. Access to mycorrhizal networks and tree roots: Importance for seedling survival and resource transfer. *Ecology* 90:2802–2822.

Tibbett, M. 2000. Roots, foraging and the exploitation of soil nutrient patches: The role of mycorrhizal symbionts. *Funct. Ecol.* 14:397–399.

Tibbett, M., K. Grantham, F. E. Sanders, and J. W. G. Cairney. 1998a. Induction of cold active acid phosphomonoesterase activity at low temperature in psychotrophic ectomycorrhizal *Hebeloma* spp. *Mycol. Res.* 102:1533–1539.

Tibbett, M., F. E. Sanders, and J. W. G. Cairney. 1998b. The effect of temperature and inorganic phosphorus supply on growth and acid phosphatase production in arctic and temperate strains of ectomycorrhizal *Hebeloma* spp., in axenic culture. *Mycol. Res.* 102:129–135.

Tibbett, M., F. E. Sanders, and J. W. G. Cairney. 2002. Low-temperature-induced changes in trehalose, mannitol and arabitol associated with enhanced tolerance to freezing in ectomycorrhizal basidiomycetes (*Hebeloma* spp.). *Mycorrhiza* 12:249–255.

Tibbett, M., F. E. Sanders, J. W. G. Cairney, and J. R. Leake. 1999. Temperature regulation of extracellular proteases in ectomycorrhizal fungi (*Hebeloma* spp.) grown in axenic culture. *Mycol. Res.* 103:707–714.

Tibbett, M., F. E. Sanders, S. J. Minto, M. Dowell, and J. W. G. Cairney. 1998c. Utilization of organic nitrogen by ectomycorrhizal fungi (*Hebeloma* spp.) of arctic and temperate origin. *Mycol. Res.* 102:1525–1532.

Tilman, D. 1993. Species richness of experimental productivity gradients: How important is colonization limitation? *Ecology* 74:2179–2191.

Tilman, D. 1997. Community invasibility, recruitment limitation, and grassland biodiversity. *Ecology* 78:81–92.

Tilman, D. 1999. The ecological consequences of changes in biodiversity: A search for general principles. *Ecology* 80:1455–1474.

Tilman, D., D. Wedin, and J. Knops. 1996. Productivity and sustainability influenced by biodiversity in grassland ecosystems. *Nature* 379:718–720.

Torres, M. S., J. F. White, X. Zhang, D. M. Hinton, and C. W. Bacon. 2012. Endophyte-mediated adjustments in host morphology and physiology and effects on host fitness traits in grasses. *Fung. Ecol.* 5:322–330.

Torpy, E. R., D. A. Morrison, and B. J. Bloomfield. 1999. The influence of fire frequency on arbuscular mycorrhizal colonization in the shrub *Dillwynia retorta* (Wendland) Druce (Fabiaceae). *Mycorrhiza* 8:289–296.

Valiela, I. and J. M. Teal. 1974. Nutrient limitation in salt marsh vegetation. In *Ecology of Halophytes*. New York: Academic Press.

van der Heijden, E. W. 2001. Differential benefits of arbuscular mycorrhizal and ectomycorrhizal infection of *Salix repens*. *Mycorrhiza* 10:185–193.

van der Heijden, M. G. A., T. Boller, A. Wiemken, and I. R. Sanders. 1998. Different arbuscular mycorrhizal fungal species are potential determinants of plant community structure. *Ecology* 79:2082–2091.

van der Heijden, M. G. A. and T. R. Horton. 2009. Socialism in soil? The importance of mycelial fungal networks for facilitation in natural ecosystems. *J. Ecol.* 97:1139–1150.

van Duin, W. E., J. Rozema, and W. H. O. Ernst. 1989. Seasonal and spatial variation in the occurrence of vesicular–arbuscular (VA) mycorrhiza in salt marsh plants. *Agric. Ecosyst. Environ.* 29:107–110.

Varma, A. and B. Hock, Eds. 1995. *Mycorrhiza: Structure, Function, Molecular Biology and Biotechnology*. Berlin: Springer-Verlag.

Villeneuve, N., M. M. Grandtner, and J. A. Fortin. 1991a. The coenological organization of ectomycorrhizal macrofungi in the Laurentide mountains of Quebec. *Can. J. Bot.* 69:2215–2224.

Villeneuve, N., F. Le Tacon, and D. Bouchard. 1991b. Survival of inoculated *Laccaria bicolor* in competition with native ectomycorrhizal fungi and effects on the growth of outplanted Douglas fir seedlings. *Plant Soil* 135:95–107.

Visser, S. 1995. Ectomycorrhizal fungal succession in jack pine stands following wildfire. *New Phytol.* 129:389–401.

Vogt, K. A., C. C. Grier, R. L. Edmonds, and C. E. Meier. 1982. Mycorrhizal role in net primary production and nutrient cycling in *Abies amabilis* (Dougl.) Forbes ecosystems in western Washington. *Ecology* 63:370–380.

von Alten, H., A. Lindemann, and F. Schonbeck. 1993. Stimulation of vesicular–arbuscular mycorrhiza by fungicides or rhizosphere bacteria. *Mycorrhiza* 2(4):167.

Walker, J. F., O. K. Miller, T. Lei, S. Semones, E. Nilsen, and B. D. Clinton. 1999. Suppression of ectomycorrhizae on canopy tree seedlings in *Rhododendron maximum* L. (Ericaceae) thickets in the southern Appalachians. *Mycorrhiza* 9:49–56.

Wallander, H., K. Arnebrant, F. Ostrand, and O. Karen. 1997. Uptake of [15]N-labelled alanine, ammonium and nitrate in *Pinus sylvestris* L. ectomycorrhiza growing in forest soil treated with nitrogen, sulphur or lime. *Plant Soil* 195:329–338.

Wallander, H. and J. E. Nylund. 1991. Effects of excess nitrogen on carbohydrate concentration and mycorrhizal development in *Pinus sylvestris* L. seedlings. *New Phytol.* 119:405–411.

Walter, L. E. F., D. C. Hartnett, A. D. Hetrick, and A. P. Schwab. 1996. Interspecific nutrient transfer in a tallgrass prairie plant community. *Am. J. Bot.* 83:180–184.

Watkinson, A. R. and R. P. Freckleton. 1997. Quantifying the impact of arbuscular mycorrhiza on plant competition. *J. Ecol.* 85:541–545.

Wells, J. M. and L. Boddy. 1990. Wood decay, and phosphorus and fungal biomass allocation, in mycelial cord systems. *New Phytol.* 116:285–295.

Wells, J. M. and L. Boddy. 1995. Phosphorus translocation by saprotrophic basidiomycete mycelial cord systems on the floor of a mixed deciduous woodland. *Mycol. Res.* 99:977–999.

Went, F. W. and N. Stark. 1968. The biological and mechanical role of soil fungi. *Proc. Natl. Acad. Sci. U.S.A.* 60:497–504.

White, J. F. and C. W. Bacon 2012. The secret world of endophytes in perspective. *Fung. Ecol.* 5:287–288.

White, J. A. and I. Charvat. 1999. The mycorrhizal status of an emergent aquatic, *Lythrum salicaria* L., at different levels of phosphorus availability. *Mycorrhiza* 9:191–197.

Wiewióra, B., G. Żurek, and M. Żurek. 2015. Endophyte-mediated disease resistance in wild populations of perennial ryegrass (*Lolium perenne*). *Fung. Ecol.* 15:1–8.

Wu, B., K. Isobe, and R. Ishii. 2004. Arbuscular mycorrhizal colonization of the dominant plant species in primary successional volcanic deserts and the Southeastern slope of Mount Fuji. *Mycorrhiza* 14:391–395.

Wu, B., K. Nara, and T. Hogetsu. 2002. Spatiotemporal transfer of carbon-14-labelled photosynthate from ectomycorrhizal *Pinus densiflora* seedlings to extraradical mycelia. *Mycorrhiza* 12:83–88.

Xiao, G. and S. M. Berch. 1999. Organic nitrogen use by salal ericoid mycorrhizal fungi from northern Vancouver Island and impacts on growth in vitro of *Gautheria shallon*. *Mycorrhiza* 9:145–149.

Yamada, A. and K. Katsuya. 2001. The disparity between the number of ectomycorrhizal fungi and those producing fruit bodies in a *Pinus desniflora* stand. *Mycol. Res.* 105:957–965.

Yamanaka, T. 1999. Utilization of inorganic and organic nitrogen in pure cultures by saprotrophic and ectomycorrhizal fungi producing sporophores on urea-treated forest floor. *Mycol. Res.* 103:811–816.

Zak, B. and D. H. Marx. 1964. Isolation of mycorrhizal fungi from roots on individual slash pines. *For. Sci.* 10:214–222.

Zettler, L. W. and T. McInnes. 1992. Propagation of *Plantathera integrilabia* (Correll) Luer, an endangered terrestrial orchid, through symbiotic seed germination. *Lindleyana* 7:154–161.

Zhu, H., B. P. Dancik, and K. O. Higginbotham. 1994. Regulation of extracellular proteinase production in an ectomycorrhizal fungus *Hebeloma crustuliniforme*. *Mycologia* 82:227–234.

Zhu, W. and J. G. Ehrenfeld. 1996. The effects of mycorrhizal roots on litter decomposition, soil biota, and nutrients in a spodosolic soil. *Plant Soil* 179:109–118.

Zhu, X-G., S. P. Long, and D. R. Ort. 2008. What is the maximum efficiency with which photosynthesis can convert solar energy into biomass? *Curr. Opin. Biotechnol.* 19:153–159.

# CHAPTER 4

# Role of Fungi in Reducing Primary Production

Plant pathogenic fungi can attack either above- or belowground and can significantly impair plant performance and reduce primary productivity (van Alfen, 2002). The effects of these pathogens are particularly seen in agroecosystems, especially in monocrop agriculture (Termorshuizen, 2014). Both plant pathogens and mycorrhizae in nature may play an important role in regulating plant community composition. The degree of impact of a fungal pathogen on its host plant depends on the fungal species and the environmental conditions in which the plant is grown. In many cases, the intensity of this effect is increased when plants are grown in suboptimal conditions and are already under some stress. Burdon (1993) categorizes plant pathogens as follows: (1) *castrators*, those with a highly significant effect on plant fecundity by affecting flowers and seed development, but little effect on vegetative growth; (2) *killers*, those that cause wilting and damping off of seedlings; and (3) *debilitators*, those that cause lesions or chronic infections. In terms of reducing primary production, all forms of pathogens can be significant, but their mode of effect is different, and the occurrence of each type may depend on the phenology of the host plant.

The most classic examples of significant reduction in plant performance due to fungal pathogen attack have been seen in the devastating potato blight in Ireland during the 1840s (Austin Bourke, 1964), oak decline in Europe (Brasier, 1996), chestnut decline in North America (Anagnostakis, 1987), and sudden oak decline in the United States (Henricot and Prior, 2004). However, most of these fungal diseases were either introduced or infected exotic plant species. The movement of plant and fungal species throughout the world is posing an increasingly serious threat to natural ecosystems (Brasier, 2001; Rossman, 2001).

The problem of discerning the effects of fungal pathogens in natural ecosystems and plant communities is much more difficult and is similar to the problem outlined earlier for the function of mycorrhizal fungi. The effects of these pathogens may be small or only occur at specific times of the year, making the effects of these fungi much less obvious than the introduced pathogens, pathogens on exotic plant species, and in particular those pathogens that attack monospecific plantations of crop plants. The literature on the effects of fungal pathogens on crop plant species is immense and beyond the scope of this book. I will refer the reader to the journals that specialize in reporting these data, such as *Phytopathology, European Journal of Pathology,* and *Plant Disease.*

## 4.1 PATHOGENIC FUNGI AND NATURAL PLANT COMMUNITIES

Harper (1990) casts some doubt on the role of pathogens in altering populations and communities of their hosts. He cites examples of dramatic negative effects of fungal pathogens on introduced or alien plants or on native plants by alien fungal pathogens. However, he suggests that where there has been evolution of communities of organisms in their natural environment, such dramatic effects of pathogens are rarely seen. It is possible that the extreme interactions have already been played out earlier in the development of the plant communities, and the current interactive responses of alien and native species of plants or fungi only represent what has happened in the past. This view is held by Anderson et al. (2010), who cite the evolutionary arms race between plants and their pathogens, in which there is constant genetic adaptation of the host plant in its defenses against a pathogen and the reciprocal development of plant antagonistic chemicals by the fungus. In addition, there are a number of factors that differ between natural and managed ecosystems that influence the success of fungal pathogens and their impacts on host plants including plant and fungal diversity, disturbance, and gene pool (Termorshuizen, 2014; Table 4.1). For example, in eastern USA, the anther-smut fungal pathogen (*Ustilago*

**Table 4.1   Examples of Fungal Root Pathogens**

|  | Species | Disease | Hosts |
|---|---|---|---|
| Plasmodiophoromycota | *Spongospora subterranea* | Powdery scab | Potato |
|  | *Polymyxa betae* | Rhizomania | Sugar beet |
|  | *Plasmodiophora brassicae* | Clubroot | Cruciferae |
| Oomycota | *Aphanomyces euteiches* | Root rot | Legumes |
|  | *Phytophthora cinnamomi* | Root rot | Many |
|  | *Phytophthora nicotianae* | Damping-off | Many |
|  | *Pythium* spp. | Root rot and damping-off | Many |
| Chytridomycota | *Synchytrium endobioticum* | Potato wart | Solanaceae |
|  | *Olpidium brassicae* | None—virus vector | Many |
| Ascomycotina | *Cylindrocarpon destructans* | Root rot | Many |
|  | *Fusarium* spp. | Wilt or root rot | Many |
|  | *Macrophomina phaseolina* | Black dot/charcoal rot | Many |
|  | *Verticillium dahlia* | Verticillium wilt | Most dicots |
| Basidiomycotina | *Thanatephorus cucumeris* | Root rot/damping-off | Many |
|  | *Armillaria mellea* | Tree root rot | Most woody species |

*Source:* Termorshuizen, A.J., in: *Interactions in Soil: Promoting Plant Growth,* ed. Dighton, J., Krumins, J.A., 119–137, Springer Science + Business Media, Dordrecht, 2014.

*violacea*) invades the stamens of *Silene alba* and replaces them with fungal structures and causes abortion of the ovary of female flowers. Even if the fungus sys-
temically infects the plant, there appears to be little effect on the survival of the
plant, other than a loss of its reproductive potential (Alexander, 1990). Some plants
within the community develop resistance to the pathogen, so the ready dispersal of
fungal spores and the patchy occurrence of resistant plants result in a fragmented
community of plants with varying degrees of fungal infection within them. It is
likely, therefore, that this heterogeneity maintains some stable equilibrium between
the abundance of host plants and the pathogenic fungus. This may be what occurs
during the evolution of plant communities, which explains why there is no evident
effect of fungal diseases on natural plant communities.

   In natural forests, the impact of a fungal pathogen can influence forest succession
and community composition. In the Pacific Northwest (USA), the laminate root rot
fungus *Phellinus weirii* impacts the growth and survival of Douglas fir, mountain
hemlock, and white fir, but western hemlock and western red cedar are either little
affected or not affected at all by the fungal pathogen. After a wildfire, dense growths
of Douglas fir are broken up by death caused by *P. weirii*, where up to 50% of trees
can be lost. The fungal pathogen slowly kills trees, and the infection spreads from
a central infected tree to neighboring trees such that, on death, gaps in the forest
are created, allowing the invasion by other plant species, such as western hemlock
(Holah et al., 1997) (Figure 4.1). Within these gaps, the diversity of vegetation during successional colonization increases in both species richness and evenness, compared with the original species composition. Changes in the resistance of trees to
the pathogen appear to be due to the nutrition of the host tree. As the infection front
advances, dead trees contribute to the nutrient pool in soil, and the elevated level of
nitrogen available to the succeeding generation of trees confers a greater resistance
to the pathogen (Hansen and Goheen, 2000). Furthermore, pathogens that do not

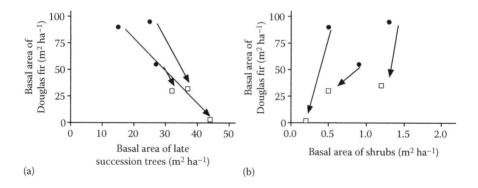

**Figure 4.1**  Changes in the relative basal area of Douglas fir trees in relation to late successional trees (a) or shrubs (b) in the H. J. Andrews forest as a result of the
root-rotting fungal pathogen, *Phelliunus weirii*. Changes are indicated by arrows
showing trends in response from plants outside infection centers (solid symbols)
to areas within infection centers (open symbols). (Data modified from Holah, J.C.
et al., *Oecologia*, 111, 429–433, 1997.)

kill host trees, such as the foliar pathogen *Phaeocryptopus gaumannii*, reduce tree vigor and allow competing tree species to increase competitiveness, again leading to increased forest diversity (Hansen and Stone, 2005). Within forest gaps, the bacterial and fungal activity was shown to be significantly different from that under closed canopy forests in subtropical forest ecosystems, and to increase plant litter decomposition in gaps, creating greater mineralization of nutrients (Zhang and Zak, 1998).

The distribution of root pathogens in soil and their ability to infect host plants may be an important determinant in the germination and survival of certain plant species. Augspurger (1990) shows how the effect of damping-off fungi (*Phytophothora*, *Rhizoctonia*, *Pythium*, and *Fusarium*) influences the development of tropical forest tree species, which develop in forest gaps. In many of these tropical tree species, seed dispersal is limited to within 100 m of the parent tree. Her research suggested that seedlings that germinated close to the parent tree had a higher percentage loss due to damping-off than seedlings developing further from the parent tree. These finding substantiate the hypothesis proposed by Janzen (1970) and Connell (1971) and referred to as the Janzen–Connell hypothesis. The essence of this hypothesis is that the effects of root pathogens and the higher incidence of herbivore grazing on seedling plants restricts the development of conspecific species under the canopy of mature tropical trees (Figure 4.2) (Clark and Clark, 1984). The hypothesis suggests that this is a mechanism that favors dispersal of seeds away from parent trees and stimulates colonization of forest gaps. Recent evidence from controlled experiments supports this hypothesis (Packer and Clay, 2000; van der Putten, 2000). In these studies, the alleviation of the pathogenic effect of soils under a parent tree was achieved by soil sterilization. This sterilization process reduced the incidence of *Pythium* damping-off of black cherry tree seedlings (Figure 4.3). However, it is difficult to argue if the pathogen alone is influencing the spatial pattern of successful seedlings. Seedlings growing in the shade of a parent tree probably exhibit signs of stress, where they are growing at suboptimal light levels. This may make them

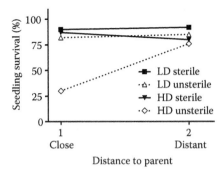

**Figure 4.2**  Effect of distance from parent tree and soil sterilization on the survival of black cherry seedlings. The regression model included density, density × distance, density × sterilization and distance × density × sterilization. Removal of any variable significantly reduced the model fit. (Data from Packer, A., Clay, K., *Nature*, 404, 278–281, 2000.)

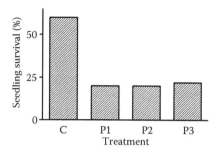

**Figure 4.3**  Black cherry seedling survival in control soil (C) of potting mix and sterilized fungal growth medium, and fungal pathogen inoculated soils (P1–P3) containing 5 ml of inoculum of *Pythium* spp. (Data from Packer, A., Clay, K., *Nature*, 404, 278–281, 2000.)

more vulnerable to root pathogens than those seedlings growing under optimal light conditions toward the center of gaps. A similar mechanism has been shown to affect recruitment of *Ocotea whitei* seedling trees under canopies of adults of the same tree species on Barro Colorado Island, Panama, because of the presence of a fungal tree canker. Seedling survival is significantly higher under the canopy of a nonsusceptible tree, *Beilschmiedia pendula* (Gilbert et al., 1994). In an experimental study of the Janzen–Connell hypothesis in a rainforest, Bagchi et al. (2014) showed that the application of fungicide significantly reduced the diversity of seedling recruits to the forest, whereas insect herbivory had much less effect on diversity (Figure 4.4). Herbivory, however, increased seedling recruitment. Suppression of fungi reduced the impact of density in reducing recruitment, confirming that fungi play a role in mediating negative density dependence.

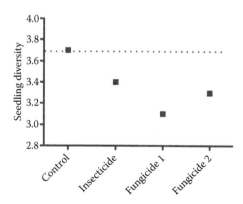

**Figure 4.4**  Comparative influence of insecticide and fungicide on seedling diversity in a tropical forest. (Data from Bagchi, R. et al., *Nature*, 506, 85–88, 2014.)

Interactions between the host plant, the pathogenic fungus, and the environment can significantly vary the outcome of the severity of the pathogenic symptoms (Paul, 1990). For example, Paul (1990) suggests that the degree of loss of photosynthetic capacity of a plant due to fungal invasion will be greater in a plant growing in the shade than one growing in full light. Similarly, he cites work to support the fact that fungal pathogen effects are greater in nutrient-poor or droughty conditions, where the fungus competes with the host plant for limited resources. Thus, the level of the impact of a pathogen may be greater on plants growing in marginal habitats than in optimal habitats. This would certainly alter the competitive abilities of plants growing in marginal conditions. This reduction in fitness of a pathogen-infected plant is significant when the host plant is grown in mixture with a nonhost plant. The reduced performance of *Senecio vulgaris* in the presence of the fungal pathogen *Puccinia lagenophorae* was shown to improve the competitive abilities of *Lactuca salvia* (Paul and Ayres, 1987), *Euphorbia peplus* (Paul, 1989), and *Capsella bursa-pastoris* with which they were grown.

The effect of seed and seedling mortality due to a fungal pathogen on plant population dynamics depends on the degree to which growth and reproduction of surviving individuals compensate for deaths. At three planting densities of the forage plant *Kummerowia stipulacea*, which has become naturalized in some parts of the United States, reduced seedling establishment and seedling size at high sowing density were noted. In in the presence of the root fungal pathogen, *Pythium* species, seed and seedling survival was low and plants were initially smaller, but at maturity the average surviving pathogen-infested plants were larger than in the other treatments (Alexander and Mihail, 2000). Both plant density and plant biomass were significantly reduced by the presence of the fungal pathogen (Figure 4.5). This suggests that the effect of the pathogen allows the surviving plants to be released from intraspecific competition. Thus, there may be a role for fungal pathogens

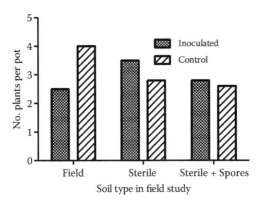

**Figure 4.5** Number of plants and plant biomass of *Kummerowia stipulacea* in the presence and absence of the plant pathogen *Pythium* in field soil, soil sterilized by microwaving, and sterile soil amended with fungal pathogen spores. (Data from Alexander, H.M., Mihail, J.D., *Oecologia*, 122, 346–353, 2000.)

in determining interplant spacing to minimize competition and increase fitness. Interactions between shade and available water levels in the competition between oak and woody shrub species in savanna ecosystems suggests that the intervention of oak wilt fungi can cause a difference in the competition between oaks and woody shrubs and facilitation of shrub layer communities (Anderson et al., 2001). Lowered water tables around healthy mature oaks reduced shrub layer community development, but where oak wilt disease reduced the growth of oak trees, shrub layer communities were able to establish, causing a change in plant community composition.

The incidence of fungal plant pathogens on grasses has been shown to increase with increasing latitude (Clay, 1997). It is assumed that this is related to the increased stresses imposed on plants by adverse climatic conditions that make them more susceptible to pathogen attack. However, much of the data underpinning this statement comes from agroecosystems, and Clay (1997) states that the number of pathogens per host plant is significantly lower in noncrop plant species than in crop species (Figure 4.6).

In natural ecosystems, the introduction of an exotic fungal pathogen is of particular concern because of the rapid genetic evolution of the introduced pathogens by their new environmental conditions (Brasier, 2001). The survival of economically important exotic crops continues to be challenged by the emergence of local diseases that adapt to new host plants. Wingfield et al. (2001) discuss the impact of exotic fungi on exotic plantation forest trees in the tropics, which can induce severe loss of forest trees with disastrous economic consequences, and Brasier (1990) reviews the devastating effects of the chestnut blight fungus, *Cryphonectria parasitica*, which was probably imported from China, on chestnuts in North America. The rapid spread of this disease (about 37 km per year) and the significant reduction in fitness of the host tree, which now exists as an understory shrub species rather than a dominant canopy tree, are witness to the effect of an introduced pathogen. In a similar manner, the fungus *Ophistoma ulmi* caused extensive decline in the elm populations of Europe and North America. However, resistance of the trees was seen to occur. This apparent resistance is partly attributed to the genetic variation in host

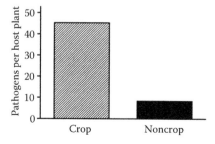

**Figure 4.6**   Mean number of fungal pathogens per host graminoid plant in relation to the agricultural status of the plant species. (Data from Clay, K., in: *Multitrophic Interactions in Terrestrial Systems.*, ed. Gange, A.C., Brown, V.K., 151–169, Blackwell Science, Oxford, 1997.)

plants producing actual resistance (Burdon et al., 1990; Crute, 1990), and also to the presence of fungal pathogenic mycoviruses (Brasier, 1990) that reduced the effectiveness of the fungal pathogen.

The loss of dominant tree species due to fungal pathogens has ecological consequences beyond the loss of that species. Alteration in canopy architecture allows subdominant trees to fill canopy gaps, and the plant community composition changes—affecting all organisms that are dependent on the lost tree species and those plants replacing it. For example, the effect of the introduced anthracnose of dogwood caused by *Discula destructans* has caused a change in the plant community structure of forest ecosystems of the Cumberland Plateau in Tennessee, USA. By selectively reducing the population of dogwood trees, the vegetation has become dominated by two bird-dispersed tree species, blackgum and spicebush. In addition to the change in the forest community, loss of the dogwood trees has reduced the cycling of calcium in the ecosystem, with consequential effects of reduced availability of calcium to birds, through their insect food, resulting in poor egg survival (Hiers and Evans, 1997).

*Phytophthora ramorum* is an emerging disease. Even a single pathogen can have devastating effects on multiple plant species as it affects members of the *Fagaceae*, *Lauraceae*, and *Ericaceae* (Appiah et al., 2004). Indeed, the devastation of oaks in California is a testament to the voracity of this disease (Condenso and Meetnemeyer, 2007; Grünwald et al., 2013) along with the decline of European beech in Europe and United States (Jung et al., 2005). The decline in oaks in southern Europe due to the destructive effects of the oomycete pathogen *Phytophthora cinnamomi* has been reviewed by Brasier (1996). In various Mediterranean locations (Spain, Portugal, Tunisia, and Morocco), this fungus has been responsible for the significant decline in the evergreen oaks *Quercus suber* and *Q. ilex*, thus significantly altering the community structure of the oak forest ecosystems in this region. Change in tree community composition after sudden oak death in California altered the ratio of tanoak, redwood, and bay laurel in the plant community, with loss of oak and increased dominance of laurel and redwood (Cobb et al., 2013). Because of the subsequent changes in leaf litter input to the forest floor, dominated by bay laurel, with a high N content, rates of N mineralization increase after oak death as litter quality (C/N) increases. The likely impact of this change in soil fertility is invasion by plant species requiring higher soil fertility.

The spread of *Phyophthora* through soil is by virtue of motile oospores that require wet or waterlogged soil for optimum dispersal. Climate change models of this area predict increasing rainfall in these regions, which would result in a potential increase in the rate of spread of the disease. However, the severity of cold winters in central and northern Europe could limit the northward spread of *Phytopthora* (Brasier, 1996). Henricot and Prior (2004) list 42 plant species from 18 families that have reportedly shown symptoms of *Phytophthora* diseases. The impact of diseases such as these can have a significant impact on forest community composition, structure, and biodiversity at the landscape level (Holdenreider et al., 2004).

Other potentially large-scale effects of fungal pathogens can relate to the global carbon cycle. The global decline in seagrass has been attributed to the fungal pathogen *Labyrinthula* spp., which causes "wasting disease" of *Zostera* (Sullivan et al., 2013). Although they cover only a small area of the globe, it is suggested that these plants account for about 20% of global C sequestration. Some 20 species of the six genera Polystigma, Didymella, Lautitia, Massarina, and Thalassoascus have been reported as weak pathogens of seaweeds (Zuccaro and Mitchell, 2005). Many seaweeds produce antifungal chemicals to a much greater degree than antibacterials, suggesting that fungal pathogens are more abundant or pose a greater threat to algae compared with bacteria (Zuccaro and Mitchell, 2005).

## 4.2 PATHOGENS AND AGROECOSYSTEMS

One of the problems with tradition monocrop agriculture and plantation forestry is the large expanse of a single plant host for fungal pathogens. Once a fungus has colonized a single plant, it is easy to spread the pathogen to adjacent plants. Long-distance spore dispersal of crop plant pathogens, particularly wheat stem rust (*Puccinia graminis*) annually moves from south to north in the United States at the rate at which crops start to grow; the "green wave" (Aylor, 2003). These rates of spread can vary between 9 and 18 km day$^{-1}$. Even for root pathogens, the aerial dispersal of spores can be important. Production and distribution of spores of the potato blight *Phytophthora infestans* can be impressive in quantity and speed of release. Sporangia release rates from the canopy of an infected potato crop can vary between 300 and 13,000 sporangia m$^{-2}$ s$^{-1}$, giving up to 26,000 sporangia m$^{-3}$ in the air directly above the crop canopy (Aylor et al., 2001). It is from these predictions that farmers can predict the optimal time for fungicide application to their crop, as spore release is controlled not just by plant phenology, but also by local microclimatic conditions. Warning systems based on instruments (e.g., the EnviroCaster) sensing local nocturnal temperature and relative humidity were compared with scheduled fungicide spraying for grapes (Madden et al., 2000). In general, however, the automated warning was no more successful in reducing disease than scheduled fungicide application. Simulation models predict significantly enhanced risks of fungal pathogen outbreaks and increased severity of these outbreaks on crop plants with predicted increases in temperature and reduced rainfall (Luo et al., 1995; Jahn et al., 1996). Similarly, it is predicted that *Phytophthora*-induced decline of oaks in Europe would also increase because of climate change (Brasier, 1996). Thus begins the race to promote more disease resistance into our crop plants, both by selecting naturally evolving resistance (Hines and Marx, 2001) and by artificially altering resistance via manipulation of genes in order to combat some of the emerging plant pathogens (Moffat, 2001; Table 4.2).

Exotic fungal species have been attributed to the severe pathogen attacks of U.S. agricultural crops resulting in economic loss for farmers and loss of tree species, such as the American chestnut and tropical trees whose wood was used for numerous purposes (Palm, 2001; Rossman, 2001; Wingfield et al., 2001). Black Sigatoka (*Mycospharella fijiensis*) is the costliest fungal disease to combat in banana crops, accounting for 27% of the maintenance costs and can cause more than 35% yield loss

**Table 4.2    Emerging Plant Pathogens for which Solutions Are Being Sought via Genetic Manipulation of Plant Species to Increase Disease Resistance**

| Disease | | Hosts | Geographic Distribution |
|---|---|---|---|
| Fungal | Late blight | Potato, tomato | Spreading worldwide |
| | Downy mildew | Corn, sorghum | Spreading out of SE Asia |
| | Rust | Soybean | Spreading from SE Asia and Russia |
| | Karnal bunt | Wheat | Pakistan, India, Nepal, Mexico, USA |
| | Monilia pod rot | Cocoa | South America |
| | Rust | Sugar cane | The Americas |
| | Blast | Rice | Asia |
| Viral | African mosaic | Cassava | Africa |
| | Streak disease | Maize, wheat, sugar cane | Africa |
| | Hoja blanca | Rice | The Americas |
| | Bunchy top | Bananas | Asia, Australia, Egypt, Pacific Islands |
| | Tungro | Rice | SE Asia |
| | Golden mosaic | Bean | Caribbean Basin, Florida, Central America |
| | Plum pox virus | Stone fruits | Europe, India, Syria, Egypt, Chile |
| | High plains virus | Cereals | Great Plains, USA |
| Bacterial | Leaf blight | Rice | Japan, India |
| | Wilt | Banana | The Americas |

*Source:* Moffat, A.S., *Science*, 292, 2270–2273, 2001.

of plantains and bananas (Marin et al., 2003). In monocrop forestry, fungal disease can significantly influence tree growth. The leaf pathogen *Teratosphaeria* spp. and rust *Puccinia psidii* at high severity significantly reduce the growth of Eucalyptus (*Eucalyptus globulus*), which is grown as a fast-growing tree crop in Uruguay (Balmelli et al., 2013; Figure 4.7).

*Phytophthora* is also an emerging disease of agricultural crops, particularly those crops depending on abundant water, such as cranberry, where spores can be

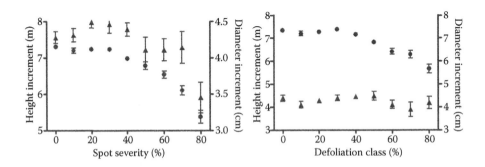

**Figure 4.7**    Effect of foliar pathogens on tree height (circles) and diameter (triangles) increment two years after damage assessment in relation to disease severity (left) and degree of defoliation (right). (Data from Balmelli, G. et al., *New For.*, 44, 249–263, 2013.)

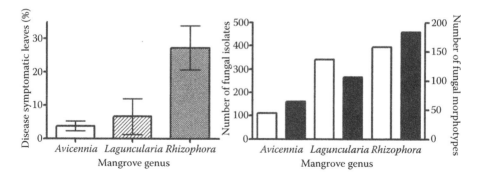

**Figure 4.8** (Left) Effect of foliar salt extrusion on fungal disease symptoms; (Right) the number of foliar epiphytes (white bars) and estimate of fungal diversity (black bars) on three genera of mangroves where *Avicennia* secretes salt from leaves and *Rhizophora* from roots. (Data from Gilbert, G.S. et al., *Oecologia*, 132, 278–285, 2002.)

easily spread through free water in soil (Polashock et al., 2005) and isolates may be tolerant of fungicides.

The natural response of plants to pathogens is to produce defensive plant hormones such as salicylic acid, jasmonates, and ethylene (Whitham and Schweitzer, 2002; Bari and Jones, 2009). However, these defense chemicals can interfere with each other and reduce efficacy. It has been shown that polysaccharide secretions of *Botrytis cinerea* elicits salicylic acid production, which in turn antagonizes the jasmonic acid signaling pathways, allowing *Botrytis* to successfully attack tomato (El Oirdi et al., 2011). Pathogens also induce the development of higher levels of plant polyphenols, especially tannins. The higher content of these chemicals reduces the palatability of dead leaves to soil fauna and, by increasing the C/N ratio of the leaf material, reduces the ability of saprotrophic and mycorrhizal fungi to decompose the leaf litter (Hättenschwiler and Vitousek, 2000). The authors conclude that with repeated or sustained high pathogen levels in plants, this positive feedback mechanism could reduce soil fertility at the local—and possibly regional—level.

Other plants can develop alternate defense systems, such as the mangrove *Avicennia germinans*, whose foliar salt exudation reduced the growth of phylloplane fungi and leaf pathogens to a greater extent than the mangrove *Rhizophora* spp., whose salt extrusion mechanism is in the roots (Gilbert et al., 2002; Figure 4.8).

## 4.3 INTERACTIONS BETWEEN MYCORRHIZAE AND PLANT PATHOGENS

A number of plant pathogenic fungi attack their host plant via roots. It is at the soil–root interface that these pathogens encounter the elevated populations of fungi and bacteria that are encouraged to grow in the rhizosphere by the presence of readily available carbohydrates in the form of root exudates, and dead root cells. Rhizospheric plant growth promoting rhizobacteria (PGPR) are part of the suppressive soil defense against root fungal pathogens (Rosas, 2007; Hol et al., 2014). A range of isolates of these bacteria have been shown to reduce the disease index of *Fusarium* wilt in chilli plants (Sundaramoorthy et al., 2012), by the production of chitinase, peroxidase, and polyphenol oxidase enzyme activity (Figure 4.9). Similar effects have been shown for combinations of the N-fixing bacterium, *Bradyrhizobium*, and PGPR isolates in protecting peanut form root rot caused by *Aspergillus niger* (Table 4.3), where disease suppression can be halved, although this is dependent on fungal spore loading (Yuttavanichakul et al., 2012), fluorescent pseudomonads protecting beans against *Rhizoctonia solani* (Ahmadzadeh and Tehrani, 2009) and *Pythium debaryanum* on alfalfa (Yanes et al., 2012). These bacterial biocontrol agents can be as effective or more effective than fungicides as has been

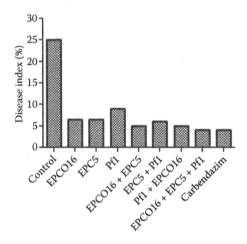

**Figure 4.9**  Effects of individual and mixtures of biocontrol agents [*Bacilus subtilis* (EPCO16 and EPC5) and *Pseudomonas fluorescens* (Pf1)] on *Fusarium* wilt disease index on chili plants. (Data from Sundaramoorthy, S. et al., *Biol. Control*, 60, 59–67, 2012.)

**Table 4.3**  Effect of Fungicide Carbendazim or Commercially Available *Bradyrhizobium* (B) in Combination with Selected PGPR Isolates (A20, A45, A62, and A106) on Root Rot Disease of Peanut Induced by Addition of *Aspergillus niger* Spores at Three Densities (spores ml$^{-1}$ seed$^{-1}$)

| | | Disease Incidence (%) | | | Disease Severity Score | | |
|---|---|---|---|---|---|---|---|
| Treatment | *A. niger* | $10^5$ | $10^6$ | $10^7$ | $10^5$ | $10^6$ | $10^7$ |
| Control | 0 | 89 | 100 | 100 | 3 | 3.7 | 4 |
| Carbendazim | 0 | 0 | 0 | 0 | 0 | 0 | 0 |
| B + A20 | 0 | 33 | 56 | 78 | 1.3 | 1.9 | 3.1 |
| B + A45 | 0 | 11 | 56 | 67 | 0.4 | 2.2 | 2.7 |
| B + A62 | 0 | 44 | 67 | 78 | 1.7 | 2.7 | 3.2 |
| B + A106 | 0 | 56 | 78 | 77 | 2.2 | 3.1 | 3.1 |

shown in potato, where isolates of *Bacillus subtilis* gave better protection against *Pythuim* leak and pink rot than phosphorus acid or hydrogen peroxide and equal to that of the systemic fungicide, azoxystrobin (Gachango et al., 2012).

In particular, the mycorrhizal fungal community associated with a plant's roots can have special significance in terms of creating a defense mechanism against root pathogenic fungi (Sylvia and Sinclair, 1983; Chakravarty et al., 1991; Quarles, 1999; Demir and Akkopru, 2007). In natural ecosystems, we know relatively little about the effect of mycorrhizae in the protection of host plants against root fungal pathogens. Most of the documented proof we have of the protective effect comes from agricultural ecosystems, where the host plant species composition is of low diversity (usually a monospecific crop), or from the forest nursery industry, where, again, a monospecific crop is being grown. However, it is of interest to note that mycorrhizae have the ability to protect host plants from pathogens even if we cannot yet put this knowledge into the context of understanding plant community dynamics in natural ecosystems.

### 4.3.1 Interactions with Arbuscular Mycorrhizae

The role of mycorrhizae in plant–pathogen interactions is outlined by Sharma et al. (2007). The effect of arbuscular mycorrhizae is probably to improve the nutrition of the host plant or alter its physiology such that the plant is better able to defend itself against the pathogen (Dehne, 1982; Smith, 1988; Volpin et al., 1994) rather than direct competition between the mycorrhizal fungal and the pathogenic fungus or nematode.

In a meta-analysis of interactions between arbuscular mycorrhizae and plant fungal pathogens, Borowicz (2001) came to the conclusion that of the published studies on this subject between 1970 and early 1998, most studies were performed on economically important plant species (agricultural crop plants), under low phosphate soil conditions in greenhouses or microplot experiments. She concludes that a much wider set of studies must be carried out before we can extrapolate the information gained to agricultural ecosystems in general, let alone to natural ecosystems. From her analyses, we can see that, in general, 50% of the studies showed that arbuscular mycorrhizae afforded some degree of protection to their host plant against plant pathogenic fungi and nematodes. The effect of the pathogenic fungi was usually to reduce the growth of the arbuscular mycorrhizal fungus, but this effect was not seen so frequently for nematodes. The interaction between the two fungal functional groups resulted in a reduction of growth of both competing fungi in only 16% of the reported cases.

The effect of arbuscular mycorrhizae and defense against plant pathogenic fungi has largely been studied in annual plants, especially in agricultural crops. Examples of this type of studies are those conducted in cotton, where the effect of inoculation with the arbuscular mycorrhizae *Glomus hoi*, *G. mosseae*, and *G. versiforme* significantly improved growth and established a significant defense against the wilt fungi, *Verticillium dahliae* (Table 4.4) (Lui, 1995). Similarly, Abdalla and Abdel-Fattah (2000) showed that peanut plants had a reduced incidence of root fungal pathogens

Table 4.4   Growth and Disease Status of Cotton Plants Grown in Association with Wilt
Pathogen *Verticillium dahliae*, Arbuscular Mycorrhizae, and Combinations
of the Two

| Treatment | Plant Height (cm) | Disease Incidence (%) |
|---|---|---|
| C | 10.5 | 0 |
| Vd$_1$ | 9.7 | 43.3 |
| Vd$_2$ | 8.3 | 35.5 |
| Gh | 12.8 | 0 |
| Gm | 13.1 | 0 |
| Gv | 13.1 | 0 |
| Gh + Vd$_1$ | 11.1 | 23.3 |
| Gm + Vd$_1$ | 11.7 | 23.3 |
| Gv + Vd$_1$ | 12.3 | 20.0 |
| Gh + Vd$_2$ | 8.9 | 23.3 |
| Gm + Vd$_2$ | 9.9 | 23.3 |
| Gv + Vd$_2$ | 12.5 | 16.7 |

*Source:* Liu, R.J., *Mycorrhiza*, 5, 293–297, 1995.
*Note:*  C, control of no fungal additions; Vd$_1$ and Vd$_2$ are two strains of *Verticillium dahlia*; Gh, Gm,
and Gv are the mycorrhizae *Glomus hoi, G. mosseae,* and *G. versiforme,* respectively.

when inoculated with the arbuscular mycorrhizal fungus *Glomus mosseae* than
in the absence of mycorrhizae. The benefit of mycorrhizal colonization of roots
was both an antagonism against the two fungal pathogens, *Fusarium solani* and
*Rhizoctonia solani*, and a growth enhancement of the host plant, leading to greater
fitness expressed in terms of pod number and seed (peanut) production (Table 4.5).
Although the presence of the pathogens reduced root colonization by the mycorrhizal fungus, propagule numbers of each pathogen isolated from a variety of plant
parts were significantly lower in mycorrhizal plants compared to nonmycorrhizal
plants. Thus, not only did *G. mosseae* protect peanut plants from infection by pod rot
fungal pathogens, it also reduced the fecundity of the fungal pathogen. Arbuscular

Table 4.5   Effect of Arbuscular Mycorrhizal Fungus *Glomus mosseae* on Growth and
Fecundity of Peanut Plants, at Maturity, Infected with Root Pathogenic Fungi
*Fusarium solani* and *Rhizoctonia solani*

| Treatment | Growth Parameters | | Yield Biomass | |
|---|---|---|---|---|
| | Shoot Weight (g) | No. of Branches | Pods per Plant | 100 Seed Weight (g) |
| Control | 9.3 | 6.3 | 9.7 | 64.5 |
| Mycorrhizal | 13.0 | 8.3 | 12.3 | 72.5 |
| *Fusarium* | 6.7 | 5.0 | 8.0 | 45.9 |
| Myco + *Fusarium* | 7.8 | 6.5 | 10.6 | 65.9 |
| *Rhizoctonia* | 6.9 | 5.0 | 7.0 | 39.8 |
| Myco + *Rhizoctonia* | 9.1 | 6.7 | 8.7 | 60.5 |
| *Fusarium* + *Rhizoctonia* | 5.8 | 5.0 | 5.3 | 38.4 |

*Source:* Abdalla, M.E., Abdel-Fattah, G.M., *Mycorrhiza*, 10, 29–35, 2000.

mycorrhizae (*Glomus etunicatum* and *Acaulospora tuberculata*) were shown to reduce the negative effects of *Phytophthora nicotianae* on citrus tree growth (Watanarojanaporn et al., 2011; Figure 4.10), and the combination of the mycorrhizal fungus *Glomus mosseae* and the plant growth-promoting fungus *Fusarium equisiti* had a synergistic effect of reducing anthracnose of cucumber (Saldajeno and Hyakumachi, 2011).

In peach tree seedlings, suppression of root rot, which is caused by *Cylindrocarpon destructans* by the arbuscular mycorrhizal fungus *Glomus aggeragatum*, has been demonstrated (Traquair, 1995; Figure 4.11).

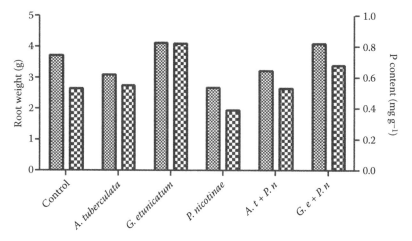

**Figure 4.10**    Effect of two arbuscular mycorrhizae (*Acaulospora tuberculata* and *Glomus etunicatum*) on citrus root growth and P content in the presence of the root pathogen *Phytophthora nicotianae*. (Data from Watanarojanaporn, N. et al., *Sci. Hortic.*, 128, 423–433, 2011.)

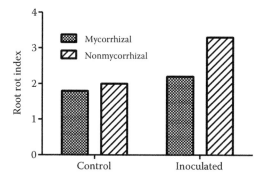

**Figure 4.11**    Degree of cylindrocarpon root rot in peach tree seedlings inoculated with the mycorrhizal fungus *Glomus aggregatum*. Inoculated trees received a conidial suspension of *Cylindrocarpon destructans*. (Data from Traquair, J.A., *Can. J. Bot.*, 73, S89–S95, 1995.)

From the data derived from mixed-species grassland, Sanders and Fitter (1992a,b) suggest that the role of arbuscular mycorrhizal colonization of plant roots may be nonnutritional. In a study of the annual winter grass (*Vulpia ciliate*) in a natural plant community, Newsham et al. (1994) used two fungicides, benomyl and prochloraz, to selectively reduce infection by arbuscular mycorrhizae and pathogenic fungi, respectively. Their determination of plant performance suggested that there was a direct interference between the mycorrhizal and pathogenic fungi and, although the plants did not show pathological symptoms, the mycorrhizal fungi induced greater fitness in the plants as a result of competition with the pathogen. It was also inferred from the data that this was the prime function of the mycorrhizal association, rather than improving phosphorus uptake by the host plant. This type of study on natural plant communities is altering the dogma of the function of mycorrhizae as being enhancement of host plant nutrition alone. However, much more research needs to be done to show the relative importance of mycorrhizae in nutrient uptake and plant defense.

A discussion of the arbuscular mycorrhizal benefits afforded to host plants in terms of nutrient acquisition and growth enhancement on one hand and the protection of the host plant against fungal root pathogens on the other, led Newsham et al. (1995) to conclude that the benefits were related to the root architecture of the host plant. In their model (Figure 4.12), Newsham et al. (1995) suggest that the derived benefit of a mycorrhizal association is predominantly nutrient acquisition if the host plant root system is poorly branched. In contrast, the benefit shifts toward pathogen prevention when the host root system is highly branched. The degree of protection afforded by arbuscular mycorrhizae may thus be related to the root morphology of the host plant. When challenged by the pathogenic fungus, *Fusarium oxycoccum*, the plant *Setaria glauca*, with a finely branched root system, benefitted more from association with mycorrhizae belonging to the Glomeraceae and the simple roots *Allium cepa* from association with Gigasporaceae (Sikes et al., 2009).

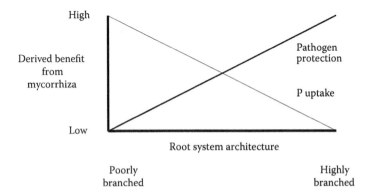

**Figure 4.12**   Nature and magnitude of the benefit derived from plant associations with arbuscular mycorrhizae depending on the branching pattern of the host plant root system. (Redrawn from Newsham, K.K. et al., *TREE*, 10, 407–441, 1995.)

The presence of mycorrhizal associations of roots can influence foliar pathogens. In rice leaves, for example, the presence of arbuscular mycorrhizae has been shown to upregulate some 120 genes associated with defense, transcription, signal transduction, protein synthesis, growth, and metabolism by more than two-fold over uninoculated plants that are considered to aid the host plant against rice blast, *Magnaporthe oryzae* (Campos-Soriano et al., 2012).

There appears to be a fine balance between the beneficial effects of a mycorrhizal fungus and the detrimental effects of a plant pathogenic fungus. In an experimental manipulation of the fungal communities associated with the annual grass, *Vulpia ciliata*, by the application of two fungicides, Newsham et al. (1994) showed that there was an interaction between the pathogenic and mycorrhizal fungal associates of the target plant. The fungicide, perchloraz, did not affect the abundance of arbuscular mycorrhizae on the plant, nor did it affect plant performance. However, benomyl significantly reduced mycorrhizal colonization of roots, although this did not significantly lower plant growth or phosphorus inflow. Benomyl, however, did increase plant fecundity, measured by seed number, in one instance. The authors suggested that the effect of the fungicide benomyl was to reduce the abundance of root pathogenic fungi as well as mycorrhizal fungi. Thus, the lack of mycorrhizal response (reduced growth with reduced colonization) was offset by the improvement in growth of the plant when relieved of the stress of root-inhabiting weak pathogens, such as *Fusarium oxysporum* and *Embellisia chlamydospora*. Much more work needs to be done to understand the role of both plant pathogens and mycorrhizae and the interaction between the two in instances such as this, where no pathological symptoms were observed. It is possible that the balance achieved between the mycorrhizae and the pathogens could alter with the phenology of the plant or changes in the edaphic and environmental conditions, leading to enhanced growth on one hand and significantly reduced growth on the other. How is the balance maintained? Has this been an evolutionary pathway to maintain a balance between two different functional groups of fungi? How much do these interactions play in the determination of plant fitness and express themselves in the community composition of plant assemblages?

In contrast to the interactions between mycorrhizae and fungal pathogens, Salonen et al. (2001) investigated the effect of arbuscular mycorrhizae on a plant parasite of clover and a grass. In greenhouse experiments, the grass *Poa annua* or clover (*Trifolium pratense*) were grown in the presence or absence of an arbuscular mycorrhizal fungus and in the presence or absence of the root hemiparasitic plants, *Odontites vulgaris* (for *Poa*) and *Rhinanthus serotinus* (for clover). Mycorrhizal colonization of roots of *P. annua* had little effect on plant growth, whereas the hemiparasite infection caused a significant reduction in host biomass. The mycorrhizal status of *P. annua* did not affect the biomass or the number of flowers produced by the parasitic *O. vulgaris* plants. In contrast, root colonization by arbuscular mycorrhizae of clover substantially increased the host plant biomass but the hemiparasite infection had no effect. The effect of the mycorrhizae on the hemiparasitic *R. serotinus* plants attached to clover was to increase the parasitic plant biomass and induce the production of more flowers than plants growing with nonmycorrhizal hosts. Salonen et al. (2001) caution, however, that improvement of the performance of the parasitic plant when attached to a mycorrhizal host depends on the degree of growth promotion afforded to the host plant by the mycorrhizae.

## 4.3.2 Interactions with Ectomycorrhizae

If we consider the ectomycorrhizal condition, we can have both a fungal–fungal interaction due to the chemical (antibiotic) interference of pathogen growth (Fravel, 1988; Chakravarty and Hwang, 1991; Duchesne, 1994) and a physical defense by virtue of the presence of a fungal sheath that envelops the short roots that are colonized by the mycorrhizal fungus (Marx, 1973). The implications of ectomycorrhizae as biocontrol agents for root pathogens of economically important trees are reviewed by Quarles (1999).

The effect of ectomycorrhizal fungi on the prevention of pathological symptoms of pine tree root pathogenic fungi has been shown, particularly in nursery conditions (Marx, 1969, 1980; Marx and Davey, 1969a,b). Similarly, Sylvia and Sinclair (1983) showed that the ectomycorrhizal fungus, *Laccaria laccata*, suppressed *Fusarium oxysporum* on Douglas fir seedlings. The same mycorrhizal fungus was also found to protect *Pinus banksiana* seedlings from *Fusarium* (Chakravarty and Hwang, 1991). Branzanti et al. (1999) showed the protective effect of four ectomycorrhizal fungi—*Laccaria laccata*, *Hebeloma crustuliniforme*, *H. sinapizans*, and *Paxillus involutus*—on pathogens of chestnut seedlings (Table 4.6). At the end of the first growing season, half of the mycorrhizal and nonmycorrhizal seedlings were challenged with *Phytophthora cambivora* or *P. cinnamomi* spores. Five months later, mycorrhizal plants infected with pathogenic fungi showed no sign of infection. The ectomycorrhizal fungi increased seedling growth and biomass in the presence of the pathogen.

Mycorrhizal fungi have been known to be effective in the prevention of root pathogen fungal attack on the host plant (Garbaye, 1991). However, as Smith and Read (1997) suggest in their review of the effects of ectomycorrhizae in pathogen resistance, much of the work has been conducted in unrealistic nursery conditions. The actual role of these mycorrhizae as antagonists to plant pathogens in nature is largely unknown. Indeed, even the recent work of Branzanti et al. (1999) that

**Table 4.6  Effect of Inoculation of Chestnut Seedlings (*Castanea sativa*) with Ectomycorrhiza against the Effects of Chestnut Ink Stain Fungal Pathogen, *Phytopthora cambivora***

| Fungal Treatment | Leaf Area (cm²) | Plant Weight (g) |
|---|---|---|
| Control | 21.3 | 9.3 |
| *Phytopthora* alone | 15.6 | 5.2 |
| *Phytopthora* + *Laccaria laccata* | 28.1 | 9.4 |
| *Phytopthora* + *Paxillus involutus* | 19.5 | 6.1 |
| *Phytopthora* + *Hebeloma crustuliniforme* | 22.0 | 5.9 |
| *Phytopthora* + *H. sinapizans* | 28.1 | 10.2 |

*Source:* Branzanti, B.M. et al., *Mycorrhiza*, 9, 103–109, 1999.

demonstrated the significant effect of inoculation of chestnut trees with the ecto-mycorrhizal fungi *Laccaria laccata*, *Hebeloma crustuliniforme*, *H. sinapizans*, and *Paxillus involutus* in preventing chestnut ink disease caused by *Phytophthora cambivora* and *P. cinnamomi*, was conducted on seedling trees (Table 4.6).

Interest in the interaction between mycorrhizae and plant pathogens has led to the use of mycorrhizal fungi as biocontrol agents (Duchesne, 1994; Quarles, 1999). The role of fungi in this regard, however, appears only to have been observed on seedling trees and in the artificial confines of nurseries. The potential role of these interactions in the maintenance of plant communities in natural ecosystems has only been speculated (Rayner, 1993). As suggested earlier in the discussion on interactions between arbuscular mycorrhizae and pathogenic fungi, we still know relatively little about the interaction between these groups of fungi in natural ecosystems. The same questions posed earlier are equally pertinent to ectomycorrhizal plant communities.

In contrast, the links between conspecific plants by mycorrhizal bridges has been shown to confer advantages to the seedlings because of interplant transfers of carbon and nutrients (Amaranthus and Perry, 1994; Rayner, 1998; Read, 1998; Simard et al., 1997a,b,c), which was discussed earlier to benefit trees developing in the shade. This mechanism has been shown to enhance the survival of tropical, ecto-mycorrhizal tree seedlings under the canopy of parent tress in Cameroon (Onguene and Kuyper, 2002; Table 4.7). However, Newbery et al. (2000) did not find such conclusive proof of the beneficial effects of ectomycorrhizae in the enhancement of tropical legume seedlings under conspecific adult trees. In the case of *Tetraberlinia bifoliata*, enhancement of seedling growth was only significant at high densities of adult trees (Table 4.8). Their data suggested that a beneficial effect could be found in

Table 4.7    Seedling Shoot Biomass of *Paraberlinia bifoliolata* When in Contact with or Isolated from the Root and Mycorrhizal System of Different Tropical Tree Species after 8 Months

| Adult Tree Species | Seedling Shoot Biomass in Contact with Adult Tree (g) | Isolated Seedling Shoot Biomass (g) |
|---|---|---|
| *Afzelia* | 1.59 | 1.16 |
| *Brachystegia* | 1.48 | 1.24 |
| *Paraberlinia* | 2.43 | 1.52 |
| *Tetraberlinia* | 2.19 | 1.78 |

*Source:* Onguene, N.A., Kuyper, T.W., *Mycorrhiza*, 12, 13–17, 2002.

Table 4.8    Survival of Seedling Tropical Ectomycorrhizal Trees after 16 Months of Growth under the Canopy of Conspecific Adults at Low and High Stem Density

| | Percent Seedling Survival | |
|---|---|---|
| Tree Species | Low Adult Stem Density | High Adult Stem Density |
| *Microberlinia bisulcata* | 25 | 33 |
| *Tetraberlinia bifoliata* | 33 | 65 |
| *Tetraberlinia moreliana* | 70 | 76 |

*Source:* Newbery, D.M. et al., *New Phytol.*, 147, 401–409, 2000.

some cases, where seedling growth was enhanced, but in other tree species this benefit was not evident. Thus, the interaction between root pathogens and mycorrhizae in the establishment of conspecific trees seedlings under the canopy of parent trees is not simple. There are probably many interactions between the mycorrhizal and pathogenic fungi that result in an advantage of one or the other depending on local environmental conditions (Connell et al., 1984).

One effect of foliar pathogens is to reduce the photosynthetic capacity of the host plant. Given that mycorrhizal fungi are dependent on a carbon supply from the host to maintain their biomass, loss of photosynthetic capacity could result in impaired mycorrhizal function. Partial defoliation of Englemann fir reduced the total abundance of ectomycorrhizal fungi on roots and altered the species composition (Cullings et al., 2001). *Inocybe* spp. were removed from the mycorrhizal community by defoliation, but Suilloid and Agaricoid mycorrhizae dominated under conditions of reduced carbon supply.

## 4.4 SAPROTROPH–PATHOGEN INTERACTIONS—BIOCONTROL

Within the phylloplane, many interactions are likely between saprotrophic fungi and bacteria and potentially pathogenic fungal species. Newton et al. (2010) call for a greater understanding of these interactions for the management of crop diseases. Many fluorescent pseudomonad bacteria are antagonistic to fungal growth on leaf surfaces. For example, isolates of these bacteria are reported to reduce the growth of sheath blight fungus (*Rhizoctonia solani*) in rice by between 50% and 70% (Akter et al., 2014).

The presence of saprotrophic fungi on plant surfaces is a long accepted fact (Last and Deighton, 1965). Leaves of terrestrial plants support extensive and diverse communities of both pathogenic and nonpathogenic fungi (Preece and Dickinson, 1971; Dickinson and Preece, 1976; Farr et al., 1989; Kenerley and Andrews, 1990; Blakeman, 1992; Donegan et al., 1996). Many saprotrophic members of the phylloplane have been shown to be antagonistic toward plant pathogens. For example, Omar and Heather (1979) showed that *Alternaria* and *Cladosporium* species were more effective inhibitors of *Melampsora larici-populina* on poplar leaves compared with *Penicillium* (Figure 4.13). Competition for nutrients among the epiphytic members of the phyllosphere of beetroot leaves was shown to negatively affect the germination of spores of plant pathogens (Blakeman and Brodie, 1977), and leaf disks (*Eucalyptus globulus*) treated with the growth staling products isolated from the leaf-inhabiting microfungi significantly reduced the development of the fungal pathogen, *Pestalotiopsis funereal* (Upadhyay and Arora, 1980). The fungi *Aspergillus niger*, *Fusarium oxysporum*, and *Penicillium citrinum* have been shown to reduce the growth of the potential leaf pathogen *Pestalotia citrinum* by more than 50% on guava leaves, partly because of the antagonistic volatiles produced by these fungi (Pandy et al., 1993).

Most studies of this type have observed interactions between a single saprotroph and a single plant pathogen; very few have looked at two or more saprotrophs in

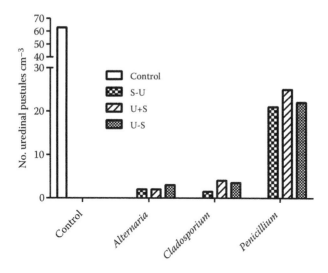

**Figure 4.13**   Effect of saprotrophic leaf surface fungi on the development of uredinal pustules of *Melampsora larici-populina*. Saprotroph conidia incubated before uredino-spores added (open bar), conidia and uredospores as a mixed inoculum (solid bar) and uredinospores added before conidia (hatched bar) compared to infection without saprotroph (control). (Data from Omar, M., Heather, W.A., *Trans. Br. Mycol. Soc.*, 72, 225–231, 1979.)

combination as antagonists. Members of the phyllosphere fungal community have been shown to coexist; however, the functional role of the organisms as a community—rather than as isolated individuals—has not been adequately investigated (Fokkema, 1991; Bills, 1995). In a study of phylloplane fungi on American cranberry, it was shown that the spore density of epiphytes had a significant effect on the weak pathogen *Pestalotia vaccinii*, especially with *Curvularia lunata* and *Alternaria alternata* as competitors, compared to *Penicillium* sp. (Nix-Stohr et al., 2007; Figures 4.14 and 4.15). Although there was little effect of differences in mixed communities of epiphytes (Stohr and Dighton, 2004), *P. vaccinii* was most suppressed with *C. lunata* in the community; however *P. vaccinii* had a larger suppressive effect on *C. lunata* than either *Alternaria* or *Penicillium*, suggesting that these interactions are strongly context-dependent.

Interactions between phylloplane pathogenic fungi and leaf surface bacterial and saprotrophic fungal communities are also important in determining the ability of the pathogen to develop pathological symptoms on the host plant (Seddon et al., 1997). The effect of these leaf surface biocontrol agents is dependent on the phenology of the plant and the infectiveness of the pathogen in relation to that phenology. If the pathogen is present in the host plant before leaf expansion or arrives at the time of leaf expansion, before a saprotrophic community can develop, the effect of the saprotrophic community to exclude the pathogen is limited. However, it has been shown that environmental conditions can significantly alter the phylloplane fungal

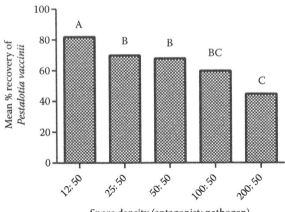

**Figure 4.14** Influence of spore density of fungal competitors to the leaf pathogenic fungus *Pestalotia vaccinii*. (Data from Nix-Stohr, S. et al., *Microb. Ecol.*, 55, 38–44, 2007.)

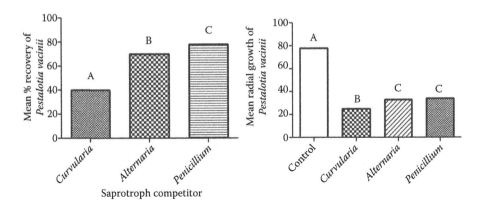

**Figure 4.15** Recovery of the leaf pathogen *Pestalotia vacinii* from leaf surfaces after competition with three saprotrophic fungal competitors (left) and effect of these fungal competitors on growth of *P. vacinii* in culture (right). (Data from Nix-Stohr, S. et al., *Microb. Ecol.*, 55, 38–44, 2007.)

community. In their study, Magan and Baxter (1996) showed that atmospheric $SO_2$ and elevated $CO_2$ can alter the community structure of both saprotrophic and pathogenic fungi on wheat flag leaves.

Interestingly, it is not only the plant whose fitness may be affected by a pathogenic fungus, but the interactions between pathogens on a plant may affect the fitness of the pathogenic fungi themselves. In a study of rust fungi on wheat leaves, Newton et al. (1997) showed that the relative fitness of a number of strains of the rust *Puccinia graminis* was controlled by density-dependent relationships. For example, the relative fitness of the fungal strain SR22 was much greater at low spore densities on the leaf than at high density. At these low densities, well below carrying capacity, the high infection efficiency of SR22 gave it a competitive edge. However, as spore density of a mixed spore inoculum on the leaf increased, the strong competitive abilities of strain SR41 allowed it to dominate in the community. Thus, in the natural ecosystem, the effect of fungal pathogens on individual plants may depend on the outcome of the competition of the fungal pathogens within their own community as much as the competition between saprotrophic fungi and pathogens.

The inhibitory attributes of phylloplane fungi have been used to develop fungal pathogen biocontrol agents. In a review of the interactions between phylloplane microorganisms and mycoherbicide efficacy, Schisler (1997) discusses only single-species interactions or the effects of microbial metabolites, without discussing the individual organisms or communities of organisms that might produce these metabolites. However, Janisiewicz (1996) evaluated the effects of multispecies combinations of yeasts and bacteria for their abilities to control blue mold (*Penicilium expansum*) on harvested apples. He suggested that the optimal species mix occurred when there was minimal niche overlap between the species. The resultant minimal competition between antagonist microbial species allowed maximal competitive interaction between the antagonist and the pathogen.

Commercial applications of fungal biocides have been shown to be important for exotic plant species. For example, the fungus *Epicoccum purpuascens* produces antifungal compounds to inhibit *Sclerotinia* head rot in sunflowers (Pieckenstain et al., 2001). In agriculture, *Trichoderma koningii* has been found to be a good biological control agent for damping-off of tomato by *Sclerotium rolfsii* (Tsahouridou and Thanassoupoulos, 2002). *Chondrostereum purpureum* has been used to suppress resprouting of exotic American bird cherry (*Prunus serotina*) and poplar (*Populus euramericana*) in parts of Europe (de Jong, 2000). This fungus had been marketed as a mycoherbicide under the name BioChon.

In the tropics, Evans (1995) suggests that it is impractical and undesirable to use herbicides in more fragile agroecosystems and natural areas, because of the unknown secondary effects of these chemicals. In contrast, biocontrol agents, such as pathogenic fungi, may be more desirable for use in reducing the abundance of exotic plant species. Although the science of fungal biocontrol of weeds has not been perfected in these ecosystems, there are indications that the fungal pathogen flora of plants changes significantly from its native range to that of its exotic range (Table 4.9). The fact there is minimal overlap of fungal pathogen species in both native and exotic ranges suggests that there is scope for the selection of effective pathogen species in the plant's exotic range to effectively reduce its fitness.

The grass endophyte *Neotyphodium occultans* has been shown to induce natural defense against another clavicipitaceous pathogenic fungus, *Claviceps purpurea*, reducing the infection of flower spikes 3-fold (Pérez et al., 2013; Figure 4.16). The toxins—ascaulitoxin, aglycone, and *trans*-4-aminoproline extracts—from the fungus *Ascochyta caulina* have been shown to be effective in controlling the noxious weed *Chenopodium album* and may prove to be a replacement for chemical

Table 4.9  Tropical Weed Plant Species and Number of Pathogenic Fungi Associated
with Them in Their Native Range and in the Range in which They Are Common
Exotics

| Plant Species | Native Range | No. Fungal Species | Exotic Range | No. Fungal Species | No. Fungal Species in Common |
|---|---|---|---|---|---|
| *Chromolaena odorata* | Neotropics | 17 | Paleotropics | 21 | 4 |
| *Mikania micrantha* | Neotropics | 29 | SE Asia | 14 | 6 |
| *Lantana acmara* | Neotropics | 28 | Paleotropics | 26 | 6 |
| *Cyperus rotundud* | Sudan, Pakistan, India | 19 | Neotropics, SE Asia, Oceania, Australia | 32 | 6 |
| *Euphorbia heterophylla* | Neotropics | 21 | Paleotropics | 33 | 7 |
| *Euphorbia hirta* | Neotropics | 15 | Paleotropics | 19 | 4 |

*Source:* Evans, H.C., *J. Bot.*, 73, S58–S64, 1995.

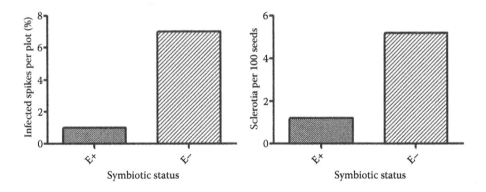

**Figure 4.16**  Effect of *Neotyphodium occultans* endophyte (E+) on *Claviceps purpureum* infection of *Lolium multiflorum* showing number of infected flower spikes (left) and number of seeds containing sclerotia (right). (Data from Pérez, L. et al., *Fung. Ecol.*, 6, 379–386, 2013.)

herbicides (Vurro et al., 2001). In greenhouse trials, extracts of toxins reduce the growth of *Chenopodium* by more than 30%.

Because of the documented inhibitory effect of leaf saprotrophs against foliar pathogens, other work has evaluated the effects that current management practices of fungicide application exert on the phylloplane community and how it might increase the pathogen's ability to initiate disease, where the saprotrophic members of the phylloplane community have been eliminated or reduced by the fungicide. Fokkema and de Nooj (1981) found that some fungicides reduced the ambient mycoflora, whereas others had no effect. Thomas and Shattock (1986) also tested this idea by applying three different fungicides (benomyl, triadimefon, and chlorothalonil) to *Lolium perenne* that had the pathogens *Drechslera siccans* and *D. dictyoides* in addition to other saprotrophic filamentous fungi. They found that the three fungicides altered the incidence of the phylloplane mycoflora in very different ways. Benomyl reduced most saprotrophs but allowed the levels of *D. siccans* and *D. dictyoides* to increase over control levels by 37% and 90%, respectively. This showed that, in the absence of saprotrophs to antagonize them, the pathogens were able to flourish beyond the established controls. Triadimefon reduced the level of pathogenic species and increased the abundance of most other common saprotrophs. Chlorothalonil removed virtually all fungi from the surface of the leaves. Thus, for agricultural purposes, there needs to be balance between encouraging natural competitors against plant pathogens and the use of traditional fungicide treatments. The importance of a protective saprotrophic fungal community on leaf surfaces, however, may only play an important part in reducing pathogenic fungal invasion during the short time the host plant is susceptible and when the spores of the pathogenic fungus are abundant for leaf inoculation.

The community interactions in the phylloplane and their ecological significance have been explored in the review by Bélanger and Avis (2002). They suggest that the diversity of fungal inhabitants on a leaf surface occur as a result of niche separation based on the temporal and spatial diversity of resources. However, Moy et al. (2000) showed that the fungal endophyte *Neotyphodium typhinum* formed epiphyllous networks of hyphae on the leaf surface of a number of grass species, particularly *Bromus setifolius* and *Poa ampla*. They suggest that these epiphyllous fungal networks could possibly act antagonistically toward fungal pathogens. The mechanism of this protection may be by direct fungal–fungal interactions or by virtue of prior space occupancy. Thus, they contend that many of the fungi may not be in competition with each other, but are utilizing unique resources. Moy et al. (2000) argue further that if this niche separation is true, evidence in the literature does not support the hypothesis of a saprotrophic fungal community affording protection to plant pathogens. Citing the experiments of Rishbeth (1963) on competition between *Peniophora giganta* and the pathogen *Fomes annosus*, they argue that the defense is merely a delay in allowing the pathogen access to its optimal resources. Whether this is defense or inadvertent competition is somewhat semantic as the result is a delay in the colonization of plant tissue by a pathogen. Similarly, Bélanger and Avis (2002) suggest that the hyperparasitism shown by *Trichoderma* spp. is probably the main mode of action of members of this genus. However, they reason that this parasitism of other fungi that occur in nature have rarely shown to be an effective means of biocontrol when the density of *Trichoderma* has been artificially increased. Jeffries (1997) reviewed the subject of mycoparasitism and came to the conclusion that this *modus operandi* is difficult to quantify in regards to its effect on populations of either fungal species. Evidence is cited in his review of positive correlations between host fungal hyphal density and that of the parasitic fungus, suggesting a direct trophic effect. He does, however, suggest that this aspect of fungal ecology could have great importance in reducing plant pathogenic fungi, although much of this information originates from the study of agricultural crops, rather than natural plant communities. Marois and Coleman (1995) also suggest that understanding the ecology of successions of phylloplane fungi could point to the appropriate species to combat pathogenic fungi. They suggest that the succession of fungi colonizing developing leaves is analogous to the colonization of freshly fallen dead leaves in the decomposer community. Their hypothesis is that an "r"-selected plant pathogenic fungus would best be controlled through competition, whereas a "K"-selected pathogen would be effectively combated by a mycoparasite.

The competitive interactions between saprotrophic and pathogenic fungi also occur in the rhizosphere. Whipps (1997) reviewed some of these fungal–fungal interactions, showing that the types of interactions could be classified as "direct antagonism" by mycoparasitism, antibiosis, or direct competition or through "indirect interactions" by the fungal induction of resistance and by plant growth promotion. An example of sustained mycoparasitism in the rhizosphere is that of the control of *Rhizoctonia* by *Verticillium biguttatum* (Van den Boogert and Velis, 1992; Van den Boogert and Deacon, 1994). The production of antibiotics by fungi in the rhizosphere has been reviewed by Lynch (1990). The introduction of nonvirulent forms of pathogenic fungal species has been shown to induce disease resistance to plants (Mandeel and Baker, 1991; Martyn et al., 1991). However, most of these studies have been conducted in agricultural settings or in artificial conditions. The importance of these interactions in natural ecosystems and their influence on plant fitness is largely unknown.

## 4.5 ALLELOPATHY

Allelopathy is usually regarded as the interactive effect of one plant species on another by the excretion of toxic metabolites. However, if we take a rather broader view of this competitive interaction we can invoke the effect of one plant on the soil microbial community, both fungi and bacteria, which may have subsequent effects on the establishment, growth, and survival of another plant species.

Lichens are particularly well known for containing plant inhibitory substances, where the foliose lichen *Cladonia cristatella* significantly inhibits germination of seeds of a number of plant species (Brown and Mikola, 1974). A number of *Cladonia* species were shown to reduce the growth of a number of both saprotrophic and ectomycorrhizal fungi, thus potentially reducing the efficacy of the mycorrhizal association, although the effect of this reduction in mycorrhizal function was not correlated to a reduction of phosphorus uptake by the plants (Brown and Mikola, 1974). The addition of *Cladonia alpestris* lichen to the soil surface significantly reduced the growth and survival of pine and spruce seedlings in a tree nursery, but had little effect on birch growth or survival (Table 4.10; Brown and Mikola, 1974), and a ground cover of more than 10% of *Cladonia alpestris* in natural ecosystems was shown to reduce Scots pine seedling growth in comparison with similar and higher ground cover of *C. rangiferina* or *Arctostaphylos uva-ursi*. In similar studies, Fisher (1979) grew white spruce and Jack pine seedlings in the greenhouse in acidic sandy soil with mulches of the lichens *Cladonia rangifera* or *Cladonia alpestris*. After 15 weeks of growth, the soil was spiked with radioactively labeled phosphorus ($^{32}P\text{-}PO_4$), and the uptake of the radiolabel measured in plants harvested at

**Table 4.10  Seedling Plant Height and Survival of Three Tree Species in a Tree Nursery Where Soil Surface Was Left Bare or Had Been Covered with Foliose Lichen, *Cladonia alpestris*, Collected from a Surrounding Forest**

| Tree Species | *Cladonia alpestris* Plots | | Control Plots | |
|---|---|---|---|---|
| | Height (cm) | Survival (%) | Height (cm) | Survival (%) |
| *Pinus sylvestris* | 22.2 | 93 | 25.8 | 100 |
| *Picea abies* | 21.4 | 84 | 31.2 | 100 |
| *Betula verrucosa* | 105.0 | 70 | 108.0 | 67 |

*Source:* Brown, R.T., Mikola, P., *Acta For. Fenn.*, 141, 1–22, 1974.

17 weeks. Lichens significantly reduced tree seedling growth and phosphate uptake. Nitrogen and phosphorus plant content were reduced, but nitrogen was reduced less than phosphorus. Potassium, calcium, and magnesium plant content was not affected (Table 4.11).

Lawrey (1986, 1989) made acetone extracts from lichen species (*Aspicilia gibbosa* and *Lasallia papulosa*) that were shown to be readily eaten by the slug *Pallifera varia*, and from lichens that were avoided by the slug (*Flavoparmelia baltimorensis* and *Xanthoparmelia cumberlandia*). The extracts were added to cultures of the bacteria *Bacillus megaterium*, *B. subtilis*, *Staphylococcus aureus*, *Escherichia coli*, and *Pseudomonas aeruginosa*. The lichens that were avoided by the herbivore had the greatest antagonistic effects against the bacteria *Bacillus megaterium*, *B. subtilis*, and *Staphylococcus aureus*, whereas the growth of *Escherichia coli* and *Pseudomonas aeruginosa* was unaffected by any lichen. Defensive chemicals produced by the two lichens that were avoided by the herbivore included norstictic, stictic, usnic, constictic, connorstic, gyrophoric, caperatic, and protocetaric acids. In a controlled experiment, Lawrey (1989) also showed that vulpinic, usnic, and stictic acids reduced the growth of the same susceptible bacterial species. Thus, there appears to be a universal defensive role of these chemicals against herbivory or microbial attack, and as an allelopathic agent (Table 4.12).

Changes in the microfungal flora of soils as a result of continuous agricultural practices have been shown to affect the ability of that soil to support plant growth and to effectively carry out the processes of leaf litter decomposition. This has been suggested to be a form of allelopathy (Golovko, 1999). In a comparative study between virgin cereal-meadow steppes of chernozem soils in the Mikhailovsky Reserve of

**Table 4.11  Allelopathic Effect of the Lichen *Cladonia* spp. on Tree Species**

| Tree Species | Lichen Species | Dry Weight | $^{32}$P Uptake | N Content | P Content |
|---|---|---|---|---|---|
| Jack pine | *Cladonia alpestris* | 56 | 62 | 85 | 81 |
| | *Cladonia rangifera* | 60 | 56 | 83 | 71 |
| White spruce | *Cladonia alpestris* | 58 | 48 | 77 | 62 |
| | *Cladonia rangifera* | 64 | 43 | 79 | 54 |

*Source:* After Fisher, R., *For. Sci.*, 25, 256–260, 1979.
*Note:* Values are expressed as percentage of the control trees, without lichen mulch.

Table 4.12   Reduction in Spore Germination of the Moss, *Funaria hygrometrica*, in the Presence of Lichen Compounds

| Lichen Compound | % Germination |
| --- | --- |
| Vulpinic acid | 37 |
| Usnic acid | 62 |
| Evernic acid | 69 |
| Psoromic acid | 72 |
| Lecanoric acid | 94 |
| Atranoric acid | 97 |
| Stictic acid | 99 |
| Fumarprotocetaric acid | 100 |

*Source:* Lawrey, J.D., *Bryologist*, 89, 111–122, 1986.
*Note:* Germination is expressed as a percentage of control.

the Ukraine and similar land in which the vegetation was periodically or frequently cut, Golovko (1999) showed that the microfungal community increased from 14 to 24 species as disturbance increased. Using cress as a bioassay plant, Golovko (1999) showed that many of the fungi associated with highly disturbed sites inhibited plant growth. Indeed, *Aspergillus fumigatus* completely inhibited cress growth. Along with changes in polyphenols in soil, he suggested that the change in fungal and bacterial communities were causal agents of allelopathy.

Increased development of plant defense chemicals in leaves can be induced by the presence of leaf pathogenic fungi. This increased concentration of polyphenols causes a significant reduction in the decomposition of plant litters (Whitham and Schweitzer, 2002). The higher content of these chemicals reduces the colonization of litter by saprotrophic and mycorrhizal fungi. Koide et al. (1998a,b) demonstrated a significant reduction in the growth of ectomycorrhizal fungi as a result of the effects of high concentrations of polyphenols. This effect could be considered analogous to an allelopathic effect if the reduction in mycorrhizal colonization of a host plant led to a reduction in the host plant fitness. However, Molofsky et al. (2000) contend that the survival and fitness of the annual plant *Cardamine pensylvanica* is related to leaf litter mass and persistence rather than litter quality. Thus, it is possible that the controls exerted by leaf litters are different for annual and perennial plants.

## 4.6 CONCLUSIONS

Although there are many books and journals devoted to the subject of fungi as plant pathogens, most of these are focused on pathogens of crop plants or plants of horticultural interest. When plants are taken from their natural environment, and particularly when planted in monoculture, the incidence of disease is increased because of the stresses placed on the plant, increased nutrient content, or proximity to other conspecifics subject to disease. In natural ecosystems, the devastation of plant biomass on the scale of that seen in agricultural settings is rare. Only in the case of an introduced exotic plant pathogen is such devastation of a natural ecosystem seen. It is, therefore, thought that the balance between plants and their pathogens has been an arms race over the course of evolution, such that neither the host plant nor the pathogen becomes an outright winner. Where imbalances do occur, because of invasions or environmental stressors altering this balance between host and pathogen, significant changes in the plant community composition can be see with knock-on effects ramifying through the populations of animals dependent on the declining plant species. Specific examples can be seen in the case of exotic fungi causing sudden oak decline, chestnut blight, and dogwood anthracnose.

# REFERENCES

Abdalla, M. E. and G. M. Abdel-Fattah. 2000. Influence of the endomycorrhizal fungus *Glomus mosseae* on the development of peanut pod rot disease in Egypt. *Mycorrhiza* 10:29–35.

Ahmadzadeh, M. and A. S. Tehrani. 2009. Evaluation of fluorescent pseudomonads for plant growth promotion, anti-fungal activity against *Rhizoctonia solani* on common bean, and biocontrol potential. *Biol. Control* 48:101–107.

Akter, S., J. Kadir, A. S. Juraimi, H. M. Saud, and S. Elmahdi. 2014. Isolation and identification of antagonistic bacteria from phylloplane of rice as biocontrol agents for sheath blight. *J. Environ. Biol.* 35:1095–1100.

Alexander, H. M. 1990. Dynamics of plant–pathogen interactions in natural plant communities. In: *Pests, Pathogens and Plant Communities*, ed. J. J. Burdon and S. R. Leather, 31–45. Oxford: Blackwell Scientific.

Alexander, H. M. and J. D. Mihail. 2000. Seedling disease in an annual legume: Consequences for seedling mortality, plant size, and population seed production. *Oecologia* 122: 346–353.

Amaranthus, M. P. and D. A. Perry. 1994. The functioning of ectomycorrhizal fungi in the field: Linkages in space and time. *Plant and Soil* 159:133–140.

Anagnostakis, S. L. 1987. Chestnut blight: The classical problem of an introduced pathogen. *Mycologia* 79:23–37.

Anderson, I. C., S. M. Chambers, and J. W. G. Cairney. 2001. Variation in nitrogen source utilization by *Pisolithus* isolates maintained in axenic culture. *Mycorrhiza* 11:53–56.

Anderson, J. P., C. A. Gleason, R. C. Foley, P. H. Thrall, J. B. Burdon, and K. B. Singh. 2010. Plants versus pathogens: An evolutionary arms race. *Funct. Plant Biol.* 37:499–512.

Appiah, A. A., P. Jennings, and J. A. Turner. 2004. *Phytophthora ramorum*: One pathogen and many diseases, an emerging threat to forest ecosystems and ornamental plant life. *Mycologist* 18:145–150.

Augspurger, C. K. 1990. Spatial patterns of damping-off disease during seedling recruitment in tropical forests. In: *Pests, Pathogens and Plant Communities*, ed. J. J. Burdon and S. R. Leather, 131–144. Oxford: Blackwell Scientific.

Aylor, D. E. 2003. Spread of plant disease on a continental scale: Role of aerial dispersal of pathogens. *Ecology* 84:1989–1997.

Aylor, D. E., W. E. Fry, H. Mayton, and J. L. Andrade-Piedra. 2001. Quantifying the rates of release and escape of *Phytopthora infestans* sporangia from potato canopy. *Phytopathol.* 91:1189–1196.

Bagchi, R., R. E. Gallery, S. Gripengerg et al. 2014. Pathogens and insect herbivores drive rainforest plant diversity and competition. *Nature* 506:85–88, doi:10.1038/nature12911.

Balmelli, G., S. Simento, N. Altier, V. Marroni, and J. J. Diez. 2013. Long-term losses caused by foliar diseases on growth and survival of *Eucalyptus globulus* in Uruguay. *New For.* 44:249–263.

Bari, R. and J. D. G. Jones. 2009. Role of plant hormones in plant defense responses. *Plant Mol. Biol,* 69:473–488.

Bélanger, R. R. and T. J. Avis. 2002. Ecological processes and interactions occurring in leaf surface fungi. In: *Phyllosphere Microbiology*, ed. S. E. Lindow, E. I. Hecht-Poinar, and V. J. Elliott, 193–207. St. Paul, MN: APS Press.

Bills, G. F. 1995. Analyses of microfungal diversity from a user's perspective. *Can. J. Bot.* 73:S33–S41.

Blakeman, J. P. 1992. Fungal interaction on plant surfaces. In *The Fungal Community: Its Organization and Role in the Ecosystem*, ed. G. C. Carroll and D. T. Wicklow, 853–867. New York: Marcel Dekker.

Blakeman, J. P. and I. D. S. Brodie. 1977. Competition for nutrients between epiphytic micro-organisms and germination of spores of plant pathogens on beetroot leaves. *Physiol. Plant Pathol.* 10:29–42.

Borowicz, V. A. 2001. Do arbuscular mycorrhizal fungi alter plant–pathogen relations? *Ecology* 82:3057–3068.

Branzanti, B. M., E. Rocca, and A. Pisi. 1999. Effect of ectomycorrhizal fungi on chestnut ink disease. *Mycorrhiza* 9:103–109.

Brasier, C. M. 1990. The unexpected element: Mycovirus involvement in the outcome of two recent pandemics, Dutch elm disease and chestnut blight. In: *Pests, Pathogens and Plant Communities*, ed. J. J. Burdon and S. R. Leather, 289–307. Oxford: Blackwell Scientific.

Brasier, C. M. 1996. *Phytopthora cinnamomi* and oak decline in southern Europe. Environmental constraints including climate change. *Ann. Sci. For.* 53:347–358.

Brasier, C. M. 2001. Rapid evolution of introduced plant pathogens via interspecific hybridization. *BioScience* 51:123–133.

Brown, R. T. and P. Mikola. 1974. The influence of fruticose soil lichens upon the mycorrhizae and seedling growth of forest trees. *Acta For. Fenn.* 141:1–22.

Burdon, J. J., A. H. D. Brown, and A. M. Jarosz. 1990. The spatial scale of genetic interactions in host–pathogen coevolved systems. In: *Pests, Pathogens and Plant Communities*, ed. J. J. Burdon and S. R. Leather, 233–247. Oxford: Blackwell Scientific.

Campos-Soriano, L., J. García-Martínez, and B. san Segundo 2012. The arbuscular mycorrhizal symbiosis promotes the systemic induction of regulatory defense-related genes in rice leaves and confers resistance to pathogen infection. *Mol. Plant Pathol.* 13:579–592.

Chakravarty, P. and S. F. Hwang. 1991. Effect of an ectomycorrhizal fungus, *Laccaria laccata* on *Fusarium* damping-off in *Pinus banksiana* seedlings. *Eur. J. For. Pathol.* 21:97–106.

Chakravarty, P., R. L. Peterson, and B. E. Ellis. 1991. Interaction between the ectomycorrhizal fungus, *Paxillus involutus*, damping-off fungi and *Pinus resinosa* seedlings. *J. Phytopathol.* 132:207–218.

Clark, D. A. and D. B. Clark. 1984. Spacing dynamics of a tropical rain forest tree: Evaluation of the Janzen–Connell model. *Am. Nat.* 124:769–788.

Clay, K. 1997. Fungal endophytes, herbivores and the structure of grassland communities. In: *Multitrophic Interactions in Terrestrial Systems*, ed. A. C. Gange and V. K. Brown, 151–169. Oxford: Blackwell Science.

Cobb, R. C., V. T. Eviner, and D. M. Rizzo. 2013. Mortality and community changes drive sudden oak death impacts on litterfall and soil nitrogen cycling. *New Phytol.* 200:422–431.

Condenso, T. E. and R. K. Meetnemeyer. 2007. Effects of landscape heterogeneity on the emerging forest disease sudden oak death. *J. Ecol.* 95:364–375.

Connell, J. H. 1971. On the role of natural enemies in preventing competitive exclusion in some marine animals and in rain forest trees. In: *Dynamics of Populations: Proceedings of the Advanced Study Institute on Dynamics of Numbers and Populations*, ed. P. J. den Boer and G. R. Gradwell, 289–310. Wageningen, The Netherlands: Center for Agricultural Publishing and Documentation.

Connell, J. H., J. G. Tracey, and L. J. Webb. 1984. Compensatory recruitment, growth and mortality as factors maintaining rain forest tree diversity. *Ecol. Monogr.* 54:141–164.

Crute, I. R. 1990. Resistance to *Bremia lactucae* (downy mildew) in British populations of *Lactuca serriola* (prickly lettuce). In: *Pests, Pathogens and Plant Communities*, ed. J. J. Burdon and S. R. Leather, 203–217. Oxford: Blackwell Scientific.

Cullings, K. W., D. R. Vogler, V. T. Parker, and S. Makhija. 2001. Defoliation effects on the ectomycorrhizal community of a mixed *Pinus contorta/Picea engelmannii* stand in Yellowstone Park. *Oecologia* 127:533–539.

de Jong, M. D. 2000. The BioChon story: Deployment of *Chondrostereum purpureum* to suppress stump sprouting in hardwoods. *Mycologist* 14:58–62.

Dehne, H. W. 1982. Interaction between vesicular–arbuscular mycorrhizal fungi and plant pathogens. *Phytopathol.* 72:1115–1119.

Demir, S. and A. Akkopru. 2007. Use of arbuscular mycorrhizal fungi for biocontrol of soilborne fungal plant pathogens. In: *Biological Control of Plant Diseases*, ed. S. B. Chincholkar and K. G. Mukerji, 17–46. New York: Haworth Press.

Dickinson, C. H. and T. F. Preece. 1976. *Microbiology of Aerial Plant Surfaces.* London: Academic Press.

Donegan, K. K., D. L. Schaller, J. K. Stone, L. M. Ganio, G. Reed, P. B. Hamm, and R. J. Seidler. 1996. Microbial populations, fungal species diversity and plant pathogen levels in field plots of potato plants expressing the *Bacillus thuringiensis* var. *tenebrionis* endotoxin. *Transgenic Res.* 5:25–35.

Duchesne, L. C. 1994. Role of ectomycorrhizal fungi in biocontrol. In *Mycorrhizae and Plant Health*, ed. F. L. Pfleger and R. G. Linderman, 27–45. St. Paul, MN: APS Press, American Phytopathological Society.

El Oirdi, M., T. A. El Rahman, L. Rigano, A. El Hadrami et al. 2011. *Botrytis cinerea* manipulates the antagonistic effects between immune pathways to promote disease development in tomato. *Plant Cell* 23:2405–2421.

Evans, H. C. 1995. Fungi as biocontrol agents of weeds: A tropical perspective. *Can. J. Bot.* 73 (Suppl. 1):S58–S64.

Farr, D. F., G. F. Bills, G. P. Chamuris, and A. Y. Rossman. 1989. *Fungi on Plant and Plant Products in the United States.* St. Paul, MN: APS Press.

Fisher, R. F. 1979. Possible allelopathic effects of reindeer-moss (*Cladonia*) on Jack pine and white spruce. *For. Sci.* 25:256–260.

Fokkema, N. J. 1991. The phyllosphere an ecologically neglected milieu: A plant pathologist's perspective. In: *Microbial Ecology of Leaves*, ed. J. H. Andrews and S. S. Hirano, 3–18. New York: Springer Verlag.

Fokkema, N. J. and M. P. de Nooij. 1981. The effect of fungicides on the microbial balance in the phyllosphere. *EPPO Bull.* 11:303–310.

Fravel, D. R. 1988. Role of antibiotics in the biocontrol of plant diseases. *Annu. Rev. Phytopathol.* 26:75–91.

Gachango, E., W. Kirk, R. Schafer, and P. Wharton. 2012. Evaluation and comparison of bio-control and conventional fungicides for control of postharvest potato tuber diseases. *Biol. Control* 63:115–120.

Garbaye, J. 1991. Biological interactions in the mycorrhizosphere. *Experientia* 47:370–375.

Gilbert, G. S., S. P. Hubbell, and R. B. Foster. 1994. Density and distance-to-adult tree effects of a canker disease of trees in a moist tropical forest. *Oecologia* 98:100–108.

Gilbert, G. S., M. Mejia-Chang, and E. Rojas. 2002. Fungal diversity and plant disease in man-grove forests: Salt excretion as a possible defense mechanism. *Oecologia* 132:278–285.

Golovko, E. A. 1999. Allelopathic soil sickness: Basic and methodological aspects. In *Biodiversity and Allelopathy: From Organisms to Ecosystems in the Pacific*, ed. C. H. Chou, G. R. Waller, and C. Reinhardt. Taipei: Academia Sinica.

Grünwald, N. J., M. Garbellotto, E. M. Goss, K. Heungens, and S. Prospero. 2013. Emergence of the sudden oak death pathogen *Phytophthora ramorum*. *Trends Microbiol.* 20:131–138.

Hansen, E. M. and E. M. Goheen. 2000. *Phelinus weirii* and other native root pathogens as determinants of forest structure and process in western North America. *Annu. Rev. Phytopathol.* 38:515–539.

Hansen, E. M. and J. K. Stone, 2005. Impacts of plant pathogenic fungi on plant communities. In: *The Fungal Community: Its Organization and Role in the Ecosystem* (3rd Edition), ed. J. Dighton, J. F. White, and P. Oudemans, 461–474. Boca Raton, FL: Taylor & Francis.

Harper, J. L. 1990. Pests, pathogens and plant communities: An introduction. In *Pests, Pathogens and Plant Communities*, ed. J. J. Burdon and S. R. Leatehr, Oxford: Blackwell Scientific.

Hättenschwiler, S. and P. M. Vitousek. 2000. The role of polyphenols in terrestrial ecosystem nutrient cycling. *TREE* 15:238–243.

Henricot, B. and C. Prior. 2004. *Phytophthora ramorum*, the cause of sudden oak death ramo-rum leaf blight and dieback. *Mycologist* 18:151–156.

Hiers, J. K. and J. P. Evans. 1997. Effects of anthracnose on dogwood mortality and forest composition of the Cumberland Plateau (USA). *Conserv. Biol.* 11:1430–1435.

Hines, P. J. and J. Marx. 2001. The endless race between plant and pathogen. *Science* 292:2269.

Hol, W. H. G, W. de Boer, and A. Medina. 2014. Beneficial interactions in the rhizosphere. In: *Interactions in Soil: Promoting Plant Growth*, ed. J. Dighton and J. A. Krumins, 59–80. Dordrecht: Springer Science + Business Media.

Holah, J. C., M. V. Wilson, and E. M. Hansen. 1997. Impacts of a native root-rotting pathogen on successional development of old-growth Douglas fir forests. *Oecologia* 111:429–433.

Holdenreider, O., M. Pautasso, P. J. Weisberg, and D. Lonsdale. 2004. Tree diseases and land-scape processes: The challenge of landscape pathology. *Trends Ecol. Evol.* 19:446–452.

Jahn, M., E. Kluge, and S. Enzian. 1996. Influence of climate diversity on fungal diseases of field crops—Evaluation of long-term monitoring data. *Aspects Appl. Biol.* 5:247–252.

Janisiewicz, W. 1996. Ecological diversity, niche overlap, and coexistence of antagonists used in developing mixtures for biocontrol of postharvest diseases of apples. *Phytopathology* 86:473–479.

Janzen, D. H. 1970. Herbivores and the number of tree species in tropical forests. *Am. Nat.* 104:501–528.

Jeffries, P. 1997. Mycoparasitism. In: *The Mycota IV: Environmental and Microbial Relationships.*, ed. D. T. Wicklow and B. Söderström, 149–164. Berlin: Springer-Verlag.

Jung, T., G. W. Hudler, S. L. Jensen-Tracy, H. M. Griffiths, F. Fleischmann, and W. Osswald. 2005. Involvement of *Phytophthora* species in the decline of European beech in Europe and the USA. *Mycologist* 19:159–166.

Kenerley, C. M. and J. H. Andrews. 1990. Interactions of pathogens on plant leaf surfaces. In: *Microbes and Microbial Products as Herbicides*, ed. R. E. Hoagland, 192–217. Washington: Am. Chem. Soc.

Koide, R. T., L. Suomi, and R. Berghage. 1998a. Tree–fungus interactions in ectomycorrhizal symbiosis. In: *Phytochemical Signals and Plant–Microbe Interactions*, ed. J. T. Romeo, K. R. Downum, and R. Verpoorte, Vol. 32, 57–70. New York: Plenum Press.

Koide, R. T., L. Suomi, C. M. Stevens, and L. McCormick. 1998b. Interactions between needles of *Pinus resinosa* and ectomycorrhizal fungi. *New Phytol.* 140:539–547.

Last, F. T. and F. C. Deighton. 1965. The non-parasitic microflora on the surfaces of living leaves. *Trans. Br. Mycol. Soc.* 48:83–99.

Lawrey, J. D. 1986. Biological role of lichen substances. *Bryologist* 89:111–122.

Lawrey, J. D. 1989. Lichen secondary compounds: Evidence for a correspondence between antiherbivore and antimicrobial function. *Bryologist* 92:326–328.

Lui, R.-J. 1995. Effect of vesicular–arbuscular mycorrhizal fungi on *Verticillium* wilt of cotton. *Mycorrhiza* 5:293–297.

Luo, Y., D. O. TeBeest, P. S. Teng, and N. G. Fabellar. 1995. Simulation studies on risk analysis of rice leaf blast epidemics associated with global climate change in several Asian countries. *J. Biogr.* 22:673–678.

Lynch, J. M. 1990. Microbial metabolites. In *The Rhizosphere*, ed. J. M. Lynch. Chichester: John Wiley & Sons.

Madden, L. V., N. Lalancette, G. Hughes, and L. L. Wilson. 2000. Evaluation of a disease warning system for downy mildew of grapes. *Plant Dis.* 84:549–554.

Magan, N. and E. S. Baxter. 1996. Effect of increased $CO_2$ concentration and temperature on the phylloplane mycoflora of winter wheat flag leaves during ripening. *Ann. Appl. Biol.* 129:189–195.

Mandeel, Q. and R. Baker. 1991. Mechanisms involved in biological control of fusarium wilt on cucumber with strains of nonpathogenic *Fusarium oxysporum*. *Phytopathology* 81:462–469.

Marin, D. H., R. A. Romero, M. Guzmán, and T. B. Sutton. 2003. Black sigatoka: An increasing threat to banana cultivation. *Plant Dis.* 87:208–222

Marois, J. J. and P. M. Coleman. 1995. Ecological succession and biological control in the phyllosphere. *Can. J. Bot.* 73 (Suppl. 1):S76–S82.

Martyn, R. D., C. L. Biles, and E. A. Dillard. 1991. Induced resistance to fusarium wilt of water melon under simulated field conditions. *Plant Dis.* 75:874–877.

Marx, D. H. 1969. The influence of ectotrophic ectomycorrhizal fungi on the resistance of pine roots to pathogenic infections: I. Antagonism of mycorrhizal fungi to pathogenic fungi and soil bacteria. *Phytopathology* 59:153–163.

Marx, D. H. 1973. Mycorrhizae and feeder root disease. In *Ectomycorrhizae: The Ecology and Physiology*, ed. G. C. Marks and T. T. Kozlowski, 351–382. New York: Academic Press.

Marx, D. H. 1980. Role of mycorrhizae in forestation of surface mines. Compact Commission and USDA Forest Service. Trees for Reclamatio; Lexington, Kentucky, USA. Interstate Mining Compact Commission and US Department of Agriculture, Forest Service; 1980: 109–116.

Marx, D. H. and C. B. Davey. 1969a. The influence of ectotrophic mycorrhizal fungi on the resistance of pine roots to pathogenic infections: III. Resistance of aseptically formed mycorrhizae to infections by *Phytopthora cinnamomi*. *Phytopathology* 59:549–558.

Marx, D. H. and C. B. Davey. 1969b. The influence of ectotrophic mycorrhizal fungi on the resistance of pine roots to pathogenic infections: IV. Resistance of naturally occurring mycorrhizae to infections by *Phytopthora cinnamomi*. *Phytopathology* 59:559–565.

Moffat, A. S. 2001. Finding new ways to fight plant diseases. *Science* 292:2270–2273

Molofsky, J., J. Lanza, and E. E. Crone. 2000. Plant litter feedback and population dynamics in an annual plant, *Cardamine pensylanica*. *Oecologia* 124:522–528.

Moy, M., F. Belanger, R. Duncan et al. 2000. Identification of epiphyllous mycelial nets on leaves of grasses infected by clavicipitaceous endophytes. *Symbiosis* 28:291–302.

Newbery, D. M., I. J. Alexander, and J. A. Rother. 2000. Does proximity to conspecific adults influence the establishment of ectomycorrhizal trees in rain forest? *New Phytol.* 147:401–409.

Newsham, K. K., A. H. Fitter, and A. R. Watkinson. 1994. Root pathogenic and arbuscular mycorrhizal fungi determine fecundity of asymptomatic plants in the field. *J. Ecol.* 82:805–814.

Newsham, K. K., A. H. Fitter, and A. R. Watkinson. 1995. Multi-functionality and biodiversity in arbuscular mycorrhizas. *TREE* 10:407–441.

Newton, A. C., C. Gravouil, and J. M. Fountaine. 2010. Managing the ecology of foliar pathogens: Ecological tolerance in crops. *Ann. Appl. Biol.* 157:343–359.

Newton, M. R., Kinkel, L. L., and K. J. Leonard. 1997. Competition and density-dependent fitness in a plant parasitic fungus. *Ecology* 78:1774–1784.

Nix-Stohr, S., R. Moshe, and J. Dighton. 2007. Effects of propagule density and survival strategies on establishment and growth: Further investigations in the phylloplane model system. *Microb. Ecol.* 55:38–44.

Omar, M. and W. A. Heather. 1979. Effect of saprophytic phylloplane fungi on germination and development of *Melampsora larici-populina*. *Trans. Br. Mycol. Soc.* 72:225–231.

Onguene, N. A. and T. W. Kuyper. 2002. Importance of the ectomycorrhizal network for seedling survival and ectomycorrhiza formation in rain forests of south Cameroon. *Mycorrhiza* 12:13–17.

Packer, A. and K. Clay. 2000. Soil pathogens and spatial patterns of seedling mortality in a temperate tree. *Nature* 404:278–281.

Palm, M. 2001. Systematics and the impact of invasive fungi on agriculture in the United States. *Bioscience* 51:141–147.

Pandy, R. P., D. K. Arora, and R. C. Dubey. 1993. Antagonistic interactions between fungal pathogens and phylloplane fungi of guava. *Mycopathology* 124:31–39.

Paul, N. D. 1989. The effect of rust (*Puccinia lagenophorae* Cooke) of groundsel (*Senecio vulgaris* L.) in competition with *Euphorbia peplus*. *J. Ecol.* 77:552–564.

Paul, N. D. 1990. Modification of the effects of plant pathogens by other components of natural ecosystems. In: *Pests, Pathogens and Plant Communities*, ed. J. J. Burdon and S. R. Leather, 81–96. Oxford: Blackwell Scientific.

Paul, N. D. and P. G. Ayres. 1987. Water stress modifies intra-specific interference between rust (*Puccinia lagenophorae*) infected and uninfected groundsel (*Senecio vulgaris*). *New Phytol.* 106:555–556.

Pérez, L. I., P. E. Gundel, C. M. Ghersha, and M. Omacini. 2013. Family issues: Fungal endophyte protects host grass from closely related pathogen *Claviceps purpurea*. *Fung. Ecol.* 6:379–386.

Pieckenstain, F. L., M. E. Bazzalo, A. M. I. Roberts, and R. A. Ugalde. 2001. *Epicoccum purpurascens* for biocontrol of *Sclerotina* head rot of sunflower. *Mycol. Res.* 105:77–84.

Polashock, J. J., J. Vaiciunas, and P. V. Oudemans. 2005. Identification of a new *Phytophthora* species causing root and runner rot of cranberry in New Jersey. *Mycology* 95:1237–1243.

Preece, T. F. and C. H. Dickinson. 1971. *Ecology of Leaf Surface Micro-organisms*. London: Academic Press.

Quarles, W. 1999. Plant disease biocontrol and ectomycorrhizae. *IPM Practitioner* 21:1–10.

Rayner, A. D. M. 1993. The fundamental importance of fungi in woodlands. *Br. Wildl.* 4:205–215.

Rayner, A. D. M. 1998. Fountains of the forest—The interconnectedness between trees and fungi. *Mycol. Res.* 102:1441–1449.

Read, D. J. 1998. Plants on the web. *Nature* 396:22–23.

Rishbeth, J. 1963. Stump protection against *Fomes annosus*: III. Inoculation with *Peniophora gigantea. Ann. Appl. Biol* 52:63–73.

Rosas, S. 2007. Role of rhizobacteria in biological control of plant diseases. In: *Biological Control of Plant Diseases*, ed. S. B. Chincholkar and K. G. Mukerji, 75–102. New York: Haworth Press.

Rossman, A. Y. 2001. A special issue on global movement of invasive plants and fungi. *BioScience* 51:93–94.

Saldajeno, M. G. B. and M. Hyakumachi. 2011. The plant growth promoting fungus *Fusarium equisiti* and the arbuscular mycorrhizal fungus *Glomus mosseae* stimulate plant growth and reduce the severity of anthracnose and damping-off diseases in cucumber (*Cucumis sativus*) seedlings. *Ann. Appl. Biol.* 159:28–40.

Salonen, V., M. Vestberg, and M. Vauhkonen. 2001. The effect of host mycorrhizal status on host plant–parasitic plant interactions. *Mycorrhiza* 11:95–100.

Sanders, I. J. and A. H. Fitter. 1992a. The ecology and functioning of vesicular–arbuscular mycorrhizas in co-existing grassland species: I. Seasonal patterns of mycorrhizal occurrence and morphology. *New Phytol.* 120:517–524.

Sanders, I. J. and A. H. Fitter. 1992b. The ecology and functioning of vesicular–arbuscular mycorrhizas in co-existing grassland species: II. Nutrient uptake and growth of vesicular–arbuscular mycorrhizal plants in a semi-natural grassland. *New Phytol.* 120:525–533.

Schisler, D. A. 1997. The impact of phyllosphere microorganisms on mycoherbicide efficacy and development. In: *The Mycota IV: Environmental and Microbial Relationships*, ed. D. T. Wicklow and B. Söderström, 219–235. Berlin: Springer Verlag.

Seddon, B., S. G. Edwards, E. Markellou, and N. E. Malathrakis. 1997. Bacterial antagonist–fungal pathogen interactions on the plant aerial surface. In: *Multitrophic Interactions in Terrestrial Systems*, ed. A. C. Gange and V. K. Brown, 5–25. Oxford: Blackwell Science.

Sharma, M. P., A. Gaur, and K. G. Mukerji. 2007. Arbuscular–mycorrhizal-mediated plant–pathogen interactions and the mechanisms involved. In: *Biological Control of Plant Diseases*, ed. S. B. Chincholkar and K. G. Mukerji, 47–74. New York: Haworth Press.

Sikes, B. A., K. Cottenie, and J. N. Klironomos 2009. Plant and fungal identity determines pathogen protection of plant roots by arbuscular mycorrhizas. *J. Ecol.* 97:1274–1280.

Simard, S. W., M. D. Jones, D. M. Durall, D. A. Perry, D. D. Myrold, and R. Molina. 1997a. Reciprocal transfer of carbon isotopes between ectomycorrhizal *Betula payrifrea* and *Pseudotsuga menziesii. New Phytol.* 137:529–542.

Simard, S. W., D. A. Perry, M. D. Jones, D. D. Myrold, D. M. Durall, and R. Molina. 1997b. Net transfer of carbon between ectomycorrhizal tree species in the field. *Nature* 338:579–582.

Simard, S. W., D. A. Perry, J. E. Sith, and R. Molina. 1997c. Effects of soil trenching on occurrence of ectomycorrhizas of *Pseudotsuga menziesii* seedlings grown in mature forests of *Betula papyrifera* and *Pseudotsuga menziesii. New Phytol.* 136:327–340.

Smith, G. S. 1988. The role of phosphorus nutrition in interactions of vesicular–arbuscular mycorrhizal fungi with soilborne nematodes and fungi. *Phytopathology* 78:371–374.

Smith, S. E. and D. J. Read. 1997. *Mycorrhizal Symbiosis*. San Diego: Academic Press.

Stohr, S. N. and J. Dighton. 2004. Effects of species diversity on establishment and coexistence: A phylloplane fungal community model system. *Microb. Ecol.* 48:431–438.

Sullivan, B. K., T. D. Shreman, V. S. Damare, O. Lilje, and F. H. Gleason. 2013. Potential roles of *Labyrinthula* spp. in global seagrass population declines. *Fung. Ecol.* 6:328–338.

Sundaramoorthy, S., T. Raguchander, N. Ragupathi, and R. Samiyappan. 2012. Combinatorial effect of endophytic and plant growth promoting rhizobacteria against with disease of *Capsicum annum* L. caused by *Fusarium solani*. *Biol. Control* 60:59–67.

Sylvia, D. M. and W. A. Sinclair. 1983. Suppressive influence of *Laccaria laccata* on *Fusarium oxysporum* and on Douglas-fir seedlings. *Phytopathology* 73:384–389.

Termorshuizen, A. J. 2014. Root pathogens. In: *Interactions in Soil: Promoting Plant Growth*, ed. J. Dighton and J. A. Krumins, 119–137. Dordrecht: Springer Science + Business Media.

Thomas, M. R. and R. C. Shattock. 1986. Effects of fungicides on *Drechslera* spp. and leaf surface filamentous saprotrophic fungi on *Lolium perenne*. *Plant Pathol.* 35:120–125.

Traquair, J. A. 1995. Fungal biocontrol of root diseases: Endomycorrhizal suppression of cylindrocarpon root rot. *Can. J. Bot.* 73 (Suppl. 1):S89–S95.

Tsahouridou, P. C. and C. C. Thanassoulopoulos. 2002. Proliferation of *Trichoderma koningii* in the tomato rhizosphere and the suppression of damping-off by *Sclerotium rolfsii*. *Soil Biol. Biochem.* 34:767–776.

Upadhyaya, R. K. and D. K. Arora. 1980. Role of fungal staling growth products in interspecific competition among phylloplane fungi. *Experientia* 36:185–186.

van Alfen, N. K. 2002. Fungal pathogens of plants. In: *Encyclopedia of Life Sciences*, 127–133. John Wiley & Sons.

van den Boogert, P. H. J. F. and J. W. Deacon. 1994. Biotrophic mycoparasitism by *Verticillium biguttatum* on *Rhizoctonia solani*. *Eur. J. Plant Pathol.* 100:137–156.

van den Boogert, P. H. J. F. and H. Velvis. 1992. Population dynamics of the mycoparasite *Verticillium biguttatum* and its host, *Rhizoctonia solani*. *Soil Biol. Biochem.* 24:157–164.

van der Putten, W. H. 2000. Pathogen-driven forest diversity. *Nature* 404:232–233.

Volpin, H. E., Y. Okon, and Y. Kapulnik. 1994. A vesicular–arbuscular mycorrhiza (*Glomus intraradix*) induces a defense response in alfalfa roots. *Plant Physiol.* 104:683–689.

Vurro, M., M. C. Zonno, A. Evidente, A. Andolfi, and P. Montemurro. 2001. Enhancement of efficacy of *Ascochyta caulina* to control *Chenopodium album* by use of phytotoxins and reduce rates of herbicides. *Biol. Control.* 21:182–19.

Watanarojanaporn, N., N. Boonkerd, S. Wongkaew, P. Prommanop, and N. Teaumroong. 2011. Selection of arbuscular mycorrhizal fungi for citris growth promotion and *Phytophthora* suppression. *Sci. Hortic.* 128:423–433.

Whipps, J. M. 1997. Interactions between fungi and plant pathogens in soil and the rhizosphere. In: *Multitrophic Interactions in Terrestrial Systems*, ed. A. C. Gange and V. K. Brown, 47–63. Oxford: Blackwell Science.

Whitham, T. G. and J. A. Schweitzer. 2002. Leaves as islands of spatial and temporal variation: Consequences for plant herbivores, pathogens, communities and ecosystems. In: *Phyllosphere Microbiology*, ed. S. E. Lindow, E. I. Hecht-Poinar, and V. J. Elliott, 279–298. St. Paul, MN: APS Press.

Wingfield, M. J., B. Slippers, J. Roux, and B. D. Wingfield. 2001. Worldwide movement of exotic forest fungi, especially in the tropics and the southern hemisphere. *BioScience* 51:134–140.

Yanes, M. L., L. de la Fuente, N. Altier, and A. Arias. 2012. Characterization of native fluorescent *Pseudomonas* isolates associated with alfalfa roots in Uruguayan agroecosytems. *Biol. Control* 63:287–295.

Yuttavanichakul, W., P. Lawongsa, S. Wonglaew et al. 2012. Improvement of peanut rhizobial inoculant by incorporation of plant growth promoting rhizobacteria (PGPR) as biocontrol against the seed borne fungus, *Aspergillus niger*. *Biol. Control* 63:87–97.

Zhang, Q. and J. C. Zak. 1998. Potential physiological activities of fungi and bacteria in rela-
    tion to plant litter decomposition along a gap size gradient in a natural subtropical forest.
    *Microb. Ecol.* 35:172–179.
Zuccaro, A. and J. I. Mitchell. 2005. Fungal communities of seaweeds In: *The Fungal
    Community: Its Organization and Role in the Ecosystem* (3rd Edition) ed. J. Dighton,
    J. F. White, and P. Oudemans, 533–579. Boca Raton, FL: Taylor & Francis.

# CHAPTER 5

# Fungi and Secondary Productivity

Fungi are an important component of the food supply to many grazing animals. How many of us have picked a mushroom in the woods, only to find it riddled with holes and full of fly larvae and other invertebrates? In many European countries, wild mushrooms are an important component of people's diet. In recent times, however, the cultivation of mushrooms by commercial growers has become more important than personal fungal forays, especially as the commercial production of mushrooms is independent of season. Indeed, the value of mushrooms as a food source for humans has driven mushroom sales to reach about $1.2 billion (USDA, 2015) in the United States in 2013–2014.

It is therefore not surprising that a number of vertebrate and invertebrate animals consume mushrooms as part of their diet (Cave, 1997). Fungi are rich in important nutrients, particularly nitrogen, phosphorus, minerals, and vitamins (Fogel, 1976; Grönwall and Pehrson, 1984) (Table 5.1). Clinton et al. (1999) measured the nutrient content of fungal fruit bodies (mushrooms of both mycorrhizal and saprotrophic basidiomycetes) of a *Nothofagus* forest floor and showed that all elements other than calcium are more concentrated in fungal tissue than the forest floor material. This suggests that fungi would be preferred food resources for many animals. However, much of the nitrogen they contain is in complex forms, such as indigestible cell walls (Cork and Kenagy, 1989a). Thus, for animals to effectively utilize the nutrients in fungi, they must have a complex community of gut symbionts to assist in the breakdown of these compounds. Indeed, experiments conducted by Cork and Kenagy (1989b) showed that the weight of ground squirrels declined when fed entirely on fruit bodies of the hypogeous ectomycorrhizal fungus, *Elaphomyces granulatus*, as more than 80% of the nitrogen was locked up in complex forms and could not be made available in the digestive tract of these animals. Fungi consist of a large amount of water (70–94%), some 8–40% proteins, 40% lipids, 28–85% carbohydrates (mainly complex), high concentrations of P, K, and Se, and high concentrations of vitamins A, B complex, C, D, and K (Fogel and Trappe, 1978; Kinnear et al., 1979; Claridge and Trappe, 2005).

Not only do animals depend on fungi for survival and growth, but they also exert effects on fungi, by dispersal of spores or reduction in fecundity of the fungi. In addition to consumption by vertebrates, the unseen grazing of fungal mycelia by soil fauna is also equally important. Many soil animals are dependent on fungi as food

Table 5.1   Protein and Mineral Content of a Range of Fungal Species Used as Food
by Red Squirrels (*Sciurus vulgaris*)

| Fungal Species | Protein | P | Ca | Mg | K | Na |
|---|---|---|---|---|---|---|
| *Amanita muscaria* | 25 | 0.48 | 0.13 | 0.04 | 7.9 | 0.02 |
| *Pholiota sqaurrosa* | 35 | 0.96 | 0.05 | 0.1 | 3.1 | 0.02 |
| *Cortinarius delibutus* | 17 | 0.55 | 0.09 | 0.07 | 5.4 | 0.02 |
| *Cortinarius armillatus* | 18 | 0.68 | 0.11 | 0.12 | 6.4 | 0.02 |
| *Gomphidius glutinosus* | 18 | 0.56 | 0.08 | 0.11 | 3.8 | 0.02 |
| *Lactarius* sp. | 16 | 0.67 | 0.06 | 0.12 | 4.9 | 0.02 |
| *Lactarius torminosus* | 17 | 0.46 | 0.12 | 0.09 | 3.0 | 0.02 |
| *Lactarius uvidus* | 22 | 0.52 | 0.06 | 0.07 | 3.5 | 0.03 |
| *Lactarius deliciosus* | 30 | 0.60 | 0.06 | 0.1 | 2.5 | 0.02 |
| *Russula flava* | 18 | 0.29 | 0.06 | 0.07 | 3.9 | 0.02 |
| *Boletus* sp. | 15 | 0.53 | 0.09 | 0.05 | 2.2 | 0.02 |
| *Boletus edulis* | 30 | 0.62 | 0.10 | 0.12 | 4.1 | 0.02 |
| *Hydnum repandum* | 26 | 0.52 | 0.16 | 0.06 | 4.1 | 0.03 |
| *Elaphomyces granulatus* | 17 | 0.21 | 0.08 | 0.12 | 0.6 | 0.10 |

*Source:* Grönwall, O., Pehrson, Å., *Oecologia*, 64, 230–231, 1984.

or as modifiers of the plant resources, making them more palatable. Leaf litter in aquatic ecosystems becomes more palatable to the "shredder" community as a result of prior colonization and activity of saprotrophic fungi (Suberkropp, 1992; Graca et al., 1993; Gessner et al., 1997). Fungi support the populations of many groups of collembola, mites, and nematodes in soil (Anderson, 2000; Edwards, 2000; Moore and de Ruiter, 2000; Ruess et al., 2000). In response, these animals exert their influence on the fungal biomass and community composition. This grazing effect can be significant in regulating the function of the fungal community in terms of modifying rates of leaf litter decomposition, competition between fungi for resources, and reducing the efficiency of mycorrhizae to effect nutrient uptake into host plants.

In a number of specific instances, close associations between animals and fungi have evolved. Leaf-cutting ants and termites rely on fungi as a food source to such a degree that they maintain cultures of specific fungal species, exclude others, and tend to the growth of their food supply as if it were an agricultural crop. Also, the close association between bark beetles and the fungi they transport is an essential relationship that permits the larvae of the beetle to obtain sufficient nitrogen from the tree that they invade (Ayres et al., 2000). Thus, it is not enough to just discuss the effects of fungi as food on the maintenance of animal growth and population size; we must also consider some of the complex interactions and feedback effects of grazing on the fungi themselves, as these feedbacks influence ecosystem processes.

## 5.1 FUNGI IN DIET OF VERTEBRATES

The abundance of fungi, particularly hypogeous (subterranean fruiting) ones, in forests can be considerable. Fogel (1976) estimated that there could be between

11,052 and 16,753 fruiting bodies produced per hectare per year in old growth Douglas fir forests in western Oregon. Fruiting accounts for about 2.3–5.4 kg ha$^{-1}$ dry mass of fungus. These fungi have a higher content of nitrogen, phosphorus, potassium, and micronutrients than epigeous fungi (fungi fruiting aboveground), making them a higher-quality food resource for mammals (Fogel and Trappe, 1978; Trappe, 1988). In comparison with available plant parts, many fungi have similar food value, but less fat content, for herbiverous small mammals (Fogel and Trappe, 1978) (Table 5.2). Thus, fungi form a significant proportion of the diet of these animals (Figure 5.1).

Reviews of mammalian mycophagy (Claridge and May, 1994; Claridge and Trappe, 2005) identified 37 species of native and four species of feral mammals as exhibiting mycophagy of some sort. The degree of dependency of each animal species on fungi as a staple or essential part of the diet is difficult to establish; however it is estimated that fungi comprise more than 25% (by volume) of the diet of brush-tailed potoroo (*Potorus longipes*) at all times of the year. Fungi occurred in the feces of these animals 90% or more during most months and never fell below 80%. Other animals, such as the smoky mouse (*Pseudomus femeus*), relied on a diet of seeds and moths during the summer months, when fungal fruiting bodies were unavailable. However, during winter, the smoky mouse along with bush rats (*Rattus fuscipes*)

**Table 5.2 Chemical Composition of Fungi in Comparison with Plant Parts and Meat**

|  |  | Protein | Fat | Carbohydrate | Ash |
|---|---|---|---|---|---|
| Fungi | Agaricus bisporus | 50 | 1.2 | – | 7 |
|  | Boletus edulis | 33 | 5 | 58 | 7 |
|  | Clavaria flava | 1927 | 2 | 47 | 5 |
|  | Lactarius deliciosus | 19 | 7 | 28 | 6 |
|  | Lentinus edodes | 40 | 5 | 54 | 3 |
|  | Marasmius oreades | 35 | 3 | 34 | 10 |
|  | Morchella esculenta | 12 | 2 | 46 | 10 |
|  | Saccharomyces cerevisiae | 14 | 1 | 21 | – |
|  | Suillus granulatus | 21 | 2 | 70 | 6 |
|  | Suillus gervillei | 20 | 2 | 64 | 6 |
|  | Suillus luteus | 17 | 4 | 53 | 6 |
|  | Trichoderma favovirens | 25 | – | 75 | 9 |
|  | Tuber melanosporum | 11 | 2 | 42 | 8 |
| Nuts | Chestnuts | 11 | 7 | 72 | 2 |
|  | Butternut | 28 | 61 | 3 | 3 |
|  | Black walnut | 30 | 57 | 6 | 2 |
|  | Pecan | 10 | 72 | 10 | 2 |
|  | Hickory nuts | 15 | 68 | 7 | 2 |
|  | Filbert nuts | 16 | 64 | 12 | 2 |
|  | Beech nuts | 22 | 52 | 19 | 3 |
| Meat | Chipped beef | 30 | 6 | 1 | – |

*Source:* Fogel, R., Trappe, J.M., *Can. J. Bot.*, 54, 1152–1162, 1978.

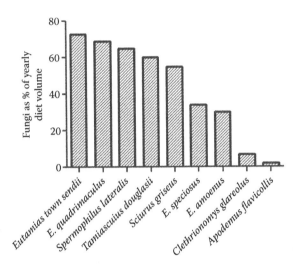

**Figure 5.1**   Percentage contribution of fungi to the annual diet of nine small mammal species in the Pacific Northwest, USA. (Data from Fogel, R., Trappe, J.M., *Northwest Sci.*, 52, 1–31, 1978.)

relied heavily on fungi. The fungi consumed are from a wide variety of taxa; however, lichenized fungi have rarely been reported to be consumed by the Australian megafaunal population. Potoroos consume the most varied fungal diet of any animal (36 fungal taxa), and most are hypogeal fungi. It would appear that body size limits the diversity of fungal species eaten, with rats and mice (<150 g body weight) feeding mainly on arbuscular mycorrhizal spores of the Endogonaceae (Cheal, 1987), whereas large animals, such as feral pigs, eat a wider variety of fungal species.

Small rodents are important vectors of fungal spores derived from mycophagy. Aboveground fruiting mycorrhizal fungi such as *Laccaria* and *Suillus* are part of the diet of mice (*Peromyscus*) (Pérez et al., 2010). However, passage of spores through the gut of these animals has different effects on spore survival, potentially altering relative survival rates. For example spores of *Suillus tomentosus* increased in activity and ability to form mycorrhizae, whereas passage through an animal gut reduced the ability of *Laccaria trichodermophora* spores to form mycorrhizae (Figure 5.2). The close association between truffles and mycophagy by small rodents is also considered to be likely a coevolution where the animals benefit from nutrition provided by the fungus, and the fungus benefits from spore dispersal from a nondehiscent fruiting structure to benefit the mycorrhizal status of forest trees (Schickmann et al., 2012) (Figure 5.3). Pigs and boar are known consumers of truffles. They, together with rodents, are potential vectors for truffle spores as passage through the gut of an animal is the only way in which truffles can disperse spores. However, the efficiency of dispersal relies on the viability of spores after passage through the animal's gut. Using pot-bellied pigs as a model, Piattoni et al. (2014) showed that although spore morphology and physical characteristics of the spore wall (determined by atomic

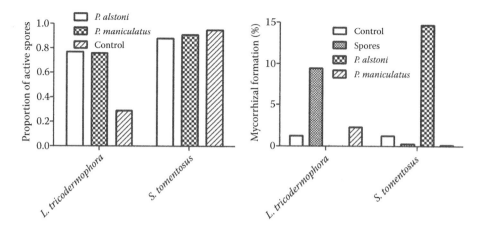

**Figure 5.2**  Influence of ectomycorrhizal spore (*Laccaria trichodermophora* and *Suillus tomentosus*) passage through the gut of two small mammals (*Peromyscus alstoni* and *P. manuiculatus*) on survival of spores (left) and ability to colonize pine tree roots (right). (Data from Perez, F. et al., *Botany*, 90, 1084–1092, 2012.)

force microscopy) were significantly altered after passage through the pig gut, spore viability and the ability to form mycorrhizae was enhanced in comparison to spores that had not passed through the animal's gut.

For example, the reindeer herds of Fennoscandia rely heavily on lichens as a food source during the winter months, without which the populations could not be sustained (Cooper and Wookey, 2001). Mathiesen et al. (2000) showed that more than 25% of the gut contents of Norwegian reindeer consist of lichens in March. The energy value of this diet is regarded as good, despite the fact that the structural carbohydrates differ significantly from plant carbohydrates. The hemicellulose in lichens contains xylan and lichen starch in $\beta$-1-4 and $\beta$-1-3 glucoside linkages. Mathiesen et al. (2000) suggest that it is this factor that induces increased bacterial fermentation in the gut, which results in an increase in the development of food absorptive papillae on the gut wall in reindeer fed exclusively with a lichen diet. On the island of Svalbard in the Barents Sea, areas that have been free of reindeer for a number of years now support large herds. As these animals rely on lichens for a major part of their diet, two aspects of their foraging are posing severe threats to the lichen community and the sustainability of a viable food reserve (Cooper and Wookey, 2001). Because of the relatively slow growth of lichens in this high arctic region, calculated as 2.5 to 10.6 mg g$^{-1}$ week$^{-1}$ relative growth rate, the density of reindeer and their grazing activity is likely to outpace lichen growth. Studies in Finland by Kumpula (2001) show that reindeer consumed up to 2.6 kg lichens day$^{-1}$ during the most intensive digging period, when snow covers the ground. With the assumption that a reindeer grazes an area of approximately 30 m$^2$ day$^{-1}$ during the period of snow cover, and calculating the energy requirements of a reindeer, Kumpula (2001) estimated that each reindeer requires some 1000 kg ha$^{-1}$ dry weight of lichens for sustenance during winter. In addition to the grazing pressure,

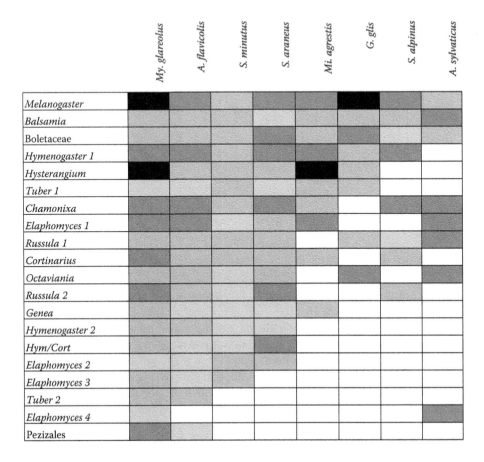

**Figure 5.3** Heat map of index of ectomycorrhizal spore egestion from small mammals showing food preferences (density of color indicates spore abundance). Animal genera: My, *Myodes*; A, *Apodemus*; S, *Sorex*; Mi, *Microtus*; G, *Glis*. (Data from Schickmann, S. et al., *Oecologia*, 170, 395–409, 2012.)

*per se*, trampling of lichens by reindeer herds is an important reason for the decline in both species diversity and biomass of lichens in areas where reindeer herd density is high. However, there are not enough data on the rates of growth of lichens to be able to predict large mammal carrying capacity based on the reliance of these animals on a predominantly lichen diet (Crittenden, 2000). It is interesting to note that, although grazing reduces lichen biomass, the presence of reindeer significantly increases the lichen nitrogen content from 0.43% to 0.91%, whereas no similar increase in N concentration is observed in Scots pine or *Empetrum* shrubs, which are the dominant vegetation of the area (Stark et al., 2000). Stark et al. (2000) also noted an increase in the abundance of bacteria and fungal-feeding nematodes in reindeer-grazed area, suggesting a more general increase in microbial activity induced by the presence of the reindeer herd.

The ranging patterns of mycophagous vertebrate may be related to the distribution of their fungal food. In Bolivia, approximately 35% of the diet of Goeldi's monkeys (*Callimico goeldii*) is fungal, and the large range of a family group of some 115–150 ha is attributable to the low density and patchy distribution of the ephemeral fungal fruiting bodies (*Auricularia* and bamboo fungi, *Ascopolyporus*), which provide a large proportion of their diet (Porter and Garber, 2010). This is especially important to the behavior of this monkey species as it devotes up to 63% of its feeding time consuming fungi, whereas most other mycophagous primates spend only less than 5% of their feeding time on fungi (Hanson et al., 2003).

Forest management can alter fungal fruit body production and, hence, influence food availability for small mammals. Control burning to reduce fuel load in *Eucalyptus* forests reduced the number of fruit bodies of primarily hypogeous fungi on which a variety of small mammals depend for food in Tasmania to between 2% and 5% of the unburned areas (Trappe et al., 2006). Forests are becoming managed for mushroom production other than truffles. Forest manipulations such as mulching significantly reduced the production of saffron milk cap mushrooms (*Lactarius deliciosus*) in radiata pine plantations for human consumption (Guerin-Laguette et al., 2014).

## 5.2 FUNGI IN DIET OF INVERTEBRATES

Reviews of invertebrate fungivory can be found in the work of Shaw (1992), McGonnigle (1997), and Ruess and Lussenhop (2005). There are interactions between a range of fungal taxa and functional groups of fungi and insects (Wilding et al., 1989). There are a number of groups of invertebrates that inhabit the mushroom fruiting structures of basidiomycete fungi. Large, fleshy mushrooms are often heavily invaded by dipteran larvae. Studies of fly larvae that consume mushrooms have shown that there is little correlation between the fungi considered poisonous to humans and those consumed by invertebrates. Indeed, Jaenike et al. (1983) found that many species of the fly *Drosophila* were tolerant of the toxic compound, α-amanitin, of *Amanatia* spp. High densities of collembola can often be found gazing the surface and spores of less fleshy species, such as *Laccaria* spp. Hanski (1989) reviewed the interactions between fungi and insects from the point of view of fungi as insect food. He suggests that the spatial distribution of fungal fruiting bodies can influence the feeding activities of aboveground fungiverous insects. In the same way, he suggests that the seasonal appearance of basidiomycete fruit bodies, in particular, can influence the growth and development of insect larvae and consequently, the fecundity of the adult insect. This may be a determining factor in why most fungal feeding insects are polyphagous, rather than monophagous; they have a greater chance of finding at least one fungal species fruiting at any time (Table 5.3).

Snow fungi (*Typhula* spp.) abundant in tundra and boreal coniferous forest provide a significant portion of the diet of microarthropods (Acari and Collembola). Using the specific isotope signature of snow fungi, Bokhorst and Wardle (2014) showed that these fungi were significant in providing food during winter.

Table 5.3    Degree of Polyphagy in Diptera Breeding in Nine Genera
of Agaricales

| Fungal Genus | Number of Species Used as Food by Diptera Larvae |
|---|---|
| Amanita | 6 |
| Hygrophorus | 4 |
| Cortinarius | 5 |
| Russula | 17 |
| Boletus | 5 |
| Tricholoma | 6 |
| Lactarius | 14 |
| Suillius | 5 |
| Leccinum | 5 |
| Small genera | 37 |

Source: Hanski, I., in: Insect–Fungus Interactions, ed. Wilding, N. et al., 25–68,
Academic Press, London, 1989.

It has been shown that not all fungi are equal in their ability to provide the necessary nutrients for adequate growth, and also that the specific secondary metabolites produced by certain fungal species act as deterrents to animal grazers. Thus, not all fungi are equally palatable to specific animals. From the limited evidence in the literature on fungal selection by a variety of faunal groups, it would appear that there is no consistent pattern in preference of specific species and avoidance of others throughout all faunal groups. Thus, the preferred fungal species varies between animal groups and even between genera and species within the same faunal taxon. Soil microfaunal feeding preferences have been determined by numerous feeding trials in the laboratory. Reddy and Das (1983) provided evidence to suggest that mites showed little food selection, when offered single or mixed cultures of *Trichoderma*, *Cladosporium*, and *Pythium*, whereas collembola preferred a mixed microfungal diet. In addition, they demonstrated that different fungi had different food values and resulted in differences in total numbers of animals at the end of a 9-week experiment. Compared to the control food (agar medium alone), they showed that some fungi were more beneficial for population growth whereas others were detrimental (Table 5.4).

Many animals have distinct preferences for certain species of fungi and dislike for other species. Their selection, however, must be based on other characteristics of the fungi than the poisons that affect humans. Parkinson et al. (1979) demonstrated that the collembolan *Onichiurus amatus* actively avoided a particular basidiomycete fungus that caused its death, even without the fungus being ingested. Shaw (1988) compared the palatability of a range of ectomycorrhizal and saprotrophic fungi to the same collembolan species and concluded that there was a consistent hierarchy of preferences (Table 5.5).

Thimm and Larink (1995) showed that each of four collembolan species (*Folsomia candida*, *Onychiurus fimatus*, *Sinella coeca*, and *Proisotoma minuta*) out

Table 5.4   Influence of Three Microfungi on Numbers of Soil Arthropods Remaining in Culture after 9 Weeks, Compared to Control Systems Containing Agar Medium Alone

| Microarthropod | Trichoderma | Cladosporium | Pythium | Control |
|---|---|---|---|---|
| Mesostigmata | 2 | 3 | 530 | 250 |
| Prostigmata | 90 | 25 | 70 | 18 |
| Cryptosigmata | 3 | 6 | 18 | 5 |
| Total mites | 95 | 34 | 618 | 273 |
| Isotomidae | 5 | 5 | 4 | 45 |
| Entomobryidae | 1 | 1 | – | 5 |
| Total collembola | 6 | 6 | 4 | 50 |

Source: Reddy, M.V., Das, P.K., J. Soil Biol. Ecol., 3, 1–6, 1983.

Table 5.5   Hierarchy of Feeding Preferences of Collembolan Onichiurus amatus When Offered a Range of Fungal Species Grown in Agar Culture

| Fungal Species | Mean% of Fungal Colony Area Consumed | Mean Fecal Count per Culture Vessel |
|---|---|---|
| Marasmius androsaceus | 72.2 | 74.8 |
| Laccaria proxima | 41.4 | 70.8 |
| Lactarius rufus | 55.7 | 64.5 |
| Suillus luteus | 50.7 | 48.9 |
| Mycena galopus | 68.2 | 19.2 |
| Suillus bovinus | 18.2 | 16.6 |
| Rhizpogon roseolus | 20.7 | 13.8 |
| Paxillus involutus | 21.3 | 13.1 |
| Mycena epiterygia | 24.2 | 10.6 |
| Piolithus tinctorius | 0.2 | 2.1 |
| Clitocybe sp. | 1.2 | 1.1 |
| Hebeloma crustuliniforme | 1.7 | 1.0 |

Source: Shaw, P.J.A., Pedobiologia, 31, 179–187, 1988.

of the five tested had a preference for a different species of arbuscular mycorrhizal fungus. *Xenylla grisea*, however, did not show a fungal feeding preference, but was observed to be feeding mainly on nonmycorrhizal root tissue.

On agar, the collembolan *Folsomia candida* grew better when fed with *Absidia* or *Cladosporium* than with *Penicillium*, but when presented with these fungi in the presence of leaf litters, *Penicillium* proved to be an equally suitable food for the animals, and litter preference changed from willow and alder to oak and willow (Heděnc et al., 2013). This suggests that although there is hierarchical food preference, this may be modified by environmental conditions. By comparing the feeding preferences of six collembolan species on eight soil fungal species, Jørgensen et al. (2003) show significant different choices by the collembolan. This is suggested to be a reason for trophic niche separation and a partial explanation for the high biodiversity in soil. Aggregations of fungal feeding macroarthropods are suggested to enhance fungal diversity in decaying wood (A'Bear et al., 2013). The outcome of competition between pairwise interactions between wood decay fungi is significantly influenced by isopod grazing, but that influence is modified by changes in temperature, suggesting that potential climate change my influence fungal community composition (A'Bear et al., 2013; Figure 5.4).

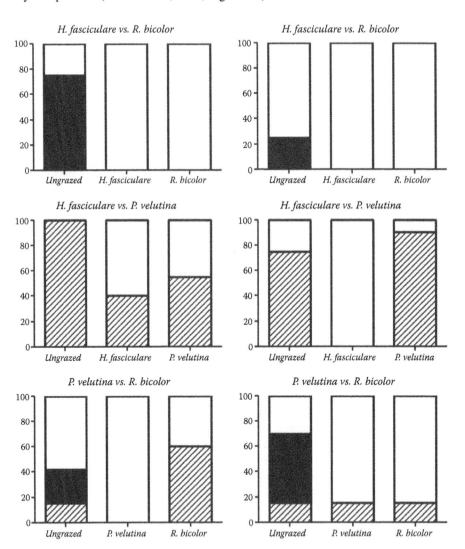

**Figure 5.4** Outcome of pairwise competition between three wood rotting fungi (percentage of outcomes) in the presence or absence of isopod grazing at ambient temperature (left series of graphs) and 4°C elevated temperature (right series of graphs). Dark bar = *R. bicolor* wins, hatched bar = *P. velutina* wins, open bar = draw and *H. fasciculare* never won. (Data from A'Bear, A.D. et al., *Fung. Ecol.*, 6, 137–140, 2013.)

Faunal grazing on arbuscular mycorrhizal fungal extraradical hyphae has been shown to reduce the efficiency of the mycorrhizae in acquiring nutrients, particularly phosphorus, for the plant. Warnock et al. (1982) showed that there was a strong interaction between collembolan density and the growth of the host plant. The effect of this grazing is likely to be more important in agroecosystems, where the diversity of soil fauna is reduced and high densities of collembola can occur in the absence of predators. In addition it has been shown that nematode feeding on mycorrhizal fungal hyphae can also reduce the effectiveness of the mycorrhizal association and has the effect of altering a plant's competitive fitness (Brussard et al., 1993).

In contrast to the suggestion that collembolan feed preferentially on saprotrophic fungi compared to arbuscular mycorrhizal fungi (Klironomos et al., 1999; Johnson et al., 2005), in a study on grassland soils, using natural abundance isotopes, Jonas et al. (2007) found that collembolan consumed both mycorrhizal and saprotrophic fungi, although saprotrophic fungi may provide higher quality food (Klironomos et al., 1999; Figure 5.5). Arbuscular mycorrhizal fungi provided less nutrition—measured by growth and fecundity of the collembolan *Folsomia candida* and *F. fimetaria*—than saprotrophic or pathogenic soil fungi (Larsen et al., 2008). The greatest fecundity was achieved with feeding on the pathogenic fungus *Fusarium culmorum*, although there was no relationship between food selection and food quality as measured by the C/N ratio.

Nematodes are ubiquitous in soils. Fungiverous nematodes may feed on a variety of fungal species, and the selection of fungi to eat may be linked to both the palatability and nutritional value of the fungi. Sutherland and Fortin (1968) provided a

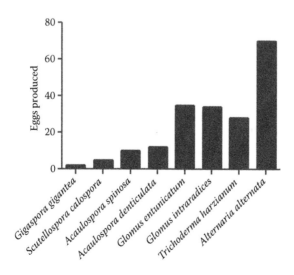

**Figure 5.5** Number of eggs produced by the collembolan, *Folsomia candida,* when feeding on arbuscular mycorrhizal and saprotrophic fungi. (Data from Klironomos, J.N. et al., *Funct. Ecol.*, 13, 756–761, 1999.)

choice of seven ectomycorrhizal fungi to the nematode *Aphelenchus avenae* and found that *Amanita rubescens* provided the best resource for nematode reproduction. Three species—*Suillus, Cenococcum geophilum*, and *Russula emetica*—provided similar nutritional value, but *Rhizopogon roseolus* had a negative impact on nematode numbers. Because of the intensity of nematode grazing on *Suillus granulatus*, the authors found that this nematode could prevent the development of mycorrhizal symbiosis with *Pinus resinosa* when present in a tripartite association in culture. It is suggested, however, that the grazing pressure of nematodes is unlikely to be intense enough to reduce the structure or function of established mycorrhizal associations in natural ecosystems. The nematode *Aphelenchoides saprophilus* is sustained at a higher population density when fed on the ectomycorrhizal fungi *Lactarius rufus* and *Laccaria laccata* than on other mycorrhizal or saprotrophic fungi tested (Ruess and Dighton, 1996). In contrast, a sustainable community could not be maintained on a diet consisting solely of *Paxillus involutus* (Table 5.6). In a comparison of ecto- and ericoid mycorrhizae, nematodes feeding on *Laccaria laccata* produced the highest proportion of females in the population as well as the largest population (Table 5.7), suggesting that this fungus was superior to the others for both population growth and potential fecundity of the population. In natural ecosystems, the diet of *Aphelenchoides* sp. was found to be not only mixed, consisting of both ectomycorrhizal and saprotrophic fungi of various higher taxa, but that the selection of the most favored fungus changed over time (Ruess et al., 2000). It was suggested that this shift in food preference might protect the nematode from an accumulation of toxic compounds from different fungal species.

Of two fungivorous nematode species (*Aphelenchoides bicaudatus* and *Aphelechus avenae*), a larger population of *A. bicaudatus* was sustained on six fungal isolates (Ikonen, 2001). Within this study, the population of both nematode species was greatest when feeding on *Fusarium chlamydosporum* and *Cladosporium herbarum* than three *Penicillium* isolates, and neither nematode species showed reproduction when fed with *Cladosporium cladosporioides*.

**Table 5.6  Numbers of Mixed Species Populations of Nematodes in Petri Plates Supporting the Growth of a Range of Saprotrophic and Ectomycorrhizal Fungal Species**

| Functional Group | Fungal Species | Number of Nematodes | (%) *A. saprophilus* |
|---|---|---|---|
| Mycorrhizal | *Lactarius rufus* | 25,890 | 99 |
| | *Laccaria laccata* | 2633 | 100 |
| | *Paxillus involutus* | 6 | 50 |
| Saprotroph | *Agrocybe gibberosa* | 349 | 95 |
| | *Chaetomium globosum* | 1427 | 99 |
| | *Mucor heimalis* | 24 | 13 |

*Source:* Ruess, L., Dighton, J., *Nematologica*, 42, 330–346, 1996.
*Note:*  The proportion of the nematode, *Aphelenchoides saprophilus*, in the population is given.

Table 5.7 Influence of Fungal Food Species (Mycorrhizal Fungi) on Population of Nematode, *Aphelenchoides saprophilus*, and Percentage of Females in the population; an Index of Population Fecundity

| Fungal Food | Population after 4 Weeks ($\times 10^5$) | Mature Females as % of Population |
|---|---|---|
| *Laccaria laccata* | 2.4 | 11.4 |
| *Cenococcum geophilum* | 1.5 | 7.3 |
| *Lactarius rufus* | 1.4 | 7.8 |
| *Hebeloma sacchariolens* | 0.8 | 2.8 |
| *Paxillus involutus* | 0.8 | 5.8 |
| *Amanita muscaria* | 0.7 | 2.0 |
| *Amanita rubescens* | 0.6 | 1.3 |
| *Hymenoscyphus ericae* | 2.0 | 6.8 |

Source: Ruess, L., Dighton, J., *Nematologica*, 42, 330–346, 1996.

Not only do fungi provide food for invertebrates, but invertebrate activity can change the physicochemical properties of resources in the decomposer system to improve their exploitation by fungi. We can see in the decomposition of plant remains that there are close interactions between soil fauna and fungi, which change as the process of decomposition progresses. In his microscopic study of pine leaf litter decomposition, Ponge (1990, 1991) showed changes in fungal species invading pine needles in concert with faunal invasions. This may have been a result of animals carrying specific fungal propagules with them, but it is more likely that the physical actions of the fauna in the comminution of the litter altered the physicochemical properties of the litter, thereby altering the competitive abilities between the fungi for access to those resources. Indeed, Anderson and Ineson (1984) showed that the decomposition of leaf litter was enhanced in the presence of isopods, which by comminution of the leaf litter increased fungal and bacterial biomass on the litter. The selective grazing on specific, preferred fungi by soil invertebrates results in changes in the competitive strengths of fungi. Thus, faunal grazing can alter the relative abundance of fungal species in the environment. A particularly good example of altered fungal competition mediated by soil arthropods is given by Newell (1984a,b). She found that of two saprotrophic basidiomycete fungi, *Onichiurus latus* preferred to feed on *Marasmius androsaceous* rather than on *Mycena galopus*. At high collembolan densities, the intensity of feeding was enough to significantly reduce the growth of *Marasmius*, such that *Mycena* dominated in the leaf litter. However, in an optimal, ungrazed system, *Marasmius* had a preferred habitat of leaf litter, whereas *Mycena* preferred a soil habitat. The effect of collembolan grazing was therefore to shift the mycelial biomass of each fungus to suboptimal niches. As noted earlier, not all fungal species are equal in their physiological attributes, and a change in species composition can have an effect on fungal-mediated processes in the ecosystem. These can be seen from examples of grazing on saprotrophic, mycorrhizal, and pathogenic fungal functional groups, although the functional effects of these interactions have only been explored to a limited degree. The impact of invertebrates grazing on fungi may have detrimental effects on the performance of fungi. For example, the host-specific ciid beetle grazing on the wood rotting fungus, *Coriolus versicolor*, reduced the reproductive potential of the fungus by up to 64% (Guevara et al., 2000), and the outcome of competition between wood rotting fungi can be significantly influenced by isopod grazing and influenced by changes in temperature (A'Bear et al., 2013).

The interaction between invertebrates and fungi can also be much more complex than a direct trophic interaction. In addition to the maintenance of a fungiverous invertebrate population, the fungus may reap benefits from the interaction. Over time, these close associations may evolve into near mutualisms. An example of this close association is found between the endophytic fungus, *Epichloë*, and *Botanophila* flies, where the fly vectors fungal spermatia to the opposite mating type to develop perithecia on which its larvae depend for development by being attracted to the fungal stroma by volatiles produced by the fungus (Bultman et al., 2000). The authors measured the reproductive output of fungi, the amount of feeding by fly larvae on fungal reproductive tissues, and the mortality of fly eggs and larvae. Contrary to the expected, the reproductive output of fungi did not decrease with increasing egg load but tended to increase as more eggs were laid on fungal stroma (Table 5.8). Larval feeding was only weakly associated with the number of eggs on fungi. The mean surface area of fungal stromata decreased as egg abundance increased, but the overall effect of the flies on *Epichloë* reproduction was positive as the number of preithecia increased with increase in egg number. The fungus, therefore, does not appear to be vulnerable to overconsumption by the fly larvae, suggesting that this could be an example of a balanced antagonism. From a population perspective, because of the heterothallic mating system of this fungus, it is essential for the flies to transfer spermatia to the opposite mating type to develop perithecia, with changes in host plant distribution, and there is a possibility of reproductive isolation with changes in host availability and fly dispersal (Bultman and Leuchtmann, 2008).

The edible wild mushroom fungus *Phlebopus portentosus* is popular in Thailand and has been found to be closely tied to insect galls on more than 20 plant species (Zhang et al., 2015). The fungal hyphae invade the epidermal and cortical cells of the root gall of Pseudococcid mealy bug species. The fungus appears to be essential

**Table 5.8  Relationship between Number of Yucca Moth Eggs Laid on Stroma of *Epichloë* and Number of Perithecia Produced by the Fungus**

| No. Eggs per Stroma | Number of *Epichloë* Perithecia |
| --- | --- |
| 0 | 10 |
| 1 | 35 |
| 2 | 57 |
| 3 | 75 |
| >3 | 70 |

*Source:* Bultman, T.L. et al., *Oecologia*, 124, 85–90, 2000.

for gall formation and may provide food for the animals. It appears likely that honeydew, secreted by the mealy bugs, may provide a resource for fungal growth. *Phlebopus portentosus*, a black bolete, forms galls in association with mealy bugs (Pseudococcidae) on the roots of *Delonix*, *Citrus*, *Coffea*, and *Arctocarpus*, where the gall provides both shelter and food for the bugs; moreover, the whole gall is consumed as human food in China and the Far East (Zhang et al., 2015). Zhang et al. (2015) cite 35 plant genera of 21 families that have fungus–insect galls associated with *Phlebopus portentosus*.

## 5.3 INFLUENCE OF FAUNAL GRAZING ON DECOMPOSITION

During decomposition, the competition for resources between microbes can be influenced by the selective grazing of fungi by soil fauna. Decomposition rates are reduced as the number of fungal species is increased, because the metabolic activity of the competing fungi is greater than the activity of an equivalent biomass of a single fungal species (Wicklow and Yocum, 1982; Robinson et al., 1993). Because of the selective grazing of fungi by soil microarthropods, the diversity of fungal species effecting decomposition is often reduced, thus probably increasing decomposition rates (Lussenhop and Wicklow, 1985). For example, the effect of the increasing complexity of the saprotrophic fungal community on rabbit dung reduced the decomposition rate and lowered fungal spore production (Lussenhop and Wicklow, 1985). This is similar to the findings of Robinson et al. (1993), who showed that a significantly higher level of respiration occurred when fungal species competed for a resource than could be predicted from the combination of respiration of each fungus. However, when Lussenhop and Wicklow (1985) introduced the mycophagous fly larvae of *Lycoriella mali*, there was a 10% increase in the rate of decomposition of rabbit feces at high fungal species diversity and a 150% increase in spore production. They suggest three possible hypotheses for this effect. First, they suggest that the larvae could directly compete with fungi for water-soluble compounds and that this competition becomes stronger as the complexity of fungal interactions increases. Second, larval grazing on mycelia could slow hyphal growth and thus reduce the chance of competitive interactions, such that the fungi can invest more resources to decomposition. Third, the larvae could concentrate enzymes as the number of fungal species increases. However, the exact effect of this interaction is unclear. In a microcosm experiment, Nieminen and Setälä (2001) showed that the presence of fungal-feeding nematodes and bacteria increased the fungal activity in soil. Each factor, nematodes or bacteria, had similar effects, but the two acting together were not additive. Nieminen and Setälä (2001) suggest that nutrient limitation and the dependence on fungi in this particular food web configuration contradicted previous studies showing that food chain length is positively correlated with rates of nutrient cycling processes.

Changes in the decomposition of plant litter in the absence of soil arthropods have been documented. For example, Beare et al. (1992) showed that the removal of soil arthropods reduced leaf litter decomposition by 5% in both conventional till and no-till treatments of an agricultural experiment. The increase in fungal biomass resulting from the alleviation of grazing pressure was correlated to an increase in the nitrogen retention (25% higher than plots with faunal populations intact). It is suggested that this increase in nitrogen content is related to N immobilization in fungal tissue. Indeed, 85% of the net immobilized nitrogen was associated with the saprotrophic fungal community. Thus, the activities of soil fauna not only moderate fungal growth, but also allow greater rates of nutrient mineralization than when animals are removed, thereby leading to greater soil fertility.

The decomposition of alfalfa residues and cellulose was increased by the presence of fungal feeding nematodes (Chen and Ferris, 1999). Where the residues were colonized by the favored fungal food (*Rhizoctonia solani*) for nematodes, both nematode populations and nitrogen mineralization were significantly higher than when the less favored fungal food (*Trichoderma* sp.) was available. This suggests that nematode feeding increased either the biomass or at least the activity of the preferred fungal food, and that this increase in fungal activity was manifested in the increase in an ecosystem function. However, in a contrasting forest ecosystem, Coleman et al. (1990) showed that reduction in microbial predators in ecosystems with high densities of forest soil fauna led to increased decomposition of litter by relief of grazing pressure.

The "top–down" regulation of processes is important in driving the ecosystem-level function of the fungal community. Thus, altering predator abundance or trophic preference of predator population can significantly influence processes effected by fungal and bacterial communities. For example, the specialist nematode feeding mite (*Parazercon radiatus*) reduced the population of both bacterial and fungal feeding nematodes by half. In contrast, the omnivorous mite (*Lysigamasus lapponicus*) increased the density of fungal feeding nematodes over the bacterial feeders. As a result, the specialist nematode predator reduced decomposition rates and nitrogen availability in soil, whereas the generalist predator caused an increase, by stimulating the microbial community as a whole (Laakso and Setälä, 1999). In another example, in the Sitka spruce plantation forests in England, the collembolan *Onichiurus latus* preferred to feed on *Marasmius androsaceous* rather than on *Mycena galopus*. At high collembolan densities, the intensity of feeding was enough to significantly reduce the growth of *Marasmius*, such that *Mycena* dominated in the leaf litter. However, it appeared that the two fungi had optimal habitats based on a vertical separation of resources. *Marasmius* was found to grow nearer the soil surface than *Mycena*, but by relieving the grazing pressure of the collembolan, *Marasmius* was found to grow readily at greater depths and *Mycena* into the less decomposed leaf litter at the soil surface (Newell, 1984a,b). It was suggested, therefore, that the vertical distribution of these two fungal species was constrained by the effects of collembolan grazing pressure. If this is indeed true, it would suggest that the fungi are growing in suboptimal habitats and are probably functioning less efficiently than if provided ideal growth conditions. These top–down controls of microbial trophic interactions have been explored by Scheu and Setälä (2002), Wardle (2002), and Krumins (2014).

Earthworms are also selective in their feeding preferences for different fungal species (Cooke, 1983; Brown, 1995). The passage of fungal spores through the gut of an earthworm selectively modifies the germinability of fungal spores (Moody et al., 1995) and influences the fungal community in worm casts. Tiwari and Mishra (1993) found greater numbers and diversity of fungi in earthworm casts than surrounding soil, thus enhancing the decomposition process. As a resource-rich environment, these earthworm feces attract entomobryid collembola, which feed on the mucus–urine mixture contained in the feces, and the casts become foci for the development of bacterial and fungal communities (Salmon and Ponge, 2001). These conditions lead to the establishment of soil microbial communities that are beneficial in forming and maintaining soil aggregates that are useful for restoration of degraded soils (Scullion and Malik, 2000; Görres et al., 2001).

The differential grazing pressure on different fungal species by soil invertebrates can have a profound effect on the distribution of fungal mycelia and probably on their function. It has often been viewed that grazing of fruiting structures by insects is inconsequential to the survival of the fungal species. Considering the mass of fungus in relation to the biomass of insects feeding on them, it was thought unlikely that the dissemination of spores, the primary purpose of a mushroom, would be impaired regardless of the intensity of invertebrate grazing pressure (Hanski, 1989; Courtney et al., 1990). However, studies of the effect of grazing of fruit bodies of the wood decomposing fungus *Coriolus versicolor* (Guevara et al., 2000) showed that the ciid beetles *Octotemnus glabriculus* and *Cis boleti* significantly reduced the fecundity of the fungus by reducing the reproductive potential by 58% and 30%, respectively, and that this reduction in fitness of the fungus may reduce the colonization potential of this fungus and decrease wood decomposition rates in the ecosystem.

Faunal grazing can change the nature of the growth pattern of the fungus. For example, Dowson et al. (1988) showed that arthropod grazing on the cord-forming fungus (*Steccherium fimbriatum*) induced the development of a fast-growing diffuse mycelium from a slow dense growth form. Hedlund et al. (1991) also showed that the collembolan *Onychiurus armatus* caused *Mortierella isabelina* to shift from appressed hyphae to aerial hyphal growth. These changes in growth form of the fungus can significantly alter the rate at which resources are colonized and utilized. Changing the outcome of fungal–fungal competition because of differential selective grazing, such as that seen in macroarthropods and wood decomposing fungi (A'Bear et al., 2013), can also alter the rates of decomposition. Thus, the indirect effect of grazing may be to alter the rate of the processes that are carried out by the fungi concerned.

Stressed ecosystems support organisms that are adapted to the stressors. Arid and semiarid regions tend to be fungal-dominated ecosystems (Zak, 1993), and as a result, the soil faunal community is dominated by fungivores (Whitford, 1989). It these dry ecosystems, Whitford (1989) suggests that there is indirect evidence that some fungiverous mites can remain inactive, in a state of cryptobiosis. It is well known that a number of nematode species can exist in a state of anhydrobiosis (Demeure and Freckman, 1981), which affords them protection during times of desiccation and is a state in which they can be dispersed by wind (Carroll and Viglierdio, 1981).

## 5.4 INFLUENCE OF FAUNAL GRAZING ON MYCORRHIZAL FUNCTION

The mycorrhizasphere is a region in the soil with high density of fungal mycelia consisting of both saprotrophic and mycorrhizal fungi. It is here that fungal–faunal interactions are likely to occur. Because collembola are one of the main fungivores in soil, collembola grazing tends to significantly reduce mycorrhizal colonization of roots (McGonnigle and Fitter, 1987; Finlay, 1985). Grazing on extraradical arbuscular mycorrhizal hyphae by *Folsomia candida* decreased the effectiveness of the mycorrhizal colonization of leek roots (Warnock et al., 1982). The severing of mycelial connections between the host plant root and soil reduced the effectiveness of the mycorrhiza to increase phosphate inflow over and above that of nonmycorrhizal plants. In a contrasting study, the addition of moderate densities of the collembolan *Folsomia candida* and *Tullbergia granulata* to field-grown soybean resulted in an increase of arbuscular mycorrhizal colonization of roots by 40% and a 5% increase in leaf nitrogen (Lussenhop, 1996). However, there were no effects on phosphorus content of the plants nor on the root nodule number. Lussenhop suggests that the density of animals in his study ($6.8 \times 10^3$ animals m$^2$) was considerably lower than that in other studies ($17 \times 10^3$ animals m$^2$), and the high phosphate availability may have induced this crop to respond differently from those in other studies. He suggests that the relationship between collembolan grazing and mycorrhizal colonization is curvilinear, rather than linear and that the intermediate densities of collembola used in his study could induce compensatory growth of fungal hyphae (Bengtsson et al., 1993) and thus cause an increase in mycorrhizae.

In another study of the effects of collembolan grazing on arbuscular mycorrhizae and consequences for plant growth, Harris and Boerner (1990) found that the growth of *Geranium robertianum* was maximal at low collembolan densities than either at high densities or in the absence of collembola. They noted that the intensity of mycorrhizal colonization of roots was inversely related to collembolan density, but there was no relationship between the intensity of root colonization and phosphorus inflow into plants, although plants with higher root colonization had the best growth. The authors suggest that the benefit of mycorrhizal association may have been through other nutrients than phosphorus (plant tissue concentrations of other nutrients were not measured). They also noted that at high collembolan densities, collembola diversified their feeding to nonfungal resources, and although they report that mycorrhizal colonization of roots was reduced at all collembolan densities, they did not suggest that compensatory growth of extraradical hyphae may have occurred at low animal density, which may have greater benefit for plant growth than the appearance of fungal structures within the root tissue. Bakonyi et al. (2002) increased the density of the collembolan, *Sinella* sp., in microcosms where maize or red fescue was grown in the presence of spores of arbuscular mycorrhizae. Significant reductions in mycorrhizal colonization were found where the collembolan density exceeded 0.2 individuals per gram of soil, but there was a significant increase in root colonization by these fungi as collembolan density increased from 0 to 0.2 animals g$^{-1}$ (Figure 5.6). Hence, it would appear that the impact of grazing

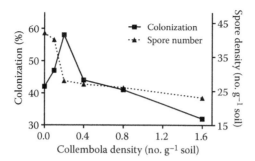

**Figure 5.6** Effect of collembolan density on arbuscular mycorrhizal spore density and colonization of maize roots by mycorrhizae. (Data from Bakonyi, G. et al., *Soil Biol. Biochem.*, 34, 661–664, 2002.)

on arbuscular mycorrhizae is a density-dependent phenomenon, and results from laboratory studies should be interpreted with caution as faunal densities may not be equivalent to those occurring in nature.

Although soil animals have been implicated in the reduction of the mycorrhizal effect of increasing plant growth and nutrient content by their feeding on extraradical hyphae, Klironomos et al. (1999) suggest that it is highly probable that arbuscular mycorrhizal fungi are rarely grazed on in natural ecosystems. By providing the collembolan *Folsomia candida* with a choice of saprotrophic and arbuscular mycorrhizal fungi, they concluded that the saprotroph *Alternaria alternata* was not only the preferred fungal food, but that a diet of exclusively arbuscular mycorrhizal fungi reduced fecundity to the point that no eggs could be produced by the second generation of animals. The assumption from their study was that there is probably little or no effect of collembolan grazing on mycorrhizal benefit to host plants, as this functional group of fungi is avoided in favor of more nutritious saprotrophic species. This finding is in contrast to an earlier study by Klironomos and Ursic (1998), in which they suggested that, despite alternate food items in the form of saprotrophic conidial fungi, collembola significantly reduced arbuscular mycorrhizal connections between the root and soil, thus reducing the beneficial effects of the mycorrhizae on plant growth. Earthworms (*Lumbricus rubellis*) appear to stimulate the growth of arbuscular mycorrhizal fungal hyphae and the growth of *Plantago lanceolata*, but less so under the influence of collembola (*Folsomia candida*), which were shown to preferentially feed on saprotrophic fungi (Gormsen et al., 2004). The degree of damage to mycorrhizal hyphae was shown to be a density-dependent function. However, these results were obtained in culture conditions, and indeed it remains to be shown if these animals can have a significant effect on mycorrhizal function in natural systems.

It has been noticed that the effect of root colonization by a range of ectomycorrhizal fungal species can alter the species composition of protozoa in the mycorrhizasphere (Ingham and Massicotte, 1994). Ingham and Massicotte (1994) showed that different bacterial communities were isolated from roots colonized by a variety of *Rhizopogon* species, *Thelephora terrestris*, and *Mycelium radicis atrovirens*, and that the communities were different on different tree hosts (Figure 5.7). It is suggested that the different mycorrhizae may encourage the growth of different bacterial flora, which, in turn, promoted a different protozoan community. However, they do not present any data to support the hypothesis of mycorrhizae inducing bacterial communities that are unique to the mycorrhizal fungal species.

In choice chamber experiments with the collembolan *Proisotoma minuta* and ectomycorrhizal fungi, collembola significantly slowed the growth rate of *Suillus luteus*, *Pisolithus tinctorius*, *Thelephora terrestris*, and *Laccaria laccata* cultures and the development of mycorrhizae of these species on roots of loblolly pine seedlings (Hiol Hiol et al., 1994). It appears that there are optimal densities of collembola to stimulate root colonization and, possibly, plant growth.

As it is believed that mycorrhizal plants direct much more of their photosynthates into the soil than nonmycorrhizal plants, it is anticipated that the growth of organisms in the detrital food web, which are energy-limited, would benefit from the flow of C through mycorrhizal plants into the belowground ecosystem. Setälä (2000) investigated the potential benefit of the presence of ectomycorrhizal fungi on the roots of Scots pine (*Pinus sylvestris*) on fungiverous and micobiverous representatives of soil mesofauna. Soil was defaunated and then reinoculated with 10 species of soil bacteria, 11 species of saprotrophic soil fungi, and pine seedlings, either infected or noninfected with four ectomycorrhizal fungi. Soil fauna were added with increasing levels of community complexity, including the omnivorous enchytraeid species *Cognettia sphagnetorum*, a collembola (*Hypogastrura assimilis*), and four species

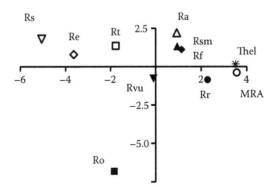

**Figure 5.7**    Principal coordinate analysis of ectomycorrhizae in protozoan species space, showing the difference in protozoan community structure of different ectomycorrhizal fungal species on roots of ponderosa pine. Four groupings appear from this analysis with similar protozoan communities in group 1 [*Rhizopogon subcaerulenscens* (Rs), *R. ellenae* (Re), and *R. truncates* (Rt)] from group 2 [*R. arctostaphylii* (Ra), *R. smithii* (Rsm), and *R. flavofibrillosus* (Rf)], from group 3 [*R. vulgaris* (Rvu), *R. rubescens* (Rr), *Thelephora terrestris* (Thel), and *Mycelium radicis atrovirens* (MRA)] from the outlier *R. occidentalis* (Ro). The data have been reworked from Ingham and Massicotte (1994) and the first two axes account for 23% and 18.2% of the variation, respectively. (Adapted from Ingham, E.R., Massicotte, H.B., *Mycorrhiza*, 5, 53–61, 1994.)

of oribatid mites. After 60 weeks, pine biomass production was significantly greater in the mycorrhizal systems, the total biomass being 1.43 times higher in the presence than absence of ectomycorrhizal (EM) fungi. Similarly, almost 10 times more fungal biomass was detected on pine roots growing in the mycorrhizal than in the nonmycorrhizal systems. However, despite the larger biomass of both pines and their associated fungal community, neither the numbers nor biomasses of the mesofauna significantly differed between the mycorrhizal and nonmycorrhizal systems (Table 5.9). The presence of collembola and *C. sphagnetorum* had a positive influence on pine growth, particularly in the absence of mycorrhizal fungi, whereas oribatid mites had no effects on pine growth. There was therefore no simple and direct relationship between the complexity of the soil faunal community and pine biomass production. For example, the complex systems with each faunal group present did not produce more pine biomass than the simple systems, where *C. sphagnetorum* existed alone. The results of this experiment suggest that the short-term role of ectomycorrhizal fungi in sustaining the detrital food web is less significant than generally considered. Schultz (1991) showed that there was selective grazing between different species of ectomycorrhizal fungi in pure culture by the collembolan *Folsomia candida*, in the same way as Shaw (1985, 1988) (Figure 5.8). The choice of fungal species is not taxonomically determined, as variation in palatability is seen between members of the same fungal taxon. In addition, Schultz's study also showed that the selection of fungi changed with time, when fungi were combined into small communities and direct selection between pairs of groups was allowed (Figure 5.9). The input of plant-derived belowground energy fuels detrital food webs. Wardle et al. (1998) suggest that negative effects on these webs could ensue from global climate change if the nature of the resources entering the system is altered as a result of increased net primary production and reduced resource quality of the litter. They suggest that this detrimental change in energy flow could be mediated through fungal–faunal interactions (Wall and Moore, 1999).

Table 5.9    **Effect of the Presence or Absence of Ectomycorrhizal Associations of Scots Pine Seedlings on the Number of Soil Fauna Supported by Experimental Systems When the Fauna Are Present as Single Species or as a Mixed Community of All Species**

| | | Faunal Density (Number per Experimental System) | |
|---|---|---|---|
| | **Faunal Group** | **No Mycorrhiza** | **With Mycorrhiza** |
| Faunal groups alone | Enchytraeid | 83 | 98 |
| | Collembola | 62 | 18 |
| | Mite | 630 | 1372 |
| Faunal groups in combination | Enchytraeid | 339 | 44 |
| | Collembola | 14 | 20 |
| | Mite | 90 | 1235 |

*Source:* Setälä, H., *Oecologia*, 125, 109–118, 2000.
*Note:* There are no statistically significant differences between the numbers of animals between mycorrhizal treatments because of the high variance around the mean values.

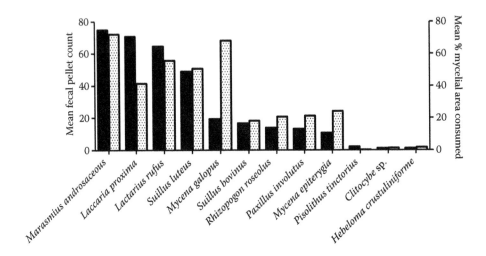

**Figure 5.8**  Hierarchical feeding preference of the collembola *Folsomia candida* on fungi as determined by mean fecal number produced by collembola and percentage consumption of the mycelial colony provided as a food source. (Data from Shaw, P.J.A., *Pedobiologia*, 31, 179–187, 1988.)

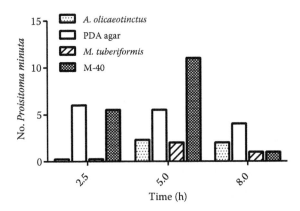

**Figure 5.9**  Changes in feeding behavior of the collembola *Proistoma minuta* over time, when offered a choice of food items of the ectomycorrhizal fungi *Alpova olicaeotinctus*, *Melanogaster tuberiformis*, an unknown isolate M-40 or PDA agar in pairwise combinations. (Data from Schultz, P.A., *Pedobiologia*, 35, 313–325, 1991.)

Indirect effects of herbivory can influence fungal communities and biomass. In a study of the effects of damage to the photosynthetic apparatus of pinyon pine forest trees by the larvae of the moth *Dioryctria albovitella*, Gehring and Whitham (1991, 1994) found that there were trees that were both susceptible and resistant to moth attack. Reduction in photosynthate supply to the roots of susceptible trees by moth larval grazing significantly reduced the number of ectomycorrhizal root tips formed on the trees, compared to the herbivore-resistant trees. When herbivore grazing pressure was artificially removed, the mycorrhizal status of susceptible trees returned to that of resistant trees. The effect of herbivory on mycorrhizal colonization of pine roots and growth of the host plant was greater in the stressed environment of an oligotrophic cinder soil than in a more nutrient-rich, neighboring sandy loam soil (Gehring and Whitham, 1994).

As we have seen, mycorrhizae are capable of altering the chemistry of their host plants, particularly in terms of their nutrient content. The selection of plant parts as food for invertebrates is often dependent on the chemistry of the plant; it is possible that there could be an influence of the mycorrhizal colonization of plant roots and the palatability of aboveground plant parts to grazing herbivores. Goverde et al. (2000) attempted to test this idea, using larvae of the common blue butterfly, *Polyommatus icarus*, that were fed with sprigs of *Lotus corniculatus* (Fabaceae) plants that had been inoculated with one of two different arbuscular mycorrhizal species, with a mixture of these mycorrhizae or with uninoculated plants. Survival of third instar larvae fed with plants colonized by both mycorrhizae was 1.6 times greater than that of larvae fed with nonmycorrhizal plants and 3.8 times higher with a single mycorrhizal species. Larvae fed with mycorrhizal plants had double the weight of those feeding on nonmycorrhizal plants, after 11 days (Table 5.10). These differences are attributable to the improved chemistry of mycorrhizal plants that had 3 times higher leaf P concentration and a higher C/N ratio than the nonmycorrhizal plants. Furthermore, larval consumption, larval food use, and adult lipid concentrations of the butterfly differed between plants inoculated with different mycorrhizal species, suggesting that herbivore performance is mycorrhizal species-specific. On the basis that our understanding of the role of mycorrhizae in natural systems is limited and that evidence indicates that there is much less effect of mycorrhizae on plant growth in natural systems than could be predicted from laboratory and greenhouse studies (Rangeley et al., 1982; Fitter, 1985; Sanders and Fitter, 1992a,b), evidence suggesting an effect of mycorrhizae on herbivores could be a reason for the maintenance of the arbuscular mycorrhizal condition in natural herbaceous ecosystems.

Table 5.10   Influence of Mycorrhizal Association of *Lotus corniculatus* on Plant Chemistry and Performance of Herbiverous Lepidopteran Larva *Polyommatus icarus*

| Mycorrhizal Treatment | Larval Mortality (%) | Leaf Chemistry (mg g$^{-1}$) | | | Larval Fresh Weight at 11 Days (mg) |
|---|---|---|---|---|---|
| | | P | N | C | |
| Nonmycorrhizal | 23 | 3.9 | 5.8 | 40.5 | 13 |
| Species 1 | 6 | 11.9 | 5.4 | 43.2 | 23 |
| Species 2 | 6 | 11.6 | 5.2 | 43.4 | 24 |
| Mixture | 14 | 10.7 | 5.1 | 43.6 | 27 |

*Source:* Goverde, M. et al., *Oecologia*, 125, 362–369, 2000.

The presence of vertically transmitted endophytic fungi in one plant has been shown to protect adjacent plants from insect herbivory. *Neotyphodium*-colonized *Lolium multiflorum* released volatiles that significantly reduced aphid abundance on the leaves of adjacent *Trifolium repens* (Parisi et al., 2014), suggesting that the effect of an endophyte may not be solely within its host plant, but may serve the plant community as a whole (Figure 5.10). The presence of arbuscular mycorrhizal colonization of roots also influences the volatiles emitted by plants in response to aphid attack. When peas became mycorrhizal before exposure to aphids, the plants became less attractive to aphids than nonmycorrhizal plants or plants that became mycorrhizal at the same time as being exposed to aphids. In all plants, the presence of aphids made the plant less attractive to aphids (Babikova et al., 2014; Figure 5.11). These changes in plant susceptibility to aphids are related to mycorrhizal-induced changes in volatile chemistry.

## 5.5 INFLUENCE OF FAUNA ON FUNGAL DISPERSAL

Soil fauna, particularly larger soil organisms, are capable of transporting fungal propagules (McGonnigle, 1997; Dighton et al., 1997). Dighton et al. (1997) suggested that the most effective distribution of microbial propagules would be by mites and collembolan. Of all soil fauna, earthworms are considered the major ecosystem engineers (Lawton and Jones, 1995; Lavelle, 1997). As endogeic species, they are able to create burrows, which facilitate both biotic and abiotic migration of propagules from the surface soil layers to depth. As epigeic species, they are capable of horizontal transport of propagules for long distances. The effect of earthworm activity on soil microbial and faunal community diversity is reviewed by Brown (1995). In this review, he suggests the different effects of the varied ecological strategies of earthworms depending on the size of the worm, its location in or on the soil

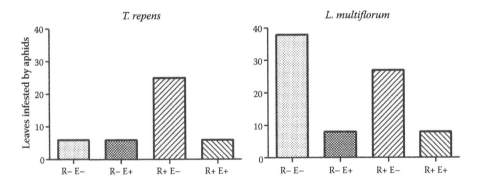

**Figure 5.10**   Influence of fungal endophytes (E) in the presence or absence of *Rhizobium* (R) on the percentage of leaves attacked by aphids. Endophyte protection in *Trifolium repens* only occurs in the presence of *Rhizobium*, but is present with or without *Rhizobium* in *Lolium multiflora*. (Data from Parisi, P.A.G. et al., *Fung. Ecol.*, 9, 61–64, 2014.)

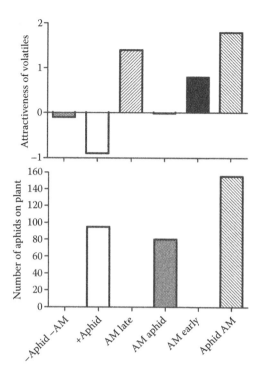

**Figure 5.11** Attractiveness of volatiles to aphids produced by pea plants in the presence or absence of arbuscular mycorrhizae (AM) and/or aphids with AM added before (early) or after (late) aphids. (Data from Babikova, Z. et al., *Funct. Ecol.*, 29, 375–385, 2014.)

surface, and the degree to which the worm is capable of altering the environmental conditions. Direct effects on fungi are through ingestion. These effects may alter the biomass of fungi by direct and indirect grazing, altering spore viability during passage through the gut of the worm, altering the environmental conditions by both physical and chemical means to improve or degrade the quality of microsites for fungal growth, and altering the dispersal patterns of fungal propagules. Dispersal can be enhanced by transport on or in the worm or may be reduced as leaf litter is buried and the sporulating fungi have less ability to disperse from depth in the soil. However, the role of earthworms in the dispersal of arbuscular mycorrhizal spores may be of importance in enhancing colonization of roots of newly emerging plants in the community (Reddell and Spain, 1991; Gange, 1993). This process is of particular importance during secondary succession, where transport of spores from surviving vegetation can be more readily moved into areas being recolonized by plants than could be achieved via physical dispersal alone. However, in clover plants, the presence of the earthworm *Aporrectodea trapezoids* reduced, rather than increased, arbuscular mycorrhizal infection of the host plant because of lateral transport of the inoculum (Figure 5.12) (Pattinson et al., 1997). Pattinson et al. (1997) suggest that the

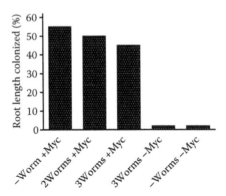

**Figure 5.12**  Colonization of subterranean roots of clover plants planted at 3-cm spacing by arbuscular mycorrhizae. The mycorrhizal inoculum and earthworms were introduced at one end of the chamber. (Data from Pattinson, G.S. et al., *Soil Biol. Biochem.*, 29, 1079–1088, 1997.)

activity of the worms disrupted the extraradical hyphal network of the arbuscular mycorrhizal fungi, preventing interplant infection by mycelial growth.

Earthworms feed preferentially on leaf material that has been previously colonized by fungi, and they are selective with respect to the fungal species colonizing the leaf material (Moody et al., 1995). Three species of earthworm had broadly similar feeding preferences between straw colonized by each of six saprotrophic fungal species, each of which had different enzymatic capabilities (Table 5.11). However, there was differential survival of the fungal spores on passage through the earthworm gut, and the effect was different between two earthworm species, *Lumbricus terrestris* and *Aporrectodea longa* (Moody et al., 1996). *Fusarium* and *Agrocybe* failed to survive passage through the gut of *Lumbricus*, and both *Fusarium* and *Mucor* failed to germinate after passage through the gut of *Aporrectodea*, but the germination of spores of *Chaetomium globosum* was enhanced after passing through this worm species. In a detailed study on the spores of *Mucor heimalis*, Moody et al. (1995)

**Table 5.11**  Mean Number of Straw Baits, Inoculated with Different Saprotrophic Fungal Species Taken by Three Earthworm Species

|  | Number of Straw Baits Taken | | |
|---|---|---|---|
| **Fungal Species** | **Lumbricus terrestris** | **Aporecttodea longa** | **Allolobophora chlorotica** |
| *Fusarium lateritium* | 12.8 | 7.9 | 7.4 |
| *Mucor heimalis* | 11.8 | 6.4 | 4.5 |
| *Trichoderma* sp. | 10.9 | 7.0 | 5.2 |
| *Chaetomium globosusm* | 9.2 | 4.4 | 0.6 |
| *Agrocybe gibberosa* | 5.8 | 2.2 | 3.3 |
| *Sphaerobolus stellatus* | 3.6 | 3.0 | 4.8 |

*Source:* Moody, S.A. et al., *Soil Biol. Biochem.*, 27, 1209–1213, 1995.

showed that the decline in spore germination on passage through an earthworm was attributable to the action of intestinal fluid, not the abrasive action of soil particles as they moved through the gut. Indeed, they established that abrasion by soil particles stimulated spore germination.

Earthworm casts are localized sites for elevated numbers of arbuscular mycorrhizal spores and soil nutrients. In an alley cropping agroecosystem in the tropics, Brussard et al. (1993) showed that earthworm casts had significantly higher contents of major plant nutrients derived from the interplanted tree species than from soil of the inter-row between the crops or from a monocrop (Table 5.12). This shows that the interaction between diverse resources and soil arthropods can stimulate leaf litter decomposition by fungi and bacteria to improve soil fertility. Gange (1993) showed that earthworm feeding activity is concentrated on dead and dying root material and, as a result, the worms ingest large amounts of arbuscular mycorrhizal spores. By depositing spores that are still viable in their casts, earthworms provide local sources of inoculum for establishing plant species. The number of spores per cast increases as vegetation succession proceeds (Figure 5.13). The effect of this process is to enhance the colonization of recruits into the plant community as succession

**Table 5.12  Combined Influence of Tree-Derived Leaf Litter, Consumption by Earthworms, and Enhancement of Decomposition by Fungi and Bacteria Significantly Improve Soil Fertility**

| Position | N | P | K | Ca | Mg |
|---|---|---|---|---|---|
| Under *Leucaena* | 401 | 8 | 42 | 191 | 23 |
| Inter-row | 72 | 1.4 | 7.5 | 27 | 3 |
| Monocrop | 46 | 2 | 5.4 | 19 | 3 |

*Source:* Brussard, L. et al., in: *Soil Organic Matter Dynamics and Sustainability of Tropical Agriculture*, ed. Mulongoy, K., Merckx, R., 241–256, Wiley-Sayce, Leuven, Belgium, 1993.

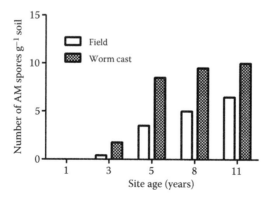

**Figure 5.13**  Mean number of arbuscular mycorrhizal spores, per unit soil weight, occurring in field soil and in earthworm casts at different ages of succession of a natural plant community. (Data from Gange, A.C., *Soil Biol. Biochem.*, 25, 1021–1026, 1993.)

proceeds by providing available spores in patches of enriched nutrient status (worm casts) where the opportunities for seedling establishment are increased. Doube et al. (1994a,b, 1995) have shown that earthworms of the genus *Aporrectodea* are important in assisting plant roots to be colonized by bacteria (especially species of *Pseudomonas*) that are antagonistic to root pathogenic fungi. They have shown that earthworms can be effective vectors for these biocontrol bacteria against the take-all fungus, *Gaumannomyces graminis*.

It is well know that mites are great pests of living fungal cultures, causing contamination between Petri plates. Fungal hyphae and spores are common in the gut of many mite species, particularly the Cryptosigmata (Mitchell and Parkinson, 1976; Price, 1976; Ponge, 1991). Collembola are also major fungal transporters in soil. Dispersal can be effected through carriage on the integument or by passage through the gut of spores and hyphal fragments (Visser, 1985). Visser et al. (1987) isolated more than 100 fungal species from collembola in an aspen woodland ecosystem. These included saprotrophic fungi as well as plant and insect pathogens. However, as Lussenhop (1992) suggests, spores of arbuscular mycorrhizal fungi are generally too large to be dispersed by microarthropods. In contrast to this suggestion, Klironomos and Moutoglis (1999) showed that the collembolan *Folsomia candida* could effect colonization of nonmycorrhizal plants from adjacent arbuscular mycorrhizal plants. However, they showed that the effect of collembola differed between fungal species. Spores of *Acaulospora denticulate* increased their dispersal distance by 10 cm in the presence of the collembola, but the dispersal of spores of *Scutellospora calospora* was reduced when collembola were present, possibly because of spore consumption. Arbuscular mycorrhizal spores may be transported by ants (Friese and Allen, 1993), which may play an important part in the colonizing of bare ground by primary plant colonizers (Allen et al., 1984). Movement of mycorrhizal spores may be greater by ants than it is by earthworms in some ecosystems (McIlveen and Cole, 1976).

Collembola of the genera *Folsomia*, *Hypogastura*, and *Proisotoma* are able to carry spores of the entomopathogenic fungi *Bauveria* and *Metathizium* on their cuticles and in their gut (Dromph, 2001). Between 8% and 78% of the spores carried on the cuticle and between 53% and 100% of the spores in feces gave rise to cultures of all three fungal species, suggesting that this type of transport of entomopathogenic fungi could be important. However, Dromph (2001) showed the ability of these fungi to form colonies on agar, not in terms of infection of insect hosts. Similarly, astigmatid mites have the ability to spread the root pathogenic fungi, *Verticillium* and *Pythium myriotylum*, by conidia and microsclerotia carried in the gut (Price, 1976; Shew and Beute, 1979).

Many insects are attracted to the glebal mass of stinkhorns (Phallales) by the volatiles produced by these fungi and the often ornate and colored fruiting structures. This attraction is undoubtedly an adaptation for spore dispersal either by insects carrying spores on their bodies or passage of spores through the animal's gut. For example, some bees have been observed to collect glebal mass, and probably disperse it (Burr et al., 1996), as well as flies and beetles (Figure 5.14). The spore mass also provides nutrition for insects, although its food value may be less than pure protein. Blowflies feeding for 1 h on either glebal mass of *Mutinus canis* or liver

**Figure 5.14** **(See color insert.)** Flies and beetles feeding on the glebal mass of spores from a stinkhorn, both deriving nourishment and effecting spore dispersal. (Photo courtesy of the author.)

only produced mature eggs on liver, but with unlimited access to fungal fruit bodies for several days, the insects produced mature eggs and first instar larvae, suggesting there is food value in the spore mass (Stoffolano et al., 1990).

Spore dispersal of ectomycorrhizal fungi has been observed with vertebrate vectors, but the association between vertebrates and fungal spore dispersal is most closely evolved in hypogeous fungi where spore dispersal is most highly dependent on animals (Trappe and Claridge, 2005). Trappe and Maser (1976) showed that spores of the arbuscular mycorrhizal fungus, *Glomus macrocarpus*, and of the hypogeous ectomycorrhizal fungus, *Hymenogaster*, were dispersed by small mammals, the Oregon vole (*Microtus oregoni*) and chickaree (*Tamiasciurus douglasi*). A proportion of the spores survived passage through the gut of the animals and germinated in the feces. The ability of these animals to effect spore dispersal assists in the colonization of bare ground by primary colonizing plant species during the initial phases of plant succession (Trappe, 1988). It has been suggested that the hypogeous habit evolved as a fungal adaptation to arid conditions, where the production of aboveground fruiting structures is limited by rapid desiccation (Trappe and Claridge, 2005). For example, the desert truffle *Phaeangium lefebvrei* fruits as a cluster of small fruit bodies that push the soil surface upward into an observable bump that can be seen by migratory birds that feed on them and disperse spores (Alsheikh and Trappe, 1983).

Effective dispersal is dependent on the spore's viability after excretion from the animal. Spores of a variety of ectomycorrhizal fungi have been found to be viable and could develop associations with Ponderosa pine after passage through the gut of the tassel-eared squirrel (*Scurius aberti*). Kotter and Farentinos (1984) provide evidence suggesting that mycophagy of fungi results in the deposition of viable spores in feces. The appearance of a variety of hypogeous and epigeous mycorrhizal fungal spores of a variety of fungal genera in the feces of pika, voles, chipmunks, marmots, mountain goats, and mule deer on the forefront of Lyman Glacier (Cázares and Trappe, 1994), strongly suggests that the deposition of these spores forms an inoculum source allowing colonization of the newly developing soils by early successional and slow-growing tree species (*Abies lasiocarpa, Larix lyalii, Tsuga mertensiana*, and *Salix* spp.). Jumpponen et al. (1999) identified "safe sites" on this glacier outwash where plant colonization was most likely to ensue. These sites consisted of concave surfaces of coarse rocky particles, which were ideal for trapping tree seeds and protecting them from desiccation. It is likely that these sites also formed foci for foraging small mammals as they were a site of abundant food in the form of seeds. Thus, the deposition of mycorrhizal spore-laden feces in these microsites would further enhance the survival of germinating tree seedlings. In these harsh environmental conditions, Jumpponen et al. (1998) showed that the dark-septate mycorrhizal fungus, *Phialocephalia fortinii*, significantly enhanced the growth of lodgepole pine (*Pinus contorta*), which is an early colonizer of the glacier forefront, but only in the presence of added nitrogen. During the succession of plants in this recent glacial till, microbial communities changed from bacterial domination to fungal-dominated communities. During this change, carbon use efficiency changes from a high rate of carbon respiration to an accumulating phase, indicating that fungi are a stabilizing force in the developing ecosystem and facilitating net carbon fixation into biomass (Ohtonen et al., 1999).

Similar results were reported by Cork and Kenagy (1989b), who demonstrated that spores of the truffle *Elaphomyces granulatus* could pass through the gut of ground squirrels and deer mouse. These spores retain viability in the feces and are thus able to colonize new seedling plants in a nutritionally favorable environment. Similarly, spores of Rhizopogon have different survival and mycorrhizal colonization abilities after passage through rodent guts (Colgan and Claridge, 2002; Figure 5.15). Consumption and dispersal of arbuscular mycorrhizal spores by the Central American spiny rat (*Proechimys semispinosus*) in neotropical forests was shown by Mangan and Adler (2002) to be maximal during the fall and winter (October to December). This dispersal was positively correlated to optimal growth and germination conditions for the host trees, but unrelated to the availability of other food sources for the rats. These authors also showed that there was either selection of mycorrhizal spore species or differential survival through the gut as the numbers of spores appearing in the feces differed between fungal species. In the same way, passage of ectomycorrhizal spores through the gut of mice (*Peromyscus*) increased the activity of *Suillus tomentosus*, but decreased the viability of *Laccaria trichodermophora* (Pérez et al., 2012). Despite significant changes in spore morphology and spore wall characteristics, spores of *Tuber aestivum* appear to survive passage through the gut of pigs, and their viability and root colonization potential are enhanced (Piattoni et al., 2014).

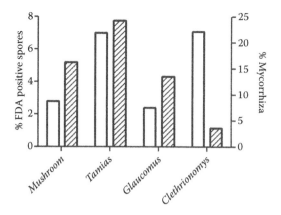

**Figure 5.15**  Viability (%FDA, open bars) and mycorrhizal formation capabilities (hatched bars) of *Rhizopogon vinicolor* spores taken from a fruit body or feces of three small mammals. (Data from Colgan, W. III, Claridge, A.W., *Mycol. Res.*, 106, 314–320, 2002.)

## 5.6 INFLUENCE OF INVERTEBRATE GRAZING ON FUNGAL PATHOGENS

As worms often have higher rates of feeding at or adjacent to roots, it is not surprising that they can be involved in the activities of root pathogenic fungi. However, these interactions are not as simple as would be predicted by logic. Clapperton et al. (2001) showed that the presence of earthworms reduced the severity of symptoms of the take-all disease (*Gaeumannomyces gramminis*) on wheat. This was not due to a reduction in the abundance of fungi, but rather a result of an increase in fungi in the earthworm colonized system. Aside from increasing fungal biomass, earthworms also stimulate an increase in bacterial populations. It is suggested that the effect of earthworms on this pathogen is indirect and mediated through changes in the microbial community by as yet unknown mechanisms. However, it could be hypothesized that an increased diversity of the microbial community would increase the abundance of bacteria that are pathogenic to fungi and increase fungal–fungal competitive interactions.

In an indirect way, soil fauna may influence the severity of a plant fungal disease by using invertebrates as vectors of biocontrol agents. Doube et al. (1994a,b, 1995) showed that the earthworms *Aporrectodea* spp. could be used to transport *Pseudomonas* bacteria to root surfaces to protect against the take-all fungus, *Gaumannomyces graminis*.

In addition to the interactions between fungi and insect pathogens, soil fauna may also modify the efficacy of plant fungal pathogens. Recently, Sabatini and Innocenti (2000) have studied the feeding preference of collembola on soil-borne plant pathogenic fungi. They determined that all of the tested species of collembola (*Onychiurus amatus*, *O. tuberculatus*, and *Folsomia candida*) preferred *Fusarium culmorum* mycelia, although mycelia of both *Gaeumannomyces graminis* and *Rhizoctonia cerealis* in the mixed culture continued to be grazed at a lower intensity and were capable of sustaining collembolan growth on their own. However, they showed that the fungus *Bipolaris sorokinaianum* was lethal to all collembolan species.

Thus, it can be seen that the dispersal of mycorrhizal spores by animal vectors can be an important component in the provision of fungal inoculum potential to sites where vegetation regeneration is occurring. Use of this information could be made in restoration sites, where the development of microhabitats suitable for small mammal refuges could enhance propagule dispersal and thus increase the rate of primary succession. In general, however, we do not know enough about the dynamics of fungal spore dispersal by invertebrates or the significance of faunal grazing on fungi belowground. From the few studies that have shown the propensity of soil animals to carry fungal spores or hyphal fragments, there appears to be possibilities of these animals carrying beneficial organisms to improve plant production by supplying inocula of mycorrhizal fungi or to deliver mycoparasitic fungi and bacteria to plant roots to reduce fungal pathogens (Doube et al., 1994a,b, 1995). Therefore, these interactions between fungi, bacteria, and soil animals warrant further investigation.

## 5.7 SPECIFIC FUNGAL–FAUNAL INTERACTIONS

### 5.7.1 Ant and Termite Fungus Gardens

Of all the close associations between fungi and animals, the interaction between leaf-cutting ants and termites and their fungus gardens is an important illustration of the role of fungi in maintenance of an insect population. This association is so close that many regard it as a true symbiosis as the ants and termites selectively allow certain fungi to colonize and grow on the leaf pieces to provide food for their colony. Indeed, the dominant mycelium in termite nests appears to be *Termitomyces* sp., which is maintained in abundance by the constant care of the termites in the face of a greater competitor, *Aspergillus*. This balance is probably actively maintained by these animals because *Aspergillus* is less palatable or has lesser food value than *Termitomyces* (Cherrett et al., 1989; Wood and Thomas, 1989).

Bass and Cherrett (1996) found that there is a close relationship between the activities of the small worker ants of the colony ("minima" workers) and the production of food rewards (staphylae) produced by the fungus. The abundance of these staphylae appears to increase in smaller passages in the colony, where only the minima workers can gain access. Indeed, the activities of ants can alter the local fungal flora. Ba et al. (2000) showed that the fire ant imported into the United States from South America develops a unique yeast flora in its brood chambers. The presence of leaf endophytes impairs leaf decomposition and production of mutualistic fungi cultivated by leaf-cutting ants. Knowing that the alkaloids of the fungi interfere with herbivory and other chemistry with other fungi, it was assumed that the chemistry of endophytes would interfere with fungus cultivated by the ants; however, by using extracts from leaves with high or low endophyte loading, Estrada et al. (2014) found that the presence of the endophytic fungus, and not its chemistry, interfered with ant mutualistic fungal growth.

The close association between Macrotermitinae and the fungus *Termitomyces* was reviewed by Wood and Thomas (1989), who showed that the digestive processes of termites is almost entirely dependent on the symbiotic association with the fungus, without which wood could not be converted into a form that could be assimilated by the termite. Similarly, Cherrett et al. (1989) described the mutualistic association in leaf-cutting ants. Here, the fungus *Attamyces bromatificus* has never been found outside the nests of leaf-cutting ants. The ants carry a fungal inoculum to new nests in an infrabuccal pocket, a small cavity at the esophageal opening to ensure colonization of new food reserves in the new colony. Korb and Linsenmair (2001) showed that the availability of food is a limiting resource for large colonies of the fungus-cultivating termite *Macrotermes bellicosus* in two habitats in the Comoé National Park (Côte d'Ivoire). The aggregation of smaller colonies in the savannah region was probably associated with the availability of trees to provide leaves for the cultivation of fungi. This patchy distribution is also related to the availability of appropriate microclimatic conditions for fungal production, which seems to be more important for young colonies. The lower density of larger colonies in high forest suggest a more stable environment and stable humidity for the cultivation of fungi, compared to the savannah ecosystem, where smaller colonies are more widely dispersed.

Symbiotic fungi in association with different termites appear to serve contrasting functions. In *Macrotermes* spp., fungi function to decompose lignin in the collected leaf material, making more palatable and higher food value, whereas in *Hypotermes makhamensis*, *Ancistrotermes pakistanicus*, and *Pseudacanthotermes militaris* the fungal biomass supported by leaf decomposition is used as a food base for the insects (Hyodo et al., 2003).

Soil fauna cause significant physical disturbance of soil as well as changes in the soil chemistry, via the introduction of feces, leaf litter, etc. Because of the aggregated distribution of most soil fauna, these activities increase the heterogeneity of the soil ecosystem. An example of such activity and its influence on fungi can be seen from the study of western harvester ants (*Pogonomyrmex occidentalis*) in arid and semi-arid ecosystems in North America. Snyder and Friese (2001) show that the activities of these ants create nests at densities up to 30 ha$^{-1}$, where each nest represent an area of soil disturbance and enrichment. They found that the density of sagebrush (*Artemesia tridentata*) roots was similar in and off nest mounds, and that the root length colonized by arbuscular mycorrhizae was similar; however, the intensity of root colonization was higher within the nest (Table 5.13). Given that nests are typically enriched in nutrients (MacMahon et al., 2000), it is surprising that there is not a greater difference in mycorrhizal colonization of roots or root length in response to this enrichment (Pregitzer et al., 1993; van Vuuren et al., 1996; Tibbett, 2000).

## 5.7.2 Bark Beetle–Fungi Interactions

The close association between wood-decomposing fungi and bark and ambrosia beetles, the evolution of the symbiotic relationship, and the physiology and

Table 5.13   Effect of Harvester Ant (*Pogonomyrmex occidentalis*) Activity on Root
Length and Arbuscular Mycorrhizal Colonization of Roots of Sagebrush
Community Plants

| | Total Root Length (cm) | | % Root Colonized (cm) | |
| Site | Mound | Off Mound | Mound | Off Mound |
|---|---|---|---|---|
| 1 | 49 | 52 | 34 | 25 |
| 2 | 23 | 118 | 20 | 11 |
| 3 | 127 | 160 | 27 | 18 |
| 4 | 150 | 180 | 16 | 12 |
| 5 | 255 | 355 | 21 | 15 |

*Source:* Snyder, S.R., Friese, C.F., *Mycorrhiza*, 11, 163–165, 2001.

behavior of the organisms involved has been reviewed by Beaver (1989). The bark beetles mainly feed on phloem, which is of relatively high food value, whereas the ambrosia beetles feed on xylem, which requires a greater dependence on the fungal symbiont to improve the resource quality of the wood by partial decomposition and incorporation of nitrogen. Beaver (1989) discusses the importance of the mycangium, a specialized appendage on the insect's leg for the transport of the fungal partner, as the adult beetles have relatively little fungal material in their gut when they hatch and disperse from the tree in which they develop. In some instances, there may a tripartite interaction between plants, animals, and fungi. One of these interactions that has recently come to light and has ecosystem-wide influence it the link between (1) mites, (2) specialized invaginations or tufts of hair on leaf surfaces providing refugia for mites (acarodomatia), and (3) plant pathogenic fungi. By manipulating access to diomatia by tydeid mites (*Orthotydea lambi*), Norton et al. (2000) observed the density of domatia per leaf and the incidence of grape mildew caused by *Uncinula necator.* They showed that the activity of mite mycophagy at high domatia densities significantly reduced the incidence of mildew on the plants.

Another complex interaction between fungi, arthropods, and plants can be seen in the blue stain fungus (*Ophiostoma minus*), the mycangial fungi (*Ceratocystiopsis ranaculosus*) and *Entomocorticium* sp., southern pine beetles (*Dendroctonus frontalis*), a phoretic mite (*Tarsonemus* spp.), and pine trees (Lombardero et al., 2000). *C. ranaculosus* is carried in specialized structures (mycangia) of the female bark beetle, and the fungus is necessary for the developing beetle larvae. Ayres et al. (2000) showed that the action of the fungus, when decomposing live phloem, doubles the nitrogen content of the wood/fungal matrix (from 0.4% to 0.86% N) (Figure 5.16). This provides the developing beetle larvae with a nutrient-enhanced food supply, allowing for faster growth rates and more rapid maturation. However, the beetle also inadvertently carries the fungus *O. minus*, but on its body surface, not in the mycangia. Similarly, the *Tarsonemus* mite is inadvertently transported between trees by the bark beetle and has the highest population growth when a feeding on *O. minus* or *C. ranaculosus*, but not on the other mycangial fungus, *Entomocorticium* sp.

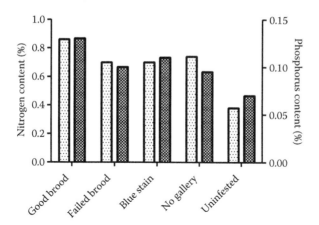

**Figure 5.16**  Concentrations of nitrogen (solid bars) and phosphorus (hatched bars) in phloem of *Pinus taeda* trees with or without infestation of *Dendroctonius frontalis* and its associated mycangial fungi. Blue stain is where the bark is also infected with the blue stain fungus, *Ophiostoma minus*, which inhibits the growth of the mycangial fungi. (Data from Ayres, M.P. et al., *Ecology*, 81, 2198–2210, 2000.)

*Ophiostoma minus* is antagonistic to the growth of the bark beetle, so the development of the two mycangial fungi into wood both provides food for developing beetle larvae, but also competes against *O. minus*. The high growth rate of the mite when feeding on *Ceratocystiopsis* reduces the growth of this fungus, allowing greater wood colonization by *O. minus*, and hence greater antagonism with the developing beetle population. Thus, there exists an indirect negative effect of a phoretic mite on the population of a bark beetle, which appears to be regulated by induced changes in competitive abilities between fungi.

Similarly, other bark beetles and wood wasps are vectors of other fungal pathogens of trees. Redfern (1989) discussed the role of the bark beetle (*Ips cembrae*) transmitting the fungal disease *Ceratocystis lariciola* and the wasp (*Urocerus gigas*) transmitting *Amylostereum chailletii*, both of which cause dieback and death of larches. The degree of damage to the tree by fungal pathogens introduced by the bark beetle and the wood wasp appears to be density-dependent. Where the population of the insects is high and causes severe attack of the tree, tree death is a likely outcome. Where insect density is low, the effect of the fungus is relatively minor (Figure 5.17).

The penetration of bark by bark beetles opens up the interior wood to wood decay fungi and the introduction of fungi carried by the beetles. In a study in Sweden, it was shown that Norway spruce wood baits attacked by beetles compared to wood punctured by hand developed significantly different saprotrophic fungal communities (Strid et al., 2014). It is possible that fungi introduced by beetles competed against selected wood-decomposing fungi.

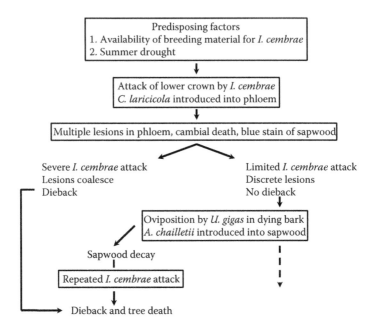

**Figure 5.17**    Interactions between the bark beetle, *Ips cembrae*, the wood wasp, *Urocerus gigas*, and the two larch fungal pathogenic fungi, *Ceratocystis lariciola* and *Amylostereum chailletti*, showing how the severity of the diseases depends on interactions between the two insects. (From Redfern, D.B., in: *Insect–Fungus Interactions*, ed. Wilding, N. et al., 195–204, Academic Press, London, 1998.)

## 5.8 FUNGAL–FAUNAL INTERACTIONS IN AQUATIC AND MARINE ECOSYSTEMS

The role of fungi in aquatic ecosystems has been reviewed by Wong et al. (1998). They suggest that some 600 fungal species are associated with aquatic ecosystems, and their function ranges from saprotrophs to pathogens of both plants and animals. In aquatic ecosystems, interactions between fungi and leaf-shredding fauna (Amphipoda, Isopoda, Diptera, Plecoptera, and Trichoptera) have been studied (Suberkropp, 1992). It has been suggested that fungal degradation of the leaf litter "conditions" the leaf material to make it more palatable for faunal grazing. These fungi alter the palatability of the litter resource, alter its chemical composition, and appear to increase its food value. Indeed, Bärlocher and Kendrick (1975) regard aquatic hyphomycetes as intermediaries in the energy flow in stream ecosystems. They showed that the amphipod *Gammarus* increased in weight faster, in relation to the amount of food ingested, when provided with fungi (*Humicola, Anguillospora, Clavariopsis, Tricladium,* or *Fusarium*), compared to a diet of elm or maple leaves.

The selection of leaves varies between animal species, and the animals are selective in the choice of fungal colonizers of the leaf material. Although the larvae of *Tipula* spp. do not select leaves that have been colonized by fungi, other shredder species select colonized leaves. The amphipod (*Gammarus pseudolimnaeus*) has been shown to be selective between fungal species colonizing leaf litter by favoring leaves colonized by some fungal species and not by others. *Gammarus pseudolimnaeus* has been shown to have greater growth on leaves colonized by four fungal species than six other fungal species. Trichopteran larvae have also been shown to have significantly improved rates of growth on leaf litter colonized by fungi than on uncolonized leaves. However, contrary to the reports of Bärlocher and Kendrick (1975), Suberkropp (1992) suggests that it is not the fungus *per se* that elicits greater productivity of animals. Suberkropp (1992) cites work showing that certain animals, such as the amphipod *Gammarus* and the isopod, *Asellus*, grew better on leaves colonized by fungi that when fed with the fungi alone or the leaf litter alone. Thus, there is a suggestion of some synergistic benefit. However, the mechanism for this is unclear as estimates show that fungal biomass accounts for between 1% and 5% of the detrital biomass in aquatic ecosystems. However, fungi make up 90–95% of the microbial biomass on decaying leaves, which suggests that bacteria are possibly less important as decomposers, at least in the initial stages of leaf decomposition. This fungal biomass has been shown to be insufficient to account for the increased rate of respiratory loss of carbon from fungal-colonized leaves. The fungal contribution to respiration accounts for between 1% and 57% of the total carbon respired. The most logical mechanism for the enhanced growth of animals grazing on fungal-colonized leaf litter is that the biochemical changes in leaf chemistry, because of fungal attack, improves the availability of essential nutrients (particularly nitrogen) and readily assimilable energy sources. In a study of the decomposition of alder and willow leaves in streams, Hieber and Gessner (2002) showed that the shredder community counted for 64% and 51% of the leaf litter mass loss for alder and willow, respectively. Fungi accounted for 15% and 18% mass loss, respectively, and that for bacteria was only 7% and 9%, respectively. Within this community of organisms, fungi were found to compose 95–99% of the total microbial biomass, which is in line with values reported in other studies (see above). Although fungi appear to dominate in these ecosystems, Wong et al. (1998) conclude that we still know relatively little about the diversity and functional role of these fungi.

In contrast to the terrestrial ecosystem, there is little evidence suggesting a faunal grazing effect on the fungal community structure or biomass. Superkropp (1992) cites work showing that leaves incubated in fine mesh bags supported a more diverse fungal community with a lower mass-specific sporulation rate than leaves incubated in coarse mesh bags. This indicates a reduced biomass of fungi under potentially higher grazing intensity, the stimulation of sporulation under a more stressed (grazed) environment, and selective feeding by fauna or differential resistance to grazing within the fungal community leading to a reduced fungal diversity in more heavily grazed conditions. However, in eutrophic streams, decomposition of leaves occurs at more than twice the rate of unpolluted streams, because of the increased abundance of aquatic hyphomycetes (Gulis et al., 2006); the abundance and richness of invertebrate "shredder" community is increased; and the resource quality of fungal-colonized leaves also increased. The interaction between aquatic hyphomycetes and other components in the aquatic ecosystem are important to nutrient cycling in these systems. Leaf litter mass loss is increased with increasing richness and diversity of aquatic hyphomycetes and with increasing detritivore richness, and the performance of detritivores increases with fungal richness (Jabiol et al., 2013). Thus, leaf litter processing rates in streams is regulated by both bottom–up and top–down processes.

In addition to fungi growing on surfaces, there is a surprisingly diverse pelagic community of fungi from Ascomycotina, Basidiomycotina, and Chytridomycotina found in aquatic systems. It is suggested that their function is related to the decomposition of suspended organic matter, where they have been suggested to colonize more efficiently than bacteria and enhance bacterial activity by fungal release of labile substrates (Jobard et al., 2010). Chytrids, in particular, have been isolated from a range of suspended materials, such as algae, pollen, seeds, and invertebrate exoskeletons, and are thought to be important decomposers. However, this is an area of mycology that needs further investigation.

Fungi are important in marine and estuarine ecosystems. Although information on their diversity, biomass, and function is limited and less than that for terrestrial and aquatic ecosystems, there is evidence to suggest that they play a role in supporting faunal populations. Fungal biomass in the decomposing material in salt marsh ecosystems is also important for sustaining invertebrate herbivore populations. The amphipod (*Ulorchestia spartinophila*) has been shown to have complex dietary requirements and appear to grow best and produce the highest number of offspring when fed on decaying leaves containing a high fungal biomass.

Kohlmeyer and Kohlmeyer (1979) show a table of 107 fungal species isolated from decomposing wood in marine habitats. These include 73 Ascomycotina, two Basidiomycotina, and 29 mitosporic species. As in terrestrial ecosystems, the low resource quality of wood appears to encourage tight linkages between fungi and fauna for its decomposition. Evidence suggests that wood-boring marine mollusks preferentially settle and feed on wood that has previously been colonized by fungi and partially decomposed, rather than invade fresh wood. The associations have become so tight that, for example, the wood-boring crustacean, the gribble (*Limnoria tripunctata*), has increased longevity when feeding on wood colonized by fungi. More importantly, it is incapable of reproduction on any substrate unless marine fungi are included as part of its diet. This may be attributable to the enhanced availability of proteins, essential amino acids, and vitamins, which are unavailable in the absence of fungi.

Fungi are major pathogens of marine animals (Hyde et al., 1998). Mitosporic fungi have been shown to cause disease in crustaceans (Smolowitz et al., 1992) and cause damage to corals (Raghukumar and Raghukumar, 1991) and juvenile clams (Norton et al., 1994). Norton et al. (1994) reported that dematiaceous fungal hyphae (*Exserohilum rostratum* and *Curvularia* spp.) isolated from the mantle and shell of juvenile *Tridacna crocea* clams is associated with their decline in health. Many oomycetes are highly destructive pathogens in finfish, mollusks, and shellfish (Noga, 1990). Very few fungi have been reported as pathogens of marine plants, although some have been observed on the leaves and roots of mangrove, despite the fact that many marine algae are susceptible to fungal pathogens, primarily with chytrids and oomycetes. The impact of these pathogens on production and fitness of their hosts has not been adequately investigated. Certainly, Hyde et al. (1998) conclude in their review that the role of fungi in this ecosystem requires considerably more research than has hitherto been devoted to it. Although numerous fungal species have been isolated from marine habitats, their function is largely unknown. Similarly, Wong et al. (1998) state that the role of fungi as plant pathogens in aquatic ecosystems appears to be minor; many fungi are pathogenic to animals. Wong et al. (1998) show that fungi are pathogens of fish, turtles, tadpoles, and adult anurans, and many invertebrates. There are suggestions that fungi may potentially be used as biocontrol agents for mosquitoes.

Like many terrestrial and aquatic arthropods, marine isopods contain fungi in their gut to assist in the breakdown of plant food. Cafaro (2000) showed that several genera and species of Patagonian intertidal isopods contained Trichomycetes of the genus *Palavascia*. A variety of gut symbionts, including the cultured fungus *Termitomyces*, are found in the gut of termites for the same reason (Wood and Thomas, 1989). In aquatic ecosystems, trichomycetes are also found as endosymbionts in black fly larval guts (Beard and Alder, 2002). Beard and Alder (2002) showed that there were significant differences in abundance of different fungal species between season, within sites, and differences between sites, although no reason was given for the between-site differences. However, Beard and Adler (2002) demonstrated that new recruits to the black fly larval population rapidly became colonized by *Harpella melusinae*, suggesting a strong dependence on this fungus as a gut symbiont.

## 5.9 CONCLUSIONS

As a component in the ecosystem, fungi are a good food resource for grazing animals. Numerous vertebrate and invertebrate animals consume either the fruiting structures or mycelia of fungi as the main component of their diet, or more usually, as part of their diet. The fact that mushroom collecting and consumption is a tradition in many European countries is testament to their food, medicinal, and cultural value (Hudler, 1998). Indeed, we point out in this chapter that the cultivation of mushrooms for human consumption is a multimillion dollar industry in the United States alone. Thus, it is not surprising that many animals use fungi as a food source.

For vertebrate animals, the consumption of fungi for food is often a part of their diet. This part is especially important at times of the year when other foods are scarce. For example, in arctic regions when the ground is snow covered, there is little vascular plant food available. At this time, reindeer forage for lichens as their main food source. In the same way, other vertebrates around the world, particularly small mammals, make use of fungi at times when other food sources are depleted or when fungi are abundant. Many small mammals also forage belowground for hypogeous fungi, whose spore dispersal is dependent on these creatures. The degree of dependence of many animals on fungi as a food source is still not entirely clear.

What would be the impact on faunal populations and communities of the removal of fungi from ecosystems?

For invertebrates, both the fruiting body and the mycelium form a food base. The nutritional value of this food item is high, especially for certain groups, such as nematodes. In the preceding discussion, we have documented the role that fungi play in the maintenance of populations of fungivorous nematodes, and that by selectively feeding on specific fungal species, the proportion of females in the population, their fecundity, and hence the size of subsequent populations can be strongly influenced (Ruess and Dighton, 1996; Ruess et al., 2000). Soil arthropods are also somewhat dependent on fungal mycelia for their growth and development. Without fungi, many collembola and mite populations would be reduced in the soil.

As animals graze on fungi, they are able to affect the rate of fungal-mediated ecosystem process. The rate of decomposition of leaf litter is affected by the rate of colonization of the litter by fungi. The rate of colonization can be helped by the invasion of soil animals by the physical breakdown of leaf litter into smaller parts (comminution) and the active transport of fungal inocula (spores of hyphal fragments) into the leaf litter by animals (Ponge, 1990, 1991). Intense grazing reduces hyphal growth, whereas moderate rates of grazing can stimulate compensatory growth and actually increase fungal growth to help fungi to grow and colonize new resources. However, as we have noted earlier, animals—like humans—are selective in what they eat. Selection of preferred fungal species can benefit the animal, by having a higher nutritional value than other species or by avoidance of poisonous secondary metabolites. So, do we really understand the importance of fungal grazers to the decomposition process? Given the observations of the feeding behavior of collembola in rhizotrons (Lussenhop, personal communication), it would appear that collembola spend a great deal of time grooming and then wandering around, "nibbling" at fungal hyphae rather than consuming large quantities at once. How important is their grazing, given that once a fungal hyphum has been severed, its translocatory function is inhibited? Much of our knowledge has been gained from studies in the greenhouse or laboratory or in the field at unrealistic densities of animals. How important is this grazing in natural ecosystems? Is it enough to cause significant decline in plant yield? Could this grazing be selective enough to affect one plant species to a greater extent than another, thus altering plant community structure?

The fact that animals are able to carry fungal propagules, either on their bodies or in their gut, is important in spreading inocula through the ecosystem. It is notable that some fungal groups have evolved to rely on faunal dispersal of their spores, and adapt their fruiting structures to encourage grazing. Hypogeous fungi, such as truffles, rely entirely on small mammals to consume the fruit body and disseminate spores in their feces. Other fungi, such as the members of the Phallales in the Gasteromycetes, have involved sticky, malodorous spore masses that attract flies, which then disperse the spores (Figure 5.14). The dispersal of fungal propagules can be important in providing, for example, mycorrhizal inoculum for plants invading bare ground during primary or secondary succession.

During evolution, the close association between animals and fungi has occasionally grown to be so close that the two are inseparable. Some species of ants and termites rely on fungi as a food source to such an extent that they actively cultivate the preferred fungal species within their colony's nest. By selective grazing, the most nutritious fungal material is fed to the developing young. The colonies of fungi are also cleaned of any contaminating fungi that are either less palatable or may compete against the preferred fungal species. In order to decompose high C/N ratio plant material of plant materials containing high levels of lignin, fungal-derived enzymes are often necessary. In addition to providing these enzymes, fungi are also able to translocate nutrients (particularly nitrogen) into such recalcitrant plant materials. It is for this reason that bark beetles carry fungi in a specialized structure, or mycangium. However, the success of these fungi in colonizing wood depends on many fungal–faunal interactions between bark beetles, phoretic, mites and multiple competing fungi (Norton et al., 2000; Lombardero et al., 2000).

## REFERENCES

A'Bear, A. D., T. W. Crowther, R. Ashfield et al. 2013. Localised invertebrate grazing moderates the effects of warming on competitive fungal interactions. *Fung. Ecol.* 6:137–140.

Allen, M. F., J. A. MacMahon, and D. C. Andersen. 1984. Reestablishment of Endogonaceae on Mount St Helens: Survival of residuals. *Mycologia* 76:1031–1038.

Alsheikh, A. M. and J. M. Trappe. 1983. Taxonomy of *Phaeangium lefebvrei*, a desert truffle eaten by birds. *Can. J. Bot.* 61:1919–1925.

Anderson, J. M. 2000. Food web functioning and ecosystem processes: Problems and perceptions of scaling. In *Invertebrates as Webmasters in Ecosystems.*, ed. D. C. Coleman and P. F. Hendrix, 3–24. Wallingford, UK: CABI.

Anderson, J. M. and P. Ineson. 1984. Interaction between microorganisms and soil invertebrates in nutrient flux pathways of forest ecosystems. In *Invertebrate–Microbial Interactions.*, ed. J. M. Anderson, A. D. M. Rayner, and D. W. H. Walton, 59–88. Cambridge: Cambridge University Press.

Ayres, M. P., R. T. Wilkins, J. J. Ruel, M. J. Lombardero, and E. Vallery. 2000. Nitrogen budgets of phloem-feeding bark beetles with and without symbiotic fungi. *Ecology* 81:2198–2210.

Ba, A. S., S. A. Jr. Phillips, and J. T. Anderson. 2000. Yeasts in mound soil of the red imported fire ant. *Mycol. Res.* 104:969–973.

Babikova, Z., L. Gilbert, T. Bruce et al. 2014. Arbuscular mycorrhizal fungi and aphids interact by changing host plant quality and volatile emission. *Funct. Ecol.* 29:375–385.

Bakonyi, G., K. Posta, I. Kiss, M. Fabin et al. 2002. Density-dependent regulation of arbuscular mycorrhiza by collembola. *Soil Biol. Biochem.* 34:661–664.

Bärlocher, F. and B. Kendrick. 1975. Hyphomycetes as intermediates of energy flow in streams. In *Recent Advances in Aquatic Mycology*, ed. E. B. G. Jones. London: Elk Sciences.

Bass, M. and J. M. Cherrett. 1996. Fungus garden structure in the leaf-cutting ant *Atta sexdens* (Formicidae, Attini). *Symbiosis* 21:9–24.

Beard, C. E. and P. H. Adler. 2002. Seasonality of trichomycetes in larval black flies from South Carolina, USA. *Mycologia* 94:200–209.

Beare, M. H., R. W. Parmelee, P. F. Hendrix et al. 1992. Microbial and faunal interactions and effects on litter nitrogen and decomposition in agroecosystems. *Ecol. Monogr.* 62:569–591.

Beaver, R. A. 1989. Insect–fungus relationships in the bark and ambrosia beetles. In *Insect–Fungus Interactions.*, ed. N. Wilding, N. M. Collins, J. F. Hammond, and J. F. Webber, 121–143. London: Academic Press.

Bengtsson, G., K. Hedlund, and S. Rundgren. 1993. Patchiness and compensatory growth in a fungus-collembola system. *Oecologia* 93:296–304.

Bokhorst, S. and D. A. Wardle 2014. Snow fungi as a food source for micro-arthropods. *Eur. J. Soil Biol.* 60:77–80.

Brown, G. G. 1995. How do earthworms affect microfloral and faunal community diversity? *Plant Soil* 170:209–231.

Brussard, L., S. Hauser, and G. Tian. 1993. Soil faunal activity in relation to the sustainability of agricultural systems in the humid tropics. In *Soil Organic Matter Dynamics and Sustainability of Tropical Agriculture*, ed. K. Mulongoy and R. Merckx, 241–256. Leuven, Belgium: Wiley-Sayce.

Bultman, T. L. and A. Leuchtman. 2008. Biology of the *Epichloë–Botanophila* interaction: An intriguing association between fungi and insects. *Fung. Biol. Rev.* 22:131–138.

Bultman, T. L., A. M. Welch, R. A. Boning, and T. I. Bowdish. 2000. The cost of mutualism in a fly–fungus interaction. *Oecologia* 124:85–90.

Burr, B., W. Barthlott, and C. Westerkamp. 1996. *Staheliomyces* (Phallales) visited by *Trigona* (Apidae): Melittophily in spore dispersal of an Amazonian stinkhorn? *J. Trop. Ecol.* 12:441–445.

Cafaro, M. J. 2000. Gut fungi of isopods: The genus *Palavascia*. *Mycologia* 92:361–369.

Carroll, J. J. and D. R. Viglierchio. 1981. On the transport of nematodes by the wind. *J. Nematol.* 13:476–483.

Cave, B. 1997. Toadstools and springtails. *The Mycologist* 11:154.

Cázares, E. and J. M. Trappe. 1994. Spore dispersal of ectomycorrhizal fungi on a glacier forefront by mammal macrophagy. *Mycologia* 86:507–510.

Cheal, D. C. 1987. The diets and dietary preferences of *Rattus fuscipes* and *Rattus lutreolus* at Walkerville in Victoria. *Aust. Wildl. Res.* 14:35–44.

Chen, J. and H. Ferris. 1999. The effects of nematode grazing on nitrogen mineralization during fungal decomposition of organic matter. *Soil Biol. Biochem.* 31:1265–1279.

Cherrett, J. M., R. J. Powell, and D. J. Stradling. 1989. The mutualism between leaf cutting ants and their fungi. In *Insect–Fungus Interactions.*, ed. N. Wilding, N. M. Collins, J. F. Hammond, and J. F. Webber, 98–143. London: Academic Press.

Clapperton, M. J., N. O. Lee, F. Binet, and R. L. Coner. 2001. Earthworms indirectly reduce the effects of take-all (*Gaeumannomyces graminis* var. *tritici*) on soft white spring wheat (*Triticum aestivum* cv. Fielder). *Soil Biol. Biochem.* 33:1531–1538.

Claridge, A. W. and T. W. May. 1994. Mycophagy among Australian mammals. *Aust. J. Ecol.* 19:251–275.

Claridge, A. W. and J. M. Trappe. 2005. Sporocarp mycophagy: Nutritional, behavioral, evolutionary, and physiological aspects. In *The Fungal Community; Its Organization and Role in the Ecosystem*, ed. J. Dighton, J. F. White, and P. Oudemans, 599–611. Boca Raton, FL: Taylor & Francis.

Clinton, P. W., P. K. Buchanan, and R. B. Allen. 1999. Nutrient composition of epigeous fungal sporocarps growing on different substrates in a New Zealand mountain beech forest. *N. Z. J. Bot.* 37:149–153.

Coleman, D. C., E. R. Ingham, H. W. Hunt, E. T. Elliott, C. P. P. Read, and J. C. Moore. 1990. Seasonal and faunal effects on decomposition in semiarid prairie meadow and lodgepole pine forest. *Pedobiologia* 34:207–219.

Colgan, W. III and A. W. Claridge. 2002. Mycorrhizal effectiveness of *Rhizopogon* spores recovered from fecal pellets of small forest-dwelling mammals. *Mycol. Res.* 106:314–320.

Cooke, A. 1983. The effects of fungi on food selection by *Lumbricus terrestris* L. In *Earthworm Ecology: From Darwin to Vermiculture*, ed. J. E. Satchell, 365–373. New York: Chapman and Hall.

Cooper, E. J. and P. A. Wookey. 2001. Field measurements of the growth rates of forage lichens and the implications of grazing by Svalbard reindeer. *Symbiosis* 31:173–186.

Cork, S. J. and G. J. Kenagy. 1989a. Nutritional value of a hypogeous fungus for a forest-dwelling ground squirrel. *Ecology* 70:577–586.

Cork, S. J. and G. J. Kenagy. 1989b. Rates of gut passage and retention of hypogeous fungal spores in two forest-dwelling rodents. *J. Mammal.* 70:512–519.

Courtney, S. P., T. T. Kibota, and T. S. Singleton. 1990. Ecology of mushroom-feeding Drosphilidae. *Adv. Ecol. Res.* 20:225–275.

Crittenden, P. D. 2000. Aspects of the ecology of mat-forming lichens. *Rangefinder* 20:127–139.

Demeure, Y. and D. Freckman. 1981. Recent advances in the study of anhydrobiosis in nematodes. In *Plant Parasitic Nematodes*, ed. B. M. Zukerman and R. A. Rohde, 204–225. New York: Academic Press.

Dighton, J., H. E. Jones, C. H. Robinson, and J. Beckett. 1997. The role of abiotic factors, cultivation practices and soil fauna in the dispersal of genetically modified microorganisms in soils. *Applied Soil Ecology* 5:109–131.

Doube, B. M., M. H. Ryder, C. W. Davoren, and T. Meyer. 1995. Earthworms: A down-under delivery service for biocontrol agents of root disease. *Acta Zool. Fenn.* 196: 219–223.

Doube, B. M., P. M. Stephens, C. W. Davoren, and M. H. Ryder. 1994a. Earthworms and the introduction and management of beneficial soil microorganisms. In *Soil Biota: Management in Sustainable Farming Systems.*, ed. C. E. Pankhurst, B. M. Doube, V. V. S. R. Gupta, and P. R. Grace. Australia: CSIRO.

Doube, B. M., P. M. Stephens, C. W. Davoren, and M. H. Ryder. 1994b. Interactions between earthworms, beneficial soil microorganisms and root pathogens. *Appl. Soil Ecol.* 1:3–10.

Dromph, K. M. 2001. Dispersal of entomopathogenic fungi by collembolans. *Soil Biol. Biochem.* 33:2047–2051.

Dowson, C. G., A. D. M. Rayner, and L. Boddy. 1998. Outgrowth patterns of mycelial cord-forming basidiomycetes into woodland soils: II. Resource capture and persistence. *New Phytol.* 109:343–349.

Edwards, C. A. 2000. Soil invertebrate controls and microbial interactions in nutrient and organic matter dynamics in natural and agroecosystems. In *Invertebrates as Webmasters in Ecosystems*, ed. D. C. Coleman and P. F. Hendrix, 141–159. Wallingford, UK: CABI.

Estrada, C., E. I. Rojas, E. T. Wcislo, and S. A. van Beal. 2014. Fungal endophyte effects on leaf chemistry alter the *in vitro* growth rates of leaf-cutting ants' fungal mutualist, *Leucocoprinus gongylophorus*. *Fung. Ecol.* 8:37–45.

Finlay, R. D. 1985. Interactions between soil microarthropods and endomycorrhizal associations of higher plants. In *Ecological Interactions in Soil*, ed. A. H. Fitter, D. Atkinson, D. J. Read, and M. B. Usher, 319–331. Oxford, UK: Blackwell.

Fitter, A. 1985. Functioning of vesicular–arbuscular mycorrhizas under field conditions. *New Phytol.* 99:257–265.

Fogel, R. 1976. Ecological studies of hypogeous fungi: II. Sporocarp phenology in a western Oregon Douglas-fir stand. *Can. J. Bot.* 54:1152–1162.

Fogel, R. and J. M. Trappe. 1978. Fungus consumption (mycophagy) by small animals. *Northwest Sci.* 52:1–31.

Friese, C. F. and M. F. Allen. 1993. The interaction of harvester ants and vesicular–arbuscular mycorrhizal fungi in a patchy semi-arid environment: The effect of mound structure on fungi dispersion and establishment. *Functional Ecology* 7:13–20.

Gange, A. C. 1993. Translocation of mycorrhizal fungi by earthworms during early succession. *Soil Biol. Biochem.* 25:1021–1026.

Gehring, C. A. and T. G. Whitham. 1991. Herbivore driven mycorrhizal mutualism in insect-susceptible pinyon pine. *Nature* 353:556–557.

Gehring, C. A. and T. G. Whitham. 1994. Comparisons of ectomycorrhizae on Pinyon pines (*Pinus edulis*; Pinaceae) across extremes of soil type and herbivory. *Am. J. Bot.* 81:1509–1516.

Gessner, M. O., K. Suberkropp, and E. Chauvet. 1997. Decomposition of plant litter by fungi in marine and freshwater ecosystems. In *The Mycota IV: Environmental and Microbial Relationships.*, ed. D. T. Wicklow and B. Soderstrom, 303–322. Berlin: Springer-Verlag.

Gormsen, D., P. A. Olsson, and K. Hedlund. 2004. The influence of collembolans and earthworms on AM fungal mycelium. *Appl. Soil Ecol.* 27:211–220.

Görres, J. F., M. C. Savin, and J. A. Amador. 2001. Soil micropore structure and carbon mineralization in burrows and casts of an anecic earthworm (*Lumbricus terrestris*). *Soil Biol. Biochem.* 33:1881–1887.

Goverde, M., M. G. A. van der Heijden, A. Wiemken, I. R. Sanders, and A. Erhardt. 2000. Arbuscular mycorrhizal fungi influence life history traits of a lepidopteran herbivore. *Oecologia* 125:362–369.

Graca, M. A. S., L. Maltby, and P. Calow. 1993. Importance of fungi in the diet of *Gammarus pulex* and *Asellus aquaticus*: II. Effects on growth, reproduction and physiology. *Oecologia* 96:304–309.

Grönwall, O. and Å. Pehrson. 1984. Nutrient content in fungi as primary food of the red-squirrel, *Sciurua vulgaris*. *Oecologia* 64:230–231.

Guerin-Laguette, A., N. Cummings, R. C. Butler, A. Willows, N. Hesom-Williams, S. Li, and Y. Wang. 2014. *Lactarius deliciosus* and *Pinus radiata* in New Zealand: Towards the development of innovative gourmet mushroom orchards. *Mycorrhiza* 24:511–523.

Guevara, R., A. D. M. Rayner, and S. E. Reynolds. 2000. Effects of fungivory by two specialist ciid beetles (*Octotemmus glabriculus* and *Cis boleti*) on the reproductive fitness of their host fungus, *Coriolus versicolor. New Phytol.* 145:137–144.

Gulis, V., V. Ferreira, and M. A. S. Graça. 2006. Stimulation of leaf litter decomposition and associated fungi and invertebrates by moderate eutrophication: Implications for stream assessment. *Freshwater Biol.* 51:1655–1669.

Hanski, I. 1989. Fungivory: Fungi, insects and ecology. In *Insect–Fungus Interactions*, ed. N. Wilding, N. M. Collins, P. M. Hammond, and J. F. Webber, 25–68. London: Academic Press.

Hanson, A. M., K. T. Hodge, and L. M. Porter. 2003. Mycophagy among primates. *Mycologist* 17:6–10.

Harris, K. K. and R. E. J. Boerner. 1990. Effects of belowground grazing by collembola on growth, mycorrhizal infection, and P uptake of *Geranium robertianum*. *Plant Soil* 129:203–210.

Heděnc, P., P, Radochová, A. Nováková, S. Kaneda, and J. Frouz. 2013. Grazing preference and utilization of soil fungi by *Folsomia candida* (Isotomidae: Collembola). *Eur. J. Soil Biol.* 55:66–70.

Hedlund, K., L. Boddy, and C. M. Preston. 1991. Mycelial response of the host fungus, *Mortierella isabellina*, to grazing by *Onichiurus armatus* (collembolan). *Soil Biol. Biochem.* 23:361–366.

Hieber, M. and M. O. Gessner. 2002. Contribution of stream detritivores, fungi and bacteria to leaf breakdown based on biomass. *Ecology* 83:1026–1038.

Hiol Hiol, F., R. K. Dixon, and E. A. Curl. 1994. The feeding preference of mycophagous Collembola varies with ectomycorrhizal symbiont. *Mycorrhiza* 5:99–103.

Hudler, G. W. 1998. *Magical Mushrooms and Mischievous Molds.* Princeton, NJ, Princeton University Press.

Hyde, K. D., E. B. Gareth Jones, E. Leano et al. 1998. Role of fungi in marine ecosystems. *Biodivers. Conserv.* 7:1147–1161.

Hyodo, F., I. T. Tayasu, T. Inoue et al. 2003. Differential role of symbiotic fungi in lignin degradation and food provision for fungus-growing termites (Macrotermitinae: Isoptera). *Funct. Ecol.* 17:186–193.

Ikonen, E. 2001. Population growth of two aphelenchid nematodes with six different fungi as a food source. *Nematology* 3:9–15.

Ingham, E. R. and H. B. Massicotte. 1994. Protozoan communities around conifer roots colonized by ectomycorrhizal fungi. *Mycorrhiza* 5:53–61.

Jabiol, J., B. G. McKie, A. Bruder et al. 2013. Trophic complexity enhances ecosystem functioning in an aquatic detritus-based model system. *J. Anim. Ecol.* 82:1042–1051.

Jaenike, J., D. Grimaldi, A. E. Shuder, and A. L. Greenleaf. 1983. Alpha-amanitin tolerance in mycophagous *Drosophila*. *Science* 221:165–167.

Jobard, M., S. Rasconi and T. Sime-Ngando. 2010. Diversity and functions of microscopic fungi: A missing component in pelagic food webs. *Aquat. Sci.* 72:255–268.

Johnson, D., M. Krsek, E. M. H. Wellington et al. 2005. Soil invertebrates disrupt carbon flow through fungal networks. *Science* 309:1047.

Jonas, J. L., G. W. T. Wilson, P. M. White, and A. Joern. 2007. Consumption of mycorrhizal and saprotrophic fungi by Collembola in grassland soils. *Soil Biol. Biochem.* 39:2549–2602.

Jørgensen, H. B., S. Elmholt, and H. Petersen. 2003. Collembolan dietary specialization on soil grown fungi. *Biol. Fertil. Soils* 39:9–15.

Jumpponen, A., K. G. Mattson, and J. M. Trappe. 1998. Mycorrhizal functioning of *Phialocephala fortinii* with *Pinus contorta* on glacier forefront soil: Interactions with soil nitrogen and organic matter. *Mycorrhiza* 7:261–265.

Jumpponen, A., J. M. Trappe, and E. Cazares. 1999. Ectomycorrhizal fungi in Lyman Lake Basin: A comparison between primary and secondary successional sites. *Mycologia* 91:575–582.

Kinnear, J. E., A. Cockson, P. E. S. Christensen, and A. R. Main. 1979. The nutritional biology of the ruminants and ruminant-like mammals: A new approach. *Comp. Biochem. Physiol.* 64A:357–365.

Klironomos, J. N., E. M. Bednarczuk, and J. Neville. 1999. Reproductive significance of feeding on saprobic and arbuscular mycorrhizal fungi by the collebolan. *Folsomia candida. Funct. Ecol.* 13:756–761.

Klironomos, J. N. and P. Moutoglis. 1999. Colonization of nonmycorrhizal plants by mycorrhizal neighbors as influenced by the collebolan. *Folsomia candida. Biol. Fertil. Soils* 29:277–281.

Klironomos, J. N. and M. Ursic. 1998. Density-dependent grazing on the extraradical hyphal network of the arbuscular mycorrhizal fungus, *Glomus intraradices*, by the collembolan, *Folsomia candida*. *Biol. Fertil. Soils.* 26:250–253.

Kohlmeyer, J. and E. Kohlmeyer. 1979. *Marine Mycology: The Higher Fungi.* New York, Academic Press.

Korb, J. and K. E. Linsenmair. 2001. The causes of spatial patterning of mounds of a fungus-cultivating termite: Results from nearest-neighbour analysis and ecological studies. *Oecologia* 127:324–333.

Kotter, M. M. and R. C. Farentinos. 1984. Tassel-eared squirrels as spore dispersal agents of hypogeous mycorrhizal fungi. *J. Mammal.* 65:684–687.

Krumins, J. A. 2014. The positive effects of trophic interactions in soil. In *Interactions in Soil: Promoting Plant Growth*, ed. J. Dighton and J. A. Krumins, 81–94. Dordrecht: Springer Science+Business Media.

Kumpula. J. 2001. Winter grazing of reindeer in woodland lichen pasture: Effect of lichen availability on condition of reindeer. *Small Rumin. Res.* 39:121–130.

Laakso, J. and H. Setala. 1999. Population- and ecosystem-level effects of predation on microbial-feeding nematodes. *Oecologia* 120:279–286.

Larsen, J., A. Johansen, S. E. Larsen, L. H. Heckmann, I. Jakobsen, and P. H. Krogh. 2008. Population performance of collembolans feeding on soil fungi from different ecological niches. *Soil Biol. Biochem.* 40:360–369.

Lavelle, P. 1997. Faunal activities and soil processes: Adaptive strategies that determine ecosystem function. *Adv. Ecol. Res.* 27:93–132.

Lawton, J. H. and C. G. Jones. 1995. Linking species and ecosystems: Organisms as ecosystem engineers. In *Linking Species and Ecosystems*, ed. C. G. Jones and J. H. Lawton, 141–150. New York: Chapman Hall.

Lombardero, M. J., K. D. Klepzig, J. C. Moser, and M. P. Ayres. 2000. Biology, demography and community interactions of *Tarsonemus* (Acari: Tarsonemidae) mites phoretic on *Dendroctonus frontalis* (coleoptera: Scolytidae). *Agric. For. Entomol.* 2:1–10.

Lussenhop, J. 1992. Mechanisms of microarthropod–microbial interactions in soil. *Adv. Ecol. Res.* 23:1–33.

Lussenhop, J. 1996. Collembola as mediators of microbial symbiont effect upon soybean. *Soil Biol. Biochem.* 28:363–369.

Lussenhop, J. and D. T. Wicklow. 1985. Interaction of competing fungi with fly larvae. *Microb. Ecol.* 11:175–182.

MacMahon, J. A., J. F. Mull, and T. O. Crist. 2000. Harvester ants (*Pogonomyrmex* spp.): Their community and influences. *Annu. Rev. Ecol. Syst.* 31:265–291.

Mangan, S. A. and G. H. Adler. 2002. Seasonal dispersal of arbuscular mycorrhizal fungi by spiny rats in a neotropical forest. *Oecologia* 131:587–597.

Mathiesen, S. D., Ø. E. Haga, T. Kaino, and N. J. C. Tyler. 2000. Diet composition, rumen papillation and maintenance of carcass mass in female Norwegian reindeer (*Rangifer tarandus tarandus*) in winter. *J. Zool. Lond.* 251:129–138.

McGonnigle, T. P. 1997. Fungivores. In *The Mycota IV*, ed. D. T. Wicklow and B. Soderstrom, 237–248. Berlin: Springer-Verlag.

McGonnigle, T. P. and A. H. Fitter. 1987. Evidence that collembola suppress plant benefit from vesicular–arbuscular mycorrhizas (VAM) in the field. Paper presented at 8th North American Conference on Mycorrhizae, at University of Florida, Gainesville, FL.

McIlveen, W. D. and H. J. Cole. 1976. Spore dispersal of Endogonaceae by worms, ants, wasps and birds. *Can. J. Bot.* 54:1486–1489.

Mitchell, M. J. and D. Parkinson. 1976. Fungal feeding oribatid mites (Acari, Cryptosigmata) in an aspen woodland soil. *Ecology* 57:302–312.

Moody, S. A., M. J. I Briones, T. G. Piearce, and J. Dighton. 1995. Selective consumption of decomposing wheat straw by earthworms. *Soil Biol. Biochem.* 27:1209–1213.

Moody, S. A., T. G. Piearce, and J. Dighton. 1996. Fate of some fungal spores associated with wheat straw decomposition on passage through the guts of *Lumbricus terrestris* and *Aporrectodea longa*. *Soil Biol. Biochem.* 28:533–537.

Moore, J. C. and P. C. de Ruiter. 2000. Invertebrates in detrital food webs along gradients of productivity. In *Invertebrates as Webmasters in Ecosystems*, ed. D. C. Coleman and P. F. Hendrix, 161–184. Wallingford, UK: CABI.

Newell, K. 1984a. Interaction between two decomposer basidiomycetes and a collembolan under Sitka spruce: Distribution, abundance and selective grazing. *Soil Biol. Biochem.* 16(3):227–233.

Newell, K. 1984b. Interactions between two decomposer basidiomycetes and a collembolan under Sitka spruce: Grazing and its potential effects on fungal distribution and litter decomposition. *Soil Biol. Biochem.* 16(3):235–239.

Nieminen, J. K. and H. Setälä. 2001. Bacteria and microbial-feeders modify the performance of a decomposer fungus. *Soil Biol. Biochem.* 33:1703–1712.

Noga, E. J. 1990. A synopsis of mycotic diseases of marine fishes and invertebrates. In *Pathology in Marine Science*, ed. F. O. Perkins, and T. C. Cheng, 143–160. New York: Academic Press.

Norton, A. P., G. English-Loeb, D. Gadoury, and R. C. Seem. 2000. Mycophagous mites and foliar pathogens: Leaf domatia mediate tritrophic interactions in grapes. *Ecology* 81:490–499.

Norton, J. H., A. D. Thomas, and J. R. Barker. 1994. Fungal infection in the cultured juvenile boring clam *Tridacna crocea*. *J. Invert. Pathol.* 64:273–275.

Ohtonen, R., H. Fritze, T. Pennanen, A. Jumpponen, and J. M. Trappe. 1999. Ecosystem properties and microbial community changes in primary succession on a glacier forefront. *Oecologia* 119:239–246.

Parisi, P. A. G., A. A. Grimoldi, and M. Omacini. 2014. Endophytic fungi of grasses protect other plants from aphid herbivory. *Fung. Ecol.* 9:61–64.

Parkinson, D., S. Visser, and J. B. Whittaker. 1979. Effects of collembolan grazing on fungal colonization of leaf litter. *Soil Biol. Biochem.* 11:75–79.

Pattinson, G. S., S. E. Smith, and B. M. Doube. 1997. Earthworm *Aporrectodea trapezoides* had no effect on the dispersal of a vesicular–arbuscular mycorrhizal fungi, *Glomus intraradices*. *Soil Biol. Biochem.* 29:1079–1088.

Pérez, F., C. Castillo-Guevara, G. Galindo-Flores, M. Cuautle, and A. Estrada-Torres. 2010. Effect of gut passage by two highland rodents on spore activity and mycorrhiza formation of two species of ectomycorrhizal fungi (*Laccaria trichodermophora* and *Suillus tomentosus*). *Botany* 90:1084–1092.

Piattoni, F., A. Amicucci, M. Iotti et al. 2014. Viability and morphology of *Tuber aestivum* spores after passage through the gut of *Sus scrofa*. *Fung. Ecol.* 9:52–60.

Ponge, J. F. 1990. Ecological study of a forest humus by observing a small volume: I. Penetration of pine litter by mycorrhizal fungi. *Eur. J. For. Pathol.* 20:290–303.

Ponge, J. F. 1991. Succession of fungi and fauna during decomposition of needles in a small area of Scots pine litter. *Plant Soil* 138:99–113.

Porter, L. M. and P. A. Garber. 2010. Mycophagy and its influence on habitat use and ranging patterns in *Callimico goeldii*. *Am. J. Phys. Anthropol.* 142:468–475.

Pregitzer, K. S., R. L. Hendrick, and R. Fogel. 1993. The demography of fine roots in response to patches of water and nitrogen. *New Phytol.* 125:575–580.

Price, D. W. 1976. Passage of *Verticillium albo-atrum* propagules through the alimentary canal of the bulb mite. *Phytopathology* 66:46–50.

Raghukumar, C. and S. Raghukumar. 1991. Fungal invasion of massive corals. *Mar. Ecol.* 12:251–260.

Rangeley, A., M. J. Daft, and P. Newbold. 1982. The inoculation of white clover with mycorrhizal fungi in unsterile hill soil. *New Phytol.* 92:89–102.

Reddell, P. and A. V. Spain. 1991. Earthworms as vectors of viable propagules of mycorrhizal fungi. *Soil Biol. Biochem.* 23:767–774.

Reddy, M. V. and P. K. Das. 1983. Microfungal food preferences of soil microarthropods in a pine plantation ecosystem. *J. Soil Biol. Ecol.* 3:1–6.

Redfern, D. B. 1989. The roles of the bark beetle *Ips cembrae*, the wood wasp *Urocerus gigas* and associated fungi in dieback and death of larches. In: *Insect–Fungus Interactions*, ed. N. Wilding, N. M. Collins, J. F. Hammond, and J. F. Webber, 195–204. London, Academic Press.

Robinson, C. H., J. Dighton, J. C. Frankland, and P. A. Coward. 1993. Nutrient and carbon dioxide release by interacting species of straw-decomposing fungi. *Plant Soil* 151:139–142.

Ruess, L. and J. Dighton. 1996. Cultural studies on soil nematodes and their fungal hosts. *Nematologica* 42:330–346.

Ruess, L., E. J. Garcia-Zapata, and J. Dighton. 2000. Food preferences of a fungal-feeding nematode *Aphelenchoides* species. *Nematologia* 2:223–230.

Ruess, L. and J. Lussenhop. 2005. Trophic interactions of fungi and animals. In *The Fungal Community; Its Organization and Role in the Ecosystem*, ed. J. Dighton, J. F. White, and P. Oudemans, 581–598. Boca Raton, FL: Taylor & Francis.

Sabatini, M. A. and G. Innocenti. 2000. Soil-borne plant pathogenic fungi in relation to some collembolan species under laboratory conditions. *Mycol. Res.* 104:1197–1201.

Salmon, S. and J.-F. Ponge. 2001. Earthworm excreta attract soil springtails: Laboratory experiments on *Heteromurus nitidus* (Collembola: Entomobryidae). *Soil Biol. Biochem.* 33:1959–1969.

Sanders, I. J. and A. H. Fitter. 1992a. The ecology and functioning of vesicular–arbuscular mycorrhizas in co-existing grassland species: I. Seasonal patterns of mycorrhizal occurrence and morphology. *New Phytol.* 120:517–524.

Sanders, I. J. and A. H. Fitter. 1992b. The ecology and functioning of vesicular–arbuscular mycorrhizas in co-existing grassland species: II. Nutrient uptake and growth of vesicular-arbuscular mycorrhizal plants in a semi-natural grassland. *New Phytol.* 120:525–533.

Scheu, S. and H. Setälä. 2002. Multitrophic interactions in decomposer food webs. In *Multitrophic Level Interactions*, ed. T. Tscharutke and B. A. Hawkins, 223–264. Cambridge: Cambridge University Press.

Schickmann, S., A., Urban, K. Kräutler, U. Nopp-Mayr, and K. Hackländer. 2012. The interrelationship of mycophagous small mammals and ectomycorrhizal fungi in primeval, disturbed and managed Central European mountainous forests. *Oecologia* 170:395–409.

Schultz, P. A. 1991. Grazing preferences of two collembolan species, *Folsomia candida* and *Proisotoma minuta*, for ectomycorrhizal fungi. *Pedobiologia* 35:313–325.

Scullion, J. and A. Malik. 2000. Earthworm activity affecting organic matter, aggregation and microbial activity in soils restored after opencast mining for coal. *Soil Biol. Biochem.* 32:119–126.

Setälä, H. 2000. Reciprocal interactions between Scots pine and soil food web structure in the presence and absence of ectomycorrhiza. *Oecologia* 125:109–118.

Shaw, P. J. A. 1985. Grazing preferences of *Onychiurus amatus* (Insecta: Collembola) for mycorrhizal and saprophytic fungi of pine plantations. In *Ecological Interactions in Soil: Plants, Microbes and Animals*, ed. A. H. Fitter, D Atkinson, D. J. Read, and M. B. Usher, 333–337. Oxford, UK: Blackwell.

Shaw, P. J. A. 1988. A consistent hierarchy in the fungal feeding preferences of the Collembola *Onychiurus armatus*. *Pedobiologia* 31:179–187.

Shaw, P. J. A. 1992. Fungi, fungivores, and fungal food webs. In *The Fungal Community: Its Organization and Role in the Ecosystem*, ed. G. C. Carrol and D. T. Wicklow, 295–310. New York: Marcel Dekker.

Shew, H. D. and M. K. Beute. 1979. Evidence for the involvement of soil-borne mites in *Pythium* pod rot of peanut. *Biol. Fertil. Soil.* 69:204–207.

Smolowitz, R. M., R. A. Bullis, and D. A. Abt. 1992. Mycotic bronchitis in the laboratory-maintained hermit crabs. *J. Crust. Biol.* 12:161–168.

Snyder, S. R. and C. F. Friese. 2001. A survey of arbuscular mycorrhizal fungal root inoculum associated with harvester ant nests (*Pogonomyrmex occidentalis*) across the western United States. *Mycorrhiza* 11:163–165.

Stark, S., D. A. Wardle, R. Ohtonen, T. Helle, and G. W. Yeates. 2000. The effect of reindeer grazing on decomposition, mineralization and soil biota in a dry oligotrophic Scots pine forest. *Oikos* 90:301–310.

Stoffolano, J. G. Jr., B.-X. Zou, and C.-M., Yin. 1990. The stinkhorn fungus, *Mutinus canis*, as a potential food for egg development in the blowfly, *Phormia regina*. *Entomol. Experimentalis Appl.* 55:267–273.

Strid, Y., M. Schroeder, B. Lindahl, K. Ihrmark, and J. Stenlid. 2014. Bark beetles have a decisive impact on fungal communities in Norway spruce stem sections. *Fung. Ecol.* 7:47–58.

Suberkropp, K. 1992. Interactions with invertebrates. In: *The Ecology of Aquatic Hyphomycetes*, ed. F. Barlocher, 118–134. Berlin, Springer-Verlag.

Sutherland, J. R. and J. A. Fortin. 1968. Effect of the nematode *Aphelenchus avenae* on some ectotrophic mycorrhizal fungi and on a red pine mycorrhizal relationship. *Phytopath.* 58:519–523.

Thimm, T. and O. Larink. 1995. Grazing preferences of some collembola for endomycorrhizal fungi. *Biol. Fertil. Soils* 19:266–268.

Tibbett, M. 2000. Roots, foraging and the exploitation of soil nutrient patches: The role of mycorrhizal symbionts. *Func. Ecol.* 14:397–399.

Tiwari, S. C. and R. R. Mishra. 1993. Fungal abundance and diversity in earthworm casts and in undigested soil. *Biol. Fertil. Soil.* 16:131–134.

Trappe, J. M. 1988. Lessons from alpine fungi. *Mycologia* 80:1–10.

Trappe, J. M. and A. W. Claridge. 2005. Hypogeous fungi: Evolution of reproductive and dispersal strategies through interactions with animals and mycorrhizal plants. In *The Fungal Community; Its Organization and Role in the Ecosystem*, ed. J. Dighton, J. F. White, and P. Oudemans, 613–623. Boca Raton, FL: Taylor & Francis.

Trappe, J. M. and C. Maser. 1976. Germination of spores of *Glomus macrocarpus* (Endogonaceae) after passage through a rodent digestive tract. *Mycologia* 68:433–436.

Trappe, J. M., A. O. Nicholls, A. W. Claridge, and S. J. Cork. 2006. Prescribed burning in a *Eucalyptus* woodland suppresses fruiting of hypogeous fungi, an important food source for mammals. *Mycol. Res.* 110:1333–1339.

USDA. 2015. Mushrooms ISSN: 1949-1530.

van Vuuren, M. M. I., D. Robinson, and B. S. Griffiths. 1996. Nutrient inflow and root proliferation during the exploitation of a temporally and spatially discrete source of nitrogen in soil. *Plant Soil* 178:185–192.

Visser, S. 1985. Role of soil invertebrates in determining composition of soil microbial communities. In *Ecological Interactions in Soil, Plants, Microbes and Animals*, ed. A. H. Fitter, D. Atkinson, D. J. Read, and M. B. Usher, 297–317. Oxford, UK: Blackwell Scientific.

Visser, S., D. Parkinson, and M. Hassall. 1987. Fungi associated with *Onichiurus subtenuis* (Collembola) in an aspen woodland. *Can. J. Bot.* 65:635–642.

Wall, D. H. and J. C. Moore. 1999. Interactions Underground. Soil biodiversity, mutualism, and ecosystem processes. *BioScience* 49(2):109–117.

Wardle, D. A. 2002. *Communities and Ecosystems: Linking Aboveground and Belowground Components*. Princeton: Princeton University Press.

Wardle, D. A., H. A. Verhoef, and M. Clarholm. 1998. Trophic relationships in the soil microfood-web: Predicting the responses to a changing global environment. *Global Change Biol.* 4:713–727.

Warnock, A. J., A. H. Fitter, and M. B. Usher. 1982. The influence of a springtail *Folsomia candida* (Insecta, Collembola) on the mycorrhizal association of leek *Allium porrum* and the vesicular–arbuscular mycorrhizal endophyte *Glomus fasciculatus*. *New Phytol.* 90:285–292.

Whitford, W. G. 1989. Abiotic controls on the functional structure of soil food webs. *Biol. Fertil. Soils* 8:1–6.

Wicklow, D. T. and D. H. Yocum. 1982. Effect of larval grazing by *Lycoriella mali* (Diptera: Sciaridae) on species abundance of coprophilous fungi. *Trans. Br. Mycol. Soc.* 78:29–32.

Wilding, N., N. M. Collins, P. M. Hammond, and J. F. Weber. 1989. *Insect–Fungus Interactions*. London, Academic Press.

Wong, M. K. M., T.-K. Goh, I. J. Hodgkiss et al. 1998. Role of fungi in freshwater ecosystems. *Biodivers. Conserv.* 7:1187–1206.

Wood, T. G. and R. J. Thomas. 1989. The mutualistic association between Macrotermitinae and *Termitomyces*. In *Insect–Fungus Interactions*, ed. N. Wilding, N. M. Collins, J. F. Hammond, and J. F. Webber, 69–92. London: Academic Press.

Zak, J. C. 1993. The enigma of desert ecosystems: The importance of interactions among the soil biota to fungi. In *Aspects of Tropical Mycology*, ed. S. Isaac, J. C. Frankland, R. Watling, and A. J. S. Whalley. Cambridge: Cambridge University Press.

Zhang, C.-X., M.-X. He, Y. Cao et al. 2015. Fungus-insect gall of *Phlebopus portentosus*. *Mycologia* 107:12–20.

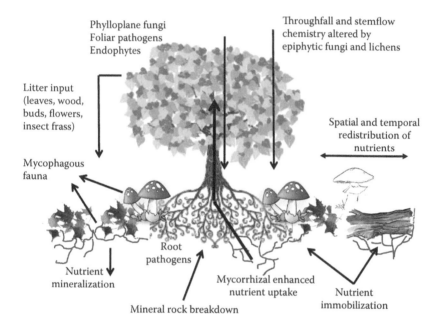

Phylloplane fungi
Foliar pathogens
Endophytes

Throughfall and stemflow
chemistry altered by
epiphytic fungi and lichens

Litter input
(leaves, wood,
buds, flowers,
insect frass)

Spatial and temporal
redistribution of
nutrients

Mycophagous
fauna

Nutrient
mineralization

Root
pathogens

Mycorrhizal enhanced
nutrient uptake

Nutrient
immobilization

Mineral rock breakdown

**Figure 1.2** Schematic representation of some of the roles of fungi in terrestrial ecosystems. (After Dighton, J., Boddy, L., *Nitrogen, Phosphorus and Sulphur Cycling in Temperate Forest Ecosystems*, ed. Boddy, L., 269–298, Cambridge University Press, Cambridge, 1989.)

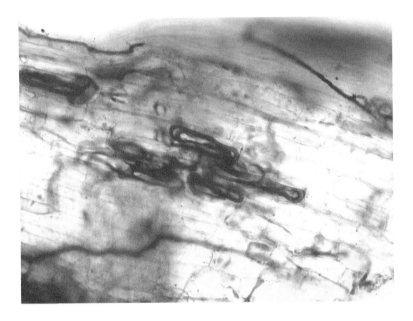

**Figure 3.2** Arbuscular mycorrhizal hyphal coils in Knieskern's beaked sedge (*Rhynchospora knieskernii*) from the New Jersey pine barrens. (From Dighton, J. et al., *Bartonia*, 66, 24–27, 2013.)

**Figure 5.14** Flies and beetles feeding on the glebal mass of spores from a stinkhorn, both deriving nourishment and effecting spore dispersal. (Photo courtesy of the author.)

© Matt Bertone 2014

**Figure 6.4** Cordyceps fruiting on a carpenter ant (*Camponotus* sp.). (Photo by Matthew Bertone, courtesy of YourWildLife.org.)

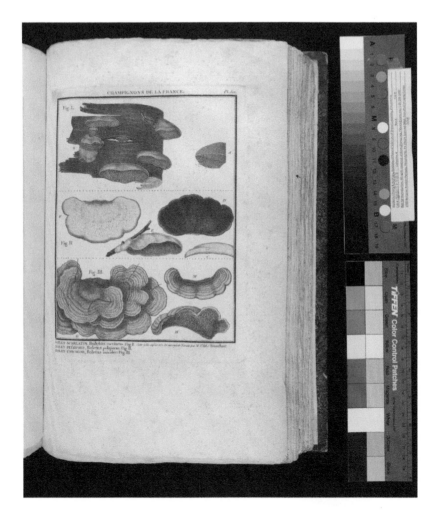

**Figure 8.2** Image of slight foxing around the edge of an old manuscript page from M. Bulliard's Champignon de France 1791. (Courtesy of the Library and Archives of the Academy of Natural Sciences, Philadelphia ANSP Call Number QK608. F8B8.)

# CHAPTER 6

# Fungi as Animal Pathogens
## *Negative Impacts on Faunal Productivity*

In addition to the positive effects of fungi on invertebrate and vertebrate animals, in terms of providing a consistent or intermittent food source, fungi can have detrimental effects on animal populations as pathogens. Within humans there are many fungal diseases that can be serious and fatal, especially to individuals with compromised immune systems. Details of some of the important human fungal pathogens can be seen in books such as those written by Blevins and Kern (1997) and Dismukes et al. (2003). However, a discussion of these pathogens is outside the scope of this book. Many of these diseases are not fatal on their own, but can exert enough influence on the health of their host to reduce growth and fecundity, which has consequential effects on the population of the animal.

## 6.1 ENTOMOPATHOGENS

Such pathogenic fungi can have negative effects on insect populations. In her presidential address to the Mycological Society of America, Blackwell (1994) and the extensive review of entomopathogens provided by Vega et al. (2009) provide a background of the evolution of insect pathogenic fungi, such as the members of the Laboulbenales that are carried by and negatively affect many arthropods. Some of these fungi have been found useful for the biocontrol of important plant insect pests and other nuisance arthropods (Charnley, 1997; Leite et al., 2000; Freimoser et al., 2000). These fungi originated from saprotophs and biotrophs having an ability to penetrate invertebrate exoskeletons, or enter through the trachea, and able to breach the animal's immune response (Charnley, 1997; Vega et al., 2009).

Although there about 100 genera of fungi that have been identified as entomopathogens, only about four genera (*Metarhizium*, *Bauvaria*, *Isaria*, and *Lecanicillium*) have been commercially used on a regular basis, out of about 12 species of current commercial value (Vega et al., 2009). In commercial applications the fungus is dispersed as spores, often in an oil carrier that increases wetting of the insect cuticle to enhance spore germination and initial growth on the cuticle surface. The application of spores of *Beauvaria bassiana* in fallow fields reduced the population of grasshoppers by 60%

and 33% within 9 and 15 days, respectively, after application (Johnson and Goettel, 1993). The incidence of fungal disease in the insects declined from 70%, 2 days after spore application, to 41% by 13 days and, subsequently, to 5% after 19 days. Similarly, the fungus, *Metarhizium anisopliae* has been used as a biocontrol agent against locusts under the trade name of Green Guard with a 90% efficiency of mortality under temperate and hot conditions (Figure 6.1), but lower kill rates under less favorable weather conditions (Miller and Hunter, 2001). The negative impact of *Metarhizium* on nontarget arthropods appears to be less significant than the pesticide fenitrothion (Table 6.1).

These two entomopathogenic fungi are, however, not restricted to infecting arthropods. Studies on the air-breathing amphibious snail (*Biomphalaria glabrata*) showed that egg masses were attacked by *Metarhizium anisopliae* and *Beauvaria bassiana*, with *M. anisopliae* being the more aggressive pathogen (Duarte et al., 2015). The

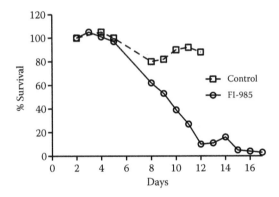

**Figure 6.1**    Effect of the entomopathogenic biocontrol agent, *Metarhizium anisopliae*, on survival of plague locust *Chortoicetes terminifera* nymphs. (Data from Miller, R.J., Hunter, D.M., *J. Orthoptera Res.*, 10, 271–276, 2001.)

**Table 6.1    Effect of Methods of Control of Plague Locust on Other Arthropods, Comparing the Biocontrol Agent *Metarhizium anisopliae* and the Insecticide Fenitrothion**

| Animal Group | Control | Metarhizium | Fenitrothion |
|---|---|---|---|
| Arachnida | 502 | 106 | 41 |
| Diptera | 82 | 11 | 3 |
| Coleoptera | 104 | 105 | 59 |
| Lepidoptera | 10 | 11 | 3 |
| Hymenoptera | 78 | 58 | 27 |
| Hemiptera | 16 | 14 | 1 |
| Collembola | 796 | 965 | 3 |
| Total | 227 | 182 | 20 |

*Source:* Miller, R.J., Hunter, D.M., *J. Orthoptera Res.*, 10, 271–276, 2001.
*Note:* Counts are mean number of animals per pitfall trap 14–21 days after application.

negative impact of the fungal pathogen on egg hatching was greater in submerged eggs compared to eggs developing in water films on rock surfaces (Figure 6.2).

In natural ecosystems, entomopathogenic fungi may be important density-dependent population regulators (Kamata, 2000). For example, in 1963 in Saskatchewan, Canada, it was shown that the naturally occurring fungus, *Entomophthora grylli*, caused an almost complete eradication of the grasshopper *Cammula pellucida* (Pickford and Riegert, 1964). The high incidence of this fungus was correlated to unusually high precipitation and high humidity in June and July of that year, compared to the long-term average and previous 2 years, which favored fungal growth and reproduction.

Kamata (2000) considered the effect of the pathogenic fungus *Cordyceps militaris*, in conjunction with avian predators, parasitoids, and abiotic factors on the population of the beech caterpillar, *Syntypistis punctatella*. He concluded that the periodic population fluctuations of these larvae were caused by delayed density-dependent regulators. The fungus was the prime suspect. This causal agent was

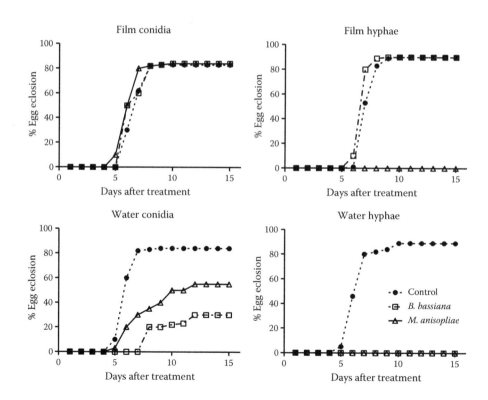

**Figure 6.2** Cumulative egg hatching (eclosion) of the snail, *Biomphalaria glabrata*, in the presence of conidia or hyphae of the pathogenic fungi *Metarhizium anisopliae* and *Beauvaria bassiana* in water films on rocks or free water. (Data from Duarte, G.F. et al., *J. Invert. Pathol.*, 125, 31–36, 2015.)

suspected as the disease started to induce population decline before it reached out-
break densities, but the delayed induced defensive response of the trees was not so
closely related to the changes in the insect population (Figure 6.3). *Cordyceps* is also
reported as frequently occurring on insects in tropical forest ecosystems (Evans,
1982). *Ophiocordyceps unilateralis* is one of a number of hypocrealen pathogens of
formicine ants that develops in the hemocoel, and by the production of nerve toxins
alters the ants' behavior to climb up a plant where the fruiting structure appears
through the ant exoskeleton for spore dispersal (Evans et al., 2011) (Figure 6.4). The
authors suggest that we still know little about the diversity of these fungi in the tropics and that our estimates of their diversity may be low. The ability of the fungus to
control the behavior of its host (extended phenotype of the parasite) is becoming an
upcoming area of research at the biochemical and molecular level (Hughes, 2013).
Indeed, new fungal species are being found in temperate zones as well, with an

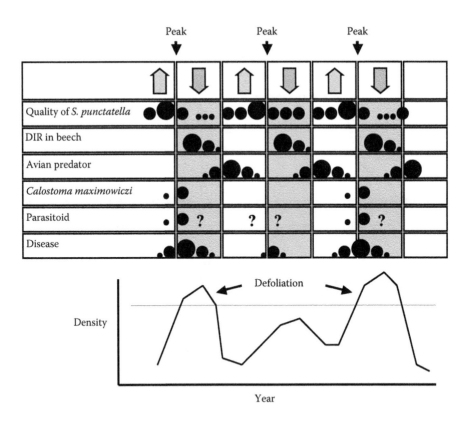

**Figure 6.3**  Model of the regulation of the population dynamics of the beech caterpillar,
*Syntypistis punctatella*, by the delayed induced defensive response (DIR) of the
beech trees and other density-dependent factors. The main correlation between
caterpillar density and a regulator appears to be with insect diseases, which are
mainly fungal. (Data from Kamata, N., *Popul. Ecol.*, 42, 267–278, 2000.)

© MATT BERTONE 2014

**Figure 6.4 (See color insert.)** Cordyceps fruiting on a carpenter ant (*Camponotus* sp.). (Photo by Matthew Bertone, courtesy of YourWildLife.org.)

example of a new species, *Ophiocordyceps myrmicarum*, found as a pathogen of the invasive European fire ant (*Myrmica rubra*) in Maine, USA (Simmons et al., 2015).

The sudden decline in the population of gypsy moth (*Lymantria dispar*) in Appalachian forests in the United States in 1996 was reported to be due to the high abundance of the fungus *Entomophaga maimaiga* (Butler and Strazanac, 2000). However, the incidence of fungal disease may be related to environmental conditions. Hicks et al. (2001) showed that there was a significant increase in entomopathogenic fungi in warm wet conditions. During these weather events, the populations of pine beauty moth in Scottish lodgepole pine forests was reduced mainly by fungal pathogens, whereas at other times predators were the main regulators of moth populations.

The isolation and culture of insect and other arthropod pathogenic fungi (Leite et al., 2000; Freimoser et al., 2000) may ultimately be useful in the discovery and use for the biocontrol of human pests. Onofre et al. (2001) have tested two isolates from each of the two species of entomopathogenic fungi against the bovine tick (*Boophilus microplus*). The fungus *Metarhizium flavoviride* proved to be more effective against adult ticks, reducing larval emergence from eggs and reducing reproductive efficiency, compared with *M. anisopliae* (Table 6.2). Different tick populations show contrasting responses to entomopathogenic fungi. Unfed larval tick mortality range from 92–99% in one population to 71–94% in another, and $LT_{90}$ values (lethal time for 90% mortality) between 19.5–27.5 and 23–37 days, respectively, when ticks were challenged with *Beauveria bassiana* (Perinotto et al., 2012). The fungus *M. anisopliae* was less effective in tick control (Table 6.3). Understanding the differential response of populations to control agents, selection of appropriately adapted isolates (climate, virulence, resistance to UV light), and the immune response characteristics of the target organism, are all challenges to the development of successful biocontrol fungal agents (Fernandez et al., 2012).

Table 6.2   Mean Lethal Dose ($LD_{50}$) and Conidial Density of Two Strains of Each of the Entomopathogenic Fungi, *Metarhizium flavoviride* and *M. anisopliae*, Required for Control of the Engorged Tick, *Boophilus microplus*

| | Mean Percent Control (C%) | | | |
| | *M. flavoviride* | | *M. anisopliae* | |
| Conidia ml$^{-1}$ | CG-291 | MF-3 | CG-30 | CG-46 |
|---|---|---|---|---|
| 0 | 0 | 0 | 0 | 0 |
| $10^4$ | 10.7 | 23.7 | 3.1 | 10.3 |
| $10^5$ | 54.6 | 50.9 | 5.1 | 32.4 |
| $10^6$ | 57.3 | 45.7 | 19.5 | 31.9 |
| $10^7$ | 77.9 | 63.2 | 51.6 | 54.8 |
| $10^8$ | 75.9 | 66.8 | 57.2 | 60.0 |
| | | $LD_{50}$ | | |
| Conidia ml$^{-1}$ | $3.4 \times 10^5$ | $4.2 \times 10^5$ | $2.3 \times 10^7$ | $1.4 \times 10^7$ |

*Source:* Onofre, S.B. et al., *Am. J. Vet. Res.*, 62, 1478–1480, 2001.

Table 6.3   Tick (*Rhipicephalus microplus*) Development Parameters at Two Locations as Affected by Biocontrol Entomopathogens *Metarhizium anisopliae* and *Bauvaria bassiana* Applied as Conidiospores at Two Densities

| | | | *M. anisopliae* | | *B. bassiana* | |
| Site | Parameter | Control | $10^7$ Conidia ml$^{-1}$ | $10^8$ Conidia ml$^{-1}$ | $10^7$ Conidia ml$^{-1}$ | $10^8$ Conidia ml$^{-1}$ |
|---|---|---|---|---|---|---|
| Pesagro | OP (days) | 17.1 | 17.2 | 13.0 | 12.6 | 9.8 |
| | %LH | 97.8 | 82.9 | 81.8 | 79.8 | 89.8 |
| | EPI (%) | 65.9 | 67.5 | 55.5 | 46.4 | 43.8 |
| Faiz | OP (days) | 19.7 | 15.4 | 14.8 | 12.7 | 9.5 |
| | % LH | 95.3 | 94.1 | 82.5 | 84.7 | 72.0 |
| | EPI (%) | 60.0 | 55.3 | 53.2 | 48.5 | 36.8 |

*Source:* Perinotto, W.M.S. et al., *Exp. Parasitol.*, 130, 257–260, 2012.
*Note:* OP = oviposition period; LH = % larval hatching; EPI = egg production index of engorged females.

## 6.2 NEMATODE PATHOGENS AND PREDATORS

The nematode-trapping fungi that occur in the rhizosphere of plants, such as *Arthrobotrys dactyloides, Dactylaria brochopaga, Monacrosporium ellipsosporum,* and *M. gehyropagum* are mentioned by Mankau (1981). These fungi, he suggests, are in the prime position to predate plant parasitic nematodes. Nematophagous fungi have adopted a variety of ways by which they are capable of trapping free-living nematodes (Nordbring-Hertz, 2004). This ability may be by the production of sticky secretions from the hyphae or from specialized structures derived from modified hyphae (sticky knobs of *Nematoctonus* sp.). In addition, a number of species (*Arthrobotrys* and *Dactylella*) produce constricting rings or nets of hyphae that close around nematodes by an almost instantaneous tactile-induced change in turgor pressure, with the hyphae cells creating the noose. The fungus secretes enzymes and digests the nematode, which is used as a source of nitrogen. *Haptocillium* spp. have similarities to the *Verticillium*-like nematode pathogens and are thought to be generalists infecting a range of soil nematode and rotifer species (Glocking and Holbrook, 2005).

Many free-living nematodes in soil have a pathogenic stage as gut intestinal parasites of ruminant animals, such as *Trichostrongylus colubriformis* and other strongyles. Conidia and chlamydospores of the nematophagous fungi *Duddingtonia* and *Arthrobotrys* are capable of surviving passage through the gut of ruminants, with *Duddingtonia flagrans* spores consistently surviving better than spores of *Arthrobotrys* spp. (Faedo et al., 1997). These spores germinate in the dung, where the free-living stage of the nematode develops high population densities and which becomes a potential source of infection for other grazing animals. The nematophagous fungi, however, offer an effective control of the population of the free-living stage of the nematode (Larsen et al., 1997; Manueli et al., 1999). By trapping and killing nematodes, these fungi are able to significantly reduce the infective potential of the nematode population (Faedo et al., 1997; Hay et al., 1998; Figure 6.5). Similarly, Ojeda-Robertos et al. (2005) found that the nematode trapping fungus

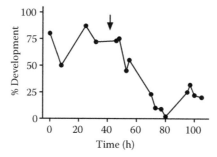

**Figure 6.5** Mean percentage development of eggs of the intestinal parasitic nematode *Trichostrongylus colubriformis* into infective larvae in sheep feces before and after oral administration of a spore inoculum of the nematophagous fungus *Duddingtonia flagrans*. (Data from Faedo, M. et al., *Vet. Parasitol.,* 72, 149–155, 1979.)

*Duddingtonia flagrans* spores could be incorporated into the diet of goats and significantly reduce gastrointestinal nematode larval populations in feces.

Two factors that influence the success of nematophagous fungi as biocontrol agents for plant pathogens are (1) survival of conidia and (2) the trapping structures produced by these fungi (Kerry and Jafee, 1997). There appears to be little information on the survival of these fungi or their role as food for other soil-inhabiting fauna. This information relates to the second factor of density-dependent parasitism, where the efficiency of nematode trapping is closely correlated to the density of nematode trapping fungal structures in soil, which is related to conidial density in soil. The ability to provide an adequate inoculum density to achieve a significant nematode kill rate may be a limiting factor to the success of a specific fungal species as a biocontrol agent.

Other fungi may also be important regulators of plant parasitic nematodes, such as *Fusarium*, *Verticillum*, and *Alternaria*, that parasitize nematode egg masses (Viaene and Abawi, 1998). Similarly, *Trichoderma* species are more effective than nematode-trapping fungi in nematode egg parasitism, and effectiveness is related to the production of chitinase enzymes regulated by the *chi18-12* gene (Szabó et al., 2012).

Populations of cereal cyst nematodes can be controlled by the fungi *Nematophthora gynophial* and *Verticillium chlamydosporium* (Kerry, 1988). These fungi, as well approximately 150 fungal species that have similar properties, parasitize female nematodes and eggs. Similarly, in the presence of arbuscular mycorrhizal fungi, populations of the plant-burrowing nematode, *Radopholus similes*, were significantly reduced (Elsen et al., 2001). The major effect of the mycorrhizae was seen in the females, suggesting potential long-term population regulation by the reduction in fecundity (Figure 6.6).

The nematicidal properties of fungi are not limited to Deuteromycetes. Barron and Thorn (1987) found that the basidiomycete fungi *Pleurotus ostreatus*, *P. strigosus*, *P. subareolatus*, and *P. cornucopiae* produced minute spatulate secretory cells that produced droplets of toxins that killed nematodes on contact within 30 seconds.

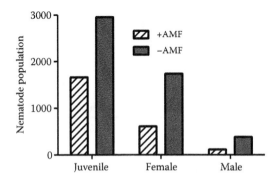

**Figure 6.6**    Effect of arbuscular mycorrhizal inoculation on the population of the root burrowing nematode, *Radopholus similis*. (Data from Elsen, A. et al., *Mycorrhiza*, 11, 49–51, 2001.)

After the death of the nematode, fungal hyphae penetrated the orifices of the nematode and destroyed it.

## 6.3 EMERGING VERTEBRATE FUNGAL PATHOGENS

Fisher et al. (2012) discuss the emerging fungal diseases that are threatening both plants and animals in our ecosystems (Table 6.4), in which they suggest that the spread of these pathogens may have significant effects on our ecosystems. For animal pathogens, they particularly cite white-nose syndrome in bats (Gargas et al., 2009), chyrid infections of amphibians (Fisher et al., 2009), coral decline (Kim and Harvell, 2004), and colony collapse of bees (Cameron et al., 2011), although this is largely attributable to a microsporidian and fungal infections of loggerhead turtles (Sarmiento-Ramírez et al., 2010), where 80% mortality of turtle eggs is attributed to infection by *Fusarium solani*. The keratinophilous fungus *Chrysosporium* and related fungi have been implicated in the disease of some mammals and chickens and has recently been reported to infect reptiles (Cabañes et al., 2014). Chlamydiosis has been found in about 50% of dead passerine birds in England as the possible causal agent of their death (Beckman et al., 2014) (Table 6.5). Many of these reports are from animals in captivity, but some have been reported in the wild, so the impact of these mainly dermal infections on populations of reptiles is still unknown.

It is suggested that there is increased pathogen virulence caused by both mutation and enhanced spore dispersal and the effects of environmental stressors increasing the susceptibility of hosts to fungal pathogens, which increases the threat imposed by these pathogens on individuals, populations, and the ecosystem services they provide (Fisher et al., 2012). In the following discussion, we highlight a few examples of emerging fungal epidemics. If climate change continues, it is likely that more animal species will become environmentally stressed and the likelihood of more epidemics is increased (Garcia-Solache and Casadevall, 2010).

**Table 6.4  Major Emerging Fungal Pathogens of Animals**

| Host | Pathogen | Disease Dynamics |
| --- | --- | --- |
| Amphibia | *Batrachochytridium dendrobatidis* (Chytridomycota) | Worldwide distribution of ultrageneralist pathogen. Influenced by biotic and abiotic factors. |
| Bats | *Geomyces destructans* (Ascomycota) | New to North America where it is more aggressive than in Europe where it probably originated. |
| Coral | *Aspergillus sydowii* (Ascomycota) | Incidence related to increased temperature and immunosuppression of corals. |
| Bees | *Nosema* spp. (Microsporidia) | Pathogen effects along with environmental stress and other pathogens. |
| Sea turtles | *Fusarium solani* (Ascomycota) | Causes hatching failure and may be related to environmental factors. |

*Source:* Adapted from Fisher, M.C. et al., *Annu. Rev. Microbiol.*, 63, 291–310, 2012.

Table 6.5    Number of Bird Carcasses in UK Tested
for Presence of Pathogenic Fungus
*Chlamydyia psittaci*

| Bird Species | Number of Cases Tested | | |
|---|---|---|---|
|  | Positive | Negative | Total |
| Great tit | 7 | 5 | 12 |
| Blue tit | 3 | 1 | 4 |
| Dunnock | 8 | 0 | 8 |
| Robin | 1 | 3 | 4 |
| Rook | 0 | 2 | 2 |
| Jackdaw | 0 | 2 | 2 |
| Chaffinch | 0 | 1 | 1 |
| Wren | 0 | 1 | 1 |
| Pied wagtail | 0 | 1 | 1 |
| Collard dove | 2 | 1 | 3 |
| Feral pigeon | 0 | 2 | 2 |
| Total | 21 | 19 | 40 |

*Source:* Beckmann, K.M. et al., *EcoHealth*, 11, 544–563,
2014.

## 6.3.1 Chytrid Induced Anuran Decline

Fungal diseases have been reported as important regulators of a number of groups of animals. Anuran populations have been shown to decline as a result of the effects of fungal pathogens (Kaiser, 1998; Morell, 1999; Lips, 1999; Reed et al., 2000; Warkentin et al., 2001; Fellers et al., 2001). Necropsies of all dead frogs showed skin abnormalities, especially around the mouth, consistent with fungal pathogens. These reports are of concern especially as the decline in frogs appears to be greatest in the tropical regions, for example, in Panama, and other tropical regions where efforts are underway to conserve biodiversity. Lips (1999) reported increasing numbers of dead frogs in her surveys over recent years. Frogs showed symptoms of fungal pathogens around the mouth and eyes. Previously, Kaiser (1998) and Morell (1999) reported an increase in the incidence of a chytrid fungal (*Batrachochytridium dendrobatidis*) disease of frogs in Panama (Kaiser, 1998) and motorbike frogs in Australia (Morell, 1999) to such an extent that Reed et al. (2000) consider this fungal group an "emerging infectious disease." These chytrids have caused oral deformities in up to 41% of larval Mountain Yellow-legged frogs (*Rana mucosa*) in the Sierra Nevada Mountains of California (Fellers et al., 2001). The distribution of *Bd* in Australia has probably been underestimated. These fungal genera also attack amphibians and fish and have recently been reviewed as potential agents of population decline and extinction of many aquatic species (Rowley et al., 2013). Revised models for suitable habitat have increased the area by more than 25% (Puschendorf et al., 2013). If this is the degree of underestimation of suitable habitat in one location, this has significant implications for the size of suitable habitat worldwide. Patchy distributions of this pathogen have also been seen in Central America in relation to environmental conditions. In the stream-dwelling amphibian *Craugastor punctariolus*, refugia in extremely arid locations appear to be chytrid-free, suggesting an intolerance of the pathogen to very dry conditions (Zumbado-Ulate et al., 2014).

Bullfrogs are raised in captivity at high densities in aquaculture facilities in places such as Brazil, for the food market. Because of the high population density, the incidence of *Bd* infection is great and of significant economic importance (McKenzie and Peterson, 2012).

Reed et al. (2000) reported the occurrence of a variety of *Chlamydia* species that cause respiratory disease in up to 90% of African clawed frogs (*Xenopus tropicalis*) imported into the United States. One breeding colony lost 90% of its individuals within 4 months owing to this fungal infection. The fungal infection is not confined to adult and larval frogs as a filamentous ascomycete fungus was found on 7% of the egg clutches of the Panamanian red-eyed tree frog (*Agalychnis callidryas*) and accounted for 40% of egg deaths (Figure 6.7) (Warkentin et al., 2001).

### 6.3.2 White-Nose Syndrome of Bats

This disease was first identified in the United States in 2006 in a cave in New York (Blehert et al., 2009) and then characterized (Gargas et al., 2009); it has subsequently spread to some 19 states. It was probably introduced to the United States from Europe (Leopaldi et al., 2015; Puechmaille et al., 2010), where it has been found in at least 12 countries. It is caused by the fungus *Pseudogymnoascus* (*Geomyces*) *destructans*, which is capable of growing at temperatures below 20°C (Blehert, 2012). The fungus attacks the bat's muzzle, ears, and wings, and can cause complete mortality within roosts, although it has been found that bats can occasionally recover from the disease (Cryan et al., 2013).

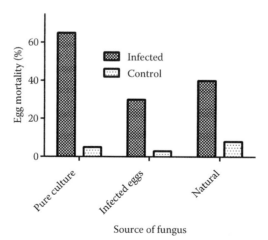

**Figure 6.7** Percentage of red-eyed treefrog egg mortality in laboratory cultured egg masses when infected by a pure culture of an ascomyete fungus, inoculum introduced by contaminated eggs, and from a natural population. (Data from Warkentin, K.M. et al., *Ecology*, 82, 2860–2869, 2001.)

Bat mortality has been attributed to wing damage by the fungus reducing electrolytes (Na concentration) and $CO_2$ partial pressure in the blood, disrupting hibernation because of dehydration and metabolic acidosis (Warnecke et al., 2013). It has been suggested that the recent introduction of the disease into bat communities in the United States is more aggressive than in Europe merely because it is more recent and that bats have yet to adapt, compared with the longer history of bat–fungal interactions in Europe (Leopaldi et al., 2015). In Europe, the fungus does not cause the mass mortality currently seen in the United States (Puechmaille et al., 2011).

The persistence of the fungus in soil and sediments in the hibernacula of bats for long periods after bats leave suggests that a source of inoculum is maintained to infect bats on their return and to newborns (Lorch et al., 2013). Indeed, the fungus possesses the ability to secrete a suite of enzymes (lipase, chitinase, cellulose, and urease) that would allow it to survive as a saprotroph, and so may have arisen as an opportunistic pathogen (Reynolds and Barton, 2014). *Geomyces* is only one genus from some 150 successfully sequenced isolates taken from tape lift samples from bat noses and wings in Illinois and Indiana prior to any symptoms of white-nose syndrome being reported. The fungal community was dominated by *Cladosporium*, *Fusarium*, *Geomyces*, *Mortierella*, *Penicillium*, and *Trichosporon* (Johnson et al., 2013), and similar ectomycota were also found in bats from New Brunswick, Canada (Vanderwolf et al., 2013). However, it is thought that many of these fungi were associated with injuries to the animal and probably represented spores as well as hyphal fragments.

With bats living in hibernacula, it has been suggested that selective culling may be a management strategy to limit the spread of this disease. However, using a modeling approach taking into consideration contact rates between social bats, it was determined that selective culling would be of very limited value (Hallam and McCracken, 2010).

Infectious diseases often have similar epidemiological characteristics, so a comparison has been made between chytrid infections of amphibians and white-nose syndrome of bats (Eskew and Todd, 2013), which is useful for suggesting questions that need to be addressed to determine the likely maximal effect of these emerging diseases on populations of affected organisms, rate of spread, and likely containment (Table 6.6). Changes in climate, particularly predictions for global warming, are expected to increase the geographical ranges of many animal species, and vector-borne diseases are expected to expand along with them. Based on results from rabbits and the disease caused by *Cryptococcus neoformans*, it is likely that fungi will adapt with increased thermotolerance to infect more mammalian hosts or hosts with more vigor (Garcia-Solache and Casadevall, 2010).

## 6.4 PATHOGENS IN AQUATIC AND MARINE ECOSYSTEMS

Wong et al. (1998) state that the role of fungi as plant pathogens in aquatic ecosystems appears to be minor, but that many fungi are pathogenic to animals. A variety

Table 6.6 Comparison of Known and Yet Unknown Information for Two Emerging Fungal Pathogens of Vertebrates, Chytridiomycosis (*Batrachochytridium dendrobatidis*) and Bat White–Nose Syndrome (*Geomyces destructans*)

| Area of Knowledge | *B. dendrobatidis* | | *G. destructans* | |
|---|---|---|---|---|
| | Current | Research Questions | Current | Research Questions |
| Disease emergence | Multiple regions of endemism and 1 hypervirulent. | Where did hypervirulent strain come from? | Genetic diversity in N America. Introduced from Europe. | Genetic differences or similarity between Europe and North America. |
| Abiotic reservoirs | Survival in soil and water. | Resistance to desiccation. Survival and reproduction as a saprotroph. | Persistence in soil and cave walls. | How widespread in the environment? Can it survive as a saprotroph? |
| Biotic reservoirs | Host generalist pathogen of amphibians. Also infects reptile, nematodes, and waterfowl. | Can pathogen complete life cycle in other hosts? | Host generalist of bats. | Can pathogen infect or exist in other vertebrate hosts? |
| Life history and host risk | Aquatic, biphasic, tropical hosts are at greatest risk. | Can life history traits of hosts predict global epidemiology? | Hibernating bat species are at greatest risk. | Are nonhibernating species susceptible? Does host life history predict species decline? |
| Host-pathogen interactions | Skin bacterial antimicrobial compounds confer some resistance. Susceptible species show no immune response. | What is the immune response? Are proteases involved in pathogenicity to evade host immune system? | Host immunity is downregulated during hibernation. | What is the immune response? Does immunity vary seasonally? How does immune response decline during hibernation? Do proteases contribute to pathogenicity? |

*Source:* Adapted from Eskew, E.A., Todd, B.D., *Emerging Infect. Dis.*, 19, 379–385, 2013.

of fungal pathogens cause significant damage to aquatic vertebrates, particularly in aquaculture facilities with about 50% mortality of coho salmon and eels (Ramaiah, 2006). *Democystidium* causes disease in oysters, fish, and amphibians; oomycetes such as *Saprolegnia* infect epidermal tissue of fish and fish eggs. *Brachiomyces* affects gills of fish; *Icthyophonus* attacks the intestine of fish and can cause kidney failure; and larval shrimp can be infected by *Lagenidium* and *Serolpidium* (Ramaiah, 2006). In his review of teleost fish pathogens and their pathology, Roberts (2012) shows a range of degrees of damage to fish from epithelial lesions and cysts by *Saprolegnia*, chytridomycetes, and *Aspergillus* to specific gill parasites attacking gill tissue and blood vessels by *Branchiomyces* to deep-seated infections of major organs (Entomophthorales and *Aspergillus*), severe necrotic erosion of skin and muscle (*Aphanomyces*), and invasion of the head, lateral line, and semicircular canal (*Phialophora*). Indeed, *Saprolegnia diclina* and *S. parasitica* not only cause population decline of freshwater fishes in natural systems but also pose a severe threat to commercial fisheries (van den Berg et al., 2013).

Aquatic fungi, particularly members of the Phycomycetes, are important freshwater fish pathogens, with some 24 fungal species isolated from fish of a Nigerian freshwater pond (Ogbonna and Alabi, 1991). With an increasing development in fish farming in Nigeria and other countries, there is a greater need for evaluating the species of fungi responsible for infecting fish, their relative pathogenicity, and the need to find either chemical or biological methods of control. This is especially important in Third World countries as this developing industry could be of importance to the country as a whole. Significant numbers of fungi, including potential fish pathogenic genera, such as *Achlya* and *Saprolegnia*, were found in commercial fishponds in Poland (Godlewska et al., 2012), with most fungi being present in clear waters and reduced in the presence of organic matter pollution. *Saprolegnia* and *Pythium* have broad host ranges infecting Trichoptera, Plecoptera, Ephemeroptera, Amhipoda, and Coleoptera as well as causing damage to fish and anuran eggs in aquatic systems (Sarowa et al., 2013). Chironomids (nonbiting midges and the biting midges (Ceratopogonidae) and black files (Simulidae) larval stages in freshwater bodies are subject to parasitism by Oomycota and Ascomycota, which may be important for population regulation (de Souza et al., 2014). The oomycte *Aphanomyces astaci* is associated with significant declines in crayfish populations where it acts as a pathogen (Kozubíková-Balcarová et al., 2014). It has been shown that crustaceans that serve as prey for fish can be a vector for fungal pathogens of fish. Some 41 species of fungi have been isolated from dead crustaceans (*Pallasea quadrispinosa*), of which 17 are known to cause mycoses of fish (Czeczuga et al., 2004). These infections cause economic loss of fish in aquaculture facilities.

Fungi are major pathogens of marine animals (Hyde et al., 1998). Mitosporic fungi have been shown to cause disease of crustaceans (Smolowitz et al., 1992) and to cause damage to corals (Raghukumar and Raghukumar, 1991) and juvenile clams (Norton et al., 1994). Norton et al. (1994) isolated dermatiatious fungal hyphae (*Exserohilum rostratum* and *Curvularia* spp.) isolated from mantle and shell of juvenile *Tridacna crocea* clams where they were associated with decline in health. Many oomycetes are highly destructive pathogens of finfish, mollusks, and shellfish (Noga, 1990).

Loss of corals owing to fungal infections has been a recent concern. Sea fan corals (*Gorgonia ventalina*) in Florida, USA, are parasitized by *Aspergillus* spp., causing more than 50% mortality or partial mortality of the fans (Kim and Harvel, 2004; Toledo-Hernandez et al., 2013). However, the authors report that over a 6-year period, fluctuations of the sea fan populations have oscillated because of interactions between recruitment and disease severity, although the authors show that there is no density-dependent interaction between population density of coral and disease prevalence. There was a significant decline in coral recruitment with increasing disease prevalence (Figure 6.8). Of considerable concern is the potential change in water quality because of water warming and pollution, which are likely to decrease the resistance of corals to fungal attack. Gorgonian corals are also susceptible to fungal diseases, particularly aspergillosus, and probably other fungal species. However, a number of other fungal species also live in the corals and appear to have significant antimicrobial activity against each other (Zhang et al., 2012). Not only are these fungi interacting with the pathogenic species but also among themselves. Novel chemistry is being sought from these fungal interactions for pharmaceutical use (Zhang et al., 2012). Diseased sea fans have yielded *Aspergillus flavus* as a potential pathogenic fungus. Zuluaga-Montero et al. (2010) investigated whether there is a separate lineage of marine-specific isolates. Their conclusion was that as *A. flavus* shows little population structure, it was not possible to molecularly determine specificity to a marine habitat.

Although fungi are implicated in the decline of corals, Work and Meteyer (2014) caution that of the nearly 500 papers published between 1965 and 2013, virtually none have established Koch's postulates and confirmed the putative pathogen with the observed disease, be it fungal, bacterial, viral, or other.

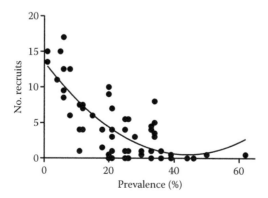

**Figure 6.8** Average number of coral recruits on Florida Keys in relation to the prevalence of aspergillosis in corals. (Data from Kim, K., Harvel, C.D., *Am. Nat.*, 164, S52–S63, 2014.)

The impact of these pathogens on production and fitness of their hosts has not been adequately investigated. Certainly, Hyde et al. (1998) conclude in their review that the role of fungi in this ecosystem requires considerably more research than has hitherto been devoted to it. Although numerous fungal species have been isolated from marine habitats, their function is largely unknown.

## 6.5 CONCLUSION

Many groups of fungi have evolved pathogenic strains, and the continuum from saprotroph to pathogen results in the same species of fungus either having isolates at either end of the spectrum or the same isolate switching trophic modes depending on local circumstances. In contrast, there are also highly specific fungal adaptations to a pathogenic mode of life with either strict host preference or behavioral adaptations (see the interactions between *Ophiocordyceps* and ants).

The degree to which fungal pathogens affect faunal populations varies considerably. We can see that many fungi are associated with animals in that they can be cultured from them or molecularly identified for animal parts. However, without obvious visual proof of death or debilitation being caused by the fungus or performing Koch's postulates, the pathogenic association may be implied rather than proven (as in the case of coral decline). Emerging diseases, however, are more easily distinguished (chytrid with frogs and white-nose syndrome in bats), but may arise as a result of species introduction and may be in a more stable state in its environment of origin.

The expansion of fungal disease of fauna is likely to increase with a combination of increased environmental stress caused by pollution and with climate change causing extensions of species range (Garcia-Solache and Casadevall, 2010). Freshwater ecosystems are described as highly dynamic because of the close interaction between the water body, surrounding terrestrial systems, their receipt of pollution from surrounding ecosystems, and because they contain a wide diversity of organisms (Johnson and Paull, 2011). Using additive models, based on the rise in publications on pathogen occurrence in the ecosystems, the authors suggest that freshwater habitats are likely to see significant increases in pathogen loading with increasing stress of climate change and pollution.

## REFERENCES

Barron, G. L. and R. G. Thorn. 1987. Destruction of nematodes by species of *Pleurotus*. *Can. J. Bot.* 65:774–778.

Beckmann, K. M., N. Borel, A. M. Pocknell et al. 2014. Chlamydiosis in British garden birds (2005–2011) retrospective diagnosis and *Chlamydyia psittaci* genotype determination. *EcoHealth* 11:544–563.

Blackwell, M. 1994. Minute mycological mysteries: The influence of arthropods on the lives of fungi. *Mycologia* 86:1–17.

Blehert, D. S. 2012. Fungal disease and the developing story of bat white-nose syndrome. *PLoS Pathol.* 8:e1002779.

Blehert, D. S., A. C. Hicks, M. Behr et al. 2009. Bat white-nose syndrome: An emerging fungal pathogen. *Science* 323:227.

Blevins, K. S. and M. E. Kern. 1997. *Medical Mycology: A Self Instructional Text*. Philadelphia, PA: F. A. Davis Company.

Butler, L. and J. Strazanac. 2000. Occurrence of Lepidoptera on selected host trees in two central Appalachian national forests. *Entomol. Soc. Am.* 93:500–511.

Cabañes, F. J., D. A. Sutton, and J. Guarro. 2014. Chrysosporium-related fungi and reptiles: A fatal attraction. *PLoS ONE Pathog.* 10:e1004367.

Cameron, S. A., J. D. Lozier, J. P. Strange, J. B. Koch, N. Cordes, L. F. Solter, and T. L. Griswold. 2011. Patterns of widespread decline in North American bumble bees. *Proc. Natl. Acad. Sci. USA* 108:662–667.

Charnley, A. K. 1997. Entomopathogenic fungi and their role in pest control. In *The Mycota IV, Environmental and Microbial Relationships*, ed. D. T. Wicklow and B. Söderstrom, 185–201. Berlin: Springer-Verlag.

Cryan, P. M., C. U. Meteyer, J. G. Boyles, and D. S. Blehert. 2013. White-nose syndrome in bats: Illuminating the darkness. *BMC Biol.* 11:47

Czeczuga, B., B. Kiziewicz, and P. Gruszka. 2004. *Pallasea quadrispinosa* G. O. Sars specimens as vectors of aquatic zoosporic fungi parasiting on fish. *Pol. J. Environ. Stud.* 13:361–366.

De Souza, J. I., F. H. Gleason, M. A. Ansari et al. 2014. Fungal and oomycete parasites of Chironomidae, Ceratopogonidae and Simulidae (Cuclicomorpha, Diptera). *Fung. Biol. Rev.* 28:13–23.

Dismukes, W. E., P. G. Pappas, and J. D. Sobel. 2003. *Clinical Mycology*. Oxford, UK: Oxford University Press.

Duarte, G. F., J. Rodrigues, E. K. K. Fernandez, R. A. Hunter, and C. Luz. 2015. New insights into the amphibious life of *Biomphalaria glabrata* and susceptibility of its egg masses to fungal infection. *J. Invertebr. Pathol.* 125:31–36.

Elsen, A., S. Declerck, and D. De Waele. 2001. Effects of *Glomus intraradices* on the reproduction of the burrowing nematode (*Radopholus similis*) in dixenic culture. *Mycorrhiza* 11:49–51.

Eskew, E. A. and B. D. Todd. 2013. Parallels in amphibian and bat declines from pathogenic fungi. *Emerg. Infect. Dis.* 19:379–385.

Evans, H. C. 1982. Entomogenous fungi in tropical forest ecosystems: An appraisal. *Ecol. Entomol.* 7:47–60.

Evans, H. C., S. L Elliot, and D. P. Hughes. 2011. *Ophiocordyceps unilateralis*: A keystone species for unravelling ecosystem functioning and biodiversity of fungi in tropical forests? *Commun. Integr. Biol.* 45:598–602.

Faedo, M., M. Larsen, and P. J. Waler. 1979. The potential of nematophagous fungi to control the free-living stages of nematode parasites of sheep: Comparison between Australian isolates of *Arthrobotrys* spp. and *Duddingtonia flagrans*. *Vet. Parasitol.* 72:149–155.

Fellers, G. M., D. E. Green, and J. E. Longcore. 2001. Oral chytridomycosis in the mountain yellow-legged frog (*Rana mucosa*). *Copea* 4:945–953.

Fernandez, E. K. K., V. R. E. P. Bittencourt, and D. W. Roberts. 2012. Perspectives on the potential of entomopathogenic fungi in the biocontrol of ticks. *Exp. Parasitol.* 130:300–305.

Fisher, M. C., T. W. J. Garner, and S. F. Walker. 2009. Global emergence of *Batrachochytridium dendrobatidis* and amphibian chytridiomycosis in space, time and host. *Ann. Rev. Microbiol.* 63:291–310.

Fisher, M. C., D. A. Henk, C. J. Briggs, J. S. Brownstein, L. C. Madoff, S. L. McCraw, and S. J. Gurr. 2012. Emerging fungal threats to animal, plant and ecosystem health. *Nature* 484:186–194.

Freimoser, F. M., A. Grundschober, M. Aebi, and U. Tuor. 2000. In vitro cultivation of the entomo-pathogenic fungus *Entomophthora thripidum*: Isolation, growth requirements and sporulation. *Mycologia* 92:208–215.

Garcia-Solache, M. A. and A. Casadevall. 2010. Global warming will bring new fungal diseases for mammals. *MBio* 1:e00061–10.

Gargas, A., M. T. Trest, M. Christensen, T. J. Volk, and D. S. Blehert. 2009. *Geomyces destructans* sp. nov. associated with bat white-nose syndrome. *Mycotaxon* 108:147–154.

Glocking, S. L. and C. P. Holbrook 2005. Endoparasites of soil nematodes and rotifers: II. The genus *Haptocillium*. *Mycologist* 19:2–9.

Godlewska, A., B. Kiziewick, E. Muszyńska, and B. Mazalaska. 2012. Aquatic fungi and heterotrophic Straminipiles from fishponds. *Pol. J. Env. Stud.* 21:615–625.

Hallam, T. G. and G. F. McCracken. 2010. Management of the panzootic white-nose syndrome through culling bats. *Conserv. Biol.* 25:189–194.

Hay, F. S., J. H. Niezen, L. Bateson, and S. Wilson. 1998. Invasion of sheep dung by nematophagous fungi and soil nematodes on hill country pasture in New Zealand. *Soil Biol. Biochem.* 30:1815–1819.

Hicks, B. J., D. A. Barbour, D. Cosens, and A. D. Watt. 2001. The influence of weather on populations of pine beauty moth and its fungal diseases in Scotland. *Scottish For.* 55:199–207.

Hughes, D. 2013. Pathways to understanding the extended phenotype of parasites in their hosts. *J. Exp. Biol.* 216:142–147.

Hyde, K. D., E. B. Gareth Jones, E. Leano, S. B. Pointing, A. D. Poonyth, and L. L. P. Vrijmoed. 1998. Role of fungi in marine ecosystems. *Biodivers. Conserv.* 7:1147–1161.

Johnson, D. L. and M. S. Goettel. 1993. Reduction in grasshopper populations following field application of the fungus *Bauvaria bassiana*. *Biocontrol Sci. Technol.* 3:165–175.

Johnson, L. J. A. N., A. N. Miller, R. A. McCleery et al. 2013. Psychrophilic and psychrotolerant fungi on bats and the presence of *Geomyces* spp. on bat wings prior to the arrival of white-nose syndrome. *Appl. Environ. Microbiol.* 79:5465–5471.

Johnson, P. T. and S. H. Paull. 2011. The ecology and emergence of diseases in fresh waters. *Freshwater Biol.* 56:638–658.

Kaiser, J. 1998. Fungus may drive frog genoside. *Science* 281:23

Kamata, N. 2000. Population dynamics of the beech catterpillar, *Syntypistis punctatella*, and biotic and abiotic factors. *Popul. Ecol.* 42:267–278.

Kerry, B. 1988. Fungal parasites of cyst nematodes. *Ag. Ecosyst. Environ.* 24:293–305.

Kerry, B. R. and B. A. Jaffee. 1997. Fungi as biocontrol agents for plant parasitic nematodes. In *The Mycota IV: Environmental and Microbial Relationships*, ed. D. T. Wicklow and B. Söderstrom. Berlin: Springer-Verlag.

Kim, K. and C. D. Harvell. 2004. The rise and fall of a six-year coral-fungal epizootic. *Am. Nat.* 164:S52–S63.

Kozubíková-Balcarová, E., O. Koukol, M. P. Martin, J. Svoboda, A. Petrusek, and J. Diéguez-Uribeondo. 2014. Reprint of: The diversity of oomycetes on crayfish: Morphological vs. molecular identification of cultures obtained while isolating the crayfish plague pathogen. *Fung. Biol.* 118:601–611.

Larsen, M., P. Nansen, J. Grønvold, J. Wolstrup, and S. A. Henriksen. 1997. Biological control of gastro-intestinal nematodes—Facts, future, or fiction? *Vet. Parasitol.* 72:479–492.

Leite, L. G., L. Smith, G. A. Morales, and D. W. Roberts. 2000. In vitro production of hyphal bodies of the mite pathogen fungus *Neozygites floridana*. *Mycologia* 92:201–207.

Leopaldi, S., D. Blake, and S. J. Puechmaille. 2015. White-nose syndrome fungus introduced from Europe to North America. *Curr. Biol.* 25:R217–R219.

Lips, K. R. 1999. Mass mortality and population declines of anurans at an upland site in western Panama. *Conservation Biol.* 13:117–125.

Lorch, J. M., L. K. Muller, R. E. Russell, M. O'Connor, D. L. Lindler, and D. S. Blehert. 2013. Distribution and environmental persistence of the causative agent of white-nose syndrome, *Geomyces destructans*, in bat hibernacula of the Eastern United States. *Appl. Environ. Microbiol.* 79:1293–1301.

Mankau, R. 1981. Microbial control of nematodes. In: *Plant Parasitic Nematodes*, ed. Zukerman, B. M. and R. A. Rohde, 475–494. New York: Academic Press.

Manueli, P. R., P. J. Waller, M. Faedo, and F. Mahommed. 1999. Biological control of nematode parasites of livestock in Fiji: Screening of fresh dung of small ruminants for the presence of nematophagous fungi. *Vet. Parasitol.* 81:39–45.

McKenzie, V. J. and A. C. Peterson. 2012. Pathogen pollution and the emergence of a deadly amphibian pathogen. *Mol. Ecol.* 21:5151–5154.

Miller, R. J. and D. M. Hunter. 2001. Recent developments in the use of fungi as biopesticides against locusts and grasshoppers in Australia. *J. Orthoptera Res.* 10:271–276.

Morell, V. 1999. Are pathogens felling frogs? *Science* 284:728–731.

Noga, E. J. 1990. A synopsis of mycotic diseases of marine fishes and invertebrates. In *Pathology in Marine Science*, ed. P. O. Perkins and T. C. Chang. New York: Academic Press.

Nordbring-Hertz, B. 2004. Morphogenesis in the nematode trapping fungus *Arthrobotrys oligospora*—An extensive plasticity of infection structures. Mycologist 18:125–13.

Norton, J. H., A. D. Thomas, and J. R. Barker. 1994. Fungal infection in the cultured juvenile boring clam *Tridacna crocea*. *J. Invertebr. Pathol.* 64:273–275.

Ogbonna, C. I. C. and R. O. Alabi. 1991. Studies on species of fungi associated with mycotic infections of fish in a Nigerian freshwater fish pond. *Hydrobiologia* 220:131–135.

Ojeda-Robertos, N. E., P. Mendoza-de Gives, J. F. J. Torres-Acosta, R. I. Rodríguez-Vivas, and A. J. Aguilar-Caballero. 2005. Evaluating the effectiveness of a Mexican strain of *Duddingtonia flagrans* as a biological control agent against gastrointestinal nematodes in goat faeces. *J. Heminthol.* 79:151–157.

Onofre, S. B., C. M. Miniuk, N. M. de Barros, and J. L. Azvedo. 2001. Pathogenicity of four strains of entomopathogenic fungi against the bovine tick *Boophilus microplus*. *Am. J. Vet. Res.* 62:1478–1480.

Perinotto, W. M. S., I. C. Angelo, P. S. Golo, S. et al. 2012. Susceptibility of different populations of ticks to entomopathogenic fungi. *Exp. Parasitol.* 130:257–260.

Pickford, R. and P. W. Riegert. 1964. The fungous disease caused by *Entomophthora grylli* Fres., and its effects on grasshopper populations in Saskatchewan in 1963. *Can. Entomol.* 96:1158–1166.

Puechmaille, S. J., P. Verdeyroux, H. Fuller, M. A. Gouilh, M. Bekaert, and E. C. Teeling. 2010. White-nose syndrome fungus (*Geomyces destructans*) in Bat, France. *Emerg. Infect. Dis.* 16:290–293.

Puechmaille, S. J., G. Wibbelt, V. Korn et al. 2011. Pan-European distribution of white-nose syndrome fungus (*Geomyces destructans*) not associated with mass mortality. *PLoS ONE* 6:e19167.

Puschendorf, R., L. Hodgson, R. A. Alford, L. F. Skerratt, and J. van der Wal. 2013. Underestimated ranges and overlooked refuges from amphibian chytridiomycosis. *Divers. Distrib.* 19:1313–1321.

Raghukumar, C. and S. Raghukumar. 1991. Fungal invasion of massive corals. *Mar. Ecol.* 12:251–260.

Ramaiah, N. 2006. A review on fungal diseases of algae, marine fishes, shrimps and corals. *Indian J. Marine Sci.* 35:380–387.

Reed, K. D., G. R. Ruth, J. A. Meyer, and S. K. Shukla. 2000. *Chlamydia pneumoniae* infection in a breeding colony of African clawed frogs (*Xenopus tropicalis*). *Emerg. Infect. Dis.* 6:196–199.

Reynolds, H. T. and H. A. Barton. 2014. Comparison of the white-nose syndrome agent *Pseudogymnoascus destructans* to cave-dwelling relatives suggests reduced saprotrophic enzyme activity. *PLoS ONE* 9:e86437.

Roberts, R. J. 2012. The Mycology of Teleosts In: *Fish Pathology* (4th Edition), ed. R. J. Roberts, 383–401. Chichester: Wiley Blackwell.

Rowley, J. J. L., F. H. Gleason, D. Andreou, W. L. Marshall, O. Lilje, and R. Gozan. 2013. Impacts of mesomycetozoan parasites on amphibian and freshwater fish populations. *Fung. Biol. Rev.* 27:100–111.

Sarmiento-Ramírez, J. M., E. Abella, M. P. Martin, M. T. Tellería, L. F. Lopez-Jurado, and J. Diéguez-Uribeondo. 2010. *Fusarium solani* is responsible for mass mortalities in nests of loggerhead sea turtle, *Caretta caretta*, in Boavista, Cape Verde. *FEMS Micob. Lett.* 321:192–200.

Sarowar, M. N., A. H. van den Berg, D. McLaggan, M. R. Young, and P. van West. 2013. Saprolegnia strains isolated from river insects and amphipods are broad spectrum pathogens. *Fung. Biol.* 117:752–763.

Simmons, D. R., J. Lund, T. Levitsky, and E. Groden. 2015. *Ophiocordyceps myrmiacarum*, a new species infecting invasive *Myrmica rubra* in Maine. *J. Invertebr. Pathol.* 125:23–39.

Smolowitz, R. M., R. A. Bullis, and D. A. Abt. 1992. Mycotic bronchitis in the laboratory-maintained hermit crabs. *J. Crust. Biol.* 12:161–168.

Szabó, M., K. Csepregi, M. Gálber, F. Virányi, and C. Fekete. 2012. Control plant-parasitic nematodes with *Trichoderma* species and nematode-trapping fungi: The role of *chi18-5* and *chi18-12* genes in nematode egg-parasitism. *Biol. Cont.* 63:121–128.

Toledo-Hernandez, C., V. Gulis, C. P. Ruiz-Diaz, A. M. Sabat, and P. Bayman. 2013. When the aspergillosis hits the fan: Disease transmission and fungal biomass in diseased versus healthy sea fans (*Gorgonia ventalina*). *Fung. Ecol.* 6:161–167.

Van den Berg, A. H., D. McLaggan, J. Diéguez-Ureondo, and P. van West. 2013. The impact of the water molds *Saprolegnia diclina* and *Saprolegnia parasitica* on natural ecosystems and the aquaculture industry. *Fung. Biol. Rev.* 27:33–42.

Vanderwolf, K. J., D. F. McAlpine, D. Malloch, and G. J. Forbes. 2013. Ectomycota associated with hibernating bats in eastern Canadian caves prior to the emergence of white-nose syndrome. *Northeast Naturalist* 20:115–130.

Vega, F. E., M. S. Goettel, M. Blackwell et al. 2009. Fungal entomopathogens: New insights on their ecology. *Fung. Ecol.* 2:149–159.

Viaene, N. M. and G. S. Abawi. 1998. Fungi parasitic on juveniles and egg masses of *Meloidogyne hapla* in organic soils from New York. *J. Nematol.* 30:632–638.

Warkentin, K. M., C. R. Currie, and S. A. Rehner. 2001. Egg-killing fungus induces early hatching of red-eyed treefrog eggs. *Ecology* 82:2860–2869.

Warnecke, L., J. M. Turner, T. K. Bollinger et al. 2013. Pathophysiology of white-nose syndrome in bats: A mechanistic model linking wing damage to mortality. *Biol. Lett.* 9:20130177.

Wong, M. K. M., T-K. Goh, I. J. Hodgkiss, K. D. Hyde, V. M. Ranghoo, C. K. M. Tsui, W-H. Ho, W. S. W. Wong, and T-K. Yuen. 1998. Role of fungi in freshwater ecosystems. *Biodiver. Conserv.* 7:1187–1206.

Work, T. and C. Meteyer. 2014. To understand coral disease, look at coral cells. *EcoHealth* 11:610–618.

Zhang, X-Y., J. Bao, G-H. Wang, F. He, X-Y. Xu, and S-H Qi. 2012. Diversity and antimicrobial activity of culturable fungi isolated from six species of the South China Sea gorgonians. *Microb. Ecol.* 64:617–627.

Zuluaga-Montero, A., L. Ramírez-Camejo, J. Rauscher, and P. Bayman. 2010. Marine isolates of *Aspergillus flavus*: Denizens of the deep or lost at sea? *Fung. Ecol.* 386–391.

Zumbado-Ulate, H., F. Bolaños, G. Gutiérrez-Espeleta, and R. Pischendorf. 2014. Extremely low prevalence of *Batrachochtridium dendrobatidis* in frog populations from neotropical dry forests of Costa Rica supports the existence of a climatic refuge from disease. *EcoHealth* 11:593–602.

# CHAPTER 7

# Fungal Interactions with Pollutants and Climate Change

As the world population increases (Silver and DeFries, 1990; Meadows et al., 1992; Brown, 1997), the effects of human activities on ecosystems escalate (Burger et al., 2012). We constantly hear about issues of climate change, acidifying pollutants (acid rain and anthropogenic nitrogen deposition), atmospheric carbon dioxide increase, stratospheric ozone depletion, and the threat of nuclear fallout. Major national and international discussions have taken place to decide the level of threat that these environmental changes may impose to human populations, their economies, and to the environment. Much of the debate has been political or driven by economic factors, but the magnitude and effects of the actual threat is still very much in debate (Lomborg, 2001), although the modified Lotka–Volterra models of Berck et al. (2012) suggest we are on track to a stable equilibrium where the increasing population will cause environmental collapse enough to drive humans to extinction.

In the discussions of the effects of human activities on the environment, the role of fungi or impacts of the pollutants on fungi rarely hit the headlines in the popular press. In this chapter, we discuss a subset of the important interactions between human-induced changes in the environment and the processes mediated through fungi. I have chosen to discuss the effects of acidifying pollutants, heavy metals, organic pollutants, radionuclides, and global carbon cycling in relation to elevated $CO_2$ levels in the atmosphere.

## 7.1 FUNGI AND ACIDIFYING POLLUTANTS

After the industrial revolution in Europe, numerous industrially related changes in ecosystems and organisms were seen to be the result of pollution of one sort or another. For example, the emergence of a black race of the peppered moth (*Biston betularia*) and its evolution toward being a new species was attributed to its increased survival by adopting a more appropriate cryptic coloration for resting, unobserved by predators, on soot-coated tree bark than its lighter counterpart (Kettlewell, 1955). The change in abundance of certain organisms has been identified as biological indicators of pollution. One such group is the lichens, whose decline in relation to increased atmospheric pollution was dramatic. They are now shown to be recovering in species abundance and diversity (Gilbert, 1992; Bates et al., 2001).

The decline in the health of central European forests in the 1970s along with the continued degradation of decorative limestone carvings on buildings and grave-stones was attributed to acid rain. The sulfur dioxide dissolved in rain resulted from combustion of high sulfur-containing coal for the energy industry in Central Europe. Observations of the declining tree canopies indicated damage to the cuticular waxes, reduction in photosynthetic capacity of the canopy, and thus a decline in tree per-formance (Bervaes et al., 1988; Oren et al., 1988). It was only in the late 1970s that soil ecologists became involved in the research on acid rain in relation to the "Waldsturben" effect of the forest dieback in Bavarian forests (Sobotka, 1964). It was the observations of Ulrich et al. (1979), Hüttermann (1982), and Blaschke et al. (1985) that alerted researchers to the fact that acid rain was affecting root growth and the mycorrhizal status of trees. However, Führer (1990) cautioned that forest decline in Europe was exacerbated by other "natural" stressors including soil nutrient deple-tion, harvest management practices, and drought. Subsequently, research moved toward the study of the primary pollutant, sulfur dioxide, rather than its solubility products. A new wave of research based on both laboratory and field fumigation experiments emerged (McLeod, 1995). At that time, it was also being recognized that the increase in ozone concentration of the lower atmosphere (rather than the loss of stratospheric ozone caused by chlorofluorocarbons) could also play a major role in damage to cuticular waxes and the photosynthetic capacity of leaves of plants. As a result, cofumigation experiments of ozone and sulfur dioxide were conducted.

Finally, $NO_x$ was recognized as another major source of pollution derived from the combustion of fossil fuels. Nitrogen deposition and the problems of nitrogen fertilization and forest soil saturation with N emerged as the most recent line of investigation (Aber et al., 1989; Aber, 1992; McNulty and Aber, 1993; Tietema et al., 1993; Gundersen et al., 1998; Reynolds et al., 1998). It is now obvious that none of these pollutants operates alone, and the reality is that there is some combination of all acidifying pollutants that influences ecosystem processes and the role of fungi within these processes. However, the logistics of teasing out the relative contribution of each pollutant is not simple, especially where they may have contrasting effects of damage to plant structure and reduction or increase in the availability of nutrients in soil.

### 7.1.1 Acidifying Pollutants, Mycorrhizae, and Plant Nutrient Uptake

#### 7.1.1.1 Acid Rain

In contrast to chemical analyses, "critical loads" of pollutants can be inferred from changes in the abundance of physiology of a biological indicator. In Europe, the changes in the composition and abundance of macrofungal fruit bodies, particularly ectomycorrhizal Basidiomycotina, have been used as a biological indicator for ter-restrial forested ecosystems (Fellner, 1988; Colpaert and Van Tichelen, 1996). The Waldsturben effect of the forest dieback in Bavarian forests suggested that part of the reason for forest tree decline was damage of roots and their ectomycorrhizae by acidifying pollutants (Sobotka, 1964; Ulrich et al., 1979; Hüttermann, 1982, 1985; Blaschke et al., 1985; Stroo and Alexander, 1985). Based on the idea that mycorrhizal formation was affected by both carbohydrate supply and nutrient levels in soil, which in turn influence hormone levels in roots (Nylund, 1989), Dighton and Jansen (1991) proposed a two-directional impact model of acidifying pollutants on the development of mycorrhizae (Figure 7.1). In this model, they suggest that two mechanisms lead to the reduction of mycorrhizal associations of the plant roots: (1) via a reduction in photosynthesis in the tree canopy, which reduces the energy supply to roots; and (2) acid-induced increase in the availability of toxic metal ions in soil, resulting in root damage. In the first scenario, pollutant-induced reduction in the photosynthetic capacity of the tree canopy reduces the allocation of photosynthate to roots and their mycorrhizae. The reduced energy supply both reduces the overall mycorrhizal colonization of roots and favors those fungal species that can survive on low carbohydrate supplies. In the second scenario, the acidifying pollutants reduce soil pH to make toxic metals (aluminum, manganese, and magnesium) more soluble, and thus, more plant-available (Skeffington and Brown, 1986; Tyler et al., 1987; Van Breemen and Van Dijk, 1988; Ruark et al., 1991). This increased toxicity leads to reduced root growth, root dieback, and reduced mycorrhizal fungal growth and root colonization. The overall results of the work on the effects of acid rain on mycorrhizae are reviewed by Jansen et al. (1988), Jansen and Dighton (1990), and Dighton and Jansen (1991).

Decline in mycorrhizal formation on roots of trees and reduced root vigor in forests were observed by Sobotka (1964), Liss et al. (1984), Meyer (1987), and Blaschke (1988), among others. Evidence for the reduction in ectomycorrhizal fruit body production comes from Arnolds (1985, 1988), Jansen and Van Dobben (1987), and Fellner (1988). Although the effect of the acidifying pollutants was different between ectomycorrhizal fungal species, there was a general trend of greater effect of reduced abundance in mycorrhizal than saprotrophic fungal species. Arnolds (1988) reported that in most healthy forest ecosystems, fruit bodies of mycorrhizal fungi formed between 45% and 50% of all fruit bodies found. In polluted stands, however, only about 10% of fruit bodies are of mycorrhizal origin. The stages of forest decline have been identified according to the macrofungal ratio of saprotrophic to mycorrhizal forms by Fellner and Pešková (1995) and are shown in Table 7.1.

The effect of acid-induced changes in soil chemistry was shown to affect the ectomycorrhizal community structure of trees roots in Finland by Markkola and Ohtonen (1988), who found that *Piloderma*, *Dermocybe*, *Hebeloma*, and a "type 03" ectomycorrhiza were significantly reduced in the presence of acidifying pollutants, whereas *Cenococcum* was found to increase. Similarly, Dighton and Skeffington (1987) found that simulated acid rain applied to *Pinus sylvestris* trees in lysimeters caused a change in the ectomycorrhizal community structure by reducing the occurrence of mycorrhizal morphotypes that were multibranched and were associated with large amounts of extraradical hyphae. It appeared that the effect of the pollutant was greatest on fungal hyphal growth rather than on the actual mycorrhizal structure. However, in a field manipulation study on Sitka spruce in Scotland,

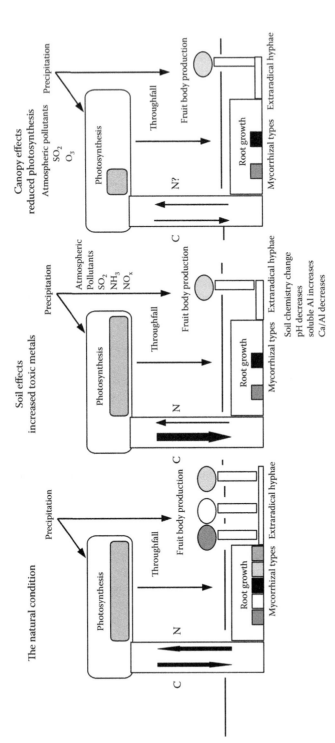

**Figure 7.1**  Schematic representation of the interactions between trees, their ectomycorrhizal symbionts, and environmental factors altered by atmospheric deposition of acidifying pollutants. The first model represents the "natural" situation, where the tree is dependent on mycorrhizal diversity and its associated function to acquire nutrients and the mycorrhizal dependency on the host tree for the supply of carbohydrates for metabolic maintenance of the symbiosis. The second model indicates effects of pollutants on soil, which increase toxic metal availability, leading to root and mycorrhizal damage and increased fertility from nitrogen sources, reducing the dependency of the plant on mycorrhizae. The third model represents damage to foliage and loss of photosynthetic capacity. This reduces the carbohydrate flux to roots, resulting in a loss of mycorrhizal diversity and function. (After Dighton, J., Jansen, A.E., *Environ. Pollut.*, 73, 179–204, 1991.)

**Table 7.1** Relationship between Saprotrophic and Ectomycorrhizal Fungal Abundance in Declining Forests, as Described by Fellner and Pešková (1995)

| Level of Disturbance | Ectotrophic Forest Stability |
| --- | --- |
| Latent | Ectomycorrhizal fungi decrease to 40%, whereas lignicolous species increase to more than 30%. |
| Acute | Ectomycorrhizal species decrease to less than 40%, lignicolous species increase to more than 40%. |
| Lethal | Ectomycorrhizal species decrease to less than 20%, whereas lignicolous species rise to more than 55%. |

Carfrae et al. (2006) observed no change in mycorrhizal fruit body production or colonization of roots with added sulfur at 50 kg ha$^{-1}$. This theory of increased solubility of Al causing root and mycorrhizal damage is supported by the research of Schier (1985) and Thompson and Medve (1984), who showed that an aluminum concentration of 146 μM suppressed the growth of hyphae of *Cenococcum*, *Pisolithus*, and *Thelephora*, but that *Suillus* showed no growth reduction below a concentration of 1000 μM. Furthermore, the Al/Ca, Al/Mg, and Al/PO$_4$ ratios were found to be important determinants of fungal growth (Jongbloed and Borst-Pauwels, 1988, 1989; Kottke and Oberwinkler, 1990). Root and mycorrhizal damage is likely caused by Al-induced increased production of superoxidase dismutase, peroxidase, and glutathione in *Pisolithus tinctorius* mycorrhizal roots of *Pinus massoniana*, leading to increased levels of damaging reactive oxygen species in the root (Kong et al., 2000). The improved growth of a variety of ectomycorrhizal fungal species at low Al/PO$_4$ ratios, even at high levels of Al concentration (Jansen et al., 1990), was explained by the absorption and immobilization of Al into polyphosphate droplets in the fungal hyphae (Kottke and Oberwinkler, 1990).

In a field fumigation experiment (McLeod et al., 1992) in which SO$_2$ and O$_3$ were released over circular plots containing the monospecific stands of Scots pine, Sitka spruce, and Norway spruce, multivariate analysis of the ectomycorrhizal community structure on root systems of the trees revealed that the only trend in change of mycorrhizal community structure was as a result of SO$_2$ fumigation on Scots pine, where a reduction in the occurrence of *Paxillus involutus* mycorrhizae occurred (Shaw et al., 1992, 1993). However, in an analysis of the fruit bodies found in the plots, *Paxillus involutus* fruited more abundantly in Scots pine plots that received high SO$_2$ loading, the converse of the root colonization. The influence of combined pollutants on mycorrhizal formation is dependent on the resistance of the plant species to the pollutant. In a study of the effects of acid rain and ozone on two provenances of Loblolly pine (*Pinus taeda*), Qui et al. (1993) showed that the ozone-tolerant provenance had less reduction in root surface area, less change in the amount of ectomycorrhizal root colonization, and less change in species composition of these mycorrhizal fungal species than the ozone-intolerant provenance. The effect of ozone alone appeared to alter the nature—and not the degree of root colonization—of the arbuscular mycorrhizal structure of sugar maple (*Acer saccharum*) by reducing the number of arbuscules, but increasing the number of vesicles and hyphal coils within the cortex of the root (Duckmanton and Widden, 1994). The authors suggest that this is a response to reduced photosynthesis of ozone-treated plants, where the production of vesicles and hyphal coils is less energy-demanding than the production of arbuscules. As vesicles are thought to be storage structures, it is possible that their production is a stress response.

### 7.1.1.2 Nitrogen Deposition

An end product of fossil fuel combustion is release of nitrogen into the atmosphere. This increased N can be deposited as both wet and dry deposition. In contrast to sulfuric acid, nitrogen is a nutrient that is in demand by plants and is often a limiting nutrient in many ecosystems. The resultant N reaching the forest floor can thus be immobilized into plant and/or microbial biomass (with different biological half-lives), and that which is in excess to plant demand may leach down the soil profile or through lateral flow into watercourses, leading to eutrophication of waterways and groundwater. The degree of N leaching is partly dependent on the limitation of other major nutrient elements (Harrison et al., 1995). When the level of N reaches a critical threshold, the change in nutrient balance (net excess of N) leads to changes in biotic components of both terrestrial and aquatic systems.

The addition of N to forests will have differing effects depending on the initial nitrogen status of the forest soil. In N-limited ecosystems, additional nitrogen will act as a fertilizer and increase plant growth (McNulty and Aber, 1993). In contrast, in N-rich and in saturating N conditions, the effect can be negative. Excess nitrate leaches from the soil into water courses, causing harm to aquatic organisms and reducing water quality for human consumption. At extreme levels of saturation there is increased $N_2O$ and $CH_4$ production, leading to increased atmospheric concentrations of these greenhouse gases (Aber, 1992; Tietema et al., 1993).

Arnolds (1989a,b, 1991, 1997) suggested that the decline in appearance of ectomycorrhizal fruit bodies and increase in saprotrophic and pathogenic fungal fruit bodies in the Netherlands is associated with a combination of acidifying pollutants, and nitrogen deposition (Termorshuizen and Schaffers, 1987, 1991; Kårén and Nylund, 1997), in particular. Although there is little experimental evidence showing the effects of acidifying pollutants on saprotrophic mushroom-forming fungi, Kuyper (1989) showed that nitrogen addition and the effects of liming, to offset the effects of acidifying pollutants, stimulate saprotrophic fungi and leaf litter decomposition where nitrogen levels in the leaf litter are low, but suppress saprotrophic activity where leaf litter nitrogen content is high. He also showed that the effect of liming on mycoflora is very similar to that of nitrogen fertilization. Although they did not observe significant changes in the ectomycorrhizal species composition, Antibus and Linkins (1992) showed that liming reduced the acid phosphatase activity of the mycorrhizal community in the litter layer of the forest floor. They did not explore how the relative availabilities of nitrogen and phosphorus in soil played a part in this, but it could be surmised that there is a synergistic activity of liming on increasing both N and P availability as pH increases in acidic soils and sorption decreases, thus reducing phosphatase activity by negative feedback mechanisms.

Changes in phosphorus availability were implicated in forest decline in the Vosges region of France (Estivalet et al., 1990). The responses of tree seedling growth to added phosphorus show that P availability was reduced in declining forest soils. Estivalet et al. (1990) attribute this to the change in the balance of rhizospheric microbial community, which, in part, suppressed the ectomycorrhizal development in declining forest soils. They showed that fungal species *Penicillium*, *Trichoderma*, *Acremonium*, and *Cylindrocarpon* were more regularly isolated from declining soils than soils from healthy forests, and suggested some antagonism—but not pathogenicity—effect between these soil fungi and the ectomycorrhizal fungi of Norway spruce.

Arnolds (1989a,b,c) catalogued the temporal changes of fungal and plant species in a nutrient-poor acidic soil of a Dutch pasture in the presence of long-term inorganic and organic N fertilizer application. The plant community changed to be dominated by plants tolerant to high levels of nitrogen. The number of fungal species found in 1974 and 1980 did not change, but the species structure did. The declining fungal species included *Hygrocybe ceracea*, *Entoloma conferendum*, *Mycena cinerella*, and *Geoglossum glutinosum*, whereas other species, including *Marasmius oreades*, *Panaeolina foenescii*, *Clitocybe amarescens*, and *Panaeolus acuminatus*, increased between 6- and 400-fold between 1974 and 1980. In addition, he showed increases in coprophytic fungal species. Although this study concentrates on the agricultural inputs of fertilizers, it suggests trends of response of the fungal communities to long-term atmospheric inputs of fertilizing pollutants, such as nitrogen. Termorshiusen and Schaffers (1987) and Rühling and Tyler (1991) showed a decrease in mycorrhizal fungal species compared to saprotrophic fungal species with increasing N. In response to simulated acid rain at two localities in Norway, Sasted and Jennsen (1993) showed a decrease in saprotrophic species richness and increase in dominance of some ectomycorrhizal fungal species. They suggested that this difference may be attributable to local conditions (soils, climate) and cautioned against the broad generalizations that have been made regarding the effects of acidifying pollutants. Arnolds' observations have led to the adoption of "red data" lists for the conservation of fungal species (Arnolds, 1989b, 1997). The call for inclusion of fungi and nonvascular plants in the lists of species for conservation has been adopted in the UK (Watling, 1999, 2005). This action stems from the fear of loss of fungal species because of anthropogenic influences.

Evidence shows that the appearance or nonappearance of fruit bodies of ectomycorrhizal fungi may bear little relation to the abundance of that mycorrhizal morphotype on the root system of the tree (Termorshuizen and Schaffers, 1989; Egli et al., 1993). Concentrating on the belowground component of the forest mycorrhizal system, it was shown that the addition of ammonium sulfate to Norway spruce forests decreased the fine root biomass, but not the degree of ectomycorrhizal colonization of roots (Kårén and Nylund, 1997). However, they suggest that there may be changes in the ectomycorrhizal community structure. Termorshuizen and Schaffers (1991) showed that the addition of nitrogen to the forest floor significantly reduced the number of fruit bodies of basidiomycete ectomycorrhizal fungal species. However, this was not reflected in a change in the mycorrhizal formation on roots. Arnebrant and Söderström (1992) showed that the total mycorrhizal colonization of roots was reduced from 70% to 55% in Scots pine forest fertilized with 1700 and 950 kg N ha⁻¹ over a period of 13 years. In contrast, in a Norway spruce forest nitrogen addition experiment (NITREX) in Sweden, Brandrud (1995) compared the macrofungal flora of a control plot receiving natural N deposition, a plot receiving an additional 35 kg h⁻¹ y⁻¹ N, and a roofed plot to exclude natural N deposition. He showed that there were changes in the ectomycorrhizal fungal flora because of the treatments (presented as a principal components analysis [PCA] in Figure 7.2). He showed that the dominant genera, *Cortinarius* and *Russula*, were reduced in abundance by additional N, *Lactarius* showed little change, and a few specific species, notably *Paxillus involutus* and *Lactarius rufus*, showed increased abundance with added N. Overall, the species diversity, number of fruiting fungi, and total number of fruit bodies did not differ significantly between treatments. Jonsson et al. (1999a) showed that there was poor correspondence between fruit bodies observed and the ectomycorrhizae on roots in both control plots and those receiving additions of nitrogen in the same NITREX experiment. Using molecular analysis of the ectomycorrhizal community structure

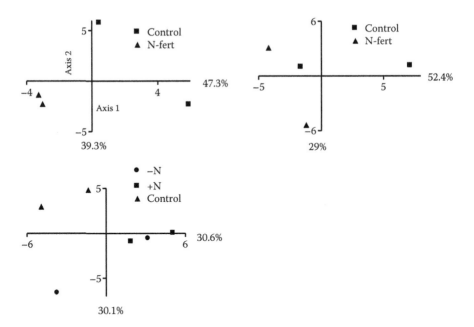

**Figure 7.2**   PCA of the effects of added nitrogen fertilizer, simulating atmospheric nitrogen deposition in a NITREX experiment in Sweden. The top two analyses are treatment plots in ectomycorrhizal fruit body species space, showing the nitrogen treated plots having a different community structure than the untreated control plots. The third analysis is similar, but uses mycorrhizal community structure based on molecular identification of root tips. This letter analysis shows less difference in mycorrhizal community structure between unfertilized and fertilized plots. (Data compiled from Brandrud, T.E., *For. Ecol. Manage.*, 71, 111–122, 1995; and Jonsson, 1998.)

on roots, she demonstrated about 1–4% correspondence between the species of fruit bodies and mycorrhizae. The occurrence of fruit bodies, however, showed a more dramatic shift in species composition than did the mycorrhizae on roots (Figure 7.2). The addition of the equivalent of three times ambient N deposition (65 and 198 kg N ha$^{-1}$ y$^{-1}$ as ammonium nitrate) in beech woodlands caused an almost complete cessation of mycorrhizal fungal fruiting (Rühling and Tyler, 1991). However, many leaf litter-inhabiting saprotrophic fungal species increased fruiting, including species of the genera *Mycena, Clitocybe, Lepista, Agaricus,* and *Lycoperdon* (Figure 7.3). A possible explanation for this is the increased availability of N to stimulate the decomposition of a recalcitrant leaf litter species. In an oligotrophic pine-dominated ecosystem, Dighton et al. (2004) identified a gradient on N deposition in the New Jersey pine barrens, USA. Along this gradient, both the abundance of ectomycorrhizal root tips and mycorrhizal species richness declined with increasing deposition (Figure 7.4). Based on root morphology, they identified a suite of mycorrhizae that were nitrophobic and declining in abundance and a suite of nitrophilic mycorrhizae that were increasing with increasing N deposition. The interesting aspect here was that these changes were observed over a small change in N deposition and at a much lower concentration of N (range between 4 and 8 kg N ha$^{-1}$ y$^{-1}$), which was much lower than the deposition rates in most European studies. This suggests that oligotrophic systems may be more responsive to chronic N addition. However, in a comparison between oligotrophic pine ecosystems in New Jersey and Florida, Adams-Krumins et al. (2009) found relatively little effects of acute N application (35 and 70 N ha$^{-1}$ y$^{-1}$ equivalent) on ectomycorrhizal community composition. Thus, the nature of the deposition of N may be more influential than the absolute loading.

Despite observations of changes in mycorrhizal communities as a result of N deposition, the function of the ectomycorrhizal community (enzyme production, rate of nutrient uptake, degree of protection of host plant to root pathogens, fungal

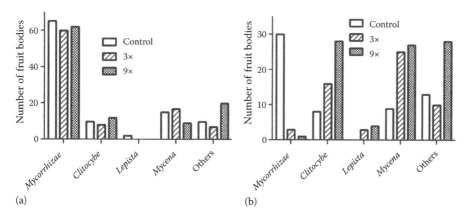

(a)                                                                    (b)

**Figure 7.3**  Mean number of mycorrhizal and saprotrophic fungal fruit bodies in experimentally N-fertilized beech forest plots at zero (control), 260 kg h$^{-1}$ N (3×) and 790 kg ha$^{-1}$ N (9×) at the start of the N addition (a) and 3 years later (b). (After Rühling, A., Tyler, G., *Ambio*, 20, 261–263, 1991.)

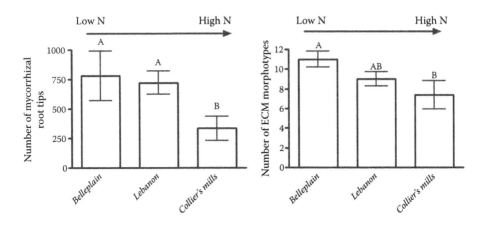

**Figure 7.4** Ectomycorrhizal abundance and morphotype richness along an N-deposition gradient in the NJ pine barrens. (After Dighton, J. et al., *For. Ecol. Manage.*, 201, 131–144, 2004.)

hyphal growth, etc.) has not been fully investigated. Some changes have been shown to occur in the growth and physiology of the mycelium of fungi in the presence of elevated N levels. Arnebrant (1994) showed that the addition of ammonium sulfate and ammonium nitrate at between 1 and 4 mg $g^{-1}$ peat, significantly reduced the mycelial growth of *Paxillus involutus* and *Suillus bovinus* in mycorrhizae synthesized on roots of lodgepole pine seedlings. In contrast, Kieliszewska-Rokicka (1992) found that the addition of small amounts of $NH_4$-N (0.17–19 mM N) increased both the growth and acid phosphatase activity of *Paxillus involutus* mycelia in agar culture. This may be a synergistic effect of the additional nitrogen causing a temporary deficiency of phosphate in the plant and stimulating phosphatase enzyme production by a positive feedback mechanism. Some ectomycorrhizal fungal species may be adapted for coping with high rates of nitrogen addition. Where fungal isolates exhibited inherently high NH4 uptake affinity, N uptake rates by these fungi were inhibited by N addition to a greater extent than fungal isolates exhibiting inherently low rates of ammonium uptake. Isolates having high uptake rates translocate a greater proportion of the assimilated N to shoots (Wallander et al., 1999), whereas low uptake rates may enable ectomycorrhizal fungi to avoid the stress of elevated nitrogen loading to the ecosystem. It has been shown that the imbalance of nutrients in soil by the addition of chronic levels of N can induce limitations of other nutrients. Potassium and phosphorus were shown to become increasingly limiting in Sitka spruce forests in the UK with increased addition of N (Harrison et al., 1995), which supports the hypothesis of Liebig's law of the minimum (Read, 1991).

As ericoid mycorrhizae are known to be capable of obtaining N from organic sources (Stribley and Read, 1980; Leake and Read, 1989, 1990), it would be expected that the addition of readily available nitrogen would have significant effects on heathland ecosystems. The heathlands of the Netherlands have suffered considerable decline as a result of nitrogen deposition. However, in her study of the effects of adding ammonium nitrate at between 0 and 75 kg ha$^{-1}$ y$^{-1}$ N to heather (*Calluna vulgaris*), Johansson (2000) found no significant effects of additional N on root production or mycorrhizal colonization of the roots. She concludes that the decline of heather under the influence of elevated nitrogen deposition is unlikely to be caused by a direct impact on the ericoid mycorrhizae.

In experiments using the interactions between misting tree canopies with a combination of sulfuric acid and ammonium nitrate, which simulates occult deposition of pollutants, Carreira et al. (2000) showed that acidifying pollutants alter the inorganic P subcycle in soil by increasing P sorption capacity and decrease the concentration of labile P. Acid phosphatase activity in soil decreased with acid misting (263, compared to 382 μg pNP h$^{-1}$ g$^{-1}$) in nonmisted soil, thus increasing the organic P and decreasing labile P in soil. They proposed that the addition of nitrogen, to the point of nitrogen saturation, also leads to a reduction in the availability of P in soil (Figure 7.5). Although they did not invoke any fungal role in this process, we can speculate that the reduction in phosphatase activity could be due to a reduction in fungal biomass and activity. However, phosphatase activity generally increases in the presence of greater amounts of organic phosphate (Dighton, 1983, 1991; Sinsabaugh et al., 1993; Antibus et al., 1997).

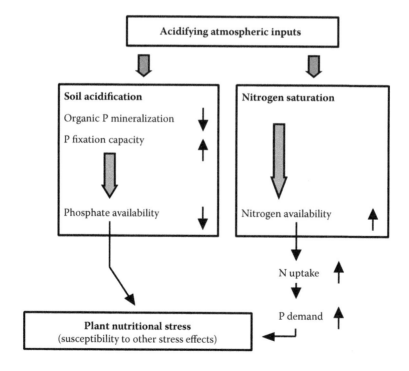

**Figure 7.5** Hypothesized methods of adicifying pollutant effects to induce phosphate nutritional stress by (1) changes in soil acidity affecting inorganic and organic P cycling and (2) increased P demand induced by N saturation. (After Carreira, J.A. et al., *Soil Biol. Biochem.*, 32, 1857–1866, 2000.)

Despite the problems of trying to tease out the effects of single and mixed pollutants, it is fair to say that a number of environmental variables, either pollutants or natural edaphic factors (Conn and Dighton, 2000; Dighton et al., 2000), result in the alteration of ectomycorrhizal community structure on root systems, although it is not fully understood (Dighton and Jansen, 1991). One of those factors was a clear understanding of the changes in ectomycorrhizal community structure. This has been addressed by the combination of fruit body, mycorrhizal morphology, and molecular analyses of these communities (Kårén and Nylund, 1997; Jonsson, 1998). In a study of the interactions between N deposition and acidification with sulfur, Carfrae et al. (2006) showed that nitrogen addition (48 kg ha$^{-1}$ y$^{-1}$) or double N concentration with high S loading (100 kg ha$^{-1}$ y$^{-1}$) had a greater effect in reducing mycorrhizal diversity and abundance of *Lactarius rufus* mycorrhizae on Sitka spruce than S or S and N applied at lower rates. There was limited effect of any treatment on the abundance of *Tylospora fibrillose*, *Inocybe* spp., or *Laccaria* spp. (Table 7.2) although *Tylospora* abundance doubles with N addition alone compared to control.

There are many suggestions that all ectomycorrhizae do not function in the same manner or the same degree of efficiency in any one physiological process (Dighton, 1983; Abuzinadah and Read, 1986, 1989; Dighton et al., 1990). It is only recently that the debate on the role of species diversity *per se* or the composition of the species assemblage and function has prompted considerable investigation using controlled experimentation (Tilman et al., 1996). The diversity/function debate regarding mycorrhizal fungi is in its infancy. Van der Heijden et al. (1998) suggested that plant productivity is at its highest when arbuscular mycorrhizal diversity is highest. Using ectomycorrhizal diversity manipulation experiments in the laboratory, where communities of one, two, or four species assemblages were constructed, Baxter and Dighton (2001, 2005) have shown that ectomycorrhizal diversity *per se* was a better determinant of improved birch seedling nutrient content than actual species composition or colonization rates. The effect of mycorrhizal diversity was greater for P than for N uptake into plant tissues. Much more work is yet to be done to fully explain the

**Table 7.2  Number of Ectomycorrhizal Root Tips and Shannon–Weiner Diversity Index (H′) of Mycorrhizae Associated with Sitka Spruce Treated with Nitrogen, Sulfur, and Combinations**

| Treatment | Control | N (48 kg ha$^{-1}$ y$^{-1}$) | S (50 kg ha$^{-1}$ y$^{-1}$) | N + S | 2× N + S |
|---|---|---|---|---|---|
| *Tylospora* | 59.4 | 136.4 | 48.6 | 82.3 | 44.6 |
| *Lactarius* | 12.1[a] | 0[b] | 9.1[a] | 11.0[a] | 6.4[a] |
| *Inocybe* | 0.75[a] | 0.44[b] | 0.75[a] | 0.44[b] | 0[b] |
| *Laccaria* | 0 | 0 | 0.75 | 0 | 0.19 |
| *Russula* | 0 | 0.81 | 0.19 | 0 | 0 |
| *Cortinarius* | 5.5 | 5.25 | 4.37 | 0.06 | 0 |
| *Thelophora* | 0.06 | 0.63 | 1.25 | 2.13 | 3.13 |
| *H′* | 0.73[a] | 0.21[d] | 0.77[a] | 0.5[d] | 0.44[a] |

*Source:* After Carfrae, J.A. et al., *Environ. Pollut.*, 141, 131–138, 2006.
*Note:* Different superscript letters indicate significant differences between treatments.

consequences of changes in both mycorrhizal and saprotrophic fungal community structure on ecosystem-level processes.

### 7.1.2 Effects of Acidifying Pollutants on Saprotrophic Fungal Activity

Plants are able to acquire nitrogen from that deposited on leaf surfaces, as well as through roots. The deposition of nitrogen onto leaf surfaces can result in the foliar uptake of the equivalent of 2–8% of the total N demand of spruce trees (Boyce et al., 1996). Not only can this change the root demand for N and significantly alter the physiology of roots and their ectomycorrhizae, but it can also affect phylloplane fungal communities and plant endophytes. In a survey of Scots pine phylloplane fungi and lichens along a transect away from a power plant in Spain, Romeralo et al. (2012) showed reduced fungal diversity with increasing proximity to the power plant. Of the lichen community, *Usnea hirta* was susceptible to the pollutant emissions, but *Hypogymnia* spp. and *Scoliciospermum umbrinum* were pollution-tolerant. Under acidic conditions, bisulfite $\left(HSO_3^-\right)$ is the main toxic product of acid rain, and in soils of pH 7 and higher, sulfite $\left(SO_3^{2-}\right)$ predominates (Dursun et al., 1996a). At the environmentally realistic levels of between 12.5 and 100 μM, sulfite had negative effects on the growth of mycelia and germination of spores of *Mycena galopus*, *Phoma exigua*, *Cladosporium cladosporioides*, and *Aureobasidium pullulans* (Boddy et al., 1996; Dursun et al., 1996a). The effects of sulfite were greatest on mycelia in terms of both growth and respiration, and spores were found to be more resistant to sulfites. The decomposition of Sitka spruce leaf litter in the presence of pure cultures of these fungi was also significantly reduced in the presence of sulfite. Moreover, litter incubated with each fungal species fumigated with sulfur dioxide at 40 nl l$^{-1}$ SO$_2$ had differential effects on the respiration of different fungal species and between leaf litter species (Dursun et al., 1996b). Although SO$_2$ reduced the respiration of *Mycena galopus* on leaf litters, this reduction was only statistically significant for Sitka spruce (*Picea sitchensis*) and after 10 weeks for hazel (*Corylus avellana*). There was effectively little reduction in respiration on Scots pine (*Pinus sylvestris*) and beech (*Fraxinus excelsior*). The greatest effect of SO$_2$ was on Sitka spruce litter decomposition, where respiration was reduced by more than one-half. Respiration of the other fungal species (*Phoma exigua*, *Cladosporium cladosporioides*, and *Aureobasidium pullulans*) was unaffected by this concentration of SO$_2$ on any of the leaf litter species. The effects of SO$_2$ fumigation at 10 to 30 nl l$^{-1}$ SO$_2$ on ash (*Fraxinus excelsior*), birch (*Betula* spp.), hazel (*Corylus avellana*), oak (*Quercus robur* and *Q. petraea*), and sycamore (*Acer pseudoplatanus*) leaf litters reduced the abundance of *Cladosporium* spp., *Epicoccum nigrum*, *Fusarium* spp., and *Phoma exigua* but increased the frequency of occurrence of *Coniothyrium quercinum*, *Cylindrocarpon ortosporum*, and *Penicillium* spp. (Newsham et al., 1992a,b). However, they suggest that there is no change in the leaf litter niches occupied by fungi and that resource utilization between these fungi is similar.

Other studies have shown reduction in litter decomposition because of sulfur dioxide fumigation (Prescott and Parkinson, 1985; Wookey and Ineson, 1991a,b), with altered fungal community composition (Wookey et al., 1991; Newsham et al., 1992a,b). However, in an experimental mesocosm study where plants and leaf litter were exposed to an ozone and acid mist (2:1 mixture of $H_2SO_4/HNO_3$), Shaw (1996) showed that the amount of fluorescein diacetate-stained (i.e., metabolically active) fungal hyphae in two leaf litters was significantly greater in acid-misted than in control systems. The stimulation in fungal activity of acid-misted leaf litter may have been attributable to the addition of nitrogen, which would help in the decomposition of high C/N litter resources (Garrett, 1963; Killham et al., 1983; Shaw and Johnston, 1993). In a comparison of decomposition of oak (*Quercus acutissima*) and pine (*Pinus massoniana*) leaf litters, He et al. (2013) showed that the decomposition was reduced with increasing acid conditions, particularly in pine litter. Fungal enzyme production (cellulose, invertase, acid and alkaline phosphatase, polyphenol oxidase, and urease) was also significantly decreased because of acidification. The effect of increasing acidity appears to favor bacterial biomass in soil. In sandy grassland soil, Aliasgharzd et al. (2010) found the fungal/bacterial ratio was lower in acid soil microsites than in more neutral sites. Similarly, in the microcosm studies of *Sorbus anifolia* litter decomposition, Cha et al. (2013) found a similar trend in fungal/bacterial ratios using phospholipid fatty acid analyses.

Because of the retardation of decomposition processes under acidifying conditions, organic nitrogen accumulates in the soil and has been shown to be released by decomposition and leaching on the reduction of acidifying pollutants over the ensuing decades (Oulehle et al., 2006). Where pollution continues, remediation by the application of lime has been carried out at ecosystem scales. In a study of liming, Veerkamp et al. (1997) identified the community structure of lignicolous fungal fruit bodies in a Scots pine forest on acidic soils. They discovered that the effect of liming increased the number of fungal species but that the resultant community was more similar to those fungi found in deciduous woodlands that coniferous woodlands. Species including *Amphinema byssoides*, *Hyphodontia breviseta*, *Hypochnicium geogenium*, and *Sitotrema octosporum* increased in abundance, whereas *Trechyspora farinacea* decreased in abundance. The rationale for the fungal community in limed sites to resemble that of deciduous woodland ecosystems was suggested to be attributable to the elevated pH of the soil increasing nitrogen availability and stimulating the decomposition of the high C/N ratio woody material. This would make the coniferous wood, which is of low resource quality, become more similar to the higher resource quality wood of angiosperms.

In aquatic ecosystems, the impact of acidifying pollutants appear to have little effects on aquatic fungal communities and their activity (Bärlocher, 1992a; Dangles and Chauvet, 2003), probably because of the large dilution factor involved. Despite this, Bärlocher (1992b) states that a reduction in the pH of water from 6 to 4 decreases fungal growth on alder leaves. Similarly, Cornut et al. (2012) show a reduction in fungal species on alder leaves in both benthic and hyporheic zones (buried at 15–30 cm in bottom substrate) in stream systems, with less effect of the acid in the hyporheic zone because of the lower sensitivity of fungi to acid in this region (Table 7.3).

Table 7.3  Number of Fungal Species and Measures of Diversity and Dominance on Decomposing Alder Leaves in Benthic and Hyporheic Zones of Streams with Contrasting Acidity

| Stream Acidity | Benthic | | | Hyporheic | | |
|---|---|---|---|---|---|---|
| | Number of Species | Simpson Diversity | Simpson Dominance | Number of Species | Simpson Diversity | Simpson Dominance |
| Acidic | 13 | 0 | 1 | 6 | 0.01 | 0.99 |
| Moderate | 18/12 | 0.69/0.03 | 0.31/0.97 | 10/11 | 0.51/0.56 | 0.49/0.44 |
| Circumneutral | 18/18 | 0.7/0.57 | 0.3/0.43 | 17/13 | 0.19/0.26 | 0.51/0.74 |

Source: Cornut, J. et al., Water Res., 46, 6430–6444, 2012.
Note: For moderate and circumneutral conditions, there are two streams.

## 7.1.3 Effects of Acidifying Pollutants on Fungal–Faunal Interactions

Given that fauna feed on fungi selectively (Shaw, 1988, 1992), it is expected that changes in the fungal community composition or fungal biomass due to acidifying pollutants will influence the faunal communities and biomass. In a 2-year study of the effects of acid precipitation (pH 4.3 and 3.6 in throughfall) in pine forests of southern United States, Esher et al. (1992) showed no effects of acidifying rain on saprotrophic microbial communities (using dilution plate counts), numbers of nematodes, or number of spores of arbuscular mycorrhizal fungi. However, they showed a significant reduction in the length and dry mass of fine roots and the number of short roots of both ectomycorrhizal loblolly pine (*Pinus taeda*) and longleaf pine (*Pinus palustris*). The numbers of oribatid, prostigmatid, and astigmatid mites increased with increasing soil acidity. Similarly, Heneghan and Bolger (1996) showed a significant increase in abundance of microphytophagous microarthropods (probably fungal feeders) with the addition of nitric acid over a water control, but a slight decrease in abundance with application of sulfuric acid (Table 7.4). This suggests that nitric acid may stimulate fungal growth, which was not seen in the ammonium treatments. Macrophytophages (leaf consumers) were eliminated by nitric acid. There was no significant difference in predator abundance with treatment. Soil microfaunal response to nitrogen additions (ammonium nitrate) at 0, 10, 20, and 50 kg ha$^{-1}$ y$^{-1}$ in semiarid Spain resulted in significant effects on abundance of pauropods and

Table 7.4  Average Abundance (Number per Soil Core) of Microarthropod Trophic Groups from Soil Treated with Acidifying Pollutants

| Treatment | Microphytophages | Predators | Macrophytophages |
|---|---|---|---|
| Nitric acid | 113 | 1.3 | 0 |
| Ammonium nitrate | 65 | 2 | 28 |
| Ammonium sulfate | 40 | 2.7 | 10 |
| Ammonium chloride | 23 | 1.3 | 8 |
| Sulfuric acid | 37 | 2.7 | 19.5 |
| Water | 54 | 1.8 | 10.5 |

Source: Heneghan, L., Bolgers, T., J. Appl. Ecol., 33, 1329–1344, 1996.

collembolan (Ochoa-Hueso et al., 2014). The decline in pauropod populations is not easy to explain from a fungal perspective as they feed mainly on highly decomposed organic matter; however, the largely fungal feeding collembolan increased abundance with increasing N deposition to a threshold between 20 and 50 kg ha$^{-1}$ y$^{-1}$, suggesting a possible change in fungal biomass at high N concentrations.

Using the maturity index (Bongers, 1990), Ruess et al. (1993) correlated the changes in nematode trophic groups (bacterial feeders and root and fungal feeders) to the changes in the bacteria/fungi ratio of forest soils subjected to experimental applications of sulfuric, nitric, and oxalic acids, showing dominance of fungal feeders with the addition of acidifying pollutants. This suggests that fungi are less sensitive to pH change than bacteria. The same general result was obtained from a 3-year acid precipitation experiment of the addition of equimolar sulfuric acid and ammonium nitrate on Sitka spruce plantation forest in Scotland. Here, Ruess et al. (1996) showed that the soil acidity decreased from pH 5.0 to 4.0 in the upper 2 cm of the soil as a result of the acid rain treatment. This induced a significant increase in fungal biomass (measured by soil ergosterol content) and ectyomycorrhizal colonization of roots, which supported an elevated population of fungiverous nematodes (particularly *Aphelenchoides* spp.) and a reduction in omnivorous and predatory nematodes (*Filenchus* spp. and *Aporcelaimellus obtusicaudatus*). The trophic interactions between forest floor fungi and fungiverous fauna requires more research for us to understand the implication of changing fungal communities under anthropogenic stress.

### 7.1.4 Effects of Acidifying Pollutants on Phylloplane Fungi

Live leaf surfaces are a resource available for fungal colonization. Many of these fungi are saprotrophs, deriving their nutrition from the wax cuticle and surface structure of the leaf, utilizing leaf exudates or the nutrients, water and carbon arriving on the leaf surface as wet or dry deposition. In addition to the saprotrophs, endophytic fungi (pathogens and nonpathogens) may spend part of their existence on the leaf surface. As the leaf surface presents a large surface area to the environment, it not only intercepts nutrients and carbon containing material from the atmosphere, it also captures atmospheric pollutants. In a 3-year study using open-air fumigation systems (McLeod et al., 1992; McLeod, 1995), Magan et al. (1995) showed that low levels of SO$_2$ markedly reduced the total phylloplane fungal population on Sitka spruce needles, but O$_3$ caused an increase in fungi on Scots pine. It was hypothesized that both pollutants would increase fungal colonization by causing damage to the leaf surface and allowing availability of more resources to the fungal community. Community changes as a result of pollutants were slight and differed between tree species, pollutants, and method (serial dilution and direct plating); a summary of results is given in Table 7.5. Again, we can see that there is no consistent trend in fungal response to pollutants and that responses are dependent on other biological factors (host tree species) and the nature of the methodology used to measure the response. In cereal crops, however, Magan and McLeod (1991) demonstrated significant reduction in the number of pink and white yeasts on the flag leaves of barley in

Table 7.5 Trends of Fungal Species Occurrence on Needle Surfaces of Scots Pine, Sitka Spruce, and Norway Spruce in Relation to Pollutant and Isolation Method

| Pollutant | Scots Pine | Sitka Spruce | Norway Spruce |
|---|---|---|---|
| $SO_2$ | | Pink yeast[dil] ↓ | *Sclerophoma pythiohila*[dp] ↑ |
| $O_3$ | *Epicoccum nigrum*[dil] ↓ *Cladosprium* spp.[dil] ↓ | *Sclerophoma pythiohila*[dil] ↑ *Rhizosphaera kalkhoffii*[dp] ↑ | *Aureobasidium pullulans*[dp] ↑ |

*Source:* Magan, N. et al., *Plant Cell Environ.*, 18, 291–302, 1995.
*Note:* Data in superscript: "dil" refers to dilution plating technique and "dp" refers to direct plating.

the presence of elevated $SO_2$, whereas *Cladosporium* spp., which are weakly parasitic, were found to increase. This suggests that in leaves with less surface waxes, the impact of $SO_2$ may increase susceptibility to the invasion of pathogenic fungi and weaken the plant. In a survey of phylloplane fungi from Scots pine needles in relation to pollutants from a power plant in Spain, Romeralo et al. (2012) found little change in isolation frequency of fungi. Most frequently, isolated fungal species were pollutant-independent, and only a few genera such as *Nectria*, *Verticillium*, *Fusarium roseum*, and a few unidentified species occurred with low frequency and are confined to the least-polluted sites.

## 7.1.5 Effects of Acidifying Pollutants on Lichens

Lichens have been known to be sensitive to pollutants, and use has been made of them as indicators of industrial effluent. In a review of human impacts on lichens, Brown (1996) outlines the ways in which lichen communities have been reported to respond to a variety of pollutants. Lichen species can vary in their tolerance to pollutant loading (Richardson, 1988; Table 7.6). Distribution maps in the United Kingdom of selected lichen species show apparent spatial relationships with areas of high industrialization and thus sulfur dioxide levels. However, there is evidence from experimental studies that the response of lichens to $SO_2$ may be less than initially thought and that some distributions are related more to intensity of collection than pollutants. However, there are strong correlations between the air pollutant loading and epiphytic lichen community classifications as shown by De Wit (1976). The degree of sensitivity of epiphytic lichens to acidifying pollutants is modified by the substrate on which they are found. De Wit (1976) showed that *Evernia prunastri* and *Parmelia physoides* were less sensitive to pollutants on stems of oak than poplar or

Table 7.6 Variation in Lichen Tolerance of Atmospheric Sulfur Dioxide

| Lichen Species | Maximum $SO_2$ Tolerance ($\mu g\ m^{-3}$) |
|---|---|
| *Lecanora conizaeoides* | 150 |
| *Parmelia caperata* | 40 |
| *Usnea* spp. | 30 |
| *Ramalina fastigiata* | 10 |

*Source:* After Richardson, D.H.S., *Bot. J. Linn. Soc.*, 96, 31–43, 1988.

elm, whereas *Pyscia tenella* was least sensitive on elm. Especially sensitive to acid rain are cyanobacterial lichens (Gilbert, 1992). However, such lichens are more tolerant than others to heavy metals and can often be found as the sole inhabitants of abandoned heavy metal-contaminated mine sites. Changes in epiphytic lichen communities have been observed in the Fraser River region in response to point source pollution, with lichens responding to bark pH and nitrogen (Coxson et al., 2014). Change in lichen communities with pollutants is likely to be a result of varied physiological responses to pollutants. Hallingbäck and Kellner (1992) observe reduced N fixation in *Peltgeria apthora* with acid rain, and particularly simulated acid rain with added ammonium nitrogen. Similarly, Kytöviita and Crittenden (1994) showed a reduction in the nitrogenase activity of *Stereocaulon paschale* with both acidity and combined acidity and nitrate-N. Rates of photosynthesis in lichens is also attributed to acidifying pollutants, such as in *Cetraria* species (Hauck, 2008). In some lichens, however, adaptations to stress may reduce the influence of acid rain. The superhydrophobic coating of *Lecanora conizaeoides* allows the lichen to shed water during very wet conditions to allow photosynthesis to occur, and the same property is suggested to protect the lichen from acidifying pollutants (Shirtcliffe et al., 2006).

Lichens have also been reported as good bioindicators of ozone pollution. In their article, Hur and Kim (2000) showed the hierarchy of the sensitivity of lichen species to ozone in the following order: *Parmotrema austrosinesnse* > *P. tinctorum* > *Certrelia braunsiana* > *Ramalina yasudae*. Recovery of lichen communities has been observed where industrialization in the greater London area has declined and pollution controls have been implemented (Bates et al., 2001). In a 70-km transect study from central London, Bates et al. (2001) recorded the revival of communities of the macrolichen *Hypogymna physoides* and the crustose lichen *Lepraria incana* due to the reduction of $SO_2$. However, the change in the range of distribution of the crustose lichen *Lecanora conizaeoides* suggests that this species needs a low level of $SO_2$ in the environment and is now found in the only remaining sites that have above ambient levels of pollution.

## 7.2 FUNGI AND HEAVY METALS

The central position of fungi in the control of pollutants in terrestrial ecosystems is shown in Figure 7.6 (Wainwright and Gadd, 1997). Heavy metal pollutants can have negative effects on the survival, fitness, and physiology of fungi, such as hyphal growth, the ability to produce extracellular enzymes, and the ability to perform their function in the ecosystem. At sublethal metal concentrations, fungi are able to immobilize heavy metals within their biomass, translocate those metals to other parts of the ecosystem, and then release heavy metals at other locations and change the chemical state of the pollutant. Thus, fungi can exert a strong influence on the fate of heavy metals in the environment. Within the mycorrhizal symbiosis, the fungus may alter the rate at which the pollutant enters the host plant or reduce the degree to which mycorrhizal fungi can colonization roots. The presence of a pollutant chemical within the mycelium of the fungus can lead to changes in the chemical

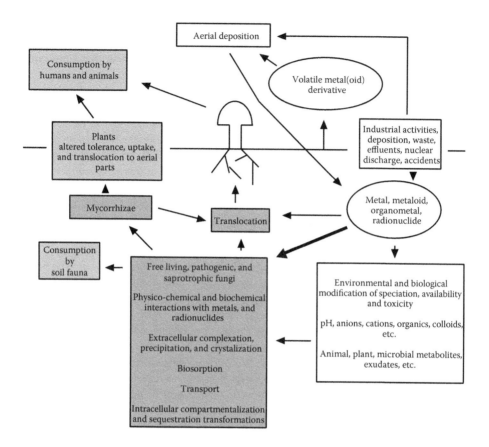

**Figure 7.6** Diagrammatic representation of the interactions between pollutant metals and fungi in a terrestrial environment. Darkly shaded boxes represent the fungal components, lighter shaded boxes the other biotic components. Open boxes represent the environmental factors that alter pollutant availability, whereas the ovals represent the pollutants and their chemical states. Thin arrows show the interactions of pollutant movement between components, but the thick arrow represents a direct effect of the pollutant on the fungi. (After Wainwright, M., Gadd, G.M., in: *The Mycota IV Environmental and Microbial Relationships*, ed. Wicklow, D.T., Soderstrom, B., 85–97. Springer-Verlag, Berlin, 1997.)

and physical state of the pollutant, making it more or less toxic to fungal consumers. The pollutant can be transferred up the food web by grazing of the mycelium or fruiting structure or the pollutant may be released through death or leakiness of the fungal mycelium.

## 7.2.1 Detrimental Effects of Metals

If heavy metals reduce the efficacy of fungal hyphal activity, it is reasonable to assume that the ecosystem-level functions carried out by fungi will be impacted. Increasing levels of chromium, copper, and arsenic wood preservatives had little effect on the biomass of the prokaryotic microbial community, but had a significant negative effect on the eukaryotic (presumed fungal) biomass, as measured by substrate-induced respiration (Bardgett et al., 1994). This suggests that fungi are more sensitive to heavy metals than bacteria. Jordan and Lechevalier (1975) noted that decomposition of cotton strips (cellulose) was significantly reduced at higher heavy metal concentrations (Nordgren et al., 1986), as cellulose decomposition in soil is mainly effected by fungi. Total and fluorescein diacetate active (FDA) fungal biomass and enzyme activity ($N$-acetylglucosaminease, $\beta$-glucosaminease, endocellulase, and acid and alkaline phosphatase) were reduced 10- to 50-fold in heavy metal-contaminated (As, Cd, Cr, Cu, Ni, Pb, and Zn) soils of the Aberdeen Proving Grounds, Maryland (Kuperman and Carreiro, 1997). Lead tends to be less available under these soil-acidifying conditions, because of its being complexed onto humic materials in soil. The abundance and diversity of fungi decreased with increasing pollutant loading (As, Cu, Cd, Pb, and Zn) in the humic soil horizon where ectomycorrhizal species were more tolerant of high metal concentrations than saprotrophs (Rühling and Söderström, 1990). For example, the ectomycorrhizal fungus *Laccaria laccata* was the most tolerant fungus to heavy metal pollutants, but the number of microfungi isolated onto agar from soil did not decrease with increasing metal loading (Rühling et al., 1984). Common species, such as *Penicillium* and *Oidiodendron*, did decrease, but some species such as *Paecilomyces* and several sterile mycelial forms were only found in the most polluted sites.

Heavy metals can have a direct and negative effect on the development and competitive abilities of mycorrhizal fungi, as has been described earlier through the increased availability of Al and Mn in soil caused by acidifying pollutants. Ectomycorrhizal colonization (total number of mycorrhizal root tips), but not plant biomass, of Loblolly pine was reduced by increasing concentrations of Pb in soil (Chappelka et al., 1991). Differential responses of mycorrhizal species to heavy metal (0 to 480 mg kg$^{-1}$ soil) have been reported, with some species being unaffected, whereas others declined in abundance (Hartley et al., 1999a), with toxicity to ectomycorrhizal Scots pine tree seedlings decreasing from Cd, through Pb, Zn and Sb to Cu. The toxicity of heavy metal combinations could not be predicted from the results of single metal applications, suggesting that the unknown complex interactions between metals may reduce the overall effect of multiple metal contamination of soil. These heavy metals also have an effect on the performance of pine seedlings and the ability of ectomycorrhizal fungi to colonize from one plant to another, previously uncolonized seedling (Hartley et al., 1999b). A general decrease in ectomycorrhizal colonization of previously uncolonized roots of Scots pine was also found in Cd- and Zn-contaminated soils (Hartley-Whittaker et al., 2000a), but this did not affect seedling growth rate. Effective concentrations reducing root colonization by 50% were recorded for cadmium as 3.7 µg g$^{-1}$ for *Paxillus involutus* and 2.3 µg g$^{-1}$ for *Suillus variegates* (Hartley-Whitaker et al., 2000b). When colonized by mycorrhizae, the fungi reduced the toxicity of heavy metals by preventing translocation of plant toxic levels into the host plant tissue (Hartley-Whitaker et al., 2000b). The root colonization of pitch pine seedlings by ectomycorrhizae showed a threshold response to mercury, with no reduction in root colonization until after 37 µg g$^{-1}$ Hg (Crane et al., 2012).

Differential tolerance to heavy metals can lead to changes in the mycorrhizal community. The grass *Arrenatherum elatius* is colonized by both arbuscular mycorrhizal and dark septate endophytes (DSEs). In a survey of plant roots in soils contaminated by a range of heavy metals, Deram et al. (2011) showed that high heavy metal loading in soil significantly reduced colonization of roots, but the presence of DSEs increased from almost nonexistent in clean soils to high levels of colonization in metal-polluted soils.

Thus, heavy metals can have negative effects on the saprotrophic and mycorrhizal community and function. However, although fungal growth and colonization of plant roots may be reduced, there are evolved mechanisms of heavy metal defense and metal accumulation by fungi that will protect plants from damaging metal concentrations and, by accumulation, possibly allow for detoxification of polluted environments.

## 7.2.2 Tolerance to Metals

Despite their toxicity, heavy metal elements have been shown to accumulate in basidiomycete fruit bodies, with concentrations often being higher than that in the environment (Byrne et al., 1979a). Not only can fungi accumulate metals such as arsenic in their fruit bodies (Byrne et al., 1979b; Byrne and Tusek-Znidaric, 1983), they are also capable of transforming the chemical nature of the element. For example the ectomycorrhizal basidiomycete *Laccaria amethystina* can methylate arsenic to dimethylarsenic acid (DMA). They discovered methylarsonic acid (MA) in *Sarcosphaera coronaria*, inorganic arsenic in *Entoloma lividum*, and a mixture of inorganic arsenic, MA, DMA, and arsenobetaine in *Sarcodon imbricatus*, *Agaricus placomyces*, and *A. haemorrhoides* (Byrne et al., 1995). Other metal transformation mechanisms have been identified in fungi (Morley et al., 1996), including the reduction of Ag and Cu, using NADPH and NADH as electron donors, as well as methylation and dealkylation of organometallic compounds by enzymes or facilitating abiotic degradation. Fischer et al. (1995) showed that the amount of methyl mercury in a fungal fruit bodies ranged from 0.2 to 8 µg Hg $g^{-1}$ dry weight, depending on the fungal species, leading to bioaccumulation factors for methyl mercury between 3 and 199, whereas those for total mercury were usually below 1 (Table 7.7). These changes in the chemical states of heavy metals can significantly affect the toxicity of the metal to other organisms in the ecosystem. The presence of a methylation process in fungi in nature could have important implications in the movement and toxicity of arsenic and mercury in the environment and its transfer within food chains.

Tolerance of ectomycorrhizal fungi to heavy metals has shown that *Pisolithus tinctorius*, *Suillus luteus*, and *Suillus variegatus* were more tolerant of the heavy metals Cu, Cd, and Zn than *Paxillus involutus*; however, *Paxillus* was more resistant to Ni (Blaudez et al., 2000). Similarly, Saxena and Bhattacharyya (2006) evaluated fungal isolates from a Ni-contaminated sludge and assessed their growth at increasing Ni concentrations. Growth of *Fusarium solani* and *Aspergillus flavus* was less affected by metal levels than that of *M. racemosus* and *Papulaspora sepedonoides*. Hartley et al. (1997a) published effective concentrations of heavy metals that caused

Table 7.7  Relative Accumulation of Mercury in Fungal Fruit Bodies in Relation to Soil Humus Concentrations

| | MeHg in Fungus ($\mu$g Hg g$^{-1}$ dry weight) | MeHg in Humus ($\mu$g Hg g$^{-1}$ dry weight) | Concentration Factor MeHg | Total Hg in Fungus ($\mu$g Hg g$^{-1}$ dry weight) | Total Hg in Humus ($\mu$g Hg g$^{-1}$ dry weight) | Concentration Factor Total Hg |
|---|---|---|---|---|---|---|
| Xercomus badius | 0.16 | 0.01 | 12.3 | 14.7 | 64.4 | 0.23 |
| Xercomus badius | 0.19 | 0.02 | 10.0 | 27.0 | 15.8 | 1.70 |
| Xercomus badius | 0.24 | 0.03 | 9.6 | 35.0 | 80.5 | 0.40 |
| Leccinum scabrum | 0.08 | 0.01 | 10.0 | 6.2 | 35.5 | 0.17 |
| Amanita muscaria | 0.55 | 0.05 | 10.4 | 67.5 | 140.2 | 0.48 |
| Amanita muscaria | 0.71 | | | 82.0 | | |
| Amanita muscaria | 0.27 | 0.09 | 3.0 | 64.0 | 61.9 | 1.03 |
| Hygrophoropsis aurantica | 0.41 | 0.02 | 19.5 | 11.2 | 29.0 | 0.39 |
| Vascellum pratense | 1.56 | 0.04 | 38.0 | 8.2 | 82.7 | 1.70 |
| Coprinus comatus | 7.94 | 0.04 | 198.5 | 144 | | |

Source: After Fischer, R.G. et al., Environ. Sci. Technol., 29, 993–999, 1995.

50% inhibition of growth ($EC_{50}$) of ectomycorrhizal isolates in pure culture. They showed that there was significant variation between species for tolerance of different heavy metals and between metals within the same species (Table 7.8). The growth of pure cultures of ectomycorrhizal species on agar showed contrasting growth trajectories with increasing concentrations of Hg (Crane et al., 2010), which could suggest changes in the outcome of fungal–fungal competition, resulting in different mycorrhizal communities developing on roots at different mercury concentrations. Hartley et al. (1997a) also showed that there were significant interactions between metals on the growth of fungi in that in combination, one heavy metal may ameliorate the negative influence of another. Examples of this are that both Pb and Sb ameliorate the toxicity of Cd to *Suillis granualtus*, and the combination of Cd + Pb + Zn was less toxic to *Lactarius deliciosus* than the individual or paired metals. They suggest two mechanisms that might cause this interaction. First, two elements with the same valency or size may compete with each other for ion transporters across the plasma membrane; second, there could be an induced physiological response to the presence of Zn that decreases sensitivity to both Zn and Cd. In their review article, Hartley et al. (1997b) suggest that there could be some evolution of resistance to heavy metal toxicity, based on prior exposure (Egerton-Warburton and Griffin, 1995). However, no relationship between metal resistance and prior exposure to fungi was found by Denny and Wilkins (1987a) or Colpaert and Van Assche (1992, 1993). They suggest, however, that in environments where heavy metals are common, such as in acidic, peaty soils, the plant species may have coevolved with their mycorrhizal fungal endophyte to develop resistance, such as ericoid mycorrhizal associations of heathland species (*Calluna* and *Vaccinium*), which have been shown to have considerable tolerance to heavy metals (Marrs and Bannister, 1978; Bradley et al., 1981, 1982).

At the biochemical level, stress reactions in plant roots to the presence of heavy metals can include increased ethylene production (especially to Cu and Zn) along with elevated reactive oxygen species (ROS), and changes in nutrient levels and membrane potentials (Khade and Adholeya, 2007; Galamero et al., 2009). Bacteria and fungi, such as *Penicillium* spp., can produce ACC deaminase that degrades precursors to ethylene. The creation of bacterial heavy metal siderophores has been implicated in reducing plant uptake of heavy metals, and Galamero et al. (2009) showed that in some arbuscular mycorrhizal fungi, gene coding for a metallothionein-like

**Table 7.8    $EC_{50}$ Levels of Heavy Metals $Cd^{2+}$, $Pb^{2+}$, $Zn^{2+}$, and $Sb^{2+}$ to Five Ectomycorrhizal Fungal Species**

|  | $EC_{50}$ (mmol $m^{-3}$) | | | |
|---|---|---|---|---|
|  | Cd | Pb | Zn | Sb |
| *Lactarius deliciosus* | 0.79 | 45 | 100 | 0.66 |
| *Paxillus involutus* | 2.3 | 25 | 188 | 50 |
| *Suillus variegatus* | 0.008 | 25 | 343 | 50 |
| *Suillus granulatus* | 12.6 | 2.5 | 336 | 5.0 |
| *Suillus luteus* | 0.42 | 25 | 284 | 50 |

*Source:* Hartley, J. et al., *Environ. Pollut.*, 106, 413–424, 1997a.

protein (GmarMT1) can be upregulated by Cd and Cu, thus reducing uptake of the metals. Metallothionein gene coding has also been seen in ectomycorrhizal fungi and is considered another mechanism that—in association with glutathione and manganese-dependent superoxide dismutase (SOD) activities—can confer metal tolerance (Lanfranco, 2007; Bellion et al., 2006). Other genes found in ectomycorrhizal fungi, reported in yeast to confer heavy metal tolerance, include transcription factors for oxidative stress tolerance and zinc transport regulation; metal influx transport regulators; intracellular binding by metallochapreones, metallothioneins, and glutathione synthesis and protection against oxidative stress induced by the metals including removal of ROS (Bellion et al., 2006). The arbuscular mycorrhiza *Rhizophagus irregularis* also stimulates the production of SOD and peroxidase in the shoots of common chicory to afford protection of the plant from heavy metal toxicity (Rozpądek et al., 2014).

In addition to biochemical defense mechanisms, physical separation of heavy metals for cytoplasm may also afford protection. Rizzo et al. (1992) showed that, despite the fact that rhizomorphs of *Armillaria* spp. have a melanized outer cortex, they are able to take up heavy metals from the environment. Some elements were 50–100 times more concentrated in fungi than in the surrounding soil, with Al, Zn, Fe, Cu, and Pb in rhizomorphs reaching up to 3440, 1930, 1890, 15, and 680 $\mu$g g$^{-1}$, respectively. X-ray dispersal electron microscopy (EDAX) showed that the ions were accumulated in the outer portion of the rhizomorph and not concentrated in the interior. It was also suggested that a coating of metal ions in the outer layers of the rhizomorphs may play a key role in the longevity and survival of rhizomorphs in soil, where the heavy metals act as an antagonist for other microorganisms or grazing fauna.

Heavy metals have been shown to be adsorbed onto microbial tissue. Tobin et al. (1984) demonstrated that dead mycelium of *Rhizopus arrhizus* was efficient at adsorbing a range of metal ions, but not alkali metals (Table 7.9). In living fungal tissue, Mg ion accumulation results from both metabolic uptake and binding to sites being positively related to the cation exchange (McKnight et al., 1990) in stipe tissue of 18 basidiomycete fungal species. Using electron microscopy coupled with EDAX, Denny and Wilkins (1987b) identified adsorption of heavy metals onto fungal hyphae in the extraradical hyphal network, the fungal sheath, and Hartig net, preventing translocation of the metal into the host cortex and, in particular, preventing movement into the vascular tissue, and identified high zinc binding on extracellular slime formed by the hyphae of the ectomycorrhizal fungus, *Pisolithus tinctorius* (Denny and Ridge, 1995). A variety of heavy metal binding sites have been reported in ectomycorrhizal fungal mycelia of *Hymenogaster* sp., *Scleroderma* sp., and *Pisolithus tinctorius*, which were tolerant of high concentrations of Al, Fe, Cu, and Zn (Tam, 1995; Table 7.10), with Cu and Zn linked to polyphosphate granules (Kottke et al., 1998; Turneau et al., 1993) and the cystine-rich proteins in the outer pigmented layer of the cell wall of *Pisolithus tinctorius* (Turneau et al., 1994).

Movement of heavy metals through fungal mycelia is likely to redistribute metals in the environment. In an experimental system, Vodnik et al. (1998) showed that lead uptake by *Pisolithus tinctorius* (1.1% of the total lead added to the experimental

Table 7.9 Adsorption of Metal Ions onto Dried Mycelium of *Rhizopus arrhizus*

| Metal Ion | Uptake ($\mu$M g$^{-1}$) |
|---|---|
| $Cr^{3+}$ | 590 |
| $La^{3+}$ | 350 |
| $Mn^{2+}$ | 220 |
| $Cu^{2+}$ | 250 |
| $Zn^{2+}$ | 300 |
| $Cd^{2+}$ | 270 |
| $Ba^{2+}$ | 410 |
| $Hg^{2+}$ | 290 |
| $Pb^{2+}$ | 500 |
| $UO_2^{2+}$ | 820 |
| $Na^+$ | 0 |
| $K^+$ | 0 |
| $Rb^+$ | 0 |
| $Ag^+$ | 500 |
| $Cs^+$ | 0 |

*Source:* Tobin, J.M. et al., *Appl. Environ. Microbiol.*, 47, 4, 821–824, 1984.

Table 7.10 50% Inhibition Concentrations (mg l$^{-1}$) of a Variety of Heavy Metals Exhibited by Five Ectomycorrhizal Fungal Species Grown in Mycelial Culture

| Metal | Ectomycorrhizal Fungal Species | | | | |
|---|---|---|---|---|---|
| | *Pisolithus tinctorius* | *Thelephora terrestris* | *Cenococcum geophilum* | *Hymeogaster* sp. | *Scleroderma* sp. |
| Al | 200 | 10 | 10 | 200 | 200 |
| Fe | 400 | 100 | 200 | 200 | 100 |
| Cu | 200 | 10 | 10 | 10 | 100 |
| Zn | 200 | 10 | 10 | 100 | 100 |
| Ni | 10 | 1 | 1 | 1 | 10 |
| Cd | 10 | 1 | 1 | 0.1 | 10 |
| Cr | 10 | 10 | 10 | 10 | 10 |
| Pb | 200 | 200 | 200 | 200 | 200 |
| Hg | 1 | 1 | 1 | 1 | 1 |

*Source:* Tam, P.C.F., *Mycorrhiza*, 5, 181–188, 1995.

system) was much less than the uptake by *Laccaria laccata* and *Suillus bovinus* (6.2% and 5.4%, respectively). About 45% of the lead taken up by *Laccaria laccata* was translocated through the mycelium and released in a different part of the mycelium, whereas with increased binding only 10% was released by *Suillius bovinus*. Although the amount of lead moved in the mycelium was low (10–45% of that applied after 30 days of incubation), the experiment showed that there could be considerable temporal and spatial redistribution of heavy metals within the fungus and redistribution of the metal in the environment by release from fungal tissue (Figure 7.7). If

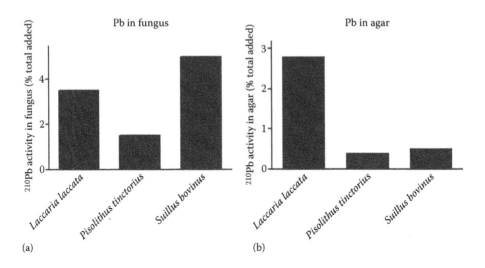

**Figure 7.7**   Lead accumulation in mycelial cultures of three ectomycorrhizal fungi (a) and the proportion of lead translocated to agar, distant from the site of lead addition, through the fungal mycelium (b). (Data from Vodnik, D. et al., *Mycol. Res.*, 102, 953–958, 1998.)

fungi are to be of importance in remediation, a deeper knowledge of the physiology of movement of heavy metals and radionuclides in relation to source–sink gradients and source-release gradients needs to be attained to adequately model the degree, duration, and strength of the retention of pollutants in the fungal thallus.

Leyval et al. (1997) suggest that there are two possible evolutionary routes that mycorrhizal fungi have taken to cope with heavy metals. One mechanism operates at low metal concentrations and is relatively metal-specific, wherein siderophores such as ferricrocin or fusigen are produced. The second approach operates at higher external concentrations of heavy metals wherein siderophore production is suppressed, but the host plant is still protected from heavy metals. Also, these fungi benefit the host plant by accumulating metals in cell walls or in vacuoles (González-Guerrero et al., 2008; Figure 7.8), or sequester metals in fungal siderophores, complex metals to metallothioneins and phytochelatins, or assist in metal extrusion by transportins associated with fungal membranes. They suggest that mixed-species communities of arbuscular mycorrhizae are better protectants than single species and that their interactions with rhizospheric bacteria perform even better.

## 7.2.3  Remediation

The role of mycorrhizae in heavy metal resistance has attracted interest because of its potential application for contaminated site restoration. Thus, research has been directed to understanding the interrelationships between metal availability, mycorrhizal infection, mycorrhizal function, and plant performance. Many of the initial findings suggesting the importance of ectomycorrhizae in protecting host plants

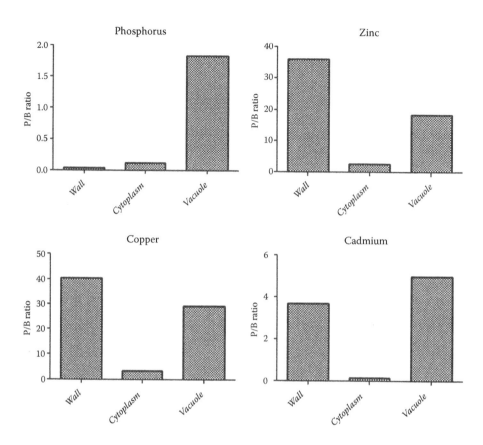

**Figure 7.8** Accumulation of phosphorus and heavy metals in hyphae of the arbuscular mycorrhizal fungus *Glomus intraradices* expressed as an EDAX peak to background ratio. (Data from González-Guerrero, M. et al., *Can. J. Microbiol.*, 54, 103–110, 2008.)

from heavy metal loadings come from the studies of Marx (1975, 1980). His observations of the survival of pine trees in mine spoil soils showed the benefit of inoculating tree seedlings with ectomycorrhizal fungi, which improved both tree survival and growth. Of particular interest was the fungal species *Pisolithus tinctorius*, which appeared to be more frequent in these polluted sites than in other habitats. The effect of inoculation with *P. tinctorius* resulted in tree volumes 250% greater than those trees assuming natural inoculum from the site or inoculation with *Thelephora terrestris* (Table 7.11). These trees also had higher foliar phosphate levels, but reduced levels of Ca, S, Fe, Mn, Zn, Cu, and Al, suggesting that the effect of this mycorrhizal fungus may reduce the uptake of heavy metals into the host tree (Denny and Wilkins 1987a,b). Bradley et al. (1982) showed that ericaceous plants of the genera *Calluna*, *Vaccinium*, and *Rhododendron* showed evidence of growth in the presence of copper and zinc at concentrations of between 0 and 150 mg l$^{-1}$ in sand culture only if

Table 7.11    Mean Survival and Growth of Transplants of Loblolly and Pitch Pine
              and Their Hybrids When Inoculated with the Ectomycorrhizal Fungus
              *Pisolithus tinctorius* (*Pt*) after Two Growing Seasons' Growth in Tennessee
              and Alabama Mine Spoil Soils

|  | Mycorrhizae at Planting | % Survival | Height (cm) | Stem Diameter (cm) | PVI ($\times 10^2$) |
|---|---|---|---|---|---|
| Tennessee | *Pt* | 85 | 79 | 3.0 | 133 |
| (overall mean) | Natural | 81 | 53 | 2.0 | 43 |
| Alabama | *Pt* | 66 | 67 | 3.0 | 96 |
| (overall mean) | Natural | 56 | 43 | 1.6 | 20 |

*Source:* Marx, D.H., *Ohio J. Sci.*, 75, 288–297, 1975.

their roots were colonized by ericoid mycorrhizae. The production of enzymes, such as polygalacturonase, which degrades pectin, has been shown to increase in the ericoid mycorrhiza *Oidiodendron maius* in the presence of increasing concentrations of Zn and Cd (Martino et al., 2000a), which coincides with their greater growth rate at higher levels of heavy metal (Martino et al., 2000b). Moreover, Hashem (1995) showed that mycorrhizal cranberry (*Vaccinium macrocarpon*) produced significantly larger plants than nonmycorrhizal plants in the presence of Mn at concentrations up to 1000 μg ml$^{-1}$. However, the effect of heavy metal availability in soil does not only affect the growth of trees—it also affects the growth, colonizing ability, and physiology of the fungal associate within this symbiosis. Endophytic (mycorrhizal) fungi isolated from these ericoid plant roots showed significantly reduced growth at 100 mg l$^{-1}$ Cu and 500 mg l$^{-1}$ zinc, indicating that although the mycorrhizal fungi appeared to be exerting a positive effect on plant growth in the presence of heavy metals, the fungi were also being inhibited.

Using a compartmentalized soil system, Joner and Leyval (1997) were able to determine the relative contribution of clover roots and their associated arbuscular mycorrhizae in the uptake of cadmium. They concluded that the presence of arbuscular mycorrhizae enhanced Cd uptake into the plant through the fungal hyphae, but that transfer from the fungus to root is reduced, retaining Cd in the fungal component of the root, thus detoxifying the system. Similar findings have been made by Tonín et al. (2001) for Cd and Zn in mycorrhizal clover and, during the second year of growth in *Salix viminalis* and *Populus × generosa*, there was no effect on plant biomass with or without inoculation with *Glomus interradices*, but mycorrhizal plants had higher concentrations of Cu in roots than uninoculated plants (Bisonnette et al., 2010).

Heavy metals (Zn, Cu, Cd, Ni) from a contaminated soil delayed the invasion of arbuscular mycorrhizal fungi into clover roots (Koomen and McGrath, 1990), although Joner and Leyval (1997) found no reduction of arbuscular mycorrhizal hyphal extension into soil with up to 20 mg extractable Cd kg$^{-1}$ soil. Notwithstanding the susceptibility of arbuscular mycorrhizal functioning in the presence of heavy metals, Call and Davis (1988) showed that the inoculation of three grass species by arbuscular mycorrhizae significantly increased grass survival, growth, and nutrient content in their attempts to restore the overburden of a surface lignite mine (Table 7.12).

Table 7.12  Survival and Growth of Three Grass Species Planted in a Restoration Project on Lignite Overburden as Influenced by the Presence of Inoculum with the Arbuscular Mycorrhizal Fungi *Glomus fasciculatum* (G. f) and *Gigaspora marginata* (G. m), or Uninoculated (NM) after 3 Years of Plant Establishment

| Grass Species | Inoculum | Survival (%) | Biomass (g plant$^{-1}$) | N Content (%) | P Content (%) |
|---|---|---|---|---|---|
| Sideoats grama | G. f | 97 | 25 | 0.82 | 0.14 |
| | G. m | 94 | 21 | 0.80 | 0.10 |
| | NM | 86 | 15 | 0.72 | 0.07 |
| Indiangrass | G. f | 92 | 36 | 1.12 | 0.12 |
| | G. m | 81 | 30 | 1.09 | 0.10 |
| | NM | 72 | 21 | 0.90 | 0.07 |
| Kleingrass | G. f | 61 | 19 | 0.75 | 0.14 |
| | G. m | 67 | 17 | 0.74 | 0.13 |
| | NM | 47 | 10 | 0.65 | 0.08 |

*Source:* Call, C.A., Davis, F.T., *Agric. Ecosyst. Environ.*, 24, 395–405, 1988.

All the factors that have just been outlined regarding the impact of heavy metals on fungi and fungi on heavy metals need to be taken into account when considering the potential role of fungi in bioremediation of contaminated soil (Skladany and Metting, 1992). Biosorption of metals by fungi relies on the ion exchange between the metal and the reactive groups of the cell wall. Some examples of degree of metal binding by a range of microfungi and yeasts suitable for industrial metal retrieval from effluent are given in Table 7.13. However, the interaction between live fungal biomass and the relative availability of carbon, nitrogen, and phosphorus may be of importance in influencing the solubility of heavy metals in the environment (Dixon-Hardy et al., 1998), showing that the role of fungi in metal binding is greatly influenced by environmental conditions.

Senior et al. (1993) and by Donnelley and Fletcher (1994) reviewed the potential role of mycorrhizal fungi in restoration and reclamation. The ability of the

Table 7.13  Ranges of Metal Uptake Capacities of a Variety of Microfungal Mycelial and Yeast Species that Could Be Used in Industrial Processes

| Metal | Uptake Capacity (mmol metal g$^{-1}$ biomass) |
|---|---|
| Copper | 0.25–1.9 |
| Silver | 0.004–0.46 |
| Zinc | 0.12–15.3 |
| Uranium | 0.12–1.3 |
| Cobalt | 0.15–0.52 |
| Thorium | 0.27–0.84 |

*Source:* Data adapted from Singleton, I., Tobin, J.M., in: *Fungi and Environmental Change*, ed. Frankland, J.C., 282–298, Cambridge University Press, Cambridge, 1996.

mycorrhizal fungal partner to protect its host plant from toxic levels of heavy metal is one of the advantages of this symbiotic association. It is also possible that the host plant may confer some enhanced survival traits for the fungus, enabling it to survive better in a metal-contaminated environment.

Arbuscular mycorrhizae can increase the heavy metal concentration of host plants as potential bioremediators of pollution. Total nickel concentration in mycorrhizal aster (*Berkheya coddii*) can be 20 times higher than nonmycorrhizal plants (Orlowska et al., 2011) with greater accumulation of Mn and Zn in the cortex and Fe, Ni, Cu, and Zn in the stele of mycorrhizal plants compared to nonmycorrhizal plants (Orlowska et al., 2013). Arsenic and lead are significantly less accumulated in the roots of mycorrhiza *Plantago lanceolata* than nonmycorrhizal plants, but the mycorrhizae significantly increase root concentrations of Zn and Cd (Orlowska et al., 2012).

In a proposal to utilize fungi for bioremediation of metal-contaminated sites, Saxena and Bhattacharyya (2006) evaluated fungal isolates from Ni-contaminated sludge and assessed their growth at increasing Ni concentrations. Growth of *Fusarium solani* and *Aspergillus flavus* was less affected by metal levels than growth of *Mucor racemosus* and *Papulaspora sepedonoides*; however, *M. racemosus* (Table 7.14) had a significant Ni accumulation potential between 0.01 and 1 mM Ni concentration.

In restoration projects, a balance must be achieved between the application of fertilizer to potentially promote plant growth and the beneficial effects of mycorrhizal fungi in detoxifying the soil. Johnson (1998) cautioned against high applications of inorganic fertilizers in taconite mine tailing restoration with *Salsola kali* and *Panicum virgatum*. She demonstrated that the addition of arbuscular mycorrhizal inoculum and organic soil amendment (papermill sludge) was more beneficial in the survival and growth of the late successional plant species (*Panicum virgatum*) and may be more cost-effective.

### 7.2.4 Lichens as Bioindicators

Lichens can accumulate heavy metals with little sign of determent to growth and survival (Richardson, 1988). However, this accumulation is not infinite, so lichens

Table 7.14    Decrease in Mycelial Weight of Fungal Isolates Grown in Increasing Ni
Concentrations, Compared to Control

| Ni Concentration (mg kg$^{-1}$) | Decrease in Mycelial Weight Compared to Control (%) | | | |
|---|---|---|---|---|
| | *Aspergillus flavus* | *Fusarium solani* | *Mucor racemosus* | *Papulaspora sepedonoides* |
| 10 | 10.5 | 13.3 | 22.2 | 4.8 |
| 100 | 26.3 | 26.7 | 31.9 | 21.0 |
| 500 | 40.4 | 35.0 | 39.3 | 35.5 |
| 1000 | 61.4 | 58.3 | 52.6 | 48.4 |
| 10,000 | 89.5 | 73.3 | 96.3 | 96.8 |

*Source:* Saxena, P., Bhattacharyya, A.K., *Geomicrob. J.*, 23, 333–340, 2006.

have been used to map pollutant plumes. This mapping has been especially useful for the determination of sulfur dioxide levels in the atmosphere because different lichen species have different tolerances. Because acidifying pollutant loading and increase in heavy metal availability are often co-occurring problems, neither Richardson (1988) nor Bates et al. (2001) considered that the changes in heavy metal availability could be contributing to the patterns of lichen distribution that associate with atmospheric $SO_2$ levels.

Fritze et al. (1989) measured soil microbial parameters along a 20,000-m transect away from a copper and nickel smelter in Finland. Within 500 m of the smelter, there was an absence of epiphytic lichens and their place was taken by an alga. *Hypogymnia physoides* and *Pseudevernia furfuracea* were more tolerant of heavy metal exposure than *Usnea hirta*, *Bryoria fuscescens*, and *Platismatia glauca*, demonstrating a differential tolerance within epiphytic lichens. Fungal hyphal length and soil respiration increased significantly, at more than 10 km from the smelter. The pattern of distribution of both lichens and mosses was found to correlate with levels of heavy metal loading in the environment (Kosta-Rick et al., 2001), where the authors found that the level of several heavy metals (Cd, Cu, Hg, Pb, Sb, Sn, and Zn) were significantly higher in lichens than in mosses.

The concentration of heavy metals in lichens can also be used as a bioindicator of heavy metal pollution. *Cetariella delisei* has shown to be a useful indicator in arctic and subarctic zones as it is widely distributed and less readily grazed by reindeer and accumulates heavy metals at similar rates to *Cladonia uncialis* and *Flavocetraria nivalis*, which have narrower ranges (Węgrzyn et al., 2013). Węgrzyn et al. (2013) suggest that lichens are easier indicators to use than mosses.

## 7.2.5 Impact of Heavy Metals in Aquatic and Marine Ecosystems

Maltby and Booth (1991) state that there is little information about the influence of pollutants on the functioning of aquatic fungi. Their study assessed the impact of coal mine effluent on the communities of fungi effecting leaf litter breakdown in stream systems. By comparing up- and downstream locations from an inflow drainage stream that emanated from a coal mine, they showed that the difference in reduction in species number downstream of the inflow was mainly attributable to a significant effect of pollutants on the hyphomycetes (Table 7.15). The rate of decomposition of leaves was significantly higher upstream than downstream, which was reflected in the more rapid decline in the C/N ratio of the leaf litter upstream.

Uptake of heavy metals into the aquatic fungi *Pythium*, *Dictyuchus*, and *Scytalidium* generally followed the order Zn > Pb > Cd. Both Cd and Zn had stimulatory effects on fungal growth at low concentrations (<1 and <10 mg l$^{-1}$, respectively), whereas lead had no effect on growth (Duddridge and Wainwright, 1980). Duddridge and Wainwright (1980) also demonstrated that there is considerable transfer of Cd from fungal hyphae to a grazing amphipod, *Gammarus pulex* (a shrimp), showing bioaccumulation and concentration through the food chain. No shrimps survived being fed on *Pythium* containing 150–170 µg g$^{-1}$ cadmium after 13 days, whereas 60% of the shrimp survived to 21 days when fed on control hyphae. These dead

Table 7.15   Presence and Absence of Hyphomycete Fungal Species
Upstream and Downstream of a Coalmine Effluent Source

| Aquatic Deteromycotina | Upstream | Downstream |
| --- | --- | --- |
| Anguillospora longissima | + | + |
| Articulospora tetracladia | + | − |
| Centrospora aquatica | + | − |
| Clavatospora longibrachiata | + | − |
| Flagellospora curvula | − | − |
| Helicodendron | + | + |
| Heliscus lugdenensis | + | + |
| Tetracaetum elegans | + | + |
| Tricladium | + | − |
| Total number of species | 8 | 4 |

Source: Maltby, L., Booth, R., Water Res., 25, 247–250, 1991.

animals contained 22 µg g$^{-1}$ Cd, demonstrating the transfer of lethal levels of heavy metal up the food chain. Bärlocher (1992a,b) showed that heavy metals can be detrimental to the growth of aquatic hyphomycetes and reduce sporulation. However, the effects of high levels of metals, such as Cd and Zn, were reduced by the presence of calcium ions in water. This Bärlocher attributes to ion complexation and suggests as a mechanism for greater heavy metal tolerance of fungi in hard water.

Aquatic fungi play a significant role in determining the decomposition of autochthonous and allochthonous materials in streams, rivers, and lakes. The activity of these organisms helps to regulate the net export of nutrients from terrestrial to marine ecosystems. The impact of pollutant metals on the fungal organisms and the processes they regulate requires deeper understanding as more of our aquatic ecosystems are being affected by industrial and agricultural effluents.

In contrast to some of the examples of negative effects of heavy metals on fungi in terrestrial ecosystems, Newell and Wall (1998) reported that they did not see any decline in fungal activity, but rather an increase in activity in saltmarsh communities contaminated by high levels of mercury (71 µg g$^{-1}$ dry weight of sediment Hg) and methylmercury (190 ng g$^{-1}$). Both fungal biomass, measured as ergosterol content of decaying leaf mass, and ascospore production were higher than those found in a less-contaminated location.

## 7.3 ORGANIC POLLUTANTS

A recent review of the role of fungi in the degradation and potential remediation of polluted ecosystems is given by Harms (2011), who identifies the phyla of fungi that have been shown to be of importance in the degradation of specific compounds as well as their appropriate functional taxonomy (Figure 7.9). Harms (2011) also highlighted the mechanisms whereby fungi metabolize organic pollutants using exoenzymes and intracellular biochemical reactions (Figure 7.10). He suggests that

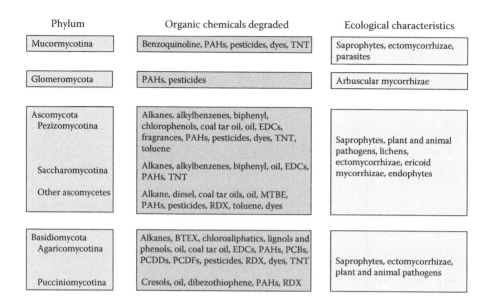

| Phylum | Organic chemicals degraded | Ecological characteristics |
|---|---|---|
| Mucormycotina | Benzoquinoline, PAHs, pesticides, dyes, TNT | Saprophytes, ectomycorrhizae, parasites |
| Glomeromycota | PAHs, pesticides | Arbuscular mycorrhizae |
| Ascomycota Pezizomycotina | Alkanes, alkylbenzenes, biphenyl, chlorophenols, coal tar oil, oil, EDCs, fragrances, PAHs, pesticides, dyes, TNT, toluene | Saprophytes, plant and animal pathogens, lichens, ectomycorrhizae, ericoid mycorrhizae, endophytes |
| Saccharomycotina | Alkanes, alkylbenzenes, biphenyl, oil, EDCs, PAHs, TNT | |
| Other ascomycetes | Alkane, diesel, coal tar oils, oil, MTBE, PAHs, pesticides, RDX, toluene, dyes | |
| Basidiomycota Agaricomycotina | Alkanes, BTEX, chloroaliphatics, lignols and phenols, oil, coal tar oil, EDCs, PAHs, PCBs, PCDDs, PCDFs, pesticides, RDX, dyes, TNT | Saprophytes, ectomycorrhizae, plant and animal pathogens |
| Pucciniomycotina | Cresols, oil, dibezothiophene, PAHs, RDX | |

**Figure 7.9** Fungal taxonomic groups and the organic pollutants degraded by them. (Data from Harms, H., *Nat. Rev. Microbiol.*, 9, 177–192, 2011.)

fungal hyphae may stimulate the activity of bacteria, through the production of exudates as an energy source, to synergize the process of pollutant degradation.

Their capabilities are limited by the appropriate suite of enzymes that any individual species can produce to break specific bonds. For example, both ericoid and ectomycorrhizal fungi produce enzymes allowing them to decompose lignin (a polymer of phenolic compounds). Because some fungi, such as the mat-forming *Hysterangium setchellii*, may account for up to 45–55% of the total soil organic biomass (Cromack et al., 1979; Fogel and Hunt, 1983), they have the potential to exert major influence on pollutant hydrocarbons in the environment. For example, 2,4-D and atrazine can be incorporated into the biomass of the ectomycorrhizal fungus *Rhizopogon vinicolor* and the ericoid mycorrhiza *Hymenoscyphus ericae*. Donnelley and Fletcher (1994) screened 21 mycorrhizal fungi for polychlorinated biphenyl (PCB) decomposition and demonstrated that 14 species could metabolize some of the PCBs by at least 20%. The ectomycorrhizal fungi *Radiigera atrogleba* and *Hysterangium gardneri* were able to degrade 80% of 2,2′-dichlorobiphenyl, but two ericoid mycorrhizae, *Hymenoscyphus ericae* and *Oidiodendron griseum*, were less effective than the ectomycorrhizal species. In contrast, the ectomycorrhizal fungus *Suillus bovinus* inoculated onto roots of scots pine seedlings was shown to impede degradation of poly-aromatic hydrocarbons in soil (Joner et al., 2006). However, ectomycorrhizal fungi alone are unlikely to complete the degradation of organic wastes, and the interaction between these fungi and saprotrophic fungi and bacteria to enhance decomposition warrants further study (Cairney and Meharg, 2002). For example, Wang et al. (2012) demonstrated that using a mixture of four bacterial

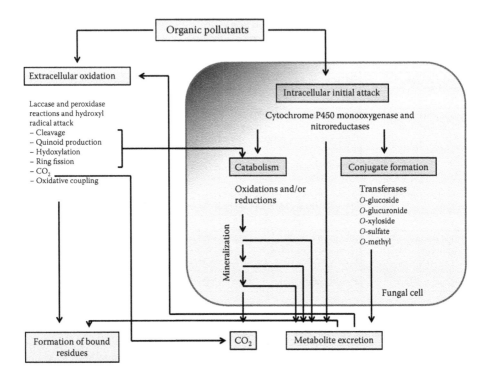

**Figure 7.10** Methods of fungal degradation of organic pollutants. (Data from Harms, H., *Nat. Rev. Microbiol.*, 9, 177–192, 2011.)

strains and seven fungal strains (ascomycetes and hyphomycetes) isolated from PAH-amended soil, the decomposition of phenanthrene, fluoranthene, and pyrene was faster than using either bacteria or fungi alone (Figure 7.11). This suggests synergistic activity within the microbial community as previously shown by Chávez-Gómez et al. (2003), wherein strains of *Penicillium*, *Trichoderma*, *Alternaria*, and *Aspergillus* were used in combination with four bacteria to enhance decomposition and removal of phenanthrene from soil to levels of between 70% and 74% compared to 35–50% by fungi alone and 20% by bacteria alone. It also appears that different species of fungi have contrasting abilities to decompose different PAHs and tolerance to PAH concentration. This results in different fungal communities developing in soils containing both type and concentration of PAH (Potin et al., 2004).

From both the aspects of metal ion accumulation and enzymatic competence of fungi, it is likely that fungi could be used to detoxify contaminated areas. Many wood-rotting fungi have been shown to be capable of degrading phthalate esters (PAEs), which are used as plasticizers of cellulose ester plastics, and are known endocrine disrupters (Luo et al., 2012). In an investigation of the fungi capable of degrading PAEs in mangrove swamps, Luo et al. (2012) found three isolates that degraded dimethyl phthalate (DMP) and 1,3-dimethyl-2-imidazzolidione (DMI) (*Aureobasidium pululans*, *Trichosporon* sp., and *Penicillium* sp.) and 10 DMT degrading isolates from two genera (eight species

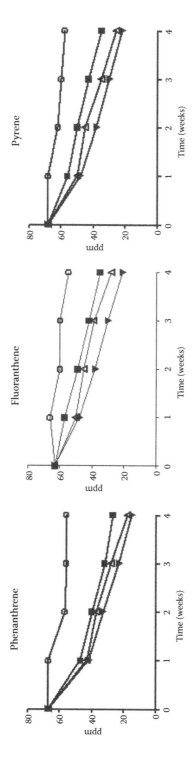

**Figure 7.11** Degradation of three PAHs in soil without amendment (open circles), with a mixture of fungi (solid squares), a mixture of bacteria (open triangles), and mixtures of fungi and bacteria (inverted solid triangles). (Data from Wang, S. et al., *J. Biosci. Bioeng.*, 113, 624–630, 2012.)

of *Fusarium* and two of *Aureobasidium*). Phenols can also be decomposed by wood-rotting fungi, such as *Trametes* (Ryan et al., 2005). When living, *Pleurotus* spp., have the capability of degrading organic pollutants such as lindane (Arisoy and Kolankaya, 1997) and treating sewage sludge because of its high production of manganese per-oxidase and laccase (Dellamatrice et al., 2005; Rigas et al., 2005). The production of these enzymes is important for all fungal species surviving in PAH soil and effecting its decomposition, including nonbasidiomycete fungal genera (D'Annibale et al., 2006).

Oleagenous fungi are capable of converting a number of complex carbon sources into lipids. This was demonstrated for the fungus *Mortierella*, which showed prolific growth on sunflower oil added to compost (Weber and Tribe, 2003). Such fungi are able to reduce pollutant phenols in oil residues to reduce toxicity of these wastes in effluents. Examples of this are the degradation of olive mill water from olive oil extraction where between 44% and 80% of the phenol content was reduced by incu-bation of waste with a variety of yeast species (Ben Sassi et al., 2008).

Numerous fungi have been found that will utilize toluene as a carbon source, and hence be important for cleanup. However, the rate of decomposition of toluene is related to concentration, wherein fungal activity will be inhibited by 50% if the concentration exceeds 2.5 mM (Prenafeta-Boldu et al., 2001).

Cleanup of aquatic systems can be achieved by the physical adsorbance prop-erties of fungi. Denizli et al. (2005) adsorbed phenol and chlorophenol from aquatic systems using dried, dead *Pleurotus sajor caju* hyphae with continuous adsorption (in mmol $g^{-1}$) rates of about 0.9 for phenol, 1.2 for *o*-chlorophenol, 1.4 for *p*-chlorophenol, and 1.86 for 2,4,6-trichlorophenol over five cycles with 90+% desorption between cycles. The use of fungi for bioadsorption can be used in con-junction with chemical methods for degradation of wastewater from industrial plants. For example, nonylphenols used as a surfactant in paper pulp production can be reduced by 90% by ultrasound-induced Fenton reaction followed by sorption into hyphae of *Paecilomyces lilacanus* (Cravotto et al., 2008).

## 7.4 FUNGI AND RADIONUCLIDES

Following the development of nuclear weapons during the Second World War and the subsequent evolution of nuclear energy-generating industries, there has been considerable concern regarding the safe storage of radionuclide waste together with the hazards of radiation pollution in the event of nuclear detonations and release from atomic energy plants and reprocessing facilities. As a result of the explosion of the Chernobyl Atomic Electric Station in the Ukraine in 1986 and, subsequently, the accident at Fukushima Daiichi, Japan, in 2011, attention has been focused on the accumulation of radioactive fallout in a variety of biotic components and the role of organisms to influence radionuclide retention in the ecosystem. In the International Commission on Radiological Protection (Coughtree, 1983), the importance of organic soil horizons and their microbial communities as potential accumulators of nutrient elements and radionuclides in terrestrial systems was highlighted by Heal and Horrill (1983). This is particularly true for forest ecosystems where the importance of fungi in radionuclide cycling in forest ecosystems has been reviewed by Steiner et al. (2002).

Fungi play an important role in the cycling of radionuclides in the ecosystem as many radioisotopes follow pathways of stable elements ($^{137}$Cs as K, $^{90}$Sr, and Ca, etc.). Saprotrophic fungi break down plant material that has accumulated radionuclides to release them in more bioavailable forms, and as mycorrhizae they influence rates of accumulation into plants. As some 90–95% of all plant species associate with mycorrhizal fungi for the uptake of nutrients, it is not surprising that mycorrhizae may be of importance in plant uptake of radionuclides. Because mushroom fruit bodies and lichens form a substantial part of the diet of animals and humans (Horyna, 1991; Table 7.16), there is an interest in the potential transfer of radionuclides through the food chain from fungi.

The renewed interest in fungi as radionuclide accumulators was based on the earlier work of Witkamp (1968) and Witkamp and Barzansky (1968), who demonstrated that fungi were capable of storing radionuclides in mushrooms. Haselwandter (1978), Eckl et al. (1986), Haselwandter et al. (1988), and Byrne (1988) also showed that lichens and mushroom forming fungi took up and accumulated radionuclides in their fruiting structures. Byrne (1988) paid special attention to members of the Cortinariaceae, which are known to be Cs accumulators. The European Community set a limit of radioactivity in foodstuffs at 600 Bq kg$^{-1}$. Byrne found that the levels of $^{134,137}$Cs radioactivity in fungi in Slovenia ranged from 0.5 kBq kg$^{-1}$ dry weight (*Cortinarius praestans*) to 43 and 44 kBq kg$^{-1}$ (*Laccaria amethystina* and *Cortinarius armillatus*, respectively), up to 80 times the limit considered safe to consume. In addition to radiocesium, fungi have been shown to take up $^7$Be, $^{60}$Co, $^{90}$Sr, $^{95}$Zr, $^{95}$Nb, $^{100}$Ag, $^{125}$Sb, $^{144}$Ce, $^{226}$Ra, and $^{238}$U (Haselwandter and Berreck, 1994), which contributes more to the radionuclide content than the natural radioisotope of potassium ($^{40}$K), where potassium may be between 0.15% and 11.7% of the dry weight of fungal tissue. They cite values of $^{137}$Cs concentration from a variety of basidiomycete fungal species at between 266 and 25,160 Bq kg$^{-1}$ dry weight before and between 95 and 947,400 Bq kg$^{-1}$ after the Chernobyl explosion. Although a

**Table 7.16   Estimated Sources of Human Internal Radiocesium Contamination due to Consumption of Foodstuffs and Fungi**

| Source | Annual Intake (Bq) | | |
|---|---|---|---|
| | **1986** | **1987** | **1988** |
| Milk | 1000 | 550 | 70 |
| Meat | 1100 | 1100 | 120 |
| Cereals | 600 | 1900 | 100 |
| Vegetables | 310 | 200 | 70 |
| Potatoes | 160 | 340 | 60 |
| Fruits | 580 | 380 | 70 |
| Mushrooms | 1400 | 1500 | 1550 |

*Source:* Horyna, J., *Isotopenpraxis*, 27, 23–24, 1991.

variety of radionuclides were released from Chernobyl, most research has focused on radiocesium.

Data collected from a variety of sources show that there is considerable geographic variation in accumulation rates and considerable variation within and between fungal species. Radiocesium accumulation into 41 species of mushrooms from sites of contrasting surface contamination in Ukraine showed that fruit body concentrations from the lowest surface contaminated regions ($<3.7 \times 10^{10}$ Bq km$^{-2}$) varied from 0 to 33 kBq kg$^{-1}$ dry weight to values between 1.4 and 3.7 MBq kg$^{-1}$ dry weight for the heavily contaminated region ($148 \times 10^{10}$ Bq km$^{-1}$ around Chernobyl and Prypyat) (Grodzinskaya et al., 1994). The ectomycorrhizal species *Suillius luteus* showed a strong positive relationship between the accumulation of both $^{137}$Cs and $^{134}$Cs in fruit bodies and the level of soil surface contamination. Accumulation into fungal fruit bodies can result in tissue concentrations significantly higher than those found in available substrates. Horyna and Randa (1988) showed concentration factors of between 0.4 and 99 (ratio of radiocesium content of mushroom to substrate) for radiocesium accumulation into basidiomycete fungal fruit bodies. Highest concentration values were found in ectomycorrhizal genera (*Boletus, Paxillus, Tylopilus, Lactarius, Leccinum, Amanita, Cortinarius*, and *Suillus*) and lowest in saprotrophic genera (*Scleroderma, Lepista*, and *Agaricus*) (Horyna and Randa, 1988; Randa and Benada, 1990), a trend also seen by Yoshida and Muramatsu (1994). These fungi also accumulated more radioactivity than other organisms in the ecosystem and can be considered good indicator species of radionuclide contamination (Fraiture et al., 1990; Giovani et al., 1990; Table 7.17).

There is considerable variation in the accumulation of radiaonuclides both between and within fungal species. In Poland, Mietelski et al. (1994) reported a difference between 300 Bq kg$^{-1}$ dry weight of $^{137}$Cs in *Macrolepiota procera* and 20,000 Bq kg$^{-1}$ in *Xercomius badius*, and a range of 300–1800 Bq kg$^{-1}$ within *Boletus edulis* in the same location. The variation in activity of the α-emitting isotopes $^{90}$Sr and $^{239+240}$Pu ranged from 0.6 to 4 Bq kg$^{-1}$ in *Leccinum* sp. for Sr and from undetectable to 90 MBq kg$^{-1}$ for *Boletus edulis*. Wide ranges of accumulation of radiocesium (<3 to 1520 Bq kg$^{-1}$) were also found in mushrooms collected in Japan (Muramatsu et al., 1991), with the lowest levels of activity (< 50 Bq kg$^{-1}$) found in the edible species *Lentinus edodes, Flammulina velutipes, Pleurotus ostreatus*, and *Pholiota nameko*. They also suggested that cesium was taken up by *Suillius granulatus* and *Lactarius hatsudake* in preference to potassium, which is in contradiction to Olsen et al. (1990), who report that the affinity of fungi to cations decreases in the order K > Rb > Cs > Na > Li. Accumulation of radionuclides into fungi is dependent on the external concentration of a number of cationic elements. Some of these have been used to reduce radionuclide bioavailability. In a laboratory study of the saprotrophic fungi *Pleurotus eryngii*, Guillén et al. (2012) varied the availability of stable potassium, cesium, strontium, and calcium on the uptake of $^{134}$Cs, $^{85}$Sr, $^{60}$Co, and $^{210}$Pb. $^{134}$Cs content in the fungus increased with increasing concentration of stable Cs, whereas $^{85}$Sr content decreased with increased concentration of stable Cs, but the transfer of $^{60}$Co and $^{210}$Pb was unaffected by changes in cation availability.

**Table 7.17    Range of Radiocesium Content of Ecosystem Components from the Same Ecosystem in Belgium in 1986 and 1987**

|  | Ecosystem Component | Radiocesium Content (Bq kg⁻¹) |
|---|---|---|
| Soil | Leaf litter | 650–1060 |
|  | Organic horizon | 240–265 |
| Trees | Spruce | 105 |
|  | Beech | 180 |
|  | Birch | 260 |
| Fern | *Pteridium aquilinum* | 137–146 |
| Ericaceous herb | *Vaccinium myrtillus* | 640 |
| Mosses | *Rhytidiadelphus* sp. | 5900 |
|  | *Dicranium* sp. | 1070–2000 |
| Lichen | *Hypogymnia physodes* | 5300 |
| Fungi | *Xercomus badius* | 8000–18,800 |
|  | *Cortinarius brunneus* | 7700–27,000 |
|  | *Cortinarius armillatus* | 8200–24,300 |
|  | *Russula ochroleuca* | 8300 |
|  | *Boletus edulis* | 1350 |
|  | *Clitocybe vibecina* | 630 |
|  | *Hypholoma* spp. | 160–660 |
| Roe deer | *Capreolus* sp. | 350–2300 |
| Wild boar | *Sus* sp. | 20–600 |
| Soil animal | Earthworms | 600 |

*Source:* After Fraiture, A. et al., in: *Transfer of Radionuclides in Natural and Semi-natural Environments*, ed. Desmet, G., 477–484, Elsevier Applied Science, London, 1990.

Variation in radionuclide acquisition by fungi is also dependent on the plant communities in the ecosystem. Recent surveys of the radiocesium content of fungal fruit bodies in contrasting ecosystems in eastern central Sweden, affected by the Chernobyl fallout, demonstrates the importance of fungi, especially ectomycorrhizal fungi, in the retention of radionuclides in the ecosystem. Vinichuk et al. (2013c) calculated the proportion of $^{137}$Cs in vegetation and fungal fruiting bodies in a *Sphagnum* peat bog, Scots pine swamp with ericaceous understory, and Scots pine forest on mineral soil (Table 7.18). They report that the fungal component of the

**Table 7.18    $^{137}$Cs in Fungal Fruit Bodies from Three Neighboring Ecosystems in Relation to Fungal Mode of Nutrition**

| | Bog | | Pine Swamp | | Forest | |
|---|---|---|---|---|---|---|
| Mode of Nutrition | $^{137}$Cs (kBq kg⁻¹) | No. spp (individuals) | $^{137}$Cs (kBq kg⁻¹) | No. spp (individuals) | $^{137}$Cs (kBq kg⁻¹) | No. spp (individuals) |
| All types | 34.8 | 16 (78) | 41.1 | 21 (51) | 54 | 21 (351) |
| Ectomycorrhizal | | | 43 | 16 (46) | 54 | 21 (351) |
| Saprotrophic | 34.8 | 16 (78) | 19.8 | 5 (5) | | |

*Source:* After Vinichuk et al., *J. Environ. Radioact.*, 90, 713–720, 2013.

total Cs loading of the ecosystem represented 0.1% in the bog, 2% in the pine swamp, and 11% in the pine forest. The greater accumulation in the pine forest was partly attributable to the soil type but mainly the greater proportion of ectomycorrhizal fungi in the forest ecosystem. The presence of bacteria and fungi in soil significantly increases retention of $^{137}$Cs or $^{85}$Sr (70% retention) compared to 10% retention in sterile soil (Parekh et al., 2008). This accumulation in the fungal fraction of the ecosystem is echoed by Shcheglov et al. (2014), whose study of the biogeochemical cycling of Chernobyl-derived radionuclides showed $^{137}$Cs predominantly residing in fungal biomass (~80%) and that of $^{90}$Sr in tree vegetation (~95%).

The Chernobyl explosion released the radiocesium isotopes $^{137}$Cs and $^{134}$Cs in a ratio of 2:1. Because $^{134}$Cs has a half-life of 2.2 years and $^{137}$Cs of 28 years, this decay-corrected ratio could be used to fingerprint Chernobyl emissions. Using this fingerprint for fruit bodies of ectomycorrhizal basidiomycete fungi, Dighton and Horrill (1988) showed that a large proportion (25–92%) of $^{137}$Cs was accumulated that originated from sources occurring before the accident at Chernobyl (Table 7.19). Similar figures (13–69%) of pre-Chernobyl accumulation of radiocesium were calculated from the data presented by Byrne (1988), and Giovani et al. (1990) commented on the deviance from the 2:1 ratio and fungal accumulation of radiocesium from atmospheric nuclear tests. This information suggests that fungi could be long-term accumulators and retainers of radionuclides in the environment. Together with an ecological half-life of 8–13 years (Vinichuk et al., 2013c), it is anticipated that fungi will continue to be important in the retention of radiocesium in the ecosystem.

Transfer of radionuclides through the food chain is partially determined by mycophagy, but consideration of the role of mycorrhizal mediation of radionuclide accumulation into plants is also of importance for transfer of radionuclides to herbivores. Haselwandter and Berreck (1994) reviewed the role of arbuscular mycorrhizae in plant uptake of radionuclides and found somewhat conflicting information. They cite an example of arbuscular mycorrhizal inoculation of sweet clover and Sudan grass, which showed slight and statistically insignificant increases in uptake of $^{137}$Cs and $^{60}$Co by mycorrhizal plants. This is in line with findings of Jackson et al. (1973),

Table 7.19   Proportion of $^{137}$Cs of Pre-Chernobyl Origin Found in Fruit Bodies of Two Ectomycorrhizal Fungal Species in Upland United Kingdom, Based on Decay-Corrected Ratios of Isotopes $^{137}$Cs and $^{134}$Cs

| Fungal Species | Location | Content of Pre-Chernobyl $^{137}$Cs (%) |
|---|---|---|
| Lactarius rufus | MH86 peat under *Pinus contorta* | 92 |
| | MH87 peat under *Pinus contorta* | 81 |
| | St humic podsol under *Picea sitchensis* | 74 |
| | S2 peat under *Picea sitchensis* | 67 |
| | SB peat under *Picea sitchensis* | 73 |
| | B humic podsol under *Picea abies* | 25 |
| Inocybe longicystis | SB peat under *Picea sitchensis* | 75 |
| | S4 peat under *Picea sitchensis* | 83 |

*Source:* After Dighton, J., Horrill, A.D., *Trans. Br. Mycol. Soc.*, 91, 335–337, 1988.

who noted that arbuscular mycorrhizal colonization of roots of soybeans by *Glomus mosseae* significantly increased [90]Sr uptake from soil. In contrast, they cite their own research that demonstrates that arbuscular mycorrhizal symbioses in the grass *Festuca ovina* reduce the uptake of radiocesium into shoots.

Information suggesting a reduced plant uptake support the findings of Clint and Dighton (1992), who showed that influx of radiocesium into mycorrhizal heather plants (*Calluna vulgaris*) with ericoid mycorrhizae was lower compared with influx of radiocesium into nonmycorrhizal plants. However, the internal redistribution of Cs within mycorrhizal plants allowed a greater proportion of the Cs taken up to be translocated to shoots in mycorrhizal plants than in nonmycorrhizal plants, especially when incubated in a high potassium environment before radiocesium exposure (Figure 7.12). Similar enhanced translocation of radiocesium into shoots of arbuscular mycorrhizal *Festuca ovina* were shown by Dighton and Terry (1996), but they did not observe enhanced shoot translocation in clover (*Trifolium repens*) (Figure 7.13). In contrast, Berreck and Hasselwandter (2001) showed a decrease in the Cs translocated to the shoots of *Agrostis tenuis* in the presence of arbuscular mycorrhizae.

Elevated levels of radionuclide in roots compared to shoots in mycorrhizal plants suggest that the mycorrhizal fungi accumulate radiocesium in the fungal tissue in a similar manner to that shown for heavy metals and ectomycorrhizae (Denny and Wilkins, 1987a,b). This concept has been reviewed for arbuscular mycorrhizal symbioses by Berreck and Haselwandter (2001), who investigated the impact of potassium fertilization as a method to reduce uptake of cesium by the mycorrhizal grass (*Agrostis tenuis*). They showed that mycorrhizal development in plant roots reduced Cs uptake by the plant at moderate nutrient levels in the soil. They further suggest that the mechanism of protection is attributable to sequestration of Cs in the

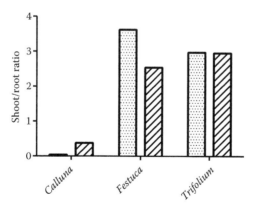

**Figure 7.12**    Histogram of shoot/root ratio of radiocesium incorporation into the ericoid mycorrhizal plant, heather (*Calluna vulgaris*), and the arbuscular mycorrhizal plants, sheep grass (*Festuca ovina*) and clover (*Trifolium repens*), in the presence (solid bars) and absence (open bars) of mycorrhizae. (Data from Dighton, J., Terry, G.M., in: *Fungi and Environmental Change*, ed. Frankland, J.C., 184–200, Cambridge University Press, Cambridge, 1996.)

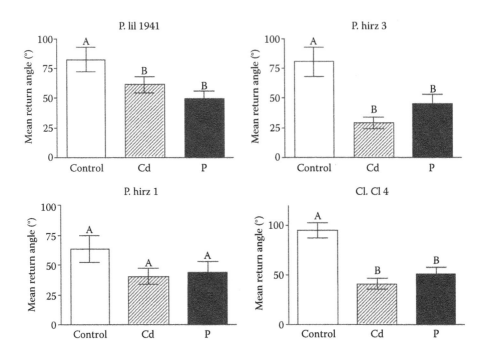

**Figure 7.13** Mean return angle (low angle means hyphal growth towards source of ionizing radiation) of four fungal isolates from Chernobyl in response to a collimated
beam of ionizing radiation from $^{109}$Cd or $^{32}$P. Fungal isolates: P. lil = *Paecilomyces
lilacinus*, P. hirz = *Penicillium hirsutum*, Cl. Cl = *Cladosporium cladosporioides*.
(Data from Dighton, J. et al., *FEMS Microb. Lett.*, 281, 109–120, 2008.)

extraradical hyphae of the mycorrhizal fungus and a reduced translocation into the
host plant. They also demonstrated that, for this fungal plant interaction, there was
no benefit of adding potassium to reduce Cs uptake. Using a method of quantitative autoradiography, Gray et al. (1995, 1996) showed that translocation of radiocesium through a fungal thallus of *Armillaria* spp. and *Schizophyllum commune* was
significantly reduced in comparison with a diffusion model of translocation. They
also demonstrated preferential movement of radiocesium to developing fruit bodies,
which were acting as nutrient sinks.

Accumulation of radionuclides by plants may also depend on their endophytic
fungal associations. DSEs inoculated onto Chinese cabbage or tomato had contrasting impacts on radiocesium accumulation depending on DSE isolate. In Chinese
cabbage, inoculation with all DSE isolates increased Cs concentration aboveground,
whereas in tomato three endophyte isolates significantly decreased the aboveground
Cs content of the plant (Diene et al., 2014).

Uptake of three radionuclides of contrasting chemistries and solubility on soil
by pine seedlings associated with the ectomycorrhizal fungus *Rhizopogon roseolus*
showed no enhanced foliar concentration of $^{137}$Cs, but an increased root concentration

and a decreased soil to the needle transfer of 95mTc (Ladeyn et al., 2008). This suggests a mycorrhiza protection for Cs but not for the more soluble and bioavailable Tc. Radiostrontium concentration was unaffected by the mycorrhizal status of the host plant.

Saprotrophic fungi are involved with the decomposition of organic resources (dead plant and animal remains) in ecosystems. Grassland soil saprotrophic fungi have been shown to have considerable potential for uptake and immobilization of radiocesium fallout (Olsen et al., 1990; Dighton et al., 1991). Assuming an average influx rate of 134 nmol Cs $g^{-1}$ dry weight of mycelium (determined from laboratory uptake studies) and an estimate of hyphal biomass of 6 g dry weight $cm^{-3}$ of soil (determined by hyphal length measurements of field-collected samples), Dighton et al. (1991) estimated that the fungal compliment of two upland grass ecosystems in northern England would have been able to take up between 350 and 804 nmol Cs $m^{-1}$ $h^{-1}$. Considering the estimates of radiocesium concentration in pore water of these soils to be at the micromolar level (Oughton, 1989), these fungi have the potential to accumulate a large percentage of the total fallout.

Much effort has been invested in the potential and actual use of plants to accumulate pollutants (mainly heavy metals) in the process of phytoremediation (Raskin and Ensley, 2000). The accumulation of radionuclides by fungal mycelia could be a useful means to attempt to effect radionuclide cleanup from both industrial processes and from contamination of natural environments. The efficacy of this is often mixed in that [137]Cs uptake by plants inoculated with arbuscular mycorrhizae show different responses between plant species and within the same plant between soil types (Vinchuk et al., 2013a). In a comparison of cucumber, ryegrass, sunflower, and nonmycorrhizal quinoa in silty clay, loamy, and loamy sand soils, the only effect of mycorrhiza was an increase in Cs uptake into inoculated sunflower plants in loamy sand soil. In an agricultural context, mycorrhizal inoculum of barley, cucumber, sunflower, and ryegrass showed no increase in radiocesium accumulation compared with plants without mycorrhizae (Vinichuk et al., 2013b).

The usefulness of fungal mycelia in environmental cleanup has been suggested by White and Gadd (1990), who developed airlift bioreactor systems containing live cultures of fungi for biosorption of radiothorium. *Rhizopus arrhizus* and *Aspergillus niger* were found to be more efficient absorbers than *Penicilium italicum* and *P. acrysogenum*. These mycelial fungi were found to be of great use as they can be pelletized to make them physically similar to commercial ion exchange materials. The ability of ectomycorrhizal basidiomycete fungal species to accumulate radionuclides could, in theory, be an important component in the restoration of radionuclide-contaminated terrestrial ecosystems. The formation of large and harvestable fruiting structures (mushrooms) provides a potential means of removal of radionuclides that have been accumulated within. Mycorrhizal fungi have been shown to be a major component of the radionuclide accumulation (radiocesium concentration) in a boreal coniferous ecosystem in Sweden (Guillette et al., 1990b) although its relative contribution to the total standing crop biomass is probably not large (Vogt et al., 1982; Fogel and Hunt, 1983). The distribution of fungal mycelia in upper soil horizons has been shown to correlate with the accumulation of radiocesium in the upper soil and reduced downward movement of the radionuclide by immobilization in fungal tissue (Guillette et al., 1990a; Rommelt et al., 1990). This attribute, together with evidence of translocation of radionuclides from soil to basidiomycete fruit bodies (Gray, 1995, 1996) points to the potential utility of fungal mycelia in soil to prevent leaching loss of radionuclides and their accumulation in mushrooms, which could be harvested and removed from the site.

The method of translocation of $^{14}$C and $^{32}$P through hyphal systems of *Rhizopus*, *Trichoderma*, and *Stemphylium* was by diffusion (Olsson and Jennings, 1991). The rate of translocation of carbon within the fungal thallus has been shown to react in real time to provide directional flow to the building phases of the hyphae (Olsson, 1995). Jennings (1990) showed that the absorption of phosphorus by rhizomorphs of *Phallus impudicus* and *Mutinus canninus* consisted of two transport systems. In contrast to the diffusion of C and P, translocation of $^{137}$Cs through hyphae of *Schizophyllum commune*, however, has been shown to be slower than diffusion, suggesting a possible mechanism for accumulation (Gray et al., 1995). In addition, there is a suggestion from this work that there is potential preferential transport to sites of basidiocarp primordium production. This observation may support the finding of that radiocesium accumulation in basidiomycete fungi could be high and long-lived (Dighton and Horrill, 1988; Yoshida and Muramatsu, 1994; Vinichuk et al., 2013c). In some cases, up to 92% of the radiocesium in mycorrhizal basidiocarps in the UK was derived from pre-Chernobyl sources of fallout, and with long ecological half-lives ectomycorrhizal fungi may significantly increase retention of radionuclides in the ecosystem.

Large numbers of fungal species can still be isolated from the walls of the reactor room (Zhdanova et al., 2000) under conditions of 1.5 to 800 mR h$^{-1}$ in the presence of alpha and beta emitting $^{239}$Pu, $^{240+241}$Pu, $^{241}$Am, and $^{244}$Cm, and mainly the gamma emitting $^{137}$Cs (Zhdanova et al., 2000). However, radionuclides exert negative effects on fungi as have been indicated by changes in the soil fungal communities because of the intense radiation doses (Zhdanova et al., 1995). Radiation results in a less diverse community dominated by melanin-containing (pigmented) fungal species at higher levels of radioactivity. Additionally, the effects of intense and sustained radiation have led to a shift in the genetic composition within some fungal species (Mironenko et al., 2000).

Initially, it was suspected that melanin could be involved in the protection of fungal tissue against damage by radionuclide emissions. Similar melanin protection is afforded to lichens, whereby the melanin-containing lichen fungi protect photosymbionts against ultraviolet light (Gauslaa and Solhaug, 2001). However, further studies revealed that fungi that have had prior exposure to ionizing radiation increase spore germination, increase hyphal length, and tend to grow toward the source of ionizing radiation (Figure 7.13; Tugay et al., 2006; Dighton et al., 2008). In other studies on melanized human skin, yeast pathogens (*Wangiella dermatidis* and *Cryptococcus neoformans*) grew up to 500 times faster and incorporated 4 times as much $^{14}$C-labeled acetate into their cells than unpigmented strains in the presence of ionizing radiation (Dadachova et al., 2007; Dadachova and Casadevall, 2008). This has led to the conclusion that yeasts are capable of using ionizing radiation for growth by incorporation of this energy into adenosine triphosphate (ATP) production (Bryan et al., 2011). By extension this is possibly also true for filamentous soil fungi.

The fungal component of the lichen symbiosis is also responsible for the elevated radionuclide content of lichens in Austria after the Chernobyl incident (Heinrich et al., 1999). Heinrich et al. (1999) report that levels of $^{137}$Cs rose from about 0.4 to more than 50 kBq kg$^{-1}$ after the explosion at Chernobyl. The $^{137}$Cs concentration of *Pseudovernia furfuracea* exceeded the natural $^{40}$K activity by 430 times. Within this foliose arboreal lichen, the biological half-life of Cs is approximately 1.3 years and that for $^{90}$Sr is 1.2–1.6 years. In contrast, the terricolous lichen (*Cetraria islandica*) had a biological half-life for $^{137}$Cs of 2.5 years. Similar biological half-lives were obtained for radiocesium in the lichens *Evernia prunastris* and *Hypogymnia physoides* by Guillette et al. (1990b). This indicates that lichens are strong accumulators of radionuclides and that the long biological half-life allows for transfer to animal grazers over an extended period. Indeed, Gaare (1990) indicates that reindeer densities of about 4 per hectare are common in parts of Norway. Lichens form the greater part of their winter feed and account for about 40% of their annual food intake. In an analysis of radionuclide content of lichens between 1986 and 1988, Gaare (1990) showed that mean activity ranged between 6700 and 24,000 Bq kg$^{-1}$ (Table 7.20). High intake of radiocesium by roe deer during the autumn in Sweden was associated with a greater proportion of fungal fruit bodies in their diet. Johanson et al. (1990) reported fungal-derived intake of 3000 Bq day$^{-1}$ in August and September and 2500 Bq day$^{-1}$ in October. Similarly, Bohac et al. (1989) demonstrated accumulation of radiocesium and radiostrontium into ectomycorrhizal fungal fruit bodies. They suggested that fungi were important in transferring radionuclides into higher trophic levels, because of the observed accumulation of radionuclides in the organic horizon of steppe soils.

In summary, fungi appear to be very resistant to radionuclides in the environment. It is possible that the presence of melanin pigment in the hyphae of mitosporic fungal species may provide some protection against ionizing radiation in the same

**Table 7.20   Radiocesium Content of Lichens in Norway Collected in 1986**

| Location | Lichen Species | $^{137}$Cs Content (Bq kg$^{-1}$) |
|---|---|---|
| Species on soil | *Alectoria ochroleuca* | 24,005 |
| | *Coelocaulon divergens* | 16,683 |
| | *Cetraria nivalis* | 24,286 |
| | *Cladina mitis* | 12,389 |
| | *Cladina stellaris* | 16,572 |
| | *Sterocaulon pashale* | 17,274 |
| Species on rocks | *Pseudephebe pubescens* | 54,080 |
| | *Ramalina polymorpha* | 9315 |
| | *Umbilicia hyperborea* | 31,767 |
| | *Umbilicia deusta* | 34,756 |

*Source:* Gaare, E., in: *Transfer of Radionuclides in Natural and Semi-natural Environments*, ed. Desmet, G., 492–501. Elsevier Applied Science, London, 1990.

Table 7.21    NADH–Ferricyanide–Melanin Reaction in Untreated and Irradiated
              (14 Gy min⁻¹) with ¹³⁷Cs *Cryptococcus neoformans* Melanin, Showing
              Increasing Transformations with Increased Radiation

| Melanin Type | Ferricyanide + Melanin Ferricyanide Reduced (nmol) | Ferricyanide + NADH + Melanin NADH Oxidized (nmol) | Ferricyanide Oxidized (nmol) |
|---|---|---|---|
| Untreated | 40 | 37 | 75 |
| Irradiated 20 min | 60 | 100 | 200 |
| Irradiated 40 min | 170 | 150 | 300 |

*Source:* After Dadachova, E. et al., *PLoS ONE*, 2, 5, e457, 2007.

way that it has been shown to protect against ultraviolet light. Because of the long-lived and extensive hyphal network, fungi appear to be very efficient in absorbing radionuclides from the environment. This is particularly true of radiocesium, which is reported to behave in a similar manner to potassium in biochemical pathways. Internal translocation of radionuclides between sources and physiological sinks occurs in the same way as essential nutrients and can account for the long-term retention of radionuclides within the fungal biomass. Adsorption of radionuclides onto ion exchange sites of fungal hyphal walls has been reported in the literature, and this attribute has been used in industrial effluent cleanup. Indeed, Entry et al. (1993) compared the uptake of ¹³⁷Cs and ⁹⁰Sr by Ponderosa and Monterey pine seedlings as both are fast-growing and potential candidate tree species for planting for site remediation. They showed that Ponderosa and Monterey respectively accumulated 6.3% and 8.3% of the radiocesium present in the growth medium and 1.5% and 4.5% of the radiostrontium. The possibility of enhancing this uptake by the addition of ectomycorrhizal symbionts showed promise (Entry et al., 1994), with 3–5 times more ⁹⁰Sr being taken up in the ectomycorrhizal seedlings. In a realistic situation, the combination of mycorrhizal enhanced uptake of radionuclides into trees together with the production of harvestable fruit bodies could prove an effective soil remediation technique (Gray, 1998). Many questions regarding the interaction of fungi and radionuclides warrant further investigation. The intriguing concept of behavioral adaptations of fungi to evolve radiotropism and the enhancement of fungal growth by melanin-mediated sequestration of ionizing radiation (Tugay et al., 2006; Dadachova et al., 2007; Dighton et al., 2008; Bryan et al., 2011) have important ramifications for fungal physiology, suggesting a photosynthesis-like attribute of melanized fungal species that may be able to grow as autotrophs (Table 7.21).

## 7.5 FUNGI AND CLIMATE CHANGE

Current debate on climate change concentrates on the need for reduction in gaseous emissions of $CO_2$ and other potential greenhouse gases. The recent International Panel on Climate Change report (IPCC, 2014) predicts an increase in land surface temperatures of between 0.3°C and 0.7°C between 2016 and 2035 with more frequent extremely hot and less frequent extremely cold temperatures. They predict changes in precipitation patterns with increased rainfall in high latitudes and in the equatorial Pacific region, a decrease in rainfall in midlatitude and subtropical dry areas and an increase in rainfall in midlatitude wet environments. Only the most stringent mitigation of greenhouse gas emissions will reduce the continued rise in atmospheric $CO_2$ concentrations over the next few decades. These predicted changes in our climate will lead to warming and increased severity of drought conditions in some areas, but an increase in precipitation and flooding in others.

In an attempt to increase the terrestrial carbon sink, international protocols have suggested that increasing plantations of forests or other crops would draw down carbon from the atmosphere. Pacala et al. (2001) suggest that land-based carbon sinks for the United States is between 0.3 and $5.8 \times 10^{15}$ g C, and plantation forests in China alone have sequestered $0.45 \times 10^{15}$ g C (Fang et al., 2001). What role do fungi have in assisting the carbon sequestration, and what effects does predicted climate change have on fungi and their ecosystem services?

## 7.5.1 Decomposition and Nutrient Availability

The inconsistency in the pattern of belowground response to elevated $CO_2$ was highlighted by Zak et al. (2000), who summarized the results of 47 publications spanning graminoid, herbaceous, and woody plant ecosystems on soil C and N cycling under elevated carbon dioxide. Zak et al. (2000) concluded that (1) there was greater plant growth under elevated $CO_2$, with more carbon entering the belowground component; and (2) there was greater metabolic activity of soil microbial communities under elevated $CO_2$. The effect of enriched atmospheric $CO_2$ on soil organisms is inconsistent (Kandeler et al., 1998; Bardgett et al., 1999). Increased carbon dioxide and elevated temperature increased microbial carbon but decreased the metabolic quotient when temperature alone was increased. However, the increase in $CO_2$ led to an increase in root biomass and an increase in C/N ratio, possibly because of a change in the balance between allocation of carbon to root growth and carbon storage.

In a review of the effects of enhanced atmospheric $CO_2$ concentrations on plant leaf chemistry, Gifford et al. (2000) showed that data were mixed and that there was little consensus about the changes in leaf chemistry. In general, they suggest that there is a trend toward an increase in the C/N ratio to a level of about 15% increase under a doubling of $CO_2$. The responses of leaves in terms of C/P ratios are also very variable, and there is no consensus for a stable trend in the data. De Angelis et al. (2000) showed that elevated $CO_2$ increased the C/N and lignin/N ratios of oak leaf litter in a Mediterranean forest ecosystem such that leaf litter decomposition was retarded. Decomposition constants for mixed leaf litter (*Quercus ilex*, *Phillyrea angustifolia*, and *Pisatcia lentiscus*) exhibited a drop of between 5% and 8% in *k* value for field experiments and a 12.5% drop in *k* value for microcosm decomposition experiments in elevated $CO_2$. Cotrufo et al. (1998) showed that the reduction in nitrogen content of leaves, grown in enhanced $CO_2$, varied from 50% in sweet chestnut (*Castanea sativa*) to 19% for sycamore (*Acer platanoides*) and that this change in nitrogen content stimulated grazing of leaves by the isopods *Oniscus asellus* and

*Pocellio scaber* (Hättenschwiler et al., 1999). The conclusion is that reduction in leaf litter resource quality in elevated $CO_2$ environments reduces the activity of the saprotrophic community, including fungi.

However, the significant increase in C/N and lignin/ratios of beech twigs grown in elevated $CO_2$ (C/N ratio change from 45.6 to 72.7 and lignin/N ratio change from 16.3 to 22.4) did not change the rate of decomposition of the twigs, nor the dynamics of nitrogen and lignin during decomposition (Cotrufo and Ineson, 2000). However, Gorisen and Cotrufo (2000) cautioned that the changes in leaf chemistry may not be correlated to the rate of decomposition processes. Although they identified an increase in the C/N ratio of grassland leaf material (*Lolium perenne*, *Agrostis capillaris*, and *Festuca ovina*) with an increase in $CO_2$ from ambient (350 µl $l^{-1}$) to double, the respired $CO_2$ during decomposition could not be attributed to the labeled C that was accumulated under the elevated $CO_2$ treatment. Their data on the internal allocation of sequestered carbon in the elevated $CO_2$ condition indicated that this additionally available carbon was translocated to the plant roots. Leaves made a contribution of about 6% and roots about 26% to the carbon remaining during decomposition over 222 days (Figure 7.14). This suggests the potential importance of plant root material to belowground storage of carbon.

In a modeling exercise, McMurtrie et al. (2000) suggest that the ability of forest systems to sequester carbon on a long-term basis is also limited by the availability

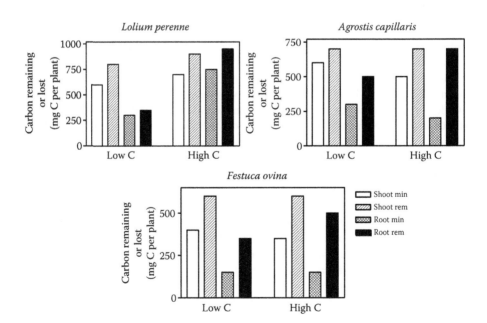

**Figure 7.14**  Amount of carbon mineralized from and remaining in leaf litter derived from three grass species grown under low (350 ml $l^{-1}$) or high (700 ml $l^{-1}$) atmospheric $CO_2$ in the presence of high nitrogen availability after 222 days. (Data from Gorisen, A., Cotrufo, M.F., *Plant Soil*, 224, 75–84, 2000.)

of nitrogen because C and N cycling are closely linked (Knicker, 2011). It may be here that mycorrhizae are of importance. Johnson et al. (2000) showed that, under elevated $CO_2$ conditions, Ponderosa pine forests accessed more nitrogen for either the surface soil horizons or from more recalcitrant forms of nitrogen in the soil to support their increased biomass. It is possible that the enzymatic capabilities of ectomycorrhizae and their ability to sequester nutrients from organic sources in soil could be of increased benefit in a high $CO_2$ world. Plant response to elevated $CO_2$ is often an increase in the C/N ratio in leaf material. The consequence of this in timothy grass occurring at 500 and 600 $\mu$mol mol$^{-1}$ $CO_2$ was also to increase sporulation of *Alternaria* to three times the number of spores, each of which contained twice the concentration of antigenic proteins (Wolf et al., 2010). Thus, the impact of climate change may be significant to human health.

As a consequence of elevated atmospheric carbon dioxide, atmospheric temperatures are predicted to increase. Increased soil temperatures raise the activity of saprotrophic microbial communities, including fungi, such that rates of decomposition are higher. In a forest simulation, Bonan and Van Cleve (1992) compared the consequences of a 5°C rise in temperature on the decomposition of forest floor residues in a variety of boreal forest ecosystems. It was concluded that the effect of air temperature rise would cause a soil temperature increase of between 300 degree days in white and black spruce forests to 500 degree days in birch forest. This would cause an increased soil respiration loss of 439 g C m$^{-2}$ in birch and 675 g C m$^{-2}$ in black spruce, a considerable loss of soil carbon store. However, the total gain in carbon as a result of increased photosynthesis was shown to effectively counteract the loss except for the white spruce ecosystem, where decomposition loss exceeded carbon gain (Table 7.22). It is likely that these changes are as a result of altered metabolic rate of the microbial community and not caused by changes in community composition. For example, Xiong et al. (2014) found no response of the soil fungal community to elevated $CO_2$, but showed a significant change in fungal communities because of soil warming with a community dominated by Ascomycota and specific general of Basidiomycota. These changes were positively related to increased soil respiration. Also, a 10-year exposure to 1.5 times ambient $CO_2$ concentration in an aspen/birch forest ecosystem had little effect on the soil or mycorrhizal fungal community but did enhance both cellulolytic and N-acetylglucosaminidase activity in the soil for the first 5 years (Edwards and Zak, 2011). In a prairie ecosystem, long-term (3 years) changes in rainfall (100% or 70%) and temperature (ambient versus 0.4–0.6°C day

**Table 7.22  Predicted Changes in Carbon Loss and Gain in Boreal Forest Ecosystems as a Consequence of a 5°C Increase in Air Temperature after 25 Years**

| | Forest Floor Mass (kg m⁻²) | | Tree Biomass (kg m⁻²) | | Ecosystem Carbon Flux (g C m⁻²) |
|---|---|---|---|---|---|
| | Control | Warming | Control | Warming | |
| Paper birch | 2.5 | 2.0 | 12.1 | 12.6 | +23 |
| White spruce | 5.4 | 5.1 | 12.6 | 13.6 | −129 |
| Black spruce | 10.7 | 9.9 | 4.3 | 5.0 | +46 |

*Source:* Bonan, G., Van Cleve, K., *Can. J. For. Res.*, 22, 629–639, 1992.

and 1.0–1.5°C night) had no effect on the soil fungal community determined by molecular techniques (Jumpponen and Jones, 2014), suggesting that the fungal community is resilient to climate change.

## 7.5.2 Mycorrhizae

Dighton and Jansen (1991) show the possible scenarios of climate change on the availability of nutrients for plant and mycorrhizal uptake and the changes in carbohydrate supply to support mycorrhizal symbioses. The suggested model for ectomycorrhizal plants (Figure 7.15) shows an enhancement of photosynthetic activity with increased $CO_2$, which would provide a larger pool of carbohydrates to support mycorrhizal development, fruiting, community diversity, and investment into extraradical hyphal exploitation of the soil. The effect of increased availability of $CO_2$ could increase the C/N ratio of plant litter, making the role of mycorrhizal fungi as saprotrophs more important. Thus, mycorrhizal diversity will be maintained and favor those species capable of producing enzymes for the acquisition of major nutrients from organic sources.

In a metadata analysis, Treseder (2004) found that elevated $CO_2$ produced a 47% increase in mycorrhizal abundance. An increase in abundance of mycorrhizae resulting from a 4°C temperature increase over ambient was seen on Douglas fir seedlings (Rygiewicz et al., 2000), whereas elevated $CO_2$ altered diversity. In a minireview of the effects of climate change on mycorrhizae, the consensus of Mohan et al. (2014) was that under elevated $CO_2$ more carbon was allocated to belowground plant and mycorrhizal parts and that warming increased mycorrhizal abundance but decreased mycorrhizal activity. In general, the literature was split over whether or not there was an effect of climate change parameters on the fungal community or its function. At extreme temperatures (up to 45°C) in thermal soils, arbuscular mycorrhizal plant growth can be suppressed by the degree of root colonization and extent of extraradical hyphal development can be significantly enhanced (Bunn et al., 2009), but the fungal community shows little adaptation to temperature. The increased C allocation to roots and mycorrhizae because of elevated atmospheric $CO_2$ may be limited to plants with particular growth strategies. In oaks in a fire-prone sandy ecosystem, most of the C allocation to roots and mycorrhizae was shown to be derived from rhizomatous roots than from newly fixed C via photosynthesis (Langley et al., 2002), reducing the potential impact of climate change in these ecosystems.

Seegmüller and Rennenberg (1994) showed that in oak, elevated $CO_2$ increased stem height, stem diameter, total plant weight, and lateral root formation, and that the effects of *Laccaria laccata* ectomycorrhizal association had more than an additive effect in the presence of elevated $CO_2$. Nylund and Wallander (1989) showed that photosynthesis was enhanced in mycorrhizal plants and that the translocation of photosynthate to roots was enhanced by the ectomycorrhizal fungus *Hebeloma crustuliniforme*, but not by *Laccaria laccata*.

Subsequent to the publication of this model, a number of research projects have looked at the role of enhanced $CO_2$ and temperature on the mycorrhizal status of plants and the physiology of their association. How well do these studies fit or refute

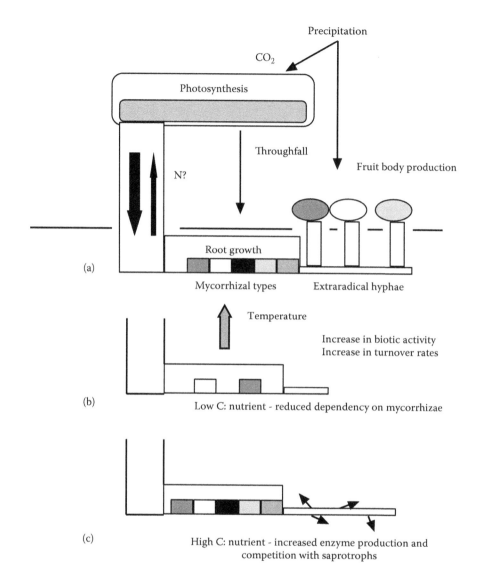

**Figure 7.15** Model of the possible effects of increased atmospheric $CO_2$ and elevated temperature on the tree–ectomycorrhizal symbiosis. (a) The top model indicates the diversity and function of mycorrhizae in an elevated $CO_2$ scenario. Increased photosynthesis leads to greater C supply to roots and fungi, which supports a greater access to nitrogen (in N-limiting systems). The effect of elevated temperature in scenario (b) suggests increased leaf litter decomposition and nutrient mineralization, leading to a reduced host dependency on the mycorrhizal symbiosis. In scenario (c), the change in leaf litter chemistry (enhanced nitrogen content) increases resource quality and the mycorrhizal fungi have to compete with the saprotrophic community for access to mineralized nutrients by increasing diversity and extraradical hyphal exploitation of soil. (Adapted from Dighton, J., Jansen, A.E., *Environ. Pollut.*, 73, 179–204, 1991.)

the model? Indeed, Vogt et al. (1993) and O'Neill (1994) discuss the importance of selecting appropriate belowground indicators of environmental change as soil organisms are sensitive to stress and may be influenced prior to any observable aboveground symptoms. They specifically suggest that fine root growth, turnover, and mycorrhizal status and function are some of the potentially most useful measures of response to environmental change. It is known that one of the effects of ectomycorrhizal colonization of root systems is an increase in root branching, so it is reasonable to assume that part of the increased carbohydrate drain to support mycorrhizal root systems is attributable to increased root branching.

Under enhanced $CO_2$, extra fixed carbon was translocated belowground and incorporated into root and fungal biomass in *Suillius bovinus* mycorrhizae, common on N-rich soils, but was respired by roots colonized by *Laccaria laccata*, which is more common on N-poor soils (Gorisen and Kuyper, 2000). The plant return for the carbon investment in *Suillus*-colonized root biomass was an increase in nitrogen uptake, resulting in plant N contents twice as high as that in *Laccaria*-colonized trees, irrespective of $CO_2$ concentration. These results lend support to the concept that carbon use for nitrogen acquisition is more important in mycorrhizal symbioses that are characteristic of N-limiting ecosystems.

The change in ectomycorrhizal community seen in Douglas fir seedlings in response to both elevated atmospheric $CO_2$ and elevated temperature (Rygiewicz et al., 2000) in a low-nitrogen soil could be a response to changes in carbon allocation to roots and nitrogen use efficiency. This study indicated that the total number of mycorrhizae increased in elevated $CO_2$ conditions, but that an increase in ectomycorrhizal diversity resulted from an increase in temperature, where *Rhizopogon* and *Cenococcum* morphotypes dominated. O'Neill et al. (1987) showed that seedling oak and pine trees produced an average of 66% (pine) and 56% (oak) increase in biomass because of the increased mycorrhizal colonization of seedlings growing in a $CO_2$-enriched environment (double ambient). It was suggested that increased photosynthesis offset the carbon drain imposed on the plant by the increased mycorrhizal colonization of the roots system. Heathland plants represented by heather (*Calluna vulgaris*) and the grass *Deschampsia flexuosa* showed increased root growth under elevated $CO_2$ levels and warming, with an increase in mycorrhizal colonization with elevated $CO_2$. However, this was not mirrored by increases in N or P uptake by the roots (Arndal et al., 2013). A similar finding in lettuce showed increased mycorrhizal colonization with $CO_2$ but no increase in plant nutrition (Baslam et al., 2012). With ectomycorrhizal plants, Deslippe et al. (2011) showed that warming significantly increased ectomycorrhizal diversity on roots of *Betula nana* in the Arctic, especially stimulating *Cortinarius* spp., which are known to produce protease enzymes. This suggests a potential increase in saprotrophic abilities enhancing nutrient cycling by priming (Phillips et al., 2012). Garcia et al. (2008) showed a 14% increase in ectomycorrhizal colonization of roots in a loblolly pine forest due to $CO_2$ enrichment, but no change in arbuscular mycorrhizal abundance. Using natural stands of Scots pine along climate gradients, Jarvis et al. (2013) showed that rainfall and soil moisture played a greater role than temperature in modifying ectomycorrhizal communities.

Treseder and Allen (2000) showed that there was some evidence suggesting that mycelial growth could be enhanced by elevated $CO_2$. For example, Sanders et al. (1998) found a 5-fold increase in arbuscular mycorrhiza extraradical mycelium with elevated $CO_2$ in experimental conditions. In a long-term exposure study, Rillig et al. (2000) compared the arbuscular mycorrhizal status of New Zealand pastures in a transect away from cold $CO_2$ springs. They found that mycorrhizal root colonization and soil hyphal length increased linearly ($r^2 = 0.47$, $P = 0.0016$ and $r^2 = 0.76$, $P < 0.0001$, respectively) along the increasing $CO_2$ gradient. However, Staddon et al. (1999) found that elevated $CO_2$ had similar growth-promoting effects on both mycorrhizal and nonmycorrhizal *Plantago lanceolata* but also noted that it had no effect on carbon partitioning between roots and shoots of either mycorrhizal status.

The increased flow of carbon through root systems in elevated $CO_2$ conditions resulted in a greater accumulation of soil organic matter, rather than plant biomass. Fitter et al. (2000) suggest that it would not be expected that there would be direct effects of increased atmospheric $CO_2$ on mycorrhizal hyphal growth because these fungi grow in a high $CO_2$ atmosphere in the soil anyway. They consider that the indirect effects via photosynthesis would be most likely. They suggest that a positive feedback loop could ensue, in which increased plant carbon fixation increased mycorrhizal growth and ability to capture P (arbuscular mycorrhizae) and N (ectomycorrhizae), to alleviate the plant from nutritional stress and increase plant growth. In the arbuscular mycorrhizal case, they suggest that high turnover of extraradical hyphae would increase soil respiration and counter the enhanced C fixation. Sanders et al. (1998) did not find any increased uptake of P through the increased extension of arbuscular mycorrhizal hyphae due to increased $CO_2$. They inferred that the increase in hyphal mass would contribute to belowground carbon stores and to the increase in soil aggregation. Klironomos et al. (1997, 1998) showed that infection and extraradical hyphal growth differed between two species of *Glomus*, *Acaulospora denticulate* and *Scutellospora calospora*, in the presence and absence of elevated $CO_2$. The two *Glomus* species only caused increased plant growth under elevated $CO_2$ conditions. The other two mycorrhizal genera caused no difference in plant growth, but only a difference in the form of the mycorrhizal association.

Carbon supply to roots can be enhanced by elevated $CO_2$ levels and cause subsequent changes in the rhizosphere community (Drigo et al., 2013). Drigo et al.'s (2013) data showed that arbuscular mycorrhizal associates of roots increased in abundance and modified the rate of release of carbohydrates into the rhizosphere, significantly reducing the bacterial phospholipid fatty acid abundance. The effects of increased $CO_2$ can also be seen in other fungal symbiotic associations. The grass endophyte (*Neotyphodoium coenophialum*) infection of tall fescue leaves was reduced by 10% because of $CO_2$ addition and showed a consequential decline of 30% in the antiherbivore compounds ergovaline and loline, but little effect of warming or altered precipitation (Brosi et al., 2011).

## 7.5.3 Rainfall Effects

Climate change is predicted to increase rainfall in some areas and decrease it in others. The ability of mycorrhizae to increase water stress tolerance of plants in droughty conditions has been well established (Cruz et al., 2000; Jany et al., 2003) and may be related to melanin in the fungal hyphae (Fernandez and Koide, 2013) and enhanced soil exploitation for water (Li et al., 2014). In contrast, where rainfall increases, waterlogging may occur. In citrus, arbuscular mycorrhizae have also been shown to ameliorate waterlogging conditions, where—despite reduced root colonization by fungi—the number of fungal entry points and vesicles has been seen to increase. This change in colonization increases the soluble protein content of the roots and leaf catalase activity, thus reducing oxidative damage (Wu et al., 2013). Altering wet and dry cycles and irrigation often cause the crystallization of salt in soil. Here, too, arbuscular mycorrhizae have been shown to improve stress tolerance by the enhanced production of antioxidant enzymes and nonenzymatic antioxidants (Evelin and Kapoor, 2014; Chandrasekaran et al., 2014; Table 7.23).

## 7.5.4 Carbon Stores

As climate changes, a potential positive feedback may build between increased soil respiration from cold highly organic soils and the $CO_2$ concentration in the atmosphere causing further warming. A similar scenario is seen with land use change, where Howard et al. (1995) document land use changes in the UK result in a predicted net loss of some $151,000 \times 10^3$ t C from soil to the atmosphere by the year 2044 (Table 7.24).

Roots and their mycorrhizal components are frequently overlooked as potential sources of carbon within ecosystems. In forested ecosystems, Vogt (1991) gives comparisons of biomass allocation within the forest for a variety of tree species and locations. Belowground biomass carbon accounts for between 16% in *Pinus menziesii* and 64% in *Pinus eliottii* of the total tree biomass carbon (Vogt, 1991). Thus, afforestation and reforestation may significantly enhance carbon sequestration in soil through roots and root death. The belowground input to the decomposer system under forested ecosystems can be considerable. Bowden et al. (1993) point out

Table 7.23 Antioxidant Enzyme Levels Induced in *Glomus intraradices* Colonized Roots of *Trigonella* by Salt Stress 14 Days after Treatment

| NsCl (mM) | AM Status | Superoxide Dismutase | Catalase | Ascorbate Peroxidase | Peroxidase | Glutathione Reductase |
|---|---|---|---|---|---|---|
| 0 | NM | 957 | 236 | 9.8 | 6.7 | 5945 |
|   | M | 1156 | 390 | 13.2 | 13 | 5973 |
| 50 | NM | 1451 | 379 | 11.1 | 12.7 | 6976 |
|   | M | 1451 | 451 | 17.5 | 12 | 6898 |
| 100 | NM | 1207 | 402 | 14.2 | 9.3 | 8470 |
|   | M | 1696 | 314 | 28.7 | 9.3 | 8737 |
| 200 | NM | 1764 | 276 | 16.5 | 11.8 | 8403 |
|   | M | 2122 | 518 | 35.8 | 17.7 | 9307 |

*Source:* After Evelin, H., Kapoor, R., *Mycorrhiza*, 24, 197–208, 2014.
*Note:* All enzymes are expressed as μkat g$^{-1}$ protein, except superoxide dismutase, which is expressed as nkat g$^{-1}$ protein.

**Table 7.24 Changes in Land Use in England and Wales (UK) during Six Years (1984–1990), the Calculated Loss of Soil Carbon, and the Predicted Loss by Year 2020**

| Cover Type | Soil C (t × 10³ km⁻²) | Area Change (km²) | Change as % of Total Area | Change in Soil C (t × 10³) | Predicted Change by 2020 (t × 10³) |
|---|---|---|---|---|---|
| Arable | 16.9 | +549 | +0.36 | +9293 | −44,137 |
| Bog | 107.4 | −111 | −0.07 | −11,921 | −69,434 |
| Coniferous | 31.8 | +267 | +0.18 | +8499 | +43,790 |
| Deciduous | 26.1 | +319 | +0.21 | +8340 | +39,553 |
| Horticulture | 19.0 | −220 | −0.14 | −4171 | −10,681 |
| Ley | 19.3 | −3099 | −2.04 | −59,891 | −74,473 |
| Lowland Heath | 24.8 | +200 | +0.13 | +4964 | +25,434 |
| Orchard | 17.0 | −200 | −0.13 | −3387 | −8338 |
| Permanent grass | 23.6 | +1854 | +1.22 | +43,759 | −8990 |
| Recreation | 21.3 | +52 | +0.03 | +1104 | +6259 |
| Rough grazing | 36.7 | +136 | +0.09 | +4979 | −28,381 |
| Scrub | 28.8 | −63 | −0.04 | −1798 | −4610 |
| Upland grass | 60.2 | −931 | −0.61 | −56,057 | −70,653 |
| Upland heath | 69.9 | +339 | +0.22 | +23,646 | +97,914 |
| Urban | 0 | +907 | +0.60 | 0 | 0 |
| Total | | | | −32,640 | −106,747 |

*Source:* Howard, D.M. et al., *J. Environ. Manage.*, 45, 287–303, 1995.

that nearly two-thirds of the soil respiration of a temperate mixed-hardwood forest comes from root activity, and the decomposition of root litter contributes 70–80% of total soil respiration in a range of forested ecosystems, of which ectomycorrhizal fungal biomass accounts for a considerable proportion. In heathland ecosystems, *Molinia* root turnover contributes 67% of the total litter production, 87% of litter nitrogen loss, and 84% of total litter phosphorus loss (Aerts et al., 1992). In contrast, *Calluna* (ericoid mycorrhizal) and *Deschampsia* (arbuscular mycorrhizal) growing in the same ecosystem contributed two to three times lower percent values. Root respiration values vary considerably from an estimated 35% in tulip tree (*Liriodendron tulipifera*) (Edwards and Harris, 1977) to 62% of soil respiration in slash pine (*Pinus eliottii*) (Nakane et al., 1983). The data support the hypothesis of Bonan (1993) that coniferous trees allocate twice as much carbon to roots than do deciduous tree species. These belowground stores of carbon may be significant in a world of increasing atmospheric $CO_2$ concentration as a potential carbon sink.

In their review paper on root system adjustments to elevated atmospheric $CO_2$, BassiriRad et al. (2001) came to the conclusion that there was no consistent pattern in the literature for compensatory adjustments such as changes in root/shoot ratios and root architecture, root nutrient absorption capacity, and nutrient use efficiency. However, numerous researchers have shown responses of both roots and their mycorrhizae to changes on atmospheric $CO_2$ concentrations (Diaz, 1996). For example, Dhillion et al. (1996) showed that a Mediterranean old field monoculture of

*Bromus madritensis* responded to an increase in atmospheric $CO_2$ from 350 to 700 µmol mol$^{-1}$ by increasing root biomass by 31%, root length by 88%, soil microbial biomass by 42%, soil fungal hyphal length by 20%, and total root colonization my arbuscular mycorrhizae by 57%. These increases in fungal biomass in soil are reflected by increases in fungal-feeding nematodes, suggesting a higher rate of energy transfer through the fungal-based food web at elevated $CO_2$ (Hungate et al., 2000).

There is a positive drain of carbohydrates from plant shoots to roots to sustain their mycorrhizal associates. In arbuscular mycorrhizal symbioses, Douds et al. (1988) showed in Carrizo citrange that there was 5–6 times greater translocation of carbon to mycorrhizal root systems than to nonmycorrhizal seedlings. Similarly, Wang et al. (1989) showed that mycorrhizal roots in a split pot experiment caused a 45% increase in photosynthate storage in leaves by *Panicum* plants and more than doubled sink activity (movement to roots) in mycorrhizal plants. The formation of different root architectures also has different demands on photosynthate translocation belowground.

Together with evidence of fruit body production of mycorrhizal fungi being a major part (15% of soil organic matter) of the forest floor biomass (Vogt et al., 1982; Vogt, 1991), it is probable that mycorrhizae form a significant carbon sink in soils. Fine roots of forest tress in eastern United States have a mean age of 3–18 years (dead roots with an age of 10–18 years) and are composed of carbon fixed between 3 and 8 years ago in organic soil horizons to 11–18 years previous in the mineral soil horizon (Gaudinski et al., 2001), suggesting that root biomass and necromass carbon is a midterm carbon store in soils. Keeping carbon belowground and in the biomass of plants and fungi is also facilitated by ectomycorrhizal fungi. Finlay and Read (1986) showed that there is significant between-plant transfer of carbon through interconnected plants via mycorrhizal bridges. This transfer is enhanced between source plants optimizing photosynthesis and sink plants, which are growing in the shade. This transfer of carbon is important both within and between tree species in the field (Simard et al., 1997a,b, 2012). These root network acts as a resource for whole plant communities to conserve both carbon and nutrients within the forest system, rather than the traditional concept of each plant and each plant species acting as a single unit in a field of competitive interactions.

Ectomycorrhizae may be important for C storage only in ecosystems where the tree species is native and the association between trees and mycorrhizal flora had coevolved. In exotic tree plantations, the interaction may be the exact opposite. Despite the suggestions that plantation forestry may be an ideal sink for excess atmospheric $CO_2$ (Fang et al., 2001), the effects of plantation forest on soil carbon dynamics is often confounded with site preparation management, such as plowing and ripping, which causes massive soil disturbance and pulses of respiratory loss of stored carbon. In a study of radiata pine plantations in Ecuador, where there was no site preparation, Chapela et al. (2001) calculates that there is greater carbon loss from soil than there is gain from photosynthesis of the forest. Stable isotope studies of the soil carbon accreted under the former paramo grassland ecosystem shows a 30% loss of this carbon during the first 20 years of forest rotation. It is suggested that the prolific fruiting (1200 kg dry weight ha$^{-1}$ y$^{-1}$ for all species), especially of the larger species (*Suillus luteus*), is a major contributor to respiratory carbon loss from the ecosystem as these mycorrhizal fungi utilize organic forms of nutrient and nonhost carbon (Dighton et al., 1987; Durall et al., 1994; Zhu et al., 1994). The concept of plantation forestry as being our salvation in creating large atmospheric carbon sinks, has to be viewed more skeptically in light of these findings.

Estimates of arbuscular mycorrhizal hyphal biomass range from 0.02 m g$^{-1}$ soil in poplar, to 38 m ml$^{-1}$ soil in shrub-steppe, to 111 m ml$^{-1}$ soil for prairie communities (Rillig and Allen, 1999), which amounts to a maximum of 457 µg ml$^{-1}$ soil of hyphal dry weight (approximately 500 kg km$^2$). However, Treseder and Allen (2000) make a contrary argument for ectomycorrhizae, whose long-lived mycelium may be resistant to microbial attack and be incorporated into long-term carbon stores in soil by entering the "slow pool" of carbon. The potential increased development of extraradical hyphae in arbuscular mycorrhizal associations, results in more fungal biomass for faunal consumption. For example, Lundquist et al. (1999) reported that the incorporation of fresh rye shoots into soil increased the number of bacterial and fungal-feeding nematodes, but that the FDA active (live) fungal hyphal length did not increase significantly. Hence, there is probably a balance between a system being able to produce fungal biomass and the rate at which it is consumed by organisms in the next trophic level. The amount of overall accretion of soil carbon, therefore, is a balance between what carbon enters and is retained in a "slow" turnover pool and that which enters a "fast" turnover pool. Mycorrhizal fungi are less favored food sources for soil fauna than saprotrophic fungi (Klironomos and Kendrick, 1996; Klironomos and Ursic, 1998; Klironomos et al., 1999), so they may have a longer hyphal residence time in soil than saprotrophic fungi. As a number of these fungi produce the glycoprotein glomalin (Wright and Upadhyaya, 1996), Treseder and Allen (2000) calculate that glomalin and glomalin-related soil protein fraction could account for 30–60% of the carbon in undisturbed soils (Schindler et al., 2007). Increase in the glomalin content of grassland soils under long-term exposure to elevated $CO_2$ (Rillig et al., 2000) suggests a secondary function of carbon storage in the increased development of soil stability because of aggregate formation. This carbon is somewhat protected in soil aggregates and could be regarded as a potential long-term carbon sink.

Jastrow (1996) suggests that the degradation of soil aggregates, as a result of soil management or mismanagement, could result in the loss of soil carbon from protected fraction in aggregates. The development of stable soil aggregates can increase the carbon sink of soils (Knicker, 2011). Hence, fungi and bacteria are of great importance in the formation and stability of aggregates. In a study of the restoration of prairie soils from agriculture, the increase in percentage of stable aggregates in the soil followed an exponential model. The rate constant ($k$) for aggregate formation was 35 times that for the accumulation of carbon into other parts of the soil, with the time to reach 99% of the equilibrium at 10.5 years for aggregates and 384 years for the whole soil. Analysis of the C/N ratio of carbon in aggregates suggests that it is of very recent origin, suggesting that it is a derivative of bacterial and fungal biomass and not highly processed as some of the carbon in whole soil. Indeed, a significant fraction of this material is of fungal cell wall residues, which contribute to binding

Table 7.25   Changes over Time in Carbon Sequestration of Different Soil Components

| Carbon Fraction | Soil under Corn | 4 Season Restoration | 10 Season Restoration | Virgin Prairie |
|---|---|---|---|---|
| Mineral-associated macroaggregate C | 1181 | 2548 | 3348 | 4692 |
| Intramacroaggregate POM C | 77 | 131 | 138 | 250 |
| Macroaggreagate associated C | 1258 | 2679 | 3485 | 4924 |
| Total C in aggregates <212 μm diameter | 1918 | 837 | 576 | 567 |
| Total C in whole soil | 3517 | 3996 | 4733 | 6106 |

Source: After Jastrow, J.D., Soil Biol. Biochem., 28, 665–676, 1996.

the microaggregates into macroaggregates. This carbon in aggregates is in a particulate or colloidal state and is physically protected from further rapid decay. The changes in partitioning of carbon over time within restored prairie soils are shown in Table 7.25. Under a $CO_2$-enriched environment, Jastrow et al. (2000) showed that soil carbon and nitrogen stocks increased under a tallgrass prairie ecosystem. They suggest that rootlike light particulate organic matter (POM) turns over more rapidly than more amorphous heavy POM, which accumulates over time under elevated $CO_2$. This they attribute to the influences in nitrogen cycling, particularly mineralization, immobilization, and asymbiotic N fixation. The first two components are very much mediated by fungal activity in the soil.

## 7.6 CONCLUSIONS

In this chapter, a small selection of human impacts on ecosystems has been explored. This coverage is far from complete or exhaustive, but serves to show that human impacts affect the community structure of fungi in the environment and their functional activities. In addition, we have seen that fungi are phenotypically plastic, allowing adaptation to their environment. In this fashion, fungi are capable of surviving in adverse conditions and altering the environment for their own growth and for the growth of other organisms. This is particularly true for the role of fungi in heavy metal and radionuclide pollution, where there is substantial evidence to show that fungi are able to accumulate, redistribute (in time and space), and alter the chemical state of elements. Industrial uses of fungi as metal accumulators have been shown to be feasible (Tobin et al., 1984), but the utility of these physiological attributes has not been fully explored in the context of potential use of fungi in bioremediation.

The discussion on the effects of atmospheric acidifying pollutants has highlighted the complexity of ecosystems. It has shown us that the effects of a single pollutant may not explain all the ecosystem and organism responses observed. Frequently, human impacts are multifactoral and the synergistic and antagonistic interactions between the impacting factors are often difficult to tease apart when studies are performed with single factors. In addition, it is often not easy to represent all of the ecosystem components in an experiment. Frequently, our information is gained from laboratory studies in highly controlled microcosms. Much more information can be gained from the study of more complex interactions between greater numbers of organisms in mesocosms (Odum, 1984), which more closely represent the natural ecosystem. However, the trade-off here is the balance between increased information and increased variation, which results in less accuracy in determining significant results. Along with excellent metadata analyses in many review articles, the use of multivariate statistics and trend analysis, however, allows us to make predictions from trends in the data, rather than just relying entirely on parametric statistics. Therefore, I see the challenge for the future as directing fungal ecology toward functional fungal ecology in a synecological setting as opposed to an autecological view.

## REFERENCES

Aber, J. D. 1992. Nitrogen cycling and nitrogen saturation in temperate forest ecosystems. *Trends Ecol. Evol.* 7:220–236.

Aber, J. D., K. Nadelhoffer, P. Steudler, and J. M. Mellillo. 1989. Nitrogen saturation in northern forest ecosystems. *Bioscience* 39:378–386.

Abuzinadah, R. A. and D. J. Read. 1986. The role of proteins in the nitrogen nutrition of ectomycorrhizal plants: III. Protein utilization by *Betula, Picea* and *Pinus* in mycorrhizal association with *Hebeloma crustuliniforme. New Phytol.* 103:507–514.

Abuzinadah, R. A. and D. J. Read. 1989. The role of proteins in the nitrogen nutrition of ectomycorrhizal plants: V. The utilization of peptides by birch (*Betula pendula* L.) infected with different mycorrhizal fungi. *New Phytol.* 112:55–60.

Adams-Krumins, J., J. Dighton, D. Gray, R. B. Franklin, P. Morin, and M. S. Roberts. 2009. Soil microbial community response to nitrogen enrichment in two scrub oak forests. *For. Ecol. Manage.* 258: 1383–1390.

Aerts, R., C. Bakker, and H. De Caluwe. 1992. Root turnover as determinant of the cycling of C, N and P in a dry heathland ecosystem. *Biogeochemistry* 15:175–190.

Aliasgharzd, N., Mårtensson, L-M. and Olsson, P. A. 2010. Acidification of a sandy grassland favours bacteria and disfavours fungal saprotrophs as estimated by fatty acid profiling. *Soil Biol. Biochem.* 42:1058–1064.

Arisoy, M. and N. Koankaya. 1997. Biodegradation of linden by *Pleurotus sajor-caju* and toxic effects of lindane and its metabolites on mice. *Bull. Environ. Toxicol.* 59:352–359.

Arndal, M. F., M. P. Merrild, A. Michelsen, I. K. Schmidt, T. N. Mikkelsen, and C. Beier. 2013. Net root growth and nutrient acquisition in response to predicted climate change in two contrasting heathland species. *Plant Soil* 369:615–627.

Antibus, R. K., D. Bower, and J. Dighton. 1997. Root surface phosphatase activities and uptake of 32P-labelled inositol phosphate in field-collected gray birch and red maple roots. *Mycorrhiza* 7:39–46.

Antibus, R. K. and A. E. Linkins. 1992. Effects of liming a red pine forest floor on mycorrhizal numbers and mycorrhizal and soil acid phosphates activities. *Soil Biol. Biochem.* 24:479–487.

Arnebrant, K. 1994. Nitrogen amendments reduce the growth of extramatrical ectomycorrhizal mycelium. *Mycorrhiza* 5:7–15.

Arnebrant K. and B. Söderström. 1992. Effects of different fertilizer treatments on the ecto-mycorrhizal colonization potential in two Scots pine forests in Sweden. *Forest Ecology and Management* 53:77–89.

Arnolds, E. 1985. *Veranderingen in de paddestoelenflora (mycoflora).* Wet. Meded. 167.

Arnolds, E. 1988. The changing macromycete flora in The Netherlands. *Trans. Br. Mycol. Soc.* 90:391–406.

Arnolds, E. 1989a. Former and present distribution of stipitate hydnaceous fungi (Basidiomycetes) in the Netherlands. *Nova Hedwigia* 1–2:107–142.

Arnolds E. 1989b. The influence of increased fertilization on the macrofungi of a sheep meadow in Drenthe, the Netherlands. *Opera Bot.* 100:7–21.

Arnolds, E. 1989c. A preliminary red data list of macrofungi in the Netherlands. *Persoonia* 14:77–125.

Arnolds, E. 1991. Decline of ectomycorrhizal fungi in Europe. *Agric. Ecosyst. Environ.* 35:209–244.

Arnolds, E. J. M. 1997. Biogeography and conservation. In: *The Mycota IV: Environmental and Microbial Relationships*, ed. D. T. Wicklow and B. Soderstrom, 115–131. Berlin: Springer-Verlag.

Baslam, M., I. Garmendina, and N. Goicoecha. 2012. Elevated $CO_2$ may impair the beneficial effect of arbuscular mycorrhizal fungi on the mineral and phytochemical quality of let-tuce. *Ann. Appl. Biol.* 161:180–191.

Bardgett, R. D., E. Kandeler, D. Tscherko et al. 1999. Below-ground microbial community development in a high temperature world. *Oikos* 85:193–203.

Bardgett, R. D., T. W. Spier, D. J. Ross, and G. W. Yeates. 1994. Impact of pasture contamina-tion by copper, chromium, and arsenic timber preservative on soil microbial properties and nematodes. *Biol. Fert. Soils* 18:71–79.

Bärlocher, F. Ed. 1992. *The Ecology of Aquatic Hyphomycetes.* Berlin: Springer-Verlag.

BassiriRad, H., V. P. Gutschick, and J. Lussenhop. 2001. Root system adjustments: Regulation of plant nutrient uptake and growth responses to elevated $CO_2$. *Oecologia* 126:305–320.

Bates, J. W., J. N. B. Bell, and A. C. Massara. 2001. Loss of *Lecanora conizaeoides* and other fluctuations of epiphytes on oak in S. E. England over 21 years of declining $SO_2$ concen-trations. *Atmos. Environ.* 35:2557–2568.

Baxter, J. W. and J. Dighton. 2001. Ectomycorrhizal diversity alters growth and nutrient acqui-sition of gray birch (*Betula populifolia* Marshall) seedlings in host-symbiont culture conditions. *New Phytol* 152:139–149.

Baxter, J. W. and J. Dighton. (2005) Diversity–functioning relationships in ectomycorrhi-zal fungal communities. In: *The Fungal Community: Its Organization and Role in the Ecosystem.* 3rd Edition. ed. J. Dighton, J. F. White Jr. and P. Oudemans, 383–398. Boca Raton, FL: Taylor & Francis.

Bellion, M., M. Courbot, C. Jaco, D. Blaudez, and M. Chalot. 2006. Extracellular and cellu-lar mechanisms sustaining metal tolerance in ectomycorrhizal fungi. *FEMS Microbiol. Lett.* 254:173–181.

Ben Sassi, A., N. Ouazzani, G. M. Walke M. et al. 2008. Detoxification of olive mill waste-waters by Moroccan yeast isolates. *Biodegradation* 19:337–346.

Berck, P., A. Levy, and K. Chowdhury. 2012. An analysis of the world's environment and population dynamics with varying carrying capacity, concerns and skepticism. *Ecol. Econ.* 73:103–112.

Berreck, M. and K. Hasselwandter. 2001. Effect of the arbuscular mycorrhizal symbiosis upon uptake of cesium and other cations by plants. *Mycorrhiza* 10:275–280.

Bervaes, J., P. Mathy, and P. Evers. 1988. *Relationships between Above and Below Ground Influences of Air Pollutants on Forest Trees.* Brussels, Belgium: Commission for European Communities (CEC).

Bisonnette, L., M. St-Arnaud, and M. Labrecque. 2010. Phytoextraction of heavy metals by two Saliciaceae clones in symbiosis with arbuscular mycorrhizal fungi during the second year of a field trial. *Plant Soil* 332:55–67.

Blaschke, H. 1988. Mycorrhizal infection and changes in fine root development of Norway spruce influenced by acid rain in the field. In: *Ectomycorrhiza and Acid Rain*, ed. A. E. Jansen, J. Dighton, and A. H. M. Bresser, 112–115. Bilthoven, The Netherlands: CEC.

Blaschke, H., U. Brehmer, and H. Schwartz. 1985. Wurtzelschäden und Waldsterben: Zur Bestimung morphometrischer Kenngrössen von Feinwurtzelsytemen mit dem IBAS— erset Ergebnisse. *Forstw. Cbl.* 104:199–205.

Blaudez, D., C. Jacob, K. Turnau, J. V. Colpaert, U. Ahonen-Jonnarth, R. Finlay, B. Botton, and M. Chalot. 2000. Differential responses of ectomycorrhizal fungi to heavy metals *in vitro. Mycol. Res.* 104:1366–1371.

Boddy, L., J. C. Frankland, S. Dursun, K. Newsham, and P. Ineson. 1996. Effects of dry-deposited SO$_2$ and sulphite on saprotrophic fungi and decomposition of tree leaf litter. In: *Fungi and Environmental Change*, ed. J. C. Frankland, N. Magan, and G. M. Gadd, 70–89. Cambridge: Cambridge University Press.

Bohac, J, D. A. Krivolutskii, and T. B. Antonova. The role of fungi in the biogenous migration of elements and in the accumulation of radionuclides. *Agric. Ecosyst. Environ.* 28:31–34. 1989.

Bonan, G. B. 1993. Physiological controls of the carbon balance of boreal forest ecosystems. *Can. J. For. Res.* 23:1453–1471.

Bonan, G. B. and K. Van Cleve. 1992. Soil temperature, nitrogen mineralization, and carbon source–sink relationships in boreal forests. *Can. J. For. Res.* 22:629–639.

Bongers, T. 1990. The maturity index: An ecological measure of environmental disturbance based on nematode species composition. *Oecologia* 83:14–19.

Bowden, R. D., K. J. Nadelhoffer, R. D. Boone, J. M. Melillo, and J. B. Garrison. 1993. Contributions of aboveground litter, belowground litter, and root respiration to total soil respiration in a temperate mixed hardwood forest. *Can. J. For. Res.* 23:1402–1407.

Boyce, R. L., A. J. Friedland, C. P. Chamberlain, and S. R. Poulson. 1996. Direct canopy nitrogen uptake from 15N labeled wet deposition by mature red spruce. *Can. J. For. Res.* 26:1539–1547.

Bradley, R., A. J. Burt, and D. J. Read. 1981. Mycorrhizal infection and resistance to heavy metal toxicity in *Calluna vulgaris. Nature* 292:335–337.

Bradley, R., A. J. Burt, and D. J. Read. 1982. The biology of the mycorrhiza in the Ericaceae: VIII. The role of mycorrhizal infection in heavy metal resistance. *New Phytol.* 91:197–209.

Brandrud, T. E. 1995. The effects of experimental nitrogen addition on the ectomycorrhizal fungal flora in an oligotrophic spruce forest at Gardsjon, Sweden. *For. Ecol. Manage.* 71:111–122.

Brosi, G. B., R. L. McCulley, L. P. Bush et al. 2011. Effects of multiple climate change factors on the tall fescue–fungal endophyte symbiosis: Infection frequency and tissue chemistry. *New Phytol.* 189:797–805.

Brown, D. H. 1996. Urban, industrial and agricultural effects on lichens. In: *Fungi and Environmental Change*, ed. Frankland, J. C., N. Magan, and G. M. Gadd, 257–281. Cambridge: Cambridge University Press.

Brown, L. R. 1997. *State of the World*. New York: W. W. Norton and Co.

Bryan, R., Z. Jiang, M. Friedman, and E. Dadachova. 2011. The effects of gamma radiation, UV and visible light on ATP levels in yeast cells depend on cellular melanization. *Fung. Biol.* 115:945–949.

Bunn, R., Y. Lekberg, and C. Zabinski. 2009. Arbuscular mycorrhizal fungi ameliorate temperature stress in thermophilic plants. *Ecology* 90:1378–1388.

Burger, J. R., C. D. Allen, J. H. Brown et al. 2012. The macroecology of sustainability. *PLoS Biol.* 10(6) e1001345. doi:10.137/journal.pbio.1001345.

Byrne, A. R. 1988. Radioactivity in fungi in Slovenia, Yugoslavia, following the Chernobyl accident. *J. Environ. Radioactivity* 6, 177–183.

Byrne, A. R., M. Dermelj, and A. Vakselj. 1979a. Silver accumulation by fungi. *Chemosphere* 10:815–821.

Byrne, A. R., Z. Slejovec, T. Stijve, L. Fay, W. Goessler, J. Gailer, and K. J. Irgolic. 1995. Arsenobetaine and other arsenic species in mushrooms. *Aust. Mycol. Newsl.* 9:305–313.

Byrne, A. R., Z. Slejkovec, T. Stijve, W. Gossler, and K. J. Irgolic. 1979b. Identification of arsenic compounds in mushrooms and evidence for mycelial methylation. *Australian Mycological Newsletter* 16:49–54.

Byrne, A. R. and M. Tusek-Znidaric. 1983. Arsenic accumulation in the mushroom *Laccaria amethystina*. *Chemosphere* 12:1113–1117.

Cairney, J. W. G. and A. A. Meharg. 2002. Interactions between ectomycorrhizal fungi and soil saprotrophs: Implications for decomposition of organic matter in soils and degradation of organic pollutants in the rhizosphere. *Can. J. Bot.* 80:803–809.

Call, C. A. and F. T. Davies. 1988. Effect of vesicular–arbuscular mycorrhizae on survival and growth of perennial grasses in lignite overburden in Texas. *Agric. Ecosyst. Environ.* 24:395–405.

Carfrae, J. A., K. R. Skene, L. J. Sheppard, K. Ingleby, and A. Crossley. 2006. Effects of nitrogen with and without acidified sulphur on an ectomycorrhizal community in a Sitka spruce (*Picea sitchensis* Bong. Carr) forest. *Environ. Pollut.* 141:131–138.

Carreira, J. A., R. Gardia-Ruiz, J. Lietor, and A. F. Harrison. 2000. Changes in soil phosphatase activity and P transformation rates induced by application of N- and S-containing acid-mist to a forest canopy. *Soil Biol. Biochem.* 32:1857–1866.

Cha, S., S-M. Lim, B. Amirasheba, and J-K. Shim. 2013. The effect of simulated acid rain on microbial community structure in decomposing leaf litter. *J. Ecol. Environ.* 36:223–233.

Chandrasekaran, M., S. Boughattas, S. Hu, S-H Oh, and T. Sa. 2014. A meta-analysis of arbuscular mycorrhizal effects on plants grown under salt stress. *Mycorrhiza* 24:611–625.

Chapela, I. H., L. J. Osher, T. R. Horton, and M. R. Henn. 2001. Ectomycorrhizal fungi introduced with exotic pine plantations induce soil carbon depletion. *Soil Biol. Biochem.* 33:1733–1740.

Chappelka, A. H., J. S. Kush, G. B. Runion, S. Meir, and W. D. Kelley. 1991. Effects of soil-applied lead on seedling growth and ectomycorrhizal colonization of loblolly pine. *Environ. Pollut.* 72:307–316.

Chávez-Gómez, B., R. Quintero, F. Esparza-Garciá et al. 2003. Removal of phenanthrene from soil by co-cultures of bacteria and fungi pregrown on sugarcane bagasse pith. *Bioresour. Technol.* 89:177–183.

Clint, G. M. and J. Dighton. 1992. Uptake and accumulation of radiocaesium by mycorrhizal and non-mycorrhizal heather plants. *New Phytol.* 122:555–561.

Colpaert, J. V. and J. A. Van Assche. 1992. Zinc toxicity in ectomycorrhizal *Pinus sylvestris* L. *Plant Soil* 143:201–211.

Colpaert, J. V. and J. A. Van Assche. 1993. The effects of cadmium on ectomycorrhizal *Pinus sylvestris* L. *New Phytol.* 123:325–333.

Colpaert, J. V. and K. K. Van Tichelen. 1996. Mycorrhizas and environmental stress. In: *Fungi and Environmental Change*, ed. J. C. Frankland, N. Magan, and G. M. Gadd, 109–128. Cambridge: Cambridge University Press.

Conn, C. and J. Dighton. 2000. Litter quality influences on decomposition, ectomycorrhizal community structure and mycorrhizal root surface acid phosphatase activity. *Soil Biol. Biochem.* 32:489–496.

Cornut, J., H. Clivot, E. Chauvet, A. Elger, C. Pagnout, and F. Guérold. 2012. Effect of acidification on leaf litter decomposition in benthic and hyporheic zones of woodland streams. *Water Res.* 46:6430–6444.

Cotrufo, M. F., B. Berg, and W. Kratz. 1998. Increased atmospheric $CO_2$ and litter quality. *Environ. Rev.* 6:1–12.

Cotrufo, M. F. and P. Ineson. 2000. Does elevated atmospheric $CO_2$ concentrations affect wood decomposition? *Plant Soil.* 224:51–57.

Coughtree, P. J. (Ed.) 1983. *Ecological Aspects of Radionuclide Release*. Oxford: Blackwell Scientific Publications.

Coxson, D., C. Björk, and M. D. Bourassa. 2014. The influence of regional gradients in climate and air pollution on epiphytes in riparian forest galleries of the upper Fraser River Watershed. *Botany* 92:23–45.

Crane, S., T. Barkay, and J. Dighton. 2012. The effect of mercury on the establishment of *Pinus rigida* seedlings and the development of their ectomycorrhizal communities. *Fungal Ecol.* 5:245–251.

Crane, S., J. Dighton, and T. Barkay. 2010. Growth responses to and accumulation of mercury by ectomycorrhizal fungi. *Fungal Biol.* 114:873–880.

Cravotto, G., S. Di Carlo, A. Binello, S. Mantegna, M. Girlanda, and A. Lazzari. 2008. Integrated sonochemical and microbial treatment for decontamination of nonylphenol-polluted water. *Water Air Soil Pollut.* 187:353–359.

Cromack, K., P. Sollins, W. C. Granstein, T. Speidel, A. W. Todd, G. Spycher, and Y-Li Ching. 1979. Calcium oxalate accumulation and soil weathering in mats of the hypogeous fungus *Hysterangium crassum. Soil Biol. Biochem.* 11:463–487.

Cruz, A. F., T. Ishii, and K. Kadoya. 2000. Effects of arbuscular mycorrhizal fungi on tree growth, leaf water potential, and levels of 1-aminocyclopropane-1-carboxylic acid and ethylene in the roots of papaya under water-stress conditions. *Mycorrhiza* 10:121–123.

D'Annibale, A., F. Rosetto, V. Leonardi, F. Federici, and M. Petruccioli. 2006. Role of autochthonous filamentous fungi on bioremediation of a soil historically contaminated with aromatic hydrocarbons. *Appl. Environ. Microbiol.* 72:28–36.

Dadachova, E., R. A. Bryan, X. Huang et al. 2007 Ionizing radiation changes the electronic properties of melanin and enhances the growth of melanized fungi. *PLoS ONE* 2(5). doi:10.1371/journal.pone.0000457.

Dadachova, E. and A. Casadevall. 2008. Ionizing radiation: How fungi cope, adapt, and exploit with the help of melanin. *Curr. Opin. Microbiol.* 11:1–7.

Dangles, O. and E. Chauvet. 2003. Effects of stream acidification on fungal biomass in decaying beech leaves and leaf palatability. *Water Res.* 37:533–538.

De Angelis, P., K. S. Chigwerewe, and G. E. Scarascia Mugnozza. 2000. Litter quality and decomposition in a $CO_2$-enriched Mediterranean forest ecosystem. *Plant Soil* 224:31–41.

De Wit, T. 1976. *Epiphytic Lichens and Air Pollution in The Netherlands*. Vaduz, Switzerland: J. Cramer.

Dellamatrice, P. M., R. T. R. Monteiro, H. M. Kamida, N. L. Nogueira, M. L. Rossi, and C. Blaise. 2005. Decolorization of municipal effluent and sludge by *Pleurotus sajor-caju* and *Pleurotus ostreatus. World J. Micribiol. Biotechnol.* 21:1363–1369.

Denizli, A., N. Cihangir, N. Tüzmen, and G. Alsancak. 2005. Removal of chlorophenols from aquatic systems using the dried and dead fungus *Pleurotus sajor caju. Bioresour. Technol.* 96:59–62.

Denny, H. J. and I. Ridge. 1995. Fungal slime and its role in the mycorrhizal amelioration of zinc toxicity to higher plants. *New. Phytol.* 130:251–257.

Denny, H. J. and D. A. Wilkins. 1987a. Zinc tolerance in *Betula* spp. I. Effects of external concentration of zinc on growth and uptake. *New Phytol.* 106:517–524.

Denny, H. J. and D. A. Wilkins. 1987b. Zinc tolerance in *Betula* spp.: IV. The mechanism of ectomycorrhizal amelioration of zinc toxicity. *New Phytol* 106:545–553.

Deram, A., F. Languereau and C. Van Haluwyn. 2011. Mycorrhizal and endophytic colonization in *Arrenatherum elatius* L. roots according to the soil contamination in heavy metals. *Soil Sediment Contam.* 20:114–127.

Deslippe, J. R., M. Hartmann, W. W. Mohn, and S. W. Simard. 2011. Long-term experimental manipulation of climate alters the ectomycorrhizal community of *Betula nana* in Arctic tundra. *Global Change Biol.* 17:1625–1636.

Dhillion, S. S., J. Roy, and M. Abrams. 1996. Assessing the impact of elevated $CO_2$ on soil microbial activity in a Mediterranean model ecosystem. *Plant Soil* 187:333–342.

Diaz, S. 1996. Effects of elevated $[CO_2]$ at the community level mediated by root symbionts. *Plant Soil* 187:309–320.

Diene, O., N. Sakagami, and K. Narisawa. 2014. The role of dark septate endophytic fungal isolates in the accumulation of cesium by Chinese cabbage and tomato plants under contaminated environments. *PLoS ONE* 9:e109233.

Dighton, J. 1983. Phosphatase production by mycorrhizal fungi. *Plant Soil* 71:455–462.

Dighton, J. 1991. Acquisition of nutrients from organic resources by mycorrhizal autotrophic plants. *Experientia* 47:362–369.

Dighton, J., G. M. Clint, and J. M. Poskitt. 1991. Uptake and accumulation of 137Cs by upland grassland soil fungi: A potential pool of Cs immobilization. *Mycol. Res.* 95:1052–1056.

Dighton, J. and A. D. Horrill. 1988. Radiocaesium accumulation in the mycorrhizal fungi *Lactarius rufus* and *Inocybe longicystis*, in upland Britain. *Trans. Br. Mycol. Soc.* 91:335–337.

Dighton, J. and A. E. Jansen. 1991. Atmospheric pollutants and ectomycorrhizas: More questions than answers? *Environ. Pollut.* 73:179–204.

Dighton, J., P. A. Mason, and J. M. Poskitt. 1990. Field use of 32P tracer to measure phosphate uptake by birch mycorrhizas. *New Phytol.* 116:655–661.

Dighton, J., A. S. Morale-Bonilla, R. A. Jimînez-Nûñez, and N. Martînez. 2000. Determinants of leaf litter patchiness in mixed species New Jersey pine barrens forest and its possible influence on soil and soil biota. *Biol. Fertil. Soils.* 31:288–293.

Dighton, J. and R. A. Skeffington. 1987. Effects of artificial acid precipitation on the mycorrhizas of Scots pine seedlings. *New Phytol.* 107:191–202.

Dighton, J. and G. M. Terry. 1996. Uptake and immobilization of caesium in UK grassland and forest soils by fungi following the Chernobyl accident. In: *Fungi and Environmental Change*, ed. J. C. Frankland, N. Magan, and G. M. Gadd, 184–200. Cambridge University Press.

Dighton, J., E. D. Thomas, and P. M. Latter. 1987. Interactions between tree roots, mycorrhizas, a saprotrophic fungus and the decomposition of organic substrates in a microcosm. *Biol. Fertil. Soils* 4:145–150.

Dighton, J. T. Tugay, and N. N. Zhdanova. 2008. Fungi and ionizing radiation from radio-nuclides. *FEMS Microb. Lett.* 281:109–120.

Dighton, J., A. R. Tuininga, D. M. Gray, R. E. Huskins, and T. Belton. 2004. Impacts of atmospheric deposition on New Jersey pine barrens forest soils and communities of ectomycorrhizae. *For. Ecol. Manage.* 201:131–144.

Dixon-Hardy, J. E., V. I. Karamushka, T. G. Gruzina, G. N. Nikovska, J. A. Sayer, and G. M. Gadd. 1998. Influence of the carbon, nitrogen and phosphorus source on the solubilization of insoluble metal compounds by *Aspergillus niger. Mycol. Res.* 102:1050–1054.

Donnelley, P. K. and J. S. Fletcher. 1994. Potential use of mycorrhizal fungi as bioremediation agents. *Am. Chem. Soc. Symp. Ser.* 563:93–99.

Douds, Jr. D. D., C. R. Johnson, and K. E. Koch. 1988. Carbon cost of the fungal symbiont relative to net leaf P accumulation in a split-root VA mycorrhizal symbiosis. *Plant Physiol* 86:491–496.

Drigo, B., G. A. Kowalchuk, B. A. Knapp, A. S. Pijl, H. T. S. Boschker, and J. A. van Veen. 2013. Impacts of 3 years of elevated atmospheric $CO_2$ on rhizosphere carbon flow and microbial community dynamics. *Global Change Biol.* 19:621–636.

Duckmanton, L. and P. Widden. 1994. Effect of ozone on the development of vesicular–arbuscular mycorrhizae in sugar maple saplings. *Mycologia* 86:181–186.

Duddridge, J. E. and M. Wainwright. 1980. Heavy metal accumulation by aquatic fungi and reduction in viability of *Gammarus pulex* fed on $Cd^{2+}$ contaminated mycelium. *Water Res.* 14:1605–1611.

Durall, D. M., A. W. Todd, and J. M. Trappe. 1994. Decomposition of $^{14}C$-labelled substrates by ectomycorrhizal fungi in association with Douglas fir. *New Phytol.* 127:725–729.

Dursun, S., J. C. Frankland, L. Boddy, and P. Ineson. 1996a. Sulphite and pH effects on $CO_2$ evolution by fungi growing on decomposing coniferous needles. *New Phytol.* 134:155–166.

Dursun, S., P Ineson, J. C. Frankland, and L. Boddy. 1996b Sulphur dioxide effects on fungi growing on leaf litter and agar media. *New Phytol.* 134:167–176.

Eckl, P., W. Hoffman, and R. Turk. 1986. Uptake of natural and man-made radionuclides by lichens and mushrooms. *Radiat. Environ. Biophys.* 25:43–54.

Edwards, N. T. and W. F. Harris. 1977. Carbon cycling in a mixed deciduous forest floor. *Ecology* 58:431–437.

Edwards, I. P. and D. R. Zak. 2011. Fungal community composition and function after long-term exposure of northern forests to elevated atmospheric $CO_2$ and tropospheric $O_3$. *Global Change Biol.* 17:2184–2195.

Egerton-Warburton, L. M. and B. J. Griffin. 1995. Differential responses of *Pisolithus tinctorius* isolates to aluminium in vitro. *Can. J. Bot.* 73:1229–1233.

Egli, S., R. Amiet, M. Zollinger, and B. Schneider. 1993. Characterization of *Picea abies* (L.) Karst. ectomycorrhizas: Discrepancy between classification according to macroscopic versus microscopic features. *TREE.* 7:123–129.

Entry, J. A., P. T. Rygiewicz, and W. H. Emmingham. 1993. Accumulation of cesium-137 and strontium-90 in Ponderosa pine and Monteray pine seedlings. *J. Environ. Qual.* 22:742–746.

Entry, J. A., P. T. Rygiewicz, and W. H. Emmingham. 1994. $^{90}Sr$ uptake by *Pinus ponderosa* and *Pinus radiata* seedlings inoculated with ectomycorrhizal fungi. *Environ. Pollut.* 86:201–206.

Esher, R. J., D. H. Marx, S. J. Ursic, R. L. Baker, L. R. Brown, and D. C. Coleman. 1992. Simulated acid rain effects on fine roots, ectomycorrhizae, microorganisms, and invertebrates in pine forests of the southern United States. *Water Air Soil Pollut.* 61:269–278.

Estivalet, D., R. Perrin, F. Le Tacon, and D. Bouchard. 1990. Nutritional and microbiological aspects of decline in the Vosges Forest area (France). *For. Ecol. Manage.* 37:233–248.

Evelin, H. and R. Kapoor. 2014. Arbuscular mycorrhizal symbiosis modulates anti-oxidant response in salt stressed *Trigonella foenum-graecum* plants. *Mycorrhiza* 24:197–208.

Fang, C., M. Radosivich, and J. J. Fuhrmann. 2001. Characterization of rhizosphere microbial community structure in five similar grass species using FAME and BIOLOG analysis. *Soil Biol. Biochem.* 33:679–682.

Fellner, R. 1988. Effects of acid deposition on the ectotrophic stability of mountain forest ecosystems in central Europe (Czechoslovakia). In: *Ectomycorrhiza and Acid Rain*, ed. A. E. Jansen, J. Dighton, and A. H. M. Bresser, 116–121. Bilthoven, The Netherlands: CEC.

Fellner, R. and V. Pesková. 1995. Effects of industrial pollutants on ectomycorrhizal relation-ships in temperate forests. *Can. J. Bot.* 73 (Suppl. 1):S1310–S1315.

Fernandez, C. W. and R. T. Koide. 2013. The function of melanin in the ectomycorrhizal fun-gus *Cenococcum geophilum* under water stress. *Fung. Ecol.* 6:479–486.

Finlay, R. D. and D. J. Read. 1986. The structure and function of the vegetative mycelium of ectomycorrhizal plants I. Translocation of $^{14}$C-labellled carbon between plants intercon-nected by a common mycelium. *New Phytol.* 103:143–156.

Fischer, R. G., S. Rapsomankis, and M. O. Andreae. 1995. Bioaccumulation of methyl mer-cury and transformation of inorganic mercury by macrofungi. *Environ. Sci. Technol.* 29:993–999.

Fitter, A. H., A. Heinemeyer, and P. L. Staddon. 2000. The impact of elevated $CO_2$ and global climate change on arbuscular mycorrhiza: A mycocentric approach. *New Phytol.* 147:179–187.

Fogel, R. and G. Hunt. 1983. Contribution of mycorrhizae and soil fungi to nutrient cycling in a Douglas-fir ecosystem. *Can. J. For. Res.* 13:219–232.

Fraiture, A., O. Guillette, and J. Lambinon. 1990. Interest of fungi as bioindicators of the radiocontamination in forest ecosystems. In: *Transfer of Radionuclides in Natural and Semi-natural Environments*, ed. G. Desmet, P. Nassimbeni, and M. Belli, 477–484. London: Elsevier Applied Science.

Fritze, H., S. Niini, K. Mikkola, and A. Makinen. 1989. Soil microbial effects of a Cu–Ni smelter in southwestern Finland. *Biol. Fertil. Soils* 8:87–94.

Führer, E. 1990. Forest decline in central Europe: Additional aspects of its cause. *For. Ecol. Manage.* 37:249–257.

Gaare, E. 1990. Lichen content of radiocesium after the Chernobyl accident in mountains in southern Norway. In: *Transfer of Radionuclides in Natural and Semi-natural Environments*, ed. G. Desmet, P. Nassimbeni, and M. Belli, 492–501. London: Elsevier Applied Science.

Galamero, E., G. Lingua, G. Berta, and B. R. Glick. 2009. Beneficial role of plant growth promoting bacteria and arbuscular mycorrhizal fungi on plant responses to heavy metal stress. *Can. J. Microbiol.* 55:501–514.

Garcia, M. O., T. Ovasapyan, M. Greas, and K. K. Treseder. 2008. Mycorrhizal dynamics under elevated $CO_2$ and nitrogen fertilization in a warm temperate forest. *Plant Soil* 303:301–310.

Garrett, S. D. 1963. *Soil Fungi and Fertility*. Oxford, UK: Pergamon Press.

Gaudinski, J. B., S. E. Trumbore, E. A. Davidson, A. C. Cook, D. Markewitz, and D. D. Richter. 2001. The age of fine-root carbon in three forests of the eastern United States measured by radiocarbon. *Oecologia* 129:420–429.

Gauslaa, Y. and K. A. Solhaug. 2001. Fungal melanins as a sun screen for symbiotic green algae in the lichen *Lobaria pulmonaria*. *Oecologia* 126:462–471.

Gifford, R. M., D. J. Barrett, and J. L. Lutze. 2000. The effects of elevated [CO$_2$] on the C:N and C:P mass ratios of plant tissues. *Plant Soil* 224:1–14.

Gilbert, O. L. 1992. Lichen reinvasion with declining air pollution. In: *Bryophytes and Lichens in a Changing Environment*, ed. J. W. Bates and A. M. Farmer, Oxford: Clarendon Press.

Giovani, C., P. L. Nimis, and R. Padovani. 1990. Investigation of the performance of macromycetes as bioindicators of radioactive contamination. In: *Transfer of Radionuclides in Natural and Semi-natural Environments*, ed. G. Desmet, P. Nassimbeni, and M. Belli, 485–491. London: Elsevier Applied Science.

González-Guerrero, M., L. H. Melville, N. Ferrol, J. A. Lott, C. Azcón-Aguilar, and R. L. Peterson. 2008. Ultrastructural localization of heavy metals in the extraradical mycelium and spores of the arbuscular mycorrhizal fungus *Glomus intrarades*. *Can. J. Microbiol.* 54:103–110.

Gorisen, A. and M. F. Cotrufo. 2000. Decomposition of leaf and root tissue of three perennial grass species grown at two levels of atmospheric CO$_2$ and N supply. *Plant Soil* 224:75–84.

Gorisen, A. and Th. W. Kutper. 2000. Fungal species-specific response of ectomycorrhizal Scots pine (*Pinus sylvestris*) to elevated [CO$_2$]. *New Phytol.* 146:163–168.

Gray, S. N., J. Dighton, and D. H. Jennings. 1996. The physiology of basidiomycete linear organs: III. Uptake and translocation of radiocaesium within differentiated mycelia of *Armillaria* spp. growing in microcosms and in the field. *New Phytol.* 132:471–482.

Gray, S. N., J. Dighton, S. Olsson, and D. H. Jennings. 1995. Real-time measurement of uptake and translocation of 137Cs within mycelium of *Schizophyllum commune* Fr. by autoradiography followed by quantitative image analysis. *New Phytol.* 129:449–465.

Grodzinskaya, A. A., M. Berreck, S. P. Wasser, and K. Haselwandter. 1994. Radiocaesium in fungi: Accumulation pattern in the Kiev district of Ukraine including the Chernobyl zone. *Beih. Sydowia*:88–96.

Guillén, J., A. Baez, and A. Salas. 2012. Influence of alkali and alkaline earth elements on the uptake of radionuclides by *Pleurotus eryngii* fruit bodies. *Appl. Radiat. Isotopes* 70:650–655.

Guillette, O., A. Fraiture, and J. Lambinon. 1990a. Soil–fungi radiocaesium transfers in forest ecosystems. In *Transfer of Radionuclides in Natural and Semi-natural Environments*, edited by Desmet, G., P. Nassimbeni, and M. Belli. London: Elsevier Applied Science.

Guillette, O., R. Kirchmann, E. van Gelder, and C. Hurtgen. 1990b. Radionuclides fallout on lichens and mosses and their leaching by rain in a forest ecosystem. In: *Transfer of Radionuclides in Natural and Semi-natural Environments*, ed. G. Desmet, P. Nassimbeni, and M. Belli, 110–117. London: Elsevier Applied Science.

Gundersen, P., B. A. Emmett, O. J. Kjønaas, C. J. Koopmans, and A. Tietma. 1998. Impact of nitrogen deposition on nitrogen cycling in forests: A synthesis of NITREX data. *For. Ecol. Manage.* 101:37–55.

Hallingbäck, T. and O. Kellner. 1992. Effects of simulated nitrogen rich and acid rain on the nitrogen fixating lichen *Peltgeria aphthosa* (L.) Wild. *New Phytol.* 120:99–103.

Harms, H. 2011. Untapped potential: Exploiting fungi in bioremediation of hazardous chemicals. *Nat. Rev. Microbiol.* 9:177–192.

Harrison, A. F., P. A. Stevens, J. Dighton, C. Quarmby, A. L. Dickinson, H. E. Jones, and D. M. Howard. 1995. The critical load of nitrogen for Sitka spruce forests on stagnopodsols in Wales: Role of nutrient limitations. *For. Ecol. Manage.* 76:139–148.

Hartley, J., J. W. G. Cairney, P. Freestone, C. Woods, and A. A. Meharg. 1999a. The effects of multiple metal contamination on ectomycorrhizal Scots pine (*Pinus sylvestris*) seedlings. *Environ. Pollut.* 106:413–424.

Hartley, J., J. W. G. Cairney, and A. A. Meharg. 1999b. Cross-colonization of Scots pine (*Pinus sylvestris*) seedlings by the ectomycorrhizal fungus *Paxillus involutus* in the presence of inhibitory levels of Cd and Zn. *New Phytol.* 142:141–149.

Hartley, J., J. W. G. Cairney, and A. A. Meharg. 1997a. Do ectomycorrhizal fungi exhibit adaptive tolerance to potentially toxic metals in the environment? *Plant Soil* 189:303–319.

Hartley, J., J. W. G. Cairney, F. E. Sanders, and A. A. Meharg. 1997b. Toxic interactions of metal ions ($Cd^{2+}$, $Pb^{2+}$, $Zn^{2+}$ and $Sb^{3-}$) on *in vitro* biomass production of ectomycorrhizal fungi. *New Phytol.* 137:551–562.

Hartley-Whitaker, J., J. W. G. Cairney, and A. A. Meharg. 2000a. Sensitivity to Cd or Zn of host and symbiont of ectomycorrhizal *Pinus sylvestris* L. (Scots pine) seedlings. *Plant Soil* 218:31–42.

Hartley-Whitaker, J., J. W. G. Cairney, and A. A. Meharg. 2000b. Toxic effects of cadmium and zinc on ectomycorrhizal colonization of scots pine (*Pinus sylvestris* L.) from soil inoculum. *Environ. Toxicol. Chem.* 19:694–699.

Haselwandter, K. 1978. Accumulation of the radioactive nuclide $^{137}$Cs in fruitbodies of basidiomycetes. *Health Phys.* 34:713–715.

Haselwandter, K. and M. Berreck. 1994. Accumulation of radionuclides in fungi. In: *Metal Ions in Fungi*, ed. G. Winkelmann and D. R. Winge, 171–176. New York: Marcel Dekker.

Haselwandter, K., M. Bereck, and P. Brunner. 1988. Fungi as bioindicators of radiocaesium contamination. Pre- and post Chernobyl activities. *Trans. Br. Mycol. Soc.* 90:171–176.

Hashem, A. R. 1995. The role of mycorrhizal infection in the resistance of *Vaccinium macrocarpon* to manganese. *Mycorrhiza* 5:289–291.

Hättenschwiler, S., S. Bühler, and C. Körner. 1999. Quality, decomposition and isopod consumption of tree litter produced under elevated $CO_2$. *Oikos* 85:271–281.

Hauck, M. 2008. Susceptibility to acid precipitation contributes to the decline of the terricolous lichens *Cetraria aculeata* and *Cetraria islandica* in central Europe. *Environ. Pollut.* 152:731–735.

Heal, O. W. and A. D. Horrill. 1983. Terrestrial ecosystems: An ecological context for radionuclide research. In: *Ecological Aspects of Radionuclide Release*, ed. P. J. Coughtree, Oxford: Blackwell Scientific Publications.

He, X., Y. Lin, G. Han, and T. Ma. 2013. Litterfall interception by understory vegetation delayed litter decomposition in *Cinnamomum camphora* plantation forest. *Plant Soil* 372:207–219.

Heinrich, G., K. Oswald, and H. J. Muller. 1999. Lichens as monitors of radiocesum and radiostrontium in Austria. *J. Environ. Radioact.* 45:13–27.

Heneghan, L. and T. Bolger. 1996. Effect of components of 'acid rain' on the contribution of soil microarthropods to ecosystem function. *J. Appl. Ecol.* 33:1329–1344.

Horyna, J. 1991. Wild mushrooms—The most significant source of internal contamination. *Isotopenpraxis* 27:23–24.

Horyna, J. and Z. Randa. 1988. Uptake of Radiocesium and Alkali Metals by Mushrooms. *J. Radionuclide Chem.* 127:107–120.

Howard, D. M., P. J. A. Howard, and D. C. Howard. 1995. A Markov model projection of soil organic carbon stores following land use change. *J. Environ. Manage.* 45:287–303.

Hungate, B. A., C. H. III Jaeger, G. Gamara, F. S. III Chapin, and C. B. Field. 2000. Soil microbiota in two annual grasslands: Responses to elevated atmospheric $CO_2$. *Oecologia* 124:589–598.

Hur, J-S. and P-G. Kim. 2000. Investigations of lichen species as a biomonitor of atmospheric ozone in 'Blackwood' mountain, Korea. *J. Korean For. Sci.* 89:65–76.

Hüttermann, A. 1982. Fruhdiagnose von Immissionsschaden im Wurzelbereich von Waldbaumen. *Landesanst. Okol Landschaftsentw. Forstpl. Nordhein-Westfalen*: 26–31.

Hüttermann, A. 1985. The effects of acid deposition on the physiology of the forest ecosystem. *Experientia* 41:585–590.

IPCC 2014. Climate Change 2014 Synthesis Report www.ipcc.ch/.

Jackson, N. E., R. H. Miller, and R. E. Franklin. 1973. The influence of vesicular–arbuscular mycorrhizae on uptake of [90]Sr from soil by soybeans. *Soil Biol. Biochem.* 5:205–212.

Jansen, A. E. and J. Dighton. 1990. Effects of air pollutants on ectomycorrhizas. *CEC Air Pollution Research Report 30*, 30.

Jansen, A. E., J. Dighton, and A. H. M. Bresser. 1988. Ectomycorrhiza and acid rain. *CEC Air Pollution Research Report 12*, Brussels.

Jansen, A. E., C. Kamminga-Van Wijk, and R. H. Jongbloed. 1990. Acid rain and ectomycorrhiza of Douglas fir. In: *Abstracts of the Fourth International Mycological Congress*, ed. A Reisinger and A. Bresinsky, 128, Regensberg, Germany.

Jany, J-L., F. Martin, and J. Garbaye. 2003. Respiration activity of ectomycorrhizas from *Cenococcum geophilum* and *Lactarius* sp. in relation to soil pore water potential in five beech forests. *Plant Soil* 255:487–494.

Jansen, A. E. and H. F. Van Dobben. 1987. Is the decline of *Cantharellus cibarius* in the Netherlands due to air pollution? *Ambio* 16:211–213.

Jarvis, S., S. Woodward, I. J. Alexander, and A. F. S. Taylor. 2013. Regional scale gradients of climate and nitrogen deposition drive variation in ectomycorrhizal fungal communities associated with native Scots pine. *Global Change Biol.* 19:1688–1696.

Jastrow, J. D. 1996. Soil aggregate formation and the accrual of particulate and mineral-associated organic matter. *Soil Biol. Biochem.* 28:665–676.

Jastrow, J. D., R. M. Miller, and C. E. Owensby. 2000. Long-term effects of elevated atmospheric $CO_2$ on below-ground biomass and transformations to soil matter in grassland. *Plant Soil* 224:85–97.

Jennings, D. H. 1990. The ability of basidiomycete mycelium to move nutrients through the soil ecosystem. In: *Nutrient Cycling in Terrestrial Ecosystems: Field Methods, Applications and Interpretation*, ed. A. F. Harrison, P. Ineson, and O. W. Heal, Elsevier.

Johanson, K. J., R. Bergstrom, S. von Bothmer, and G. Karlen. 1990. Radiocaesium in wildlife of a forest ecosystem in central Sweden. In: *Transfer of Radionuclides in Natural and Semi-natural Environments*, ed. G. Desmet, P. Nassimbeni and M. Belli, 183–193. London: Elsevier Applied Science.

Johansson, M. 2000. The influence of ammonium nitrate on the root growth and ericoid mycorrhizal colonization of *Calluna vulgaris* (L.) Hull from a Danish heathland. *Oecologia* 123:418–424.

Johnson, D. W., W. Cheng, and J. T. Ball. 2000. Effects of $CO_2$ and N fertilization on decomposition and N immobilization in ponderosa pine litter. *Plant Soil* 224:115–122.

Johnson, N. C. 1998. Responses of *Salsola kali* and *Panicum virgatum* to mycorrhizal fungi, phosphorus and soil organic matter: Implications for reclamation. *J. Appl. Ecol.* 35:86–94.

Joner, E. J. and C. Leyval. 1997. Uptake of [109]Cd by roots and hyphae of a *Glomus mosseae/Trifolium subterraneum* mycorrhiza from soil amended with high and low concentrations of cadmium. *New Phytol.* 135:353–360.

Joner, E. J., C. Leyval, and J. V. Colpaert. 2006. Ectomycorrhizas impede phytoremediation of poly aromatic hydrocarbons (PAHs) both within and beyond the rhizosphere. *Environ. Pollut.* 142:34–38.

Jongbloed, R. H. and G. W. F. H. Borst Pauwels. 1988. Efects of $Al^{3+}$ and $NH_4^+$ on growth and uptake of $K^+$ and $H_2PO_4^-$ by three ectomycorrhizal fungi in pure culture. In: *Ectomycorrhiza and Acid Rain*, ed. Jansen, A. E., J. Dighton, and A. H. M. Bresser, 47–52. Bilthoven, The Netherlands: CEC.

Jongbloed, R. H. and G. W. F. H. Borst Pauwels. 1989. Effects of ammonium and pH on growth and potassium uptake by the ectomycorrhizal fungus *Laccaria bicolor* in pure culture. *Agric. Ecosyst. Environ.* 28:207–212.

Jonsson, L. 1998. Community Structure of Ectomycorrhizal Fungi in Swedish Boreal Forests. PhD Thesis, Swedish University of Agriculture, *Silvestria* 75, Uppsala, Sweden.

Jonsson, L., A. Dahlberg, M-C. Nilsson, O. Karen, and O. Zackrisson. 1999a. Continuity of ectomycorrhizal fungi in self-regulating boreal *Pinus sylvestris* forests studied by comparing mycobiont diversity on seedlings and mature trees. *New Phytol.* 142:151–162.

Jonsson, L., A. Dahlberg, M-C. Nilsson, O. Zackrisson, and O. Karen. 1999b. Ectomycorrhizal fungal communities in late-successional Swedish boreal forests, and their composition following wildfire. *Mol. Ecol.* 8:205–215.

Jordan, M. J. and M. P. Lechevalier. 1975. Effects of zinc smelter emissions on forest soil microflora. *Can. J. Microbiol.* 21:1855–1865.

Jumpponen, A. and K. L. Jones. 2014. Tallgrass prairie soil fungal communities are resilient to climate change. *Fung. Ecol.* 10:44–57.

Kandeler, E., D. Tscherko, R. D. Bardgett, P. J. Hobbs, C. Kampichler, and T. H. Jones. 1998. The response of soil microorganisms and roots to elevated $CO_2$ and temperature in a terrestrial model ecosystem. *Plant Soil* 202:251–262.

Kårén, O. and J-E. Nylund. 1997. Effects of ammonium sulphate on the community structure and biomass of ectomycorrhizal fungi in a Norway spruce stand in southwestern Sweden. *Can. J. Bot.* 75:1628–1642.

Kettlewell, H. B. D. 1955. Selection experiments on industrial melanism in the Lepidoptera. *Heredity* 10:287–301.

Khade, S. W. and A. Adholeya. 2007. Feasible bioremediation through arbuscular mycorrhizal fungi imparting heavy metal tolerance: A retrospective. *Bioremed. J.* 11:33–43.

Kieliszewska-Rokicka, B. 1992. Effect of nitrogen level on acid phosphatase activity of eight isolates of the ectomycorrhizal fungus *Paxillus involutus* cultured in vitro. *Plant Soil* 139:229–238.

Killham, K., M. K. Firestone, and J. G. McColl. 1983. Acid rain and soil microbial activity: Effects and their mechanisms. *J. Environ. Qual.* 12:133–137.

Klironomos, J. N., E. M. Bednarczuk, and J. Neville. 1999. Reproductive significance of feeding on saprobic and arbuscular mycorrhizal fungi by the collembolan *Folsomia candida*. *Funct. Ecol.* 13:756–761.

Klironomos, J. N. and W. B. Kendrick. 1996. Palatability of microfungi to soil arthropods in relation to the functioning of arbuscular mycorrhizae. *Biol. Fertil. Soils* 21:43–52.

Klironomos, J. N., M. C. Rillig, M. F. Allen, D. R. Zak, M. Kubiske, and K. S. Pregitzer. 1997. Soil fungal–arthropod responses to *Populus tremuloides* grown under enriched atmospheric $CO_2$ under field conditions. *Global Change Biol.* 3:473–478.

Klironomos, J. N. and M. Ursic. 1998. Density dependent grazing on the extraradical hyphal network of the arbuscular mycorrhizal fungus *Glomus intraradices* by the collembolan *Folsomia candida*. *Biol. Fertil. Soils* 26:250–253.

Klironomos, J. N., M. Ursic, M. Rillig, and M. F. Allen. 1998. Interspecific differences in the response of arbuscular mycorrhizal fungi to *Artimissia tridentata* grown under elevated atmospheric $CO_2$. *New Phytol.* 138:599–605.

Knicker, H. 2011. Soil organic N—An under-rated player for C sequestration in soils? *Soil Biol. Biochem.* 43:1118–1129.

Kong, F. X., Y. Liu, W. Hu, P. P. Shen, C. L. Zhou, and L. S. Wang. 2000. Biochemical responses of the mycorrhizae in *Pinus massoniana* to combined effects of Al, Ca and low pH. *Chemosphere* 40:311–318.

Koomen, I. and S. P. McGrath. 1990. Mycorrhizal infection of clover is delayed in soils contaminated with heavy metals from past sewage sludge applications. *Soil Biol. Biochem.* 22:871–873.

Kosta-Rick, R., U. S. Leffler, B. Markert, U. Herpin, M. Lusche, and J. Lehrke. 2001. Assessing the pollution impact on terrestrial ecosystems by plant and soil monitoring: Conception, implementation and assessment scales. *Umweltwiss. Schadst-Forsch.* 13:5–12.

Kottke, I. and F. Oberwinkler. 1990. Pathways of elements in ectomycorrhizae in respect to Hartig net development and endodermis differentiation. In: *Abstracts of the Fourth International Mycological Congress*, ed. A. Reisinger and A. Bresinsky. Regensberg, Germany.

Kottke, I., X. M. Quian, K. Pritsch, I. Haug, and F. Oberwinkler. 1998. *Xercomus badius–Picea abies*, an ectomycorrhiza of high activity and element storage capacity in acidic soils. *Mycorrhiza* 7:267–275.

Kuperman, R. G. and M. M. Carreiro. 1997. Soil heavy metal concentrations, microbial biomass and enzyme activities in a contaminated grassland ecosystem. *Soil Biol. Biochem.* 29:179–190.

Kuyper, Th. W. 1989. Effects of forest fertilization on the mycoflora. *Beitr. Kenntis Pilze Mitteleurpas.* 5:5–20.

Kytöviita, M.-M. and P. D. Crittenden. 1994. Effects of simulated acid rain on nitrogenase activity (acetylene reduction) in the lichen *Sterocaulon paschale* (L.) Hoffm., with special reference to nutritional aspects. *New Phytol.* 128:263–271.

Ladeyn, I., C. Plassard, and S. Staunton. 2008. Mycorrhizal association of maritime pine, *Pinus pinaster*, with *Rhizopogon roseolus* has contrasting effects on the uptake from soil and root-to-shoot transfer of $^{137}$Cs, $^{85}$Sr and $^{95m}$Tc. *J. Environ. Radioact.* 99:853–863.

Lanfranco, L. 2007. The fine tuning of heavy metals in mycorrhizal fungi. *New Phytol.* 174:3–6.

Langley, J. A., B. G. Drake, and B. A. Hungate. 2002. Extensive belowground carbon storage supports roots and mycorrhizae in regenerating scrub oaks. *Oecologia* 131:542–548.

Leake, J. R. and D. J. Read. 1989. The biology of mycorrhiza in the Ericaceae: XIII. Some characteristics of the extracellular proteinase activity of the ericoid endophyte *Hymenoscyphus ericae*. *New Phytol.* 112:69–76.

Leake, J. R. and D. J. Read. 1990. Proteinase activity in mycorrhizal fungi: I. The effect of extracellular pH on the production and activity of proteinase by the ericoid endophytes of soils of contrasted pH. *New Phytol.* 115:243–250.

Leyval, C., K. Turnau, and K. Hasselwandter. 1997. Effect of heavy metal pollution on mycorrhizal colonization and function: Physiological, ecological and applied aspects. *Mycorrhiza* 7:139–153.

Li, T., G. Lin, X. Zhang, Y. Chen, S. Zhang, and B. Chen. 2014. Relative importance of an arbuscular mycorrhizal fungus (*Rhizophagus intraradicis*) and root hairs in plant drought tolerance. *Mycorrhiza* 24:595–602.

Liss, B., H. Blaschke, and P. Schutt. 1984. Verleichende Feinwurzeluntersuchungen an gesunden und erkrankter Altfichten auf zwei Standorten in Bayern—ein Beitrag zur Waldsterbenforschung. *Eur. J. For. Pathol.* 14:90–102.

Lomborg, B. 2001. *The Skeptical Environmentalist: Measuring the Real State of the World.* Cambridge, UK: Cambridge University Press.

Lundquist, E. J., L. E. Jackson, K. M. Scow, and C. Hsu. 1999. Changes in microbial biomass and community composition, and soil carbon and nitrogen pools after incorporation of rye into three California agricultural soils. *Soil Biol. Biochem.* 31:221–236.

Luo, Z-H., K-L. Pang, Y-R. Wu, J-D. Gu, R. K. K. Chaow, and L. L. P. Vrijmoed. 2012. Degradation of phthalate esters by *Fusarium* sp. DMT-5-3 and *Trichosporon* sp. DMI-5-1 isolated from mangrove sediments. In: *Biology of Marine Fungi*, ed. C. Raghukumar, 299–328. Berlin: Springer-Verlag.

Magan, N., I. A. Kirkwood, A. R. McLeod, and M. K. Smith. 1995. Effect of open air fumigation with sulphur dioxide and ozone on phyllosphere and endophytic fungi of conifer needles. *Plant Cell Environ.* 18:291–302.

Magan, N. and A. R. McLeod. 1991. Effect of open air fumigation with sulphur dioxide on the occurrence of phylloplane fungi on winter barley. *Agric. Ecosyst. Environ.* 33:245–261.

Maltby, L. and R. Booth. 1991. The effect of coal-mine effluent on fungal assemblages and leaf breakdown. *Water Res.* 25:247–250.

Markkola, A. M. and R. Ohtonen. 1988. The effect of acid deposition on fungi in forest humus. In: *Ectomycorrhiza and Acid Rain*, ed. A. E. Jansen, J. Dighton, and A. H. M. Bresser, 122–126. Bilthoven, The Netherlands: CEC.

Marrs, R. H. and P. Bannister. 1978. The adaptation of *Calluna vulgaris* (L.) Hull. to contrasting soil types. *New Phytol.* 81:753–761.

Martino, E., J. D. Coisson, I. Lacourt, F. Favaron, P. Bonfante, and S. Perotto. 2000a. Influence of heavy metals on production and activity of pectolytic enzymes in ericoid mycorrhizal fungi. *Mycol. Res.* 104:825–833.

Martino, E., K. Turnau, M. Girlanda, P. Bonfante, and S. Perotta. 2000b. Ericoid mycorrhizal fungi from heavy metal polluted soils: Their identification and growth in the presence of zinc ions. *Mycol. Res.* 104:338–344.

Marx, D. H. 1975. Mycorrhiza and establishment of trees on strip-mined land. *Ohio J. Sci.* 75:288–297.

Marx, D. H. 1980. Role of mycorrhizae in forestation of surface mines. Paper presented at Trees for Reclamation. Reprint.

McKnight, K. B., K. H. McKnight, and K. T. Harper. 1990. Cation exchange capacities and mineral element concentrations of macrofungal stipe tissue. *Mycologia* 82:91–98.

McLeod, A. R. 1995. An open-air system for exposure of young forest trees to sulphur dioxide and ozone. *Plant Cell Environ.* 18:215–225.

McLeod, A. R., P. J. A. Shaw, and M. R. Holland. 1992. The Liphook forest fumigation project: Studies of sulphur dioxide and ozone effects on coniferous trees. *For. Ecol. Manage.* 51:121–127.

McMurtrie, R. E., R. C. Dewar, B. E. Medlyn, and M. P. Jeffreys. 2000. Effects of elevated [$CO_2$] on forest growth and carbon storage: A modelling analysis of the consequences of change in litter quality/quantity and root exudation. *Plant Soil* 224:135–152.

McNulty, S. G. and J. D. Aber. 1993. Effects of chronic nitrogen additions on nitrogen cycling in a high-elevation spruce-fir stand. *Can. J. For. Res.* 23:1252–1263.

Meadows, D. H., D. L. Meadows, and J. Randers. 1992. *Beyond the Limits: Confronting Global Collapse, Envisioning a Sustainable Future.* White River Junction, VT: Chelsea Green Publishing Co.

Meyer, F. H. 1987. Das Wurzelsystem geschadigter Waldbestande. *Allg. Forst Zeitscr.* 27/28/29:754–757.

Mietelski, J. W., M. Jasinska, B. Kubica, K. Kozak, and P. Macharski. 1994. Radioactive contamination of Polish mushrooms. *Sci. Total Environ.* 157:217–226.

Mironenko, N. V., I. A. Alekhina, N. N. Zhdanova, and S. A. Bulat. 2000. Intraspecific variation in gamma-radiation resistance and genomic structure in the filamentous fungus *Alternaria alternata*: A case study of strains inhabiting Chernobyl Reactor No. 4. *Ecotoxicol. Environ. Saf.* 45:177–187.

Mohan, J. E., C. C. Cowden, P. Baas et al. 2014. Mycorrhizal fungi mediation of terrestrial ecosystem responses to global change: Mini-review. *Fung. Ecol.* 10:3–19.

Morley, G. F., J. A. Sayer, S. C. Wilkinson, M. M. Gharieb, and G. M. Gadd. 1996. Fungal sequestration, mobilization and transformation of metals and metalloids. In: *Fungi and Environmental Change*, ed. J. C. Frankland, N. Magan, and G. M. Gadd, 235–256. Cambridge: Cambridge University Press.

Muramatsu, Y., S. Yoshida, and M. Sumia. 1991. Concentrations of radiocesium and potassium in basidiomycetes collected in Japan. *Sci. Total Environ.* 105:29–39.

Nakane, K., M. Yamamoto, and H. Tsuboto. 1983. Estimation of root respiration in a mature forest ecosystem. *Jpn. J. Ecol.* 33:397–408.

Newell, S. Y. and J. W. Wall. 1998. Response of saltmarsh fungi to the presence of mercury and polychlorinated biphenyls at a Superfund site. *Mycologia* 90:777–784.

Newsham, K. K., J. C. Frankland, L. Boddy, and P. Ineson. 1992a. Effects of dry-deposited sulphur dioxide on fungal decomposition of angiosperm tree leaf litter I. Changes in communities of fungal saprotrophs. *New Phytol.* 122:97–116.

Newsham, K. K., P. Ineson, L. Boddy, and J. C. Frankland. 1992b. Effects of dry deposited sulphur dioxide on fungal decomposition of angiosperm leaf litter: II. Chemical content of leaf litters. *New Phytol.* 122:111–125.

Nordgren, A., T. Kauri, E. Baath, and B. Soderstrom. 1986. Soil microbial activity, mycelial lengths and physiological groups of bacteria in a heavy metal polluted area. *Environ. Pollut. Ser. A* 41:89–100.

Nylund, J-E. 1989. Nitrogen, carbohydrate and ectomycorrhiza—The classical theories crumble. *Agriculture, Ecosystems and Environment* 28:361–364.

Nylund, J-E. and H. Wallander. 1989. Effects of ectomycorrhiza on host growth and carbon balance in a semi-hydroponic cultivation system. *New Phytol.* 112:389–398.

O'Neill, E. G. 1994. Responses of soil biota to elevated atmospheric carbon dioxide. *Plant Soil* 165:55–65.

O'Neill, E. G., R. J. Luxmoore, and R. J. Norby. 1987. Increases in mycorrhizal colonization and seedling growth in *Pinus echinata* and *Quercus alba* in an enriched $CO_2$ atmosphere. *Can. J. For. Res.* 17:878–883.

Ochoa-Hueso, R., I. Rocha, C. J. Stevens, E. Manrique, and M. J. Luciañez. 2014. Simulated nitrogen deposition affects soil fauna from a semiarid Mediterranean ecosystem in central Spain. *Biol. Fertil. Soils* 50:191–196.

Odum, E. P. 1984. The mesocosm. *BioScience* 34:558–562.

Olsen, R. A., E. Joner, and L. R. Bakken. 1990. Soil fungi and the fate of radiocaesium in the soil ecosystem—A discussion of possible mechanisms involved in the radiocaesium accumulation in fungi, and the role of fungi as a Cs-sink in the soil. In: *Transfer of Radionuclides in Natural and Semi-natural Environments*, ed. G. Desmet, P. Nassimbeni, and M. Belli, 657–663. London: Elsevier Applied Science.

Olsson, S. 1995. Mycelial density profiles of fungi on heterogenous media and their interpretation in terms of nutrient reallocation patterns. *Mycol. Res.* 99:143–153.

Olsson, S. and D. H. Jennings. 1991. Evidence for diffusion being the mechanism of translocation in the hyphae of three moulds. *Exp. Mycol.* 15:302–309.

Oren, R., E. D. Schulze, K. S. Werk, J. Meyer, B. U. Schneider, and H. Heilmeir. 1988. Performance of two *Picea abies* (L.) Karst. stands at different stages of decline: I. Carbon relations and stand growth. *Oecologia* 1877:25–37.

Orlowska, E., B. Godzik, and K. Turnau. 2012. Effect of different arbuscular mycorrhizal fungal isolates on growth and arsenic accumulation in *Plantago lanceolata* L. *Environ. Pollut.* 168:121–130.

Orlowska, E., W. Przybylowicz, D. Orlowski, N. P. Mongwaketsi, K. Turnau, and J. Mesjasz-Przybyłowicz. et al. 2013. Mycorrhizal colonization affects the elemental distribution in roots of Ni-hyperaccumulator *Berkheya coddii* Roessler. *Environ. Pollut.* 175:100–109.

Orlowska, E., W. Przybylowicz, D. Orlowski, K. Turnau, and J. Mesjasz-Przylowicz. 2011. The effect of mycorrhiza on the growth and elemental composition on Ni-hyperaccumulating plant *Berkheya coddii* Roessler. *Environ. Pollut.* 159:3730–3738.

Oughton, D. H. 1989. The environmental chemistry of radiocaesium and other nuclides. PhD thesis, University of Manchester.

Oulehle, F., J. Hofmeister, P. Cudlin, and J. Hruška. 2006. The effect or reduced atmospheric deposition on soil and soil solution chemistry at a site subjected to long-term acidification, Načetín, Czech Republic. *Sci. Total Environ.* 370:532–544.

Pacala, S. W., G. C. Hurtt, D. Baker et al. 2001. Consistent land- and atmospheric-based US carbon sink estimates. *Science* 292:2316–2319.

Parekh, N. R., J. M. Poskitt, B. A. Dodd, E. D. Potter, and A. Sanchez. 2008. Soil microorganisms determine the sorption of radionuclides within organic soil systems. *J. Environ. Radioact.* 99:841–852.

Phillips, R. P., I. C. Meier, E. S. Bernhard, A. S. Grandy, K. Wickings, and A. C. Finzi 2012. Roots and fungi accelerate carbon and nitrogen cycling in forests exposed to elevated $CO_2$. *Ecol. Lett.* 15:1042–1049.

Potin, O., C. Rafin, and E. Veignie. 2004. Bioremediation of an aged polycyclic aromatic hydrocarbons (PAHs)-contaminated soil by filamentous fungi isolated from the soil. *Int. Biodeter. Biodegr.* 54:45–52.

Prenafeta-Boldu, F. X., A. Kuhn, D. M. A. M. Luykx, H. Anke, J. W. van Groenstijn, and J. A. M. de Bont. 2001. Isolation and characterization of fungi growing on volatile aromatic hydrocarbons as their sole source of carbon and energy source. *Mycol. Res.* 105:477–484.

Prescott, C. E. and D. Parkinson. 1985. Effects of sulphur pollution on rates of litter decomposition in a pine forest. *Can. J. Bot.* 63:1436–1443.

Qui, Z., A. H. Chappelka, Somers, G. L., B. G. Lockaby, and R. S. Meldahl. 1993. Effects of ozone and simulated acidic precipitation on ectomycorrhizal formation on loblolly pine seedlings. *Environ. Exp. Bot.* 33:423–431.

Randa, Z. and J. Benada. 1990. Mushrooms—Significant source of internal contamination by radiocaesium. In: *Transfer of Radionuclides in Natural and Semi-natural Environments*, ed. G. Desmet, P. Nassimbeni, and M. Belli, 169–178. London: Elsevier Applied Science.

Raskin, I. and B. D. Ensley (Eds). 2000. *Phytoremediation of Toxic Metals: Using Plants to Clean Up the Environment.* New York: John Wiley and Sons, Inc.

Read, D. J. 1991. Mycorrhizas in ecosystems—Nature's response to the "Law of the Minimum". In: *Frontiers in Mycology*, ed. D. L. Hawksworth, Wallingford, U. K.: CAB International.

Reynolds, B., E. J. Wilson, and B. A. Emmett. 1998. Evaluating critical loads of nutrient nitrogen and acidity for terrestrial systems using ecosystem-scale experiments (NITREX). *For. Ecol. Manage.* 101:81–94.

Richardson, D. H. S. 1988. Understanding the pollution sensitivity of lichens. *Bot. J. Linn. Soc.* 96:31–43.

Rigas, F., V. Drista, R. Marchant, K. Papadopoulou, E. J. Avramides, and I. Hatzianestis. 2005. Biodegradation of lindane by *Pleurotus ostreatus* via central composite design. *Environ. Int.* 31:191–196.

Rillig, M. C. and M. F. Allen. 1999. What is the role of arbuscular mycorrhizal fungi in plant-to-ecosystem responses to elevated atmospheric $CO_2$? *Mycorrhiza* 9:1–8.

Rillig, M. C., G. Y. Hernandez, and P. C. D. Newton. 2000. Arbuscular mycorrhizae respond to elevated atmospheric $CO_2$ after long-term exposure: Evidence from a $CO_2$ spring in New Zealand supports the resource balance model. *Ecol. Lett.* 3:475–478.

Rizzo, D. M., R. A. Blanchette, and M. A. Palmer. 1992. Biosorption of metals by *Armillaria* rhizomorphs. *Can. J. Bot.* 70:1515–1520.

Romeralo, C., J. J. Diez, and N. F. Santiago. 2012. Presence of fungi in Scots pine needles found to correlate with air quality as measured by bioindicators in northern Spain. *For. Pathol.* 42:443–453.

Rommelt, R., L. Hiersche, G. Schaller, and E. Wirth. 1990. Influence of soil fungi (Basidiomycetes) on the migration of $Cs^{134 + 137}$ and $Sr^{90}$ in coniferous forest soils. In: *Transfer of Radionuclides in Natural and Semi-natural Environments*, ed. G. Desmet, P. Nassimbeni, and M. Belli, 152–160. London: Elsevier Applied Science.

Rozpądek, P., K. Wężowicz, A. Stojakowska et al. 2014. Mycorrhizal fungi modulate phytochemical production and antioxidant activity of *Cichorium intybus* L. (*Asteraceae*) under metal toxicity. *Chemosphere* 112:217–224.

Ruark, G. A., F. C. Thornton, A. E. Tiarks, B. G. Lockarby, A. H. Chappelka, and R. S. Meldahl. 1991. Exposing loblolly pine seedlings to acid precipitation and ozone: Effects on soil rhizosphere chemistry. *J. Environ. Qual.* 20:828–832.

Ruess, L., W. Funke, and A. Breunig. 1993. Influence of experimental acidification on nematodes, bacteria and fungi: Soil microcosms and field experiments. *Zool. Jahrb. Syst.* 120:189–199.

Ruess, L., P. Sandbach, P. Cudlin, J. Dighton, and A. Crossley. 1996. Acid deposition in a spruce forest soil: Effects on nematodes, mycorrhizas and fungal biomass. *Pedobiologia* 40:51–66.

Rühling, A. E. Bååth, A. Nordgren, and B. Söderström. 1984. Fungi in metal-contaminated soil near the Gusum brass mill, Sweden. *Ambio* 13:34–36.

Rühling, A. and B. Söderström. 1990. Changes in fruitbody production of mycorrhizal and litter decomposing macromycetes in heavy metal polluted coniferous forests in North Sweden. *Water Air Soil Pollut.* 49:375–387.

Rühling, A. and G. Tyler. 1991. Effects of simulated nitrogen deposition to the forest floor on the macrofungal flora of a beech forest. *Ambio* 20:261–263.

Ryan, D. R., W. D. Leukes, and S. G. Burton. 2005. Fungal bioremediation of phenolic wastewaters in an airlift reactor. *Biotechnol. Prog.* 21:1068–1074.

Rygiewicz, P. T., K. J. Martin, and A. R. Tuininga. 2000. Morphotype community structure of ectomycorrhizas on Douglas fir (*Pseudotsuga menziesii* Mirb. Franco) seedlings grown under elevated atmospheric $CO_2$ and temperature. *Oecologia* 124:299–308.

Sanders, I. R., R. Strietwolf-Engel, M. G. A. van der Heijden, T. Boller, and A. Wiemken. 1998. Increased allocation to external hyphae of arbuscular mycorrhizal fungi under $CO_2$ enrichment. *New Phytol.* 117:496–503.

Sasted, S. M. and H. B. Jenssen. 1993. Interpretation of regional differences in the fungal biota as effects of atmospheric pollution. *Mycol. Res.* 97:1451–1458.

Saxena, P. and A. K. Bhattacharyya. 2006. Nickel tolerance and accumulation by filamentous fungi from sludge of metal finishing industry. *Geomicrobiol.* 23:333–340.

Schier, G. A. 1985. Response of red spruce and balsam fir seedlings to aluminium toxicity in nutrient solutions. *Can. J. For. Res.* 15:29–33.

Schindler, F. V., E. J. Mercer, and J. A. Rice. 2007. Chemical characteristics of glomalin-related soil protein (GRSP) extracted from soils of varying organic matter content. *Soil Biol. Biochem.* 39:320–329.

Seegmüller, S. and H. Rennenberg. 1994. Interactive effects of mycorrhization and elevated carbon dioxide on growth of young pedunculate oak (*Quercus robur* L.) trees. *Plant Soil* 167:325–329.

Senior, E, J. E. Smith, I. A. Watson-Craik, and J. E. Tosh. 1993. Ectomycorrhizae and landfill site reclamations: Fungal selection criteria. *Lett. Appl. Microbiol.* 16:142–146.

Shaw, P. J. A. 1988. A consistent hierarchy in the fungal feeding preferences of the Collembola *Onychiurus armatus*. *Pedobiologia* 31:179–187.

Shaw, P. J. A. 1992. Fungi, fungivores, and fungal food webs. In: *The Fungal Community: Its Organization and Role in the Ecosystem*, ed. G. C. Carrol and D. T. Wicklow, 295–310. New York: Marcel Dekker.

Shaw, P. J. A. 1996. Influences of acid mist and ozone on the fluorescein diacetate activity of leaf litter. In: *Fungi and Environmental Change*, ed. J. C. Frankland, N. Magan, and G. M. Gadd, 102–108, Cambridge University Press.

Shaw, P. J. A., J. Dighton, and J. M. Poskitt. 1993. Studies on the mycorrhizal community infecting trees in the Liphook forest fumigation experiment. *Agric. Ecosyst. Environ.* 47:185–191.

Shaw, P. J. A., J. Dighton, J. M. Poskitt, and A. R. McLeod. 1992. The effects of sulphur dioxide and ozone on the mycorrhizas of Scots pine and Norway spruce in a field fumigation system. *Mycol. Res.* 96:785–791.

Shaw, P. J. A. and J. P. N. Johnston. 1993. Effects of $SO_2$ and $O_3$ on the chemistry and FDA activity of coniferous leaf litter in an open air fumigation experiment. *Soil Biol. Biochem.* 5:897–908.

Shcheglov, A., O. Tsetnova, and A. Klyashtorin. 2014. Biogeochemical cycles of Chernobyl-born radionuclides in the contaminated forest ecosystems. Long-term dynamics of the migration process. *J. Geochem. Explor.* 144:260–266.

Shirtcliffe, N. J., F. B. Pyatt, M. I. Newton, and G. McHale. 2006. A lichen protected by a super-hydrophobic breathable structure. *Plant Physiol.* 163:1193–1197.

Silver, C. S. and R. S. DeFries. 1990. *One Earth, One Future: Our Changing Global Environment*. Washington, D.C.: National Academy Press.

Simard, S. W., K. J. Beiler, M. A. Bingham, J. R. Deslippe, L. J. Philip, and F. P. Teste. 2012. Mycorrhizal networks: Mechanisms, ecology and modelling. *Fung. Biol. Rev.* 26:39–60.

Simard, S. W., M. D. Jones, D. M. Durall, D. A. Perry, D. D. Myrold, and R. Molina. 1997a. Reciprocal transfer of carbon isotopes between ectomycorrhizal *Betula payrifrea* and *Pseudotsuga menziesii*. *New Phytol.* 137:529–542.

Simard, S. W., D. A. Perry, M. D. Jones, D. D. Myrold, D. M. Durall, and R. Molina. 1997b. Net transfer of carbon between ectomycorrhizal tree species in the field. *Nature* 338:579–582.

Singleton, I. and J. M. Tobin. 1996. Fungal interactions with metals and radionuclides for environmental bioremediation. In: *Fungi and Environmental Change*, ed. J. C. Frankland, N. Magan and G. M. Gadd, 282–298. Cambridge: Cambridge University Press.

Sinsabaugh, R. L., R. K. Antibus, A. E. Linkins, and C. A. McClaughery. 1993. Wood decomposition: Nitrogen and phosphorus dynamics in relation to extracellular enzyme activity. *Ecology.* 74(5):1586–1593.

Skeffington, R. A. and K. A. Brown. 1986. The effect of five years of acid treatment on leaching, soil chemistry and weathering of a humo-ferric podslol. *Water Air Soil Pollut.* 31:981–990.

Skladany, G. J. and B Metting F. 1992. Bioremediation of contaminated soil. In *Soil Microbial Ecology: Applications in Agriculture and Environmental Management*, ed. F. B. Metting, 483–513. New York: Marcel Dekker.

Sobotka, A. 1964. Effects of industrial exhalations on soil biology of Norway spruce stands in the Ore mountains. *Lesnicky Casopis* 37:987–1002.

Staddon, P. L., A. H. Fitter, and D. Robinson. 1999. Effects of mycorrhizal colonization and elevated atmospheric carbon dioxide on carbon fixation and below-ground carbon partitioning in *Plantago lanceolata. J. Exp. Bot.* 50:853–860.

Steiner, M., I. Linkov, and S. Yoshida. 2002. The role of fungi in the transfer and cycling of radionuclides in forest ecosystems. *J. Environ. Radioactivity.* 58:217–241.

Stribley, D. P. and D. J. Read. 1980. The biology of mycorrhiza in the Ericaceae: VII. The relationship between mycorrhizal infection and the capacity to utilize simple and complex organic nitrogen sources. *New Phytol.* 86:365–371.

Stroo, H. F. and M. Alexander. 1985. Effect of simulated acid rain on mycorrhizal infection of *Pinus strobus* L. *Water Air Soil Pollut.* 25:107–114.

Tam, P. C. F. 1995. Heavy metal tolerance by ectomycorrhizal fungi and metal amelioration by *Pisolithus tinctorius. Mycorrhiza* 5:181–188.

Termorshuizen, A. J. and A. P. Schaffers. 1987. Occurrence of carpophores of ectomycorrhizal fungi in selected stands of *Pinus sylvestris* L. in the Netherlands in relation to stand vitality and air pollution. *Plant Soil* 104:209–217.

Termorshuizen, A. J. and A. P. Schaffers. 1989. The relation in the field between fruitbodies of mycorrhizal fungi and their mycorrhizas. *Agric. Ecosyst. Environ.* 28:509–512.

Termorshuizen, A. J. and A. P. Schaffers. 1991. The decline of carpophores of ectomycorrhizal fungi in stands of *Pinus sylvestris* L. in the Netherlands: Possible causes. *Nova Hedwigia* 53:267–289.

Thompson, G. W. and R. J. Medve. 1984. Effects of aluminum and manganese on the growth of ectomycorrhizal fungi. *Appl. Environ. Microbiol.* 48:556–560.

Tietema, A., L. Riemer, J. M. Verstraten, M. P. van der Maas, A. J. van Wijk, and I. van Voorthuyzen. 1993. Nitrogen cycling in acid forest soils subject to increased atmospheric nitrogen input. *For. Ecol. Manage.* 57:29–44.

Tilman, D., D. Wedin, and J. Knops. 1996. Productivity and sustainability influenced by biodiversity in grassland ecosystems. *Nature* 379:718–720.

Tobin J.M., D. G. Cooper, and R. J. Neufeld 1984. Uptake of metal ions by *Rhizopus arrhizus* biomass. *Appl. Environ. Microbiol.* 47 4:821–824.

Tonín, C., P. Vandenkoornhuyse, E. J. Joner, J. Straczek, and C. Leyval. 2001. Assessment of arbuscular mycorrhizal fungi diversity in the rhizosphere of *Viola calaminaria* and effect of these fungi on heavy metal uptake by clover. *Mycorrhiza* 10:161–168.

Treseder, K. K. 2004. A meta-analysis of mycorrhizal responses to nitrogen, phosphorus, and atmospheric $CO_2$ in field studies. *New Phytol.* 164:347–355

Treseder K. K. and M. F. Allen. 2000. Mycorrhizal fungi have a potential role in soil carbon storage under elevated $CO_2$ and nitrogen deposition. *New Phytol.* 147:189–200.

Tugay, T., N. N. Zhdanova, V. Zheltonozhsky, L. Sadovnikov, and J. Dighton. 2006. The influence of ionizing radiation on spore germination and emergent hyphal growth response reactions of microfungi. *Mycologia* 98:521–527.

Turneau, K., I. Kottke, J. Dexheimer, and B. Botton. 1994. Element distribution in *Pisolithus tinctorius* mycelium treated with cadmium dust. *Ann. Bot.* 74:137–142.

Turneau, K., I. Kottke, and F. Oberwinkler. 1993. *Paxillus involutus–Pinus sylvestris* mycorrhizae from heavily polluted forest. *Bot. Acta* 106:213–219.

Tyler, G., D. Berggren, B. Bergkvist, U. Falkengren-Grerup, L. Folkeson, and A. Ruhling. 1987. Soil acidification and metal solubility in forests of southern Sweden. In: *Effects of Pollutants on Forests, Wetlands and Agricultural Ecosystems*, ed. T. C. Hutchinson and K. M. Meema, 347–359. Berlin: Springer Verlag.

Ulrich, B., R. Mayer, and P. K. Khanna. 1979. Deposition von Luftverunreinigungen und ihre Auswirkungen in Waldökosystemen im Solling. J. D. Sauerlanders Verlag, Frankfurt, Schriften aus der Forstlichen Fakultät der Üniversität Göttingen. 58.

Van Breemen, N. and H. F. G. Van Dijk. 1988. Ecosystem effects of atmospheric deposition of nitrogen in the Netherlands. *Environ. Pollut.* 54:249–274.

van der Heijden, M. G. A., T. Boller, A. Wiemken, and I. R. Sanders. 1998. Different arbuscular mycorrhizal fungal species are potential determinants of plant community structure. *Ecology* 79:2082–2091.

Veerkamp, M. T., B. W. L. De Vries, and Th. W. Kuyper. 1997. Shifts in species composition of lignicloous macromycetes after application of lime in a pine forest. *Mycol. Res.* 101:1251–1256.

Vinichuk, M., A. Mårtensson, T. Ericsson, and K. Rosén. 2013a. Effect of arbuscular mycorrhizal (AM) fungi on [137]Cs uptake by plants on different soils. *J. Environ. Radioact.* 115:151–156.

Vinichuk, M., A. Mårtensson, and K. Rosén. 2013b. Inoculation with arbuscular mycorrhizae does not improve [137]Cs uptake in crops grown in the Chernobyl region. *J. Environ. Radioact.* 126:14–19.

Vinichuk, M., K. Rosén, and A. Dahlberg. 2013c. [137]Cs in fungal sporocarps in relation to vegetation in a bog, pine swamp and forest along a transect. *J. Environ. Radioact.* 90:713–720.

Vodnik, D., A. R. Byrne, and N. Gogala. 1998. The uptake and transport of lead in some ectomycorrhizal fungi in culture. *Mycol. Res.* 102:953–958.

Vogt K. 1991. Carbon budgets of temperate forest ecosystems. *Tree Physiol.* 9:69–86.

Vogt, K. A., C. C. Grier, R. L. Edmonds, and C. E. Meier. 1982. Mycorrhizal role in net primary production and nutrient cycling in *Abies amabilis* (Dougl.) Forbes ecosystems in western Washington. *Ecology* 63:370–380.

Vogt, K. A., D. A. Publicover, J. Bloomfiled, J. M. Perez, D. J. Vogt, and W. L. Silver. 1993. Belowground responses as indicators of environmental change. *Environ. Exp. Bot.* 33:189–205.

Wainwright, M. and G. M. Gadd. 1997. Fungi and industrial pollutants. In: *The Mycota IV Environmental and Microbial Relationships*, ed. D. T. Wicklow and B. Soderstrom, 85–97. Berlin: Springer-Verlag.

Wallander, H., K. Arnebrant, and A. Dahlberg. 1999. Relationships between fungal uptake of ammonium, fungal growth and nitrogen availability in ectomycorrhizal *Pinus sylvestris* seedlings. *Mycorrhiza* 8:215–223.

Wang, G. M., D. C. Coleman, D. W. Freckman, M. I. Dyer, S. J. McNaughton, M. A. Acra, and J. D. Goeschl. 1989. Carbon partitioning patterns of mycorrhizal versus non-mycorrhizal plants: Real-time dynamic measurements using 11CO2. *New Phytol.* 112:489–493.

Wang, S., N. Nomura, T. Nakajima, and H. Uchiyama. 2012. Case study of the relationship between fungi and bacteria associated with high-molecular-weight polycyclic aromatic hydrocarbon degradation. *J. Biosci. Bioeng.* 113:624–630.

Watling, R. 1999. Launch of the UK Biodiversity Action Plan for Lower Plants. *Mycologist* 13:158.

Watling, R. 2005. Fungal conservation: Some impression—A personal view. In: *The Fungal Community: Its Organization and Role in the Ecosystem*. 3rd Edition, ed. J. Dighton, J. F. White, and P. Oudemans 881–896, Baton Rouge, CRC Press.

Weber, R. W. S. and H. T. Tribe. 2003. Oil as a substrate for *Mortierella* species. *Mycologist* 17:134–139.

Węgrzyn, M., M. Lisowska, and P. Nicia. 2013. The value of the terricolous lichen *Cetrariella delisei* in the biomonitoring of heavy-metal levels in Svalbard. *Polish Polar Res.* 34:375–382.

White, C. and G. M. Gadd. 1990. Biosorption of radionuclides by fungal biomass. *J. Chem. Technol. Biotechnol.* 49:331–343.

Witkamp, M. 1968. Accumulation of $^{137}$Cs by *Trichoderma viride* relative to $^{137}$Cs in soil organic matter and soil solution. *Soil Sci* 106:309–311.

Witkamp, M. and B. Barzansky. 1968. Microbial immobilization of $^{137}$Cs in forest litter. *Oikos* 19:392–395.

Wolf, J., N. R. O'Neill, A. Rogers, M. L. Muilenberg, and L. H. Ziska. 2010. atmospheric carbon dioxide concentrations amplify *Alternaria alternata* sporulation and total antigen production. *Environ. Health Perspect.* 118:1223–1228.

Wookey, P. A. and P. Ineson. 1991a. Chemical changes in decomposing forest litter in response to atmospheric sulphur dioxide. *J. Soil Sci.* 42:615–628.

Wookey, P. A. and P. Inseon. 1991b. Combined use of open air and indoor fumigation systems to study effects of $SO_2$ on leaching processes in Scots pine litter. *Environ. Pollut.* 74:325–343.

Wookey, P. A., P. Ineson, and T. A. Mansfield. 1991. Effects of atmospheric sulphur dioxide on microbial activity in decomposing forest litter. *Agric. Ecosyst. Environ.* 33:263–280.

Wright, S. F. and A. Upadhyaya. 1996. Extraction of an abundant and unusual protein from soil and comparison with hyphal protein from arbuscular mycorrhizal fungi. *Soil Sci.* 161:575–586.

Wu, Q.-S., Y-N. Zou, and Y-M. Huang. 2013. The arbuscular mycorrhizal fungus *Diversispora spurca* ameliorates effects of waterlogging on growth, root system architecture ad antioxidant enzyme activities of citrus seedlings. *Fung. Ecol.* 6:37–43.

Xiong, J., F. Peng, H. Sun, X. Xue, and H. Chu. 2014. Divergent responses of soil functional groups to short-term warming. *Microb. Ecol.* 68:708–715.

Yoshida, S. and Y. Muramatsu. 1994. Accumulation of radiocesium in basidiomycetes collected from Japanese forests. *Sci. Total Environ.* 157:197–205.

Zak, D. R., K. S. Pregitzer, J. S. King, and W. E. Holmes. 2000. Elevated atmospheric $CO_2$, fine roots and the response of soil microorganisms: A review and hypothesis. *New Phytol.* 147:201–222.

Zhdanova, N. N, A. I. Vasilevskaya, L. V. Artyshkova, Yu S., Sadovnikov, V. I Gavrilyuk, and J. Dighton. 1995. Changes in the micromycete communities in soil in response to pollution by long-lived radionuclides emitted by in the Chernobyl accident. *Mycol. Res.* 98:789–795.

Zhdanova, N. N., V. A. Zakharchenko, V. V. Vember, and L. T. Nakonechnaya. 2000. Fungi from Chernobyl: Mycobiota of the inner regions of the containment structures of the damaged nuclear reactor. *Mycol. Res.* 104:1421–1426.

Zhu, H., B. P. Dancik, and K. O. Higginbotham. 1994. Regulation of extracellular proteinase production in an ectomycorrhizal fungus *Hebeloma crustuliniforme*. *Mycologia* 82:227–234.

# CHAPTER 8

# Fungi in the Built Environment

As humans developed from a hunter–gatherer and somewhat nomadic lifestyle, they created structures to provide shelter. Over time, and with our social behavior, aggregations of these shelters became hamlets, villages, towns, and now the modern cities and urban sprawl we see today. These large urban areas are now a major part of the landscape and have their own environment and ecology, which can be considered distinct from other more natural environments. As fungi are ubiquitous, it is no surprise that they play a significant role in this "built environment" as both saprotrophs and pathogens. The use of natural materials for the construction of buildings, the fact that we still gather (from the supermarket) our food supplies and store them in our homes, and that we shed body parts (skin and hair), from ourselves and pets, means that there are plenty of resources available in this built environment for fungi to grow (Shadzi et al., 2002). As a result of this fungal growth, our food supply can be reduced by spoilage and the production of fungal spores, and volatile secondary metabolites can affect our health through pathogenicity, allergies, and reduced function because of "sick building syndrome."

The desire for humans to keep warm and hydrated results in our sustained efforts to maintain our living environment at temperatures and relative humidity levels that are often optimum for fungal growth. This frequently leads to unwanted consequences of fungal growth on commonly used work surfaces, bathroom walls, and fixtures and items of art or historical value that we wish to preserve. In our attempts to reduce the abundance of fungi in our living environment, by the use of cleaning agents, we are likely selecting for more and more resistant strains, leading to the evolution of a new fungal community that is best adapted for life in the built environment (Gostinčar et al., 2011).

## 8.1 DECOMPOSITION OF FABRIC OF BUILDINGS

In their review articles, Singh (1999) and Schmidt (2007) discuss the role of predominantly the brown rot fungi *Serpula lachrymans* and *Meruliporia incrassata*, which degrade coniferous wood within buildings. They identify the optimum moisture (35–100%) and temperature (20–27°C) regimes for growth of the fungi. These fungi can be aggressive and cause extensive damage to the structure of wooden or wood framed houses, reducing the structural properties of wood and leading to potential collapse of buildings. Singh (1999) discusses the biology of the fungi showing that ingress of the fungus into a building is attributable to foraging hyphae and rhizomorphs, through which the fungus is able to transport water into the wood to maintain fungal growth and activity. Remediation includes removal and replacement of all fungal colonized wood, treating remaining wood with preservatives, severing any rhizomorphs at the exterior of the property, and ensuring there is no direct contact between wood and soil. Apart from removal of damaged wood, other suggested effective treatments are high-frequency electromagnetic radiation and microwave radiation. Additionally, chemical treatments can be used to either suppress fungal growth or to interfere with essential physiology. Aminoisobutyric acid interferes with trehalose metabolism and translocation processes that are important for water movement through the fungus and polyoxin inhibits chitin production. There is evidence that some *Trichoderma* species can be used as biocontrol agents, but there is concern that sporulation may increase sick building syndrome. The best strategy is avoidance by preventing leaking water pipes and ensuring adequate ventilation (Schmidt, 2007). It is interesting to note that strains of *Meruliporia incrassata* (e.g., strain TFFH 294) have been tested for their ability to decompose CCA (chromium copper arsenate)-treated wood in landfills as a way to reduce the bulk of these materials in the waste stream (Yang and Illman, 1999).

Historic buildings are similarly prone to fungal attack and degradation even under extreme environments. Blanchette et al. (2010) found that Shackleton's historic buildings and wooden artifacts in Antarctica were being attacked by fungi with dominant fungal species of the genera *Cadophora* (44%) (see also Arenz and Blanchette, 2009), *Thielavia* (17%) and *Geomyces* (15%). Extensive decay was revealed by scanning electron microscopy. They suggest that climatic constraints, high UV irradiation, and the abundance of penguin guano as a nutrient source selected for specific groups of wood decay fungi in these wooden structures, which are in close contact with soil. Additionally, the interior of huts had relative humidity levels constantly between 75% and 80%, and although temperatures rarely exceeded 0°C, this was a suitable environment for fungal growth (Held et al., 2005). In cold climates, growth of fungi can be maintained by physiological adaptations where the production of trehalose and other sugars and polyols (glycerol, erythritol, mannitol) act as cryporotectants allowing fungi to grow at temperatures of 5°C and survive freeze–thaw cycles (Robinson, 2001; Ludley and Robinson, 2008).

Both the surfaces of building materials and the integral structure of buildings can be attacked by fungi when relative humidity or water content is high. Painted surfaces are susceptible to fungal degradation (Shirakawa et al., 2002) with a defined succession of fungi colonizing newly painted surfaces. Initial colonizers are thought to exist in the paint in the can, probably surviving as spores. These are then replaced by *Alternaria, Curvularia, Epicoccum, Helminthosporium, Pestalotia, Monascus*, and *Nigrospora* after a few weeks. *Pithomyces, Tripospermum, Trichoderma*, and yeasts appear as part of the fungal community after about 30 weeks. However, the incorporation of the biocides carbamate, *N*-octyl-2*H*-isothiazolin-3-one and

*N*-(3,4-dichlorophenyl)*N-N*-dimethyl urea at 0.25% into acrylic paint significantly reduces fungal growth on painted surfaces (Shirakawa et al., 2002). Other protective measures include the incorporation of nanoparticles of copper or silver into compressed particle board, which has been shown to be successful as a fungal retardant, when exposed to inoculum of *Trametes versicolor* (Taghiyari et al., 2014).

## 8.2 AIR SPORA, HEALTH PROBLEMS, AND MOLDS ON STRUCTURES

As fungi grow in our dwellings and offices, they produce spores and volatile organic chemicals that can be harmful to our health, and this is called sick building syndrome. The abundance of fungal colonization of buildings can seriously affect human health (Cabral, 2010; Crook and Burton, 2010; OSHA, 2013). The incidence of asthma increases with exposure to these agents and, in extreme cases, the fungal spores can lead to pathogenic diseases such as aspergillosis and pulmonary hemorrhage (Singh, 2002; Bennett and Klich, 2003). Allergies and asthmas can be aggravated by volatile secondary metabolites (mycotoxins) emitted by the fungi (Miller, 1992; Bennett and Klich, 2003). Examples are aflatoxin B (*Aspergillus flavus*), sterigmatocystin (*A. versicolor*) and zeralenon, and T-2 toxin (*Fusarium graminarum*) and diacetoscirpenol (*Trichoderma viride*) (Gravesen, 1979; Cvetnić and Pepeljnjak, 1997; see review by Bennett and Klich, 2003).

Sick building syndrome is exacerbated in buildings where dampness is prevalent or after environmental catastrophies such as major storms or hurricanes (Hyvärinen et al., 2002). Different materials in a wet building are resources for different fungal species. Hyvärinen et al. (2002) showed a variety of fungi-colonized wooden materials, with *Penicillum* species being most prevalent, but noted that *Cladosporium* and *Stachybotris* dominated on paper products and *Stachybotris* on gypsum wall boards (Figure 8.1). A specific example of this is what occurred after hurricane Katrina in 2005, which left numerous New Orleans houses inundated with water and subsequently abandoned. With wet fabric, high relative humidity, and warm temperatures, many buildings became colonized by fungi, resulting in health hazards from spores in the air and contaminated debris as a problem for cleanup crews (Solomon et al., 2006; Rao et al., 2007; Weinhold, 2007; Bloom et al., 2009; Cho et al., 2011).

Much of the indoor contamination of structural surfaces comes from spores and dust-containing spores entering the building. Condiospores, in general, have smaller spore volume (100–300 $\mu m^3$) compared with basidiospores (65–600 $\mu m^3$), which tend to have a 95% dispersal distance of tens of centimeters (Li, 2005; Galante et al., 2011) (Table 8.1). Thus, conidial fungi are more likely to invade interiors of buildings than basidiomyces. The exception to this is *Stachybotris*, which is an important fungus for human health that has a spore volume of about 600 $\mu m^3$ (Bennett and Klich, 2003). In sensitive environments, protocols for reducing the ingress of dust and spores may be important for human health. For example, Berthelot et al. (2006) advocated protocols to restrict construction dust entering hospital wards to reduce the risk of aspergillosis in patients. However, a discussion of fungal diseases

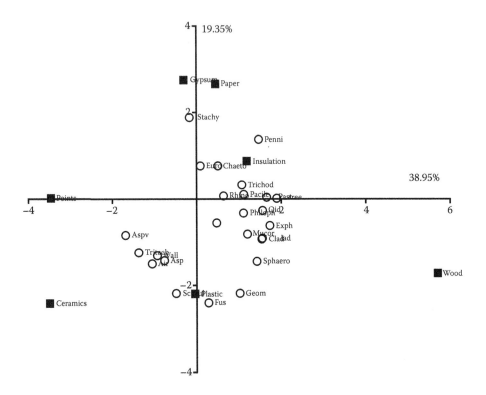

**Figure 8.1**   PCA analysis of fungal communities associated with different building materials in water damaged buildings. Square symbols represent construction materials and circles represent fungal genera. (Analysis based on data from Hyvärinen, A. et al., *J. Biodeter. Bioremed.*, 49, 27–37, 2002.)

**Table 8.1    Spore Dispersal Distance from Basidiomes of Six Fungal Species**

| Species | Spore Volume (μm³) | 95% Dispersal Distance (cm) |
|---|---|---|
| *Inocybe lacera* | 208 | 3 |
| *Thelephora americana* | 211 | 7 |
| *Laccaria laccata* | 279 | 14 |
| *Suillus tomentosus* | 84 | 16 |
| *Suillus brevipes* | 65 | 25 |
| *Lactarius rufus* | 171 | 58 |

*Source:* Galante, T.E. et al., *Mycologia*, 103, 1175–1183, 2011.

and medical mycology is beyond the scope of this book, but readers may refer to Dismukes et al. (2003) and Singh (2005).

The accumulation of fungal spores and generation of harmful volatile metabolites is not restricted to human dwellings but may also be an important health threat in other confined areas where access by humans is necessitated. For example, freight containers and containers carrying food stuff, which may be colonized by fungi causing spoilage, may induce high fungal spore loading in air to which humans are exposed upon opening. For example, Hill et al. (1995) found that wooden floors of general-purpose freight containers harbored a wide range of fungal species that increased airborne chlorophenol concentrations of more than 5 orders of magnitude. Nineteen of the 38 isolated fungal species were known to biomethylate chlorophenols in wood to chloranisoles, which are known to produce strong odors, although posing limited health risks (Gunschera et al., 2004).

Attention has been focused on the density of spores in indoor air as an index of potential health risk, and various health organizations have tried to provide guidelines for maximum desired spore loading. Using Andersen spore traps and culturing on malt extract agar (MEA), Salonen et al. (2007) measured the fungal spore community in mold-damaged and nondamaged offices in a subarctic winter in southern Finland. The mold-infected buildings had significantly higher fungal spore loadings than the undamaged buildings. Medrela-Kuder (2003) compared the seasonal air spora within and outside of a building in Craćow, Poland. During summer, highest levels of spores were found both inside and outside with a dominance of *Cladosporium*. However, during winter, spore density was less than 40% of summer values and there were 3 times the spore density inside the building than outside (dominated by *Penicilium* and *Aspergillus*). Spore density was measured as colony forming units on Petri plates with a Chirana aeroscope (volumetric sampling device).

Where abundant substrates are available for fungal growth, spore loading in buildings can be significantly higher than that found outdoors. In contrast, the outdoor air density in Tianan City, Taiwan, was similar to that inside residential houses (Wu et al., 2009), but spore density was twice as high in urban than in suburban areas in summer because of the higher abundance of *Cladosporium* and *Penicillium* spores. In a study of fungal spore densities in sawmills, Simeray et al. (1997) found that indoor spore densities were up to 6.5 times greater than that found outside, using MEA as a growth medium and up to 21 times higher when using sucrose-amended MEA. Additionally, sawmills cutting hardwoods had 42% higher spore loading than mills cutting softwood.

However, just when you thought it was safer to go outside, other studies have shown that outdoor spore density can be significantly higher than that indoors. In a tropical environment in a library in Singapore, Goh et al. (2000) found that the indoor air fungal spore density was 50 times lower than that outside, which they attribute to the lower humidity produced by the use of air conditioners (Table 8.2). In contrast, bacterial spore density was 10 times higher inside than out. The abundance of fungal spores in outdoor air is related to both seasonality and climate. O'Gorman and Fuller (2008) showed that in Dublin city air *Aspergillus fumigatus* was present at a low spore density throughout the year with sporadic high concentrations (300–400 CFU m$^{-3}$); *Cladosporium* reached allergenic levels of 3200 CFU m$^{-3}$ during summer (August), whereas *Penicillium* never reached allergenic levels. The abundance of *Cladosporium* spores was positively correlated with temperature, whereas *Penicillium* and *Aspergillus* spore densities were positively correlated to relative humidity. Similarly, Lee et al. (2010), using T-RFLP (terminal restriction fragment length polymorphism) and real-time polymerase chain reaction (PCR) analysis

Table 8.2    Fungal Spore Density in Air Taken Inside and Outside the Central Library
             at the National University of Saigon

| Time | Relative Humidity (%) | | Spore Count (CFU m$^{-3}$) | |
| | Inside | Outside | Inside | Outside |
| --- | --- | --- | --- | --- |
| 0800 | 58.7 | 91.2 | 41.2 | 2868.1 |
| 1200 | 58.1 | 85.4 | 64.4 | 864.5 |
| 1500 | 59.8 | 86.1 | 34.2 | 1332.2 |
| 1800 | 55.8 | 96.2 | 53.0 | 2378.1 |
| 2000 | 57.8 | 97.9 | 43.6 | 4969.4 |
| Average | 58.0 | 91.4 | 47.3 | 2482.5 |

Source: Goh, L. et al., *Acta Biotechnol.*, 1, 67–73, 2000.

of fungal community of air spora of outdoor air in Seoul, Korea, measured ranges of spore density from 10 to $400 \times 10^4$ CFU m$^{-3}$. The fungal community changed significantly with time during the year, probably related to climatic conditions, phenology of plants, and other environmental factors. Thus, the abundance of spores in air is very much dictated by the response of specific fungal genera to season and local climate.

## 8.3 HOT SPOTS OF FUNGAL ACTIVITY RELATED TO ENVIRONMENT

Using Real time PCR, Adams et al. (2013) identified 966 OUTs (observational taxonomic units or putative species) of fungi from window sills, basin drains, and skin from the forehead of occupants of a university dormitory. Differences exist between the fungal communities in drains compared to dust, window sills, or skin samples, and may reflect the smaller niche dimension here, leading to reduction in diversity and dominance by *Exophilia* (Eurotiomycetes). A similar dominance by this group of fungi was found in dishwashers and water residing in dishwashers (Zalar et al., 2011; Hamada and Abe, 2013) (Tables 8.3 and 8.4). This fungal genus

Table 8.3    Fungal Contamination (Colony Forming Units
             [CFU]) of Internal Parts of Dishwashers
             Determined by Swab Cultures

| Dishwasher Part | Average CFU cm$^{-2}$ |
| --- | --- |
| Rubber seal | 5.3 |
| Lid of drain | 4.6 |
| Interior wall | 5.5 |
| Dish rack | 2.7 |
| Bottom | 1.3 |
| Interior of door | 28.4 |
| Spoon basket | 6.4 |
| Edge of tab | 4.1 |
| Others | 13.0 |

Source: Hamada, N., Abe, N., *J. Antibact. Antifung. Agents*,
        41, 527–534, 2013.

Table 8.4   Fungi Isolated from 189 Dishwashers around the World, of Which 117 (61.9%) Yielded Positive for Fungal Contamination

| Fungus | Number of Dishwashers | % Dishwashers |
|---|---|---|
| Exophilia dermatidis/phaeomuriformis | 66 | 34.9 |
| Candida parapsilopsis | 16 | 8.5 |
| Magnusiomyces capitatus | 9 | 4.8 |
| Fusarium dimerum | 5 | 2.6 |
| Pichia guilliermondii | 8 | 4.2 |
| Rhodotorula mucilaginosa | 4 | 2 |
| Other | 12 | 63 |

Source: Zalar, P. et al., Fungal Biol., 115, 997–1007, 2011.

appears to exhibit extreme thermo-, halo-, and pH tolerance (Zalar et al., 2011). The adaptation of stress tolerance in fungal species in hot and humid environments within houses and especially to fluctuations in temperature and humidity has been proposed as an evolutionary force leading to the development of higher abundance of pathogens in buildings (Gostinčar et al., 2011).

Even human-inhabited spaces in extreme environments harbor a fungal flora. Fungal isolates were obtained from numerous sites within the International Space Station, including *Penicillium*, *Aspergillus*, *Curvularia*, *Trichphyton*, *Microsporium*, and *Streptomyces* species (Castro et al., 2004). Indeed, De Vera (2012) demonstrated that 90% of lichens could survive and grow on return to Earth conditions after a period under simulated space conditions of vacuum and enhanced UV radiation.

## 8.4 DEGRADATION OF ARTIFACTS

Our cultural heritage is largely contained in museum pieces. Preservation of these artifacts is an important part of the curatorial duties of museum staff by regulation of the climate in which materials are held. However, despite these efforts, either the location of climate monitoring systems or local (microsite) conditions are such that humidity control is insufficient to prevent fungal growth on these substrates (Sterflinger, 2010). As a result, fungal staining of paper and nonpaper materials (silk, hemp, etc.), referred to as "foxing," appears as brown stains (Choi, 2007) (Figure 8.2). This is common on old manuscripts and paintings and is caused by a number of fungi, particularly *Aspergillus* spp. (Arai, 2000; Rakotonirainy et al., 2007) (Figure 8.3). In a specific study of eighteenth and nineteenth century water-damaged manuscripts in Maryland, USA, Szczepanowska and Cavaliere (2000) found evidence of damage caused by *Chaetomium* spp. that had pigmented the documents magenta/orange to yellow or olive. In her review, Sterflinger (2010) identified some 29 fungal genera associated with paintings, paper, parchment, or keratinous substrates in museums.

Abrusci et al. (2005) identified 17 fungal species causing damage to film in three archives in Spain. Of 18 isolates of 12 species tested, they found all but one

**Figure 8.2** **(See color insert.)** Image of slight foxing around the edge of an old manuscript page from M. Bulliard's Champignon de France 1791. (Courtesy of the Library and Archives of the Academy of Natural Sciences, Philadelphia ANSP Call Number QK608.F8B8.)

exhibiting significant levels of gelatin hydrolysis. Degradation of wooden artifacts is also attributable to fungal attack. For example, a South American Jesuit sculpture, *The Trinity*, in a museum in Buenos Aires has been shown to be affected by the soft rot fungus *Chaetomium globosum* along with *Nigrospora sphaerica*, despite the fact that covering pigment and plaster contained up to 6% Al and 8% Pb (Fazio et al., 2010). It appears to be only recently that museums have adopted greater steps to monitor for microbial decay and taken appropriate actions beyond climate (mainly humidity) control by introducing air filtering and biocide treatments to individual artifacts (Sterflineger, 2010).

Fungi are involved in the degradation of rocks (see reference on geomycology in Chapter 2; Gadd 2004, 2007), so it is not surprising that they are involved in the degradation of structures constructed from stone or artifacts that have similar properties. Stone and brick buildings are subjected to attack by a variety of hyphomycetes, particularly black fungi such as *Hortea*, *Sarcinomyces*, and *Exophiala* within the stone and in association with lichens (Sterflinger, 2010). These fungi have caused

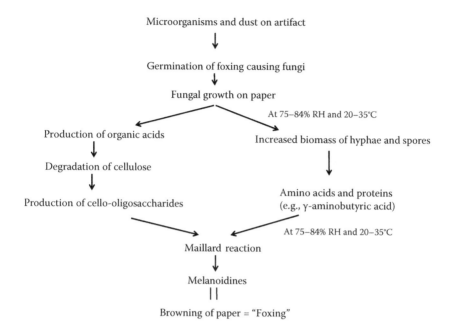

**Figure 8.3**  The process of developing foxing on paper artifacts. (After Aria, H., *Int. Biodeter. Biodegr.*, 46, 181–188, 2000.)

extensive deterioration of important historical buildings including the Acropolis of Athens, temples of Delos, and marble monuments in Crimea. Ivory is chemically similar to natural apatite (a natural phosphate and calcium mineral) and has been used for carving sculptures and ornaments. In the inoculation of ivory tusks of wild boar and walrus with *Aspergillus niger* and *Serpula himantiodes*, Pinzari et al. (2013) observed similar degradation characteristics shown by the Lewis Chessmen (medieval sculptures), suggesting that the surface channels and tunneling within the pieces could be due to fungal activity.

Other rock-derived substrates are also susceptible to fungal attack and degradation. Rodrigues et al. (2014) isolated five ascomycetes (*Alternaria*, *Chaetomium*, *Cladosporium*, *Didymella*, and *Penicilium*) and one basidiomycete (*Sistotrema*) from stained glass in a Portuguese palace (Table 8.5). They inoculated reproduction of colorless and stained glass fragments with either *Penicilium* or *Cladosporium* isolate and found significant damage to the glass after 25 months. Damage, as revealed by scanning electron microscopy EDS and Raman Fourier transform infrared spectroscopy, included pitting, staining, leaching, and deposition of elements. Although these fungi were inoculated along with a small amount of medium onto the glass in the experiments, it is well known that oligotrophic fungi are capable of growing on glass and other substrates by scavenging nutrients from the atmosphere (Wainwright et al., 1991, 1997).

TABLE 8.5  Fungi Isolated from and Doing Damage to Stained Glass Windows from Samples Collected in Germany and Switzerland

| Germany | Switzerland |
| --- | --- |
| Alternaria tenuissima | Cladosporium coarctatum |
| Cladosporium sphaerospermum | Cladosporium langeronii |
| Didymella phacae | Cladosporium sphaerospermum |
| Penicillium sp. | Penicillium sp. |
| Penicillium citreonigrum | Penicillium citreonigrum |
| Sistotrema sp. | Penicillium roseopurpureum |

Source: Rodrigues, A. et al., Int. Biodeter. Biodegr., 90, 152–160, 2014.

## 8.5 POSSIBLE PROTECTION OF ARTIFACTS

Remedial treatment of paper foxed by fungi requires the use of nonaqueous solvents. Szczepanowska and Lovett (1992) demonstrated that the use of 1,4-dioxane, *N,N*-dimethyl-formamide or pyridine was successful in removing stains from a variety of fungi. Laser treatment of paper with a 1.6-W, 532-nm pulsed laser was successful for damage caused by *Penicillium notatum*, less so for damage caused by *Alternaria solani*, but not for damage caused by *Chaetomium* or *Fusarium*, because of the extensive damage in the fabric of the paper (Szczepanowska and Moomaw, 1994). Treating paper artifacts with a 1.5% concentration of deacetylated chitosan, chitosan salts, and cellulose ethers has been shown to significantly reduce the growth of *Aspergillus*, *Penicillium*, *Trichoderma*, and *Mucor* species on paper, despite the slight changes in color and brightness changes in the paper (Ponce-Jiménez et al., 2002; Shiah, 2009). In a survey of conservators' prevention measures, Sequeira et al. (2014) found that improving air circulation and controlling relative humidity and temperature were the main factors. Very few conservators had done more than drying or alcohol wipes of paper that had been damaged by water.

Nanoparticle colloids of antifungal chemicals (mainly heavy metals) have been investigated for possible use as protectants of a variety of materials. Gómez-Ortíz et al. (2014) report that use of zinc oxide nanoparticles inhibited the growth of *Aspergillus niger* and *Penicilium oxalicum* on limestone from the Chichén Itza region of Mexico, where the rock is being used in restoration efforts. In the presence of the protectant, they reported no fungal growth within 21 days. Similarly, Mantanis et al. (2014) showed that zinc oxide, zinc borate, and copper oxide nanoparticles applied as a vacuum treatment to pine wood could protect wood from decay by some fungal species. Wood degradation by the white rot fungus *Trametes versicolor* was inhibited by the zinc treatments, but they were ineffective against the brown rot fungus *Tyromyces palustris*.

As wood/polyvinylchloride materials are becoming more commonly used as construction materials, Kositchaiyong et al. (2014) examined their ability to withstand biodegradation under UV light and with soil burial. They showed that antifungal performance was decreased from 81.4% to 28.3% when composites were exposed to 32 days of UV light, but significantly increased loss of performance (down to 4.4%

Table 8.6  **Percent Inhibition of Growth of *Aspergillus niger* on PVC (Polybinylchloride) or WPVC (Wood/Polyvinylchloride Composite) in the Presence or Absence of the Preservative IPBC (3-Iodopropinyl-*N*-Butylcarbamate) under UV Exposure (Aging) or Burial in Soil**

|  | IPBC (ppm) | UV Accelerated Aging (days) | | | Soil Burial (months) | |
|---|---|---|---|---|---|---|
|  |  | 0 | 8 | 32 | 3 | 6 |
| PVC | 0 | −31 | −53 | −90 | +26 | n/a |
|  | 10,000 | −47 | −39 | −45 | +28 | +11 |
| WPVC | 0 | −40 | −74 | −96 | +7 | n/a |
|  | 10,000 | +81 | +73 | +28 | +24 | +4 |

*Source:* Kositchaiyong, A. et al., *Int. Biodeter. Biodegr.*, 91, 128–137, 2014.
*Note:* Negative numbers indicate no growth inhibition. n/d, bacterial contamination and no reading. % inhibition calculated from diameter of inoculum ($D_i$) placed between two samples 30 mm apart ($D_s$) as (($D_s − D_i)/D_s$) × 100.

antifungal performance) when they were buried in soil for the same time interval. The addition of 3-iodopropinyl-*N*-butylcarbamate at 10,000 ppm significantly increased antifungal performance (Table 8.6). Selection of decay-resistant wood is also a strategy that can be adopted. Olivera et al. (2010) showed that out of five selected woods, *Caesalpinia ferrea* (jucá) supported the greatest growth of *Phanerochaete chrysosporium*, but *Anadenanthera colubria* (Angico-branco) and *Manilkara huberi* (sapotilla) supported little or no fungal growth because of their high lignin and phenolic content. In the same way, Atlantic white cedar (*Chamaecyparis thyoides*) wood was commonly used for roofing shingles as its high polyphenol content provided resistance to decay. Plaschkies et al. (2014) showed that oak wood had greater resistance to decay by *Coniophora puteana* than did beech, pine, or larch wood.

Surface corrosion and discoloration of metal objects and artwork exposed to the environment has led to a number of efforts to provide protection using organic coatings of waxes or acrylics. However, recently bioprotective coatings of fungal origin have been shown to have potential to prevent decay and retain more natural patinas to the metal surfaces (Joseph et al., 2011). Of five fungal species, *Beauveria bassiana* was found to have the greatest oxalate formation, which would develop as a metal-oxalate film on the surface of structures. This deposition as a film imparts far better protection than the aggregated concretions of oxalates formed by the other fungal species tested (*Aspergillus niger*, *A. alliacea*, *Penicillum* sp., and *Fusarium* spp.) (Table 8.7; Figure 8.4). This method of protection is especially good for copper metal and is suggested to be a potential protective treatment for copper artifacts (Joseph et al., 2012).

## 8.6 FOOD SPOILAGE

The environmental requirements for fungal growth on stored food products and a list of fresh and processed or preserved foods and their fungal contaminants can be found in the work of Pitt and Hocking (1997a,b,c). Not only do fungi degrade the fabric of our buildings, but they also degrade the products we bring into these buildings. The spoilage of food products by fungi has been well documented, with *Penicillium*

**Table 8.7 Selection of Fungal Species for Production of Oxalates on a Variety of Growth Media**

| Medium | Aspergillus niger | Penicillium sp. | Aspergillus alliaceus | Beauvaria bassiana | Fusarium sp. |
|---|---|---|---|---|---|
| Malt agar (MA) | −/+ | − | − | −/+ | − |
| MA + calcite | ++ | ++ | − | +++ | −/+ |
| MA + gypsum | + | + | − | ++ | ++ |
| MA + brochanite | − | − | − | +++ | − |
| MA + atacamite | − | − | − | ++ | − |
| MA + cuprite | − | − | − | ++ | − |
| Milk whey agar | + | +/− | − | + | − |

*Source:* Joseph, E. et al., *Anal. Bioanal. Chem.*, 399, 2899–2907, 2011.

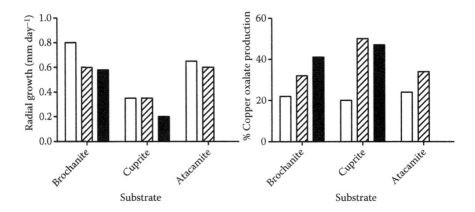

**Figure 8.4** Ability of *Beauvaria bassiana* to grow (left) and produce protective copper oxide (right) on three solid media amended with Cu(II) at thee concentrations (open bar = 10 g l⁻¹, hatched bar = 20 g l⁻¹ and solid bar = 30 g l⁻¹). (Data from Joseph, E. et al., *Front. Microbiol.*, 2, 1–8, 2012.)

growth on cheeses (Kure and Skaar, 2000) (Table 8.8) and *Fusarium*, *Aspergillus*, and *Penicillium* on stored grains, producing their associated mycotoxins (Miller, 1995). Fresh vegetables and fruits are especially susceptible to damage by fungi. Tournas (2005) and Tournas and Katsoudas (2005) reported that 33–100% of soft fruit (raspberries, blackberries, and strawberries) had fungal growth, and many fresh vegetables in supermarkets had up to $4 \times 10^8$ CFU g⁻¹ of yeasts and $4 \times 10^4$ CFU g⁻¹ of filamentous fungi on their surfaces (Table 8.9). How much of a health hazard this may be is not qualified, but there is great potential for loss of revenue when these fungi cause actual damage to the food or the mycotoxins that fungi produce (Taniwaki et al., 2009), and considerable effort has been exerted to address ways in which food products are stored before and after sale to minimize these effects (Dagnas and Membré, 2013). For example, increasing the carbon dioxide level in storage reduces both the growth of fungi and the mycotoxin production of these fungi (Taniwaki et al., 2009), and bread spoilage could be reduced significantly by the addition of very

**Table 8.8   Frequency of Isolation of Contaminant Fungi from Norwegian Cheeses**

|                | Isolation Frequency (%) | |
| Fungal Genus   | Norvegia | Jarlsberg |
| -------------- | -------- | --------- |
| *Alternaria*   | 0.6      | –         |
| *Aureobasidium* | 1.3     | –         |
| *Cladosporium* | 1.3      | –         |
| *Epicoccum*    | 0.6      | –         |
| *Geotrichum*   | –        | 1.3       |
| *Mucor*        | 3.2      | 0.7       |
| *Penicillium*  | 89.2     | 98        |
| *Phoma*        | 2.5      | –         |
| *Ulocladium*   | 1.3      | –         |

Source: Kure, C.F., Skaar, I., *Int. J. Food Microbiol.*, 62, 133–137, 2000.

**Table 8.9   Fungal Contamination of Fruits Purchased from a U.S. Supermarket**

| Fruits  | Number of Samples | % Samples Contaminated |
| ------- | ----------------- | ---------------------- |
| Berries | 90                | 98                     |
| Grapes  | 69                | 39                     |
| Citrus  | 92                | 81                     |

Source: Tournas, V.H., Katsoudas, E., *Int. J. Food Microbiol.*, 105, 11–17, 2005.

low concentrations (0.8–2%) in the package (Berni and Scaramuzza, 2013). In addition to their appearance on fruits and vegetables, fungi have been shown to appear in less likely places. Cabral and Pinto (2002) found *Penicillium*, *Cladosporium*, and *Alternaria* in bottled mineral water form a variety of commercial sources. Some of these caused actual spoilage, but some were just detected as being present in the water (Table 8.10). Sørensen et al. (2008) report the occurrence of fungi growing on meat and meat products in a number of meat processing facilities. The occurrence of

**Table 8.10   Percent of Commercial Water Bottles Yielding Fungal Contaminants with Visible Contamination and No Visible Contamination**

| Visible Contamination | | | No Visible Contamination | |
| Fungus | % with Visible Hyphae | % No Visible Hyphae | Fungus | % Samples |
| ------------- | --- | --- | ------------- | --- |
| *Penicillium*  | 46 | 33 | *Penicillium*  | 46 |
| *Cladosporium* | 50 | 17 | *Cladosporium* | 32 |
| *Alternaria*   | 21 | 0  | *Rhizopus*     | 8  |
|                |    |    | *Aspergillus*  | 3  |
|                |    |    | *Phoma*        | 3  |

Source: Cabral, D., Pinto, V.E.F., *Int. J. Food Microbiol.*, 72, 73–76, 2002.

fungi taints meat and is visually unappealing, thus reducing the market potential for the product. A wide range of fungal genera was found on sausage products, but liver paté was mainly colonized by *Penicillium*.

Not only do human food stuffs decompose by fungi, but also foodstock for our animals may be degraded and contaminated by fungi-producing mycotoxins (Bennett and Klich, 2003). For example, patulin production from fungi (mainly *Penicillium*) in corn silage significantly reduced digestion of silage in artificial rumen conditions, thus reducing the food quality (Tapia et al., 2005).

## 8.7 CONCLUSION

The impact of humans on the landscape has been huge. The evolution of our towns and cities has opened up new or different opportunities for microorganisms to colonize novel or altered substrates. As opportunists, many fungi have readily adapted to these new environments, and in their proliferation on the substrates have either been a nuisance or caused heath-related issues. With potential climate change and an alteration of weather patterns, more opportunities for fungal growth on man-made or manipulated materials are likely. Understanding the diversity and function of fungi in the built environment is a relatively new area of study in mycology, and it is likely that many more studies will be forthcoming.

## REFERENCES

Abrusci, C., A. Martín-González, A. Del Amo, F. Catalina, J. Collado, and G. Platas. 2005. Isolation and identification of bacteria and fungi from cinematographic films. *Int. Biodeter. Biodegr.* 56:58–68.

Adams, R. I., M. Miletto, J. W. Taylor, and T. D. Bruns. 2013. The diversity and distribution of fungi on residential surfaces. *PLoS ONE* 8(11):e78866. doi:10.1371/journal .pone.0078866.

Arenz, B. E. and R. A. Blanchette. 2009. Investigations of fungal diversity in wooden structures and soils at historic sites on the Antarctic Peninsula. *Can. J. Microbiol.* 55:46–56.

Arai, H. 2000. Foxing caused by fungi: Twenty-five years of study. *Int. Biodetr. Biodegr.* 46:181–188.

Bennett, J. W. and M. Klich. 2003. Mycotoxins. *Clin. Microbiol. Rev.* 16:497–516.

Berni, E. and N. Scaramuzza. 2013. Effect of ethanol on growth of *Chrysonilia sitophila* ('the red bread mould') and *Hyphopicia burtonii* ('the chalky mould') in sliced bread. *Lett. Appl. Microbiol.* 57:344–349.

Berthelot, P., P. Loulerge, H. Raberin et al. 2006. Efficacy of environmental measures to decrease the risk of hospital-acquired aspergillosis in patients hospitalized in haematological wards. *Clin. Microbiol. Infect.* 12:738–744.

Blanchette, R. A., B. W. Held, B. E. Arenz et al. 2010. An Antarctic hot spot for fungi at Shackleton's historic hut on Cape Royds. *Microb. Ecol.* 60:29–38.

Bloom, E., L. F. Grimsley, C. Pehrson, J. Lewis, and L. Larsson. 2009. Molds and mycotoxins in dust from water-damaged homes in New Orleans after hurricane Katrina. *Indoor Air* 19:153–158.

Cabral, D. and V. E. F. Pinto. 2002. Fungal spoilage of bottled water. *Int. J. Food Microbiol.* 72:73–76.

Cabral, J. P. S. 2010. Can we use indoor fungi as bioindicators of indoor air quality? Historical perspectives and open questions. *Sci. Total Environ.* 408:4285–4295.

Castro, V. A., A. N. Thrasher, M. Healy, C. M. Ott, and D. L. Pierson. 2004. Microbial characterization during the early habitation of the International Space Station. *Microb. Ecol.* 47:119–126.

Cho, S. J., J-H. Park, K. Kreiss, and J. M. Cox-Ganser. 2011. Levels of microbial agents in floor dust during remediation of a water-damaged office building. *Indoor Air* 21:417–426.

Choi, S. 2007. Foxing on paper: A literature review. *J. Am. Inst. Conserv.* 46:137–152.

Crook, B. and N. V. Burton. 2010. Indoor moulds, Sick Building Syndrome and building related illness. *Fung. Biol. Rev.* 24:106–113.

Cvetnić, Z. and S. Pepeljnjak. 1997. Distribution and mycotoxin-producing ability of some fungal isolates from the air. *Atmosph. Envir.* 31:491–495.

Dagnas, S. and J.-M. Membré. 2013. Predicting and preventing mold spoilage of food products. *J. Food Protect.* 76:538–551.

De Vera, J.-P. 2012. Lichens as survivors in space and on Mars. *Fung. Ecol.* 5:475–479.

Dismukes, W. E., P. G. Pappas, and J. D. Sobel. 2003. *Clinical Mycology.* New York: Oxford University Press, 519 pp.

Fazio, A. T., L. Papinutti, B. A. Gómez et al. 2010. Fungal deterioration of a Jesuit South American polychrome wood sculpture. *Int. Biodeter. Biodegr.* 64:694–701.

Gadd, G. M. 2004. Mycotransformation of organic and inorganic substrates. *Mycologist* 18:60–70.

Gadd, G. M. 2007. Geomycology: Biogeochemical transformation of rocks, minerals and radionuclides by fungi, bioweathering and bioremediation. *Mycol. Res.* 111:3–49.

Galante, T. E., T. R. Horton, and D. P. Swaney. 2011. 95% of basidiospores fall within 1 m of the cap: A field- and modeling-based study. *Mycologia* 103:1175–1183.

Goh, L., J. P. Obbard, S. Viswanathan, and Y. Huang. 2000. Airborne bacteria and fungal spores in the indoor environment: A case study in Singapore. *Acta Biotechnol.* 1:67–73.

Gómez-Ortíz, N. M., W. S. González- Gómez, S. C. De la Rosa-García et al. 2014. Antifungal activity of Ca[Zn(OH)$_3$]$_2$·2H$_2$O coatings for the prevention of limestone monuments: An *in vitro* study. *Int. Biodeter. Biodegr.* 91:1–8.

Gostinčar, C., M. Grube, and N. Gune-Cimerman. 2011. Evolution of fungal pathogens in domestic environments? *Fungal Biol.* 115:1008–1018.

Gravesen, S. 1979. Fungi as a cause of allergenic disease. *Allergy* 34:135–154.

Gunschera, J., F. Fuhrmann, T. Salthammer, A. Schulze, and E. Unde. 2004. Formation and emission of chloranisoles as indoor pollutants. *Environ. Sci. Pollut. Res. Int.* 11:147–151.

Hamada, N. and N. Abe. 2013. Fungal contamination in dishwashers. *J. Antibact. Antifung. Agents* 41:527–534.

Held, B. W., J. A. Jurgens, B. E. Arenz, S. M. Duncan, R. L. Farrell, and R. A. Blanchette. 2005. Environmental factors influencing microbial growth inside historic expedition huts of Ross Island, Antarctica. *Int. Biodeter. Biodegr.* 55:45–53.

Hill, J. L., A. D. Hocking, and F. B. Whitfield. 1995. The role of fungi in the production of chloranisoles in general purpose freight containers. *Food Chem.* 54:161–166.

Hyvärinen, A., T. Meklin, A. Vepsäläinen, and A. Nevalainen. 2002. Fungi and actinobacteria in moisture-damaged materials—Concentration and diversity. *J. Biodeter. Bioremed.* 49:27–37.

Joseph, E., S. Cario, A. Simon, M. Wörle, R. Mazzeo, P. Junier, and D. Job. 2012. Protection of metal artefacts with the formation of metal-oxalates complexes by *Beauvaria bassiana*. *Front. Microbiol.* 2:1–8.

Joseph, E., A. Simon, S. Prati, M. Wörle, D. Job, and R. Mazzeo. 2011. Development of an analytical procedure for evaluation of the protective behavior of innovative fungal patinas on archeological and artistic metal artefacts. *Anal. Bioanal. Chem.* 399:2899–2907.

Kositchaiyong, A., V. Rosarpitak, H. Hamada, and N. Sombatsompop. 2014. Anti-fungal performance and mechanical–morphological properties of PVC and wood/PVC composites under UV-weathering ageing and soil-burial exposure. *Int. Biodeter. Biodegr.* 91:128–137.

Kure, C. F. and I. Skaar. 2000. Mould growth on the Norwegian semi-hard cheeses Norvegia and Jarlsberg. *Int. J. Food Microbiol.* 62:133–137.

Lee, S.-H., H.-J. Lee, S.-J. Kim, H. M. Lee, H. Kang, and Y. P. Kim. 2010. Identification of airborne bacterial and fungal community in an urban area by T-RFLP analysis and quantitative real-time PCR. *Sci. Total Environ.* 408:1349–1357.

Li, D.-W. 2005. Release and dispersal of basidiospores from *Amanita muscaria* var. *alba* and their infiltration into a residence. *Mycol. Res.* 109:1235–1242.

Ludley, K. E. and C. H. Robinson. 2008. Decomposer Basidiomycota in Arctic and Antarctic ecosystems. *Soil Biol. Biochem.* 40:11–29.

Mantanis, G., E. Terzi, S. N. Kartal, and A. N. Papadopoulos. 2014. Evaluation of mold, decay and termite resistance of pine wood treated with zinc- and copper-based nanocompounds. *Int. Biodeter. Biodegr.* 90:140–144.

Medrela-Kuder, E. 2003. Seasonal variations in the occurrence of culturable airborne fungi in outdoor and indoor air in Craćow. *Int. Biodeter. Biodegr.* 52:203–205.

Miller, J. D. 1992. Fungi as contaminants in indoor air. *Atmos. Environ.* 26A:2163–2172.

Miller, J. D. 1995. Fungi and mycotoxins in grain: Implications for stored product research. *J. Stored Prod. Res.* 31:1–16.

O'Gorman, C. M. and H. T. Fuller. 2008. Prevalence of culturable airborne spores of selected allergenic and pathogenic fungi in outdoor air. *Atmos. Environ.* 42:4355–4368.

Olivera, L. S., A. L. B. D. Santana, C. A. Maranhão et al. 2010. Natural resistance of five woods to *Phanerochaete chrysosporium* degradation. *Int. Biodeter. Biodegr.* 64:711–715.

OSHA 2013. A brief guide to mold in the workplace. *OSHA Safety and Health Information Bulletin* 03-10-10 (http://www.osha.gov/dts/shib/shib101003.html).

Pinzari, F., J. Tate, M. Bicchieri, Y. J. Rhee, and G. M. Gadd. 2013. Biodegradation of Ivory (natural apatite): Possible involvement of fungal activity in biodeterioration of the Lewis Chessmen. *Environ. Microbiol.* 15:1050–1062.

Pitt, J. J. and A. D. Hocking. 1997a. The ecology of fungal food spoilage. In: *Fungi and Food Spoilage* eds. J.J. Pitt and A. D. Hocking, 3–9. Springer Science.

Pitt, J. J. and A. D. Hocking. 1997b. Fresh and perishable foods. In: *Fungi and Food Spoilage* eds. J.J. Pitt and A. D. Hocking, 383–400. Springer Science.

Pitt, J. J. and A. D. Hocking. 1997c. Spoilage of stored, processed and preserved foods. In: *Fungi and Food Spoilage* eds. J.J. Pitt and A. D. Hocking, 401–421. Springer Science.

Plaschkies, K., K. Jacobs, W. Scheiding, and E. Melcher. 2014. Investigations on natural durability of important European species against wood decay fungi: Part 1. Laboratory tests. *Int. Biodeter. Biodegr.* 90:52–56.

Ponce-Jiménez, M. D. P., F. A. L.-D. Toral, and H. Gutierrez-Pulido. 2002. Antifungal protection and sizing of paper with chitosan salts and cellulose ethers: Part 2. Antifungal effects. *J. Am. Inst. Conserv.* 41:255–268.

Rao, C.Y., M. A. Riggs, G. L. Chew et al. 2007. Characterization of airborne molds, endotoxins, and glucans in homes in New Orleans after hurricane Katrina and Rita. *Appl. Environ. Microbiol.* 73:1630–1634.

Rakotonirainy, M. S., E. Heude, and B. Lavédrine. 2007. Isolation and attempts of biomolecular characterization of fungal strains associated to foxing on a 19th century book. *J. Cult. Heritage* 8:126–133.

Robinson, C. H. 2001. Cold adaptation in Arctic and Antarctic fungi. *New Phytol.* 151:341–353.

Rodrigues, A., S. Gutierrez-Patricio, A. Z. Miller et al. 2014. Fungal deterioration of stained-glass windows. *Int. Biodeter. Biodegr.* 90:152–160.

Salonen, H., S. Lappalainen, O. Lindroos, R. Harju, and K. Reijula. 2007. Fungi and bacteria in mould-damaged and non-damaged office environments in a subarctic climate. *Atmos. Environ.* 41:6797–6807.

Schmidt, O. 2007. Indoor wood-decay basidiomycetes: Damage, causal fungi, physiology, identification and characterization, prevention and control. *Mycol. Progress* 6:261–279.

Sequeira, S. O., E. J. Cabrita, and M. F. Macedo. 2014. Fungal biodeterioration of paper: How are paper and book conservators dealing with it? An international survey. *Restaurator* 35:181–199.

Shadzi, S., M. Chadeganipour, and M. Alimoradi. 2002. Isolation of keratinophilic fungi from elementary schools and public parks in Isfahan, Iran. *Mycoses* 45:496–499.

Shiah, T.-C. 2009. Applying chitosan to increase fungal resistance of paper-based cultural relics. *Taiwan J. For. Sci.* 24:285–294.

Shirakawa, M. A., C. C. Gaylarde, P. M. Gaylard, V. John, and W. Gambale. 2002. Fungal colonization and succession on newly painted buildings and the effect of biocide. *FEMS Micribiol. Ecol.* 39:165–173.

Simeray, J., D. Mandin, and J. P. Chaumont. 1997. An aeromycological study of sawmills: Effects of type of installation and timber on mycoflora and inhalation hazards for workers. *Int. Biodeter. Biodegr.* 40:11–17.

Singh, J. 1999. Dry rot and other wood-destroying fungi: Their occurrence, biology, pathology and control. *Indoor Built Environ.* 8:3–20.

Singh, J. 2002. Fungi in buildings, holistic conservation and health, environmental management of fungal problems in our cultural heritage. Environmental Building Solutions Ltd., http://www.ebssurvey.co.uk/downloads.html.

Singh, J. 2005. Toxic molds and indoor air quality. *Indoor Built Environ.* 14:229–234.

Solomon, G. M., M. Hjelmroos-Koski, M. Rotkin-Ellman, and S. K. Hammond. 2006. Airborne mold and endotoxin concentrations in New Orleans, Louisiana after flooding, October through November 2005. *Environ. Health Perspect.* 114:1381–1386.

Sørensen, L. M., T. Jacobsen, P. V. Nielsen, J. C. Frisvad, and A. G. Koch. 2008. Mycobiota in the processing areas of two different meat products. *Int. J. Food Microbiol.* 124:58–64

Sterflinger, K. 2010. Fungi: Their role in deterioration of cultural heritage. *Fung. Biol. Rev.* 24:47–55.

Szczepanowska, H. and A. R. Cavaliere. 2000. Fungal deterioration of 18th and 19th century documents: A case study of the Tilghman family collection, Wye House, Easton, Maryland. *Int. Biodeter. Biodegr.* 46:245–249.

Szczepanowska, H. and C. M. Lovett. 1992. A study of the removal and prevention of fungal stains on paper. *J. Am. Inst. Conserv.* 31:147–160.

Szczepanowska, H. and W. R. Moomaw. 2004. Laser stain removal of fungal-induced stains from paper. *J. Am. Inst. Conserv.* 33:25–32.

Taghiyari, H. R., B. Moradi-Malek, M. G. Kookandeh, and O. F. Bibalan. 2014. Effects of silver and copper nanoparticles in particleboard to control *Trametes versicolor* fungus. *Int. Biodeter. Biodegr.* 94:69–72.

Taniwaki, M. H., A. D. Hocking, J. I. Pitt, and G. H. Fleet. 2009. Growth and mycotoxin pro-
    duction by food spoilage fungi under high carbon dioxide and low oxygen atmospheres.
    *Int. J. Food Microbiol.* 132:100–108.

Tapia, M. O., M. D. Stern, A. L. Soraci et al. 2005. Patulin-producing molds in corn silage
    and high moisture corn and effects of patulin on fermentation by ruminal microbes in
    continuous culture. *Anim. Food Sci. Technol.* 119:147–258.

Tournas, V. H. 2005. Moulds and yeasts in fresh and minimally processed vegetables and
    sprouts. *Int. J. Food Microbiol.* 99:71–77.

Tournas, V. H. and E. Katsoudas. 2005. Mould and yeast flora in fresh berries, grapes and
    citrus fruits. *Int. J. Food Microbiol.* 105:11–17.

Wainwright, M., K. Al-Wajeeh, and S. Grayston. 1997. Effect of silicic acid and other sili-
    con compounds on fungal growth in oligotrophic and nutrient-rich media. *Mycol. Res.*
    101:933–938.

Wainwright, M., F. Barakah, I. al-Turk, and T. A. Ali. 1991. Oligotrophic micro-organisms in
    industry, medicine and the environment. *Sci. Progress* 75:313–322.

Weinhold, R. 2007. A spreading concern: Inhalation health effects of mold. *Environ. Health
    Perspect.* 115:A300–A305.

Wu, P.-C., H.-J. Su, and C.-Y. Lin. 2009. Characteristics of indoor and outdoor airborne fungi
    at suburban and urban homes in two seasons. *Sci. Total Environ.* 253:111–118.

Yang, V. W. and B. L. Illman. 1999. Optimum growth conditions for the metal-tolerant wood
    decay fungus, *Merulioporia incrassata* TFFH 294. *Inter. Res. Group on Wood Preserv.
    Document No.* IRG/WP 99-50142, 8 pp.

Zalar, P., M. Novak, G. S. deHoog, and N. Gunde-Cimeran. 2011. Dishwashers—A man-made
    ecological niche accommodating human opportunistic fungal pathogens. *Fung. Biol.*
    115:997–1007.

# Index

Page numbers followed by f and t indicate figures and tables, respectively.

Printed and bound by CPI Group (UK) Ltd, Croydon, CR0 4YY

17/10/2024

01775709-0011